Abstract Algebra
A Computational Approach

Abstract Algebra
A Computational Approach

Charles C. Sims
Rutgers University

JOHN WILEY & SONS

New York   Chichester   Brisbane   Toronto   Singapore

Library of Congress Cataloging in Publication Data

Sims, Charles C.
Abstract algebra.

Includes index.
1. Algebra, Abstract–Data processing. I. Title.
QA162.S54 1984 512'.02'028542 83-6715
ISBN 0-471-09846-9

Printed in the United States of America

10 9 8 7 6 5 4 3 2 1

In memory of my father,
Ernest M. Sims
(1883-1973)

PREFACE

This book is intended as a text for a one-year introductory course in abstract algebra in which algorithmic questions and computation are stressed. A significant amount of computer usage by students is anticipated. My decision to write the book grew out of my interest in group-theoretic algorithms and my observation that learning the definitions, the theorems, and even the proofs of algebra too often fails to equip students adequately to solve computational algebraic problems. The goals of the book are to:

1 Introduce students to the basic concepts of algebra and to elementary results about them.
2 Present the concept of an algorithm and to discuss certain fundamental algebraic algorithms.
3 Show how computers can be used to solve algebraic problems and to provide a library, *CLASSLIB*, of computer programs with which students can investigate interesting computational questions in algebra.
4 Describe the APL computer language to the extent needed to achieve the other goals.

To help meet these goals, two additional manuals have been prepared, an instructor's manual and a *CLASSLIB* user's manual. There is more material here than can be covered in a one-year course. The instructor's manual contains suggested course outlines, hints on how to use this book, and the answers to selected exercises, including all that involve APL. The *CLASSLIB* user's manual contains detailed information about the library with complete listings of all programs. Generally, students should not need to acquire the user's manual. However, anyone wishing to make extensive use of *CLASSLIB* will probably want to have a copy. Both manuals are available from the publisher. The library can be obtained in machine-readable form from the author.

Computation in algebra is not really new. In some areas, such as number theory, there is a tradition of hand calculation going back hundreds of years. However, the development of the digital computer has inspired new interest in the subject. More and more research effort is being devoted to the existence and efficiency of algebraic algorithms. In addition, the use of computers to solve problems in algebra is growing steadily. Many algebraic algorithms can be understood by students in an introductory course.

The choice of the computer language to be used was very important. Of the languages normally available at college computer centers, only one is really suited for use in the teaching of algebra to students with little or no prior computing experience. That language is APL. The superiority of APL stems as much from the way the language is implemented as it does from the nature of the language itself. Here are some of the features of APL that make it the natural choice for this book.

1 APL is implemented in an interactive mode.
2 Arrays exist independent of programs.
3 One-line statements can be entered and executed immediately. In effect, beginning students do not have to write programs in the traditional sense.
4 The language contains many powerful primitive operations for manipulating arrays that are very useful in describing algebraic algorithms.

Even with the power of the APL language, most of the algorithms discussed are too complicated to be coded efficiently by beginning programmers. Therefore I decided to provide a library of programs that would allow students to use the algorithms on nontrivial problems while developing their skill in using APL. Students should not need to acquire a separate APL text; the appendices provide an adequate introduction to the language.

At this point, it would be good to mention several things that the book is not. It is not a text in applied algebra, which emphasizes the use of algebraic techniques to solve problems that arise outside of mathematics. Neither is this a book on numerical linear algebra, which deals with the numerical analysis aspects of linear algebra over the real and complex fields. Although the difficulties of performing computer calculations with real and complex numbers are discussed, the emphasis is on exact computation. For this reason, many of the computational exercises in linear algebra involve the fields $GF(p)$, p a prime.

There are two reasons for recommending that this book be used for a one-year course. First, the time required to introduce students to APL is too great to leave sufficient time to cover a reasonable amount of algebra in a one-semester course. In order to be able to understand and reproduce the dialogues in the text, one needs to know the material in Sections 1 to 3 and 5 to 7 of Appendix 1 and Sections 2 and 3 of Appendix 2, as well as certain topics discussed in Sections A1.4 and A2.1. Approximately three weeks are necessary to cover this material, and even more time must be spent if significant original programming is to be required of students. The second reason for suggesting a one-year course is that the most interesting algorithms, at least to me, come in the second half of the course. All of the chapters contain computational topics, but it was my desire to describe the material in Sections 6.5, 7.4, and 8.2, which provided the main motivation for this book. There are several additional topics that I would have liked to include. Some, such as factorization in $\mathbf{Z}[X]$ and a study of the ideals in $\mathbf{Z}[X]$, were omitted for lack of space. Others, such as some of the recent developments in computational group theory, could not be included because they involve concepts that do not fit easily into an introductory algebra course. Galois theory has been left out because practical algorithms for computing Galois groups are too involved to be presented at this level.

In the text, the lemmas, theorems, and corollaries are numbered consecutively within a section. Theorem 3 of Section 4 of Chapter 6 is referred to as Theorem 4.3 in other sections of Chapter 6 and as Theorem 6.4.3 outside of Chapter 6. A similar numbering system is used for examples and for exercises. Names occurring in brackets are references to the bibliography. In the contents, sections marked with an asterisk may be omitted without affecting the logical development. Exercises of greater than average difficulty are also flagged with asterisks. The ends of proofs are marked with the symbol □. A □ at the end of a theorem indicates that the proof of that theorem will be omitted.

Much of the writing of this book was done in college libraries. I am indebted to the library staffs at Rutgers University, Princeton University, Monmouth College, Southern Methodist University, and the Australian National University for the facilities they made available. Many individuals provided assistance throughout the eight years during which this book took shape. I wish to thank James England, Eugene Klotz, and Don Orth for many useful conversations concerning the use of APL in expostion. Michael O'Nan and Hale Trotter provided assistance on various mathematical topics. In particular, some of the material in Section 7.4 is based on a talk by Trotter. Certainly thanks are due to Kenneth Iverson. Without his development of APL, this book, at least in its present form, would not have been possible. Finally, I wish to record my deep gratitude to my wife, Annette, who typed an early version of this text and then typed the entire manuscript into a homegrown word processor. Her assistance made the preparation of the book much easier than it would otherwise have been.

Charles C. Sims

CONTENTS

SIGNIFICANT DEPENDENCIES BETWEEN SECTIONS

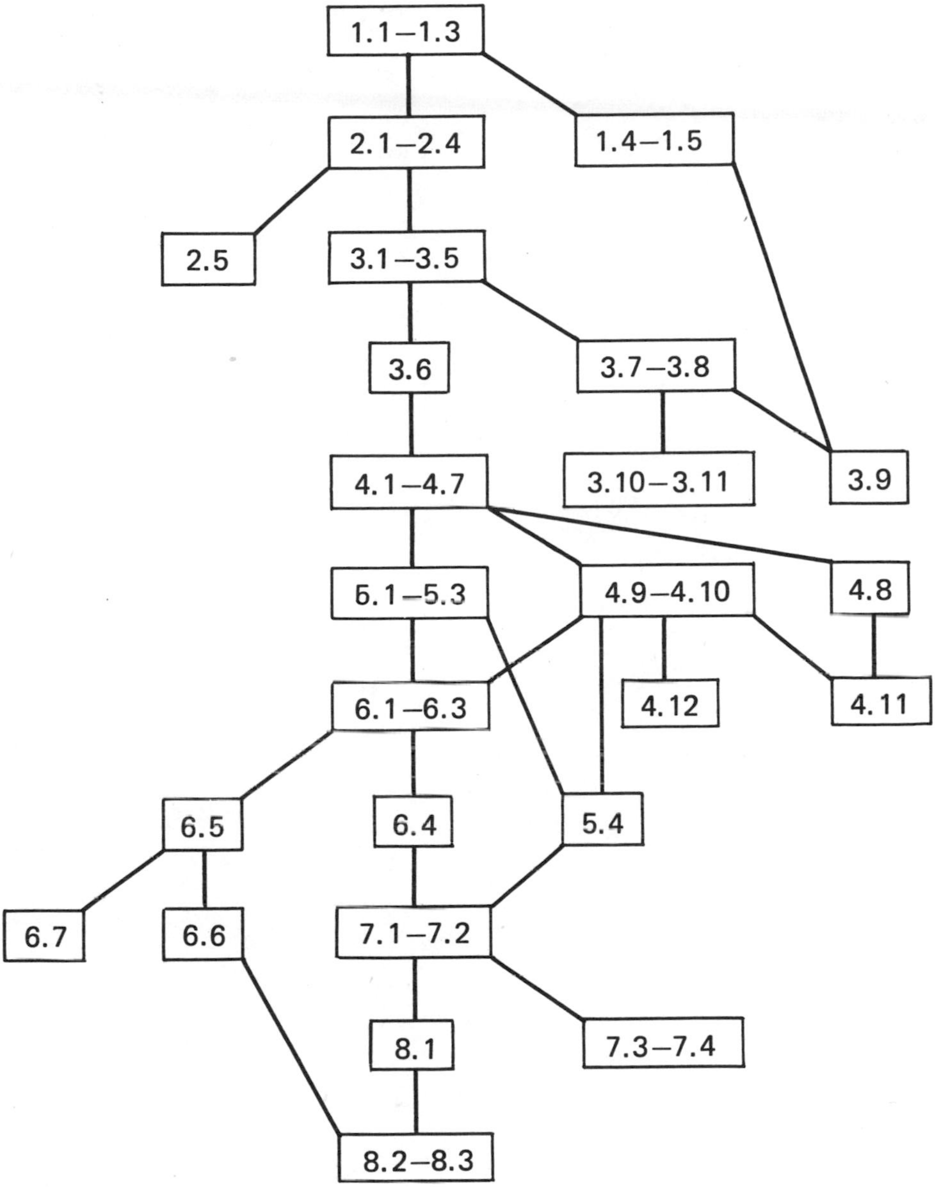

INTRODUCTION

This book is an introduction to the area of mathematics called abstract algebra or, simply, algebra. It differs from most books on the subject in that computation plays a central role throughout. A substantial portion of the text is devoted to algorithms for solving algebraic problems. Loosely defined, an *algorithm* is a sequence of instructions for solving a particular problem or class of problems. The instructions must be unambiguous, with no room for different interpretations by different individuals, and must lead to the solution of the problem in a finite number of steps. The emphasis here will be on algorithms that can be carried out, or executed, by a computer.

It is difficult to establish a specific date for the beginning of any branch of mathematics. Nevertheless, it is widely agreed that the work of the French mathematician Evariste Galois (1811-1832) set the stage for the development of algebra into one of the major areas of mathematical activity. It was not until later in the nineteenth century, however, that abstraction became an important part of algebra. Abstraction is the process by which similarities are recognized between apparently dissimilar mathematical objects and by which these similarities are shown to be consequences of a few basic properties (axioms) that are possessed by all of the objects being studied. It is to this process that the word "abstract" in the phrase "abstract algebra" refers.

Introductory algebra courses are often referred to as courses on groups, rings, and fields. Algebra involves more than the study of groups, rings, and fields, but these three types of algebraic structures, together with one additional type, modules, form the subject matter of this text. The term "group" was coined by Galois, but the first formal definition was not given until 1849 and the value of the concept of an "abstract group" was not recognized for nearly 30 years more. The idea of a field is present in Galois' work, but the term was introduced by the German mathematician

Richard Dedekind (1831-1916) and the definition was not standardized until late in the nineteenth century. Although many examples of rings were known in the nineteenth century, the abstract theory was developed during the present century. The term "ring" was formulated by David Hilbert (1862-1943), a very important German mathematician.

The formal prerequisites for the study of abstract algebra are minimal. However, it will be assumed that readers are familiar with certain concepts normally covered in lower-level undergraduate mathematics courses. These concepts include proofs by induction and the elementary properties of sets, the integers, and rational numbers. Some acquaintance with real and complex numbers will also be assumed.

In this text, as in any text on abstract algebra, a considerable amount of space is devoted to the formal development of the subject. As axioms are stated, definitions made, and theorems proved, readers are encouraged to study particular examples in detail. It is only through the study of examples that one can see how the abstract theory provides an efficient method of deriving useful information about many different mathematical objects. The investigation of examples can be facilitated with the help of a computer. The computer makes it possible to look at more complicated and, one hopes, more interesting examples by removing the drudgery of time-consuming hand calculation.

In order to communicate with a computer we must use a computer language. The language chosen for use in this book is APL. The APL language is extremely powerful, which means that complicated calculations can be described with a few symbols. APL also possesses a high degree of internal consistency and, in many ways, is more logical than traditional mathematical notation. No prior knowledge of APL will be assumed. The appendices contain a description of the aspects of APL that are important for using this text. Appendix 1 describes the APL language in sufficient detail to permit readers to follow the computer examples in this book. However, it is very important that readers be able to work out these and other examples on the computer. Appendix 2 contains further information about APL systems that can help readers use their local systems efficiently.

In order to make the best possible use of this book, readers should have access to two APL workspaces that have been specifically created to supplement the text. The workspace *CLASSLIB* contains procedures for carrying out many types of algebraic computations. The workspace *EXAMPLES* contains arrays that represent various kinds of algebraic objects. The arrays in *EXAMPLES* are used in the computational examples in the text. A more complete description of the contents of these two workspaces can be found in Appendix 3. It is suggested that the naming conventions described in Section A3.1 be read before making extensive use of *CLASSLIB*. All of the computer examples in the text assume that the contents of both *CLASSLIB* and *EXAMPLES* are present in the active workspace.

The algorithms used in most of the procedures in *CLASSLIB* are discussed in the text. As each procedure is introduced, readers should concentrate on learning what it does and on understanding the basic algorithm involved. Once some familiarity with a procedure has been achieved, it is an extremely valuable exercise to write one's own version of the procedure and compare it to the version in the library. I would appreciate being informed about possible improvements to the procedures in *CLASSLIB*.

Readers having some experience with APL may proceed immediately to Chapter 1. Those not familiar with APL should begin by reading through the first seven sections of Appendix 1. It is not necessary to become an expert in the use of APL before starting to learn algebra using this book. Once the fundamentals of the language have been grasped, the study of the real subject matter—abstract algebra—should be begun. The appendices can then be used for reference, as needed.

We will be using two different systems of symbolic notation, APL and traditional, and it is important to be able to recognize which system is being used in a particular expression. APL expressions are printed in a special type font used only for APL. Thus *C←A ZGCD B* is an APL expression. When any other type font is used, as in the statement "let $c = \gcd(a, b)$", traditional mathematical notation is assumed. Some care is required to distinguish between the two commas that occur in this book. The ordinary comma (,) is a mark of punctuation, but the APL comma (,) represents one of two APL operations that are described in Section A1.5. Occasionally, we will borrow certain aspects of APL notation for use with traditional notation. For example, we will sometimes denote the entry in the ith row and jth column of the matrix A by $A[i;j]$, even though A is not an APL array. These borrowings should not present any serious problems.

1
SETS

For nearly 100 years the formal exposition of mathematics has been based on the concept of a set. Readers are no doubt familiar with sets from previous courses in mathematics. In this chapter we will summarize the basic definitions, notation, and operations of set theory. We will also discuss ways of representing sets by APL arrays and techniques for manipulating these arrays to perform set-theoretic operations. The APL index origin, described in Section A1.2, is normally assumed to be 1.

1. SETS

One of the most important ideas in the development of mathematics is the use of the axiomatic method, in which all of the theorems in a particular branch of mathematics are obtained as logical consequences of a few axioms that state the basic properties that are assumed to hold for the objects under study. This approach is probably most familiar in the area of plane geometry. In an axiomatic treatment of plane geometry, no attempt is made to say what points and lines really are. Instead, one writes down axioms such as "through any two distinct points there passes eaactly one line" in an attempt to formalize our intuitive notions about the kinds of pictures we can draw using a straightedge and a very sharp pencil. The axiomatic method is universally agreed to be the proper approach to the study of abstract algebra.

The idea of a set has been found to be of fundamental importance not only in algebra but in most of present-day mathematics. All of the algebraic objects we will study will be sets. It would seem reasonable, therefore, to begin our study of algebra with an axiomatic treatment of sets. This approach seems even more essential when we learn, as we will at the end of this section, that our intuition concerning sets can lead us to logical contradictions. However, we will follow the accepted practice in introductory algebra texts and omit a formal treatment of sets. There are two reasons for this. First, the exposition of axiomatic set theory would delay too long our study of the main subject matter of this book: the basic properties of algebraic systems such as groups, rings, and fields and the algorithms for solving problems related to them. Second, our intuition leads us astray only when we try to consider sets that are "too big". The sets we will encounter in our study of algebra will be "small enough" that our intuition can be trusted not to get us into trouble.

We define a *set* to be any collection of objects called the *elements* or *members* of the set. A commonly used synonym for "set" is "family". The word "group" should not be used as a synonym for "set" because a group in mathematical terminology is a particular kind of algebraic object that we will study in Chapter 3. If x is an element of the set X, we write $x \in X$ and say that x *belongs to* X or that x *is in* X. If x is not an element of X, we write $x \notin X$. Two sets are equal if they have the same elements. Thus the statement $X = Y$ is equivalent to the following pair of assertions:

1. If $x \in X$, then $x \in Y$.
2. If $y \in Y$, then $y \in X$.

To show that X and Y are not equal, we must exhibit an element of X that is not an element of Y or an element of Y that is not an element of X.

There are two standard ways of describing a set. The first is to list the elements of the set separated by commas and enclosed in braces. Thus

$$S = \{1, 2, 3, 5, 8, 13\}, \qquad U = \{2, 3, 5, 7, 11\},$$
$$A = \{2, 4, 8, 16\}, \qquad B = \{16, 8, 4, 2\},$$
$$C = \{2, 4, 8, 16, 8, 4, 2\}$$

are all sets whose elements are positive integers. Since neither the order in which the elements are listed nor the fact that some elements are repeated has any significance, the sets A, B, and C are all equal.

The easiest sets to represent in APL are finite sets of real numbers. Since the entries of an APL array are real numbers, we may simply define a vector whose components list the elements of the set. Thus, if

```
S←1 2 3 5 8 13
U←2 3 5 7 11
A←2 4 8 16
B←16 8 4 2
C←2 4 8 16 8 4 2
```

then the vectors S, U, A, B, C correspond naturally to the definition of the sets S, U, A, B, C. (To save space, APL dialogues such as the preceding one are printed in two columns. At a terminal, they would appear as one long column.)

The notation $\{a_1, \ldots, a_n\}$ and the use of an APL vector to list the elements of a set both have the drawback that the representation for a particular set is not unique. In general, there is no natural order on the elements of a set but, if the elements of the set are real numbers, we do get a unique representation if we assume the elements are listed in increasing order and without repetitions. The procedure *SSORT* in *CLASSLIB* produces this standard list of the elements in the set described by an arbitrary vector.

```
      B                          C
16 8 4 2                  2 4 8 16 8 4 2
      SSORT B                    SSORT C
2 4 8 16                  2 4 8 16
```

If X is an APL vector, we will often speak of "the set X instead of "the set represented by X". In particular, we will often refer to the set ⍳N, which can, of course, mean either $\{1,2,\ldots,N\}$ or $\{0,1,\ldots,N-1\}$, depending on the index origin.

Sets whose elements are not real numbers are more difficult to represent in APL. We will have to represent sets of sets of real numbers, sets of polynomials, and many other types of sets. Techniques for doing this will be discussed as the need arises.

The second way to describe a set is to specify a property that characterizes the elements of that set. The statement

$$X = \{x|P(x)\}$$

is read "X is the set of all x such that the property P holds for x." For example, we may define two sets L and M as follows:

$$L = \{x|x \text{ is a positive real number}\},$$

$$M = \{t|t \text{ is an even integer}\}.$$

There are a few sets that will come up so frequently that it is convenient to have special symbols for them. The set of *integers* will be denoted by **Z** and the set of positive integers or *natural numbers* by **N**. The symbols **Q**, **R**, and **C** will stand for the set of *rational numbers*, the set of *real numbers*, and the set of *complex numbers*, respectively.

Suppose we define a set E by

$$E = \{x|x \in \mathbf{R},\quad x^2 = -1\}.$$

Since every real number has a nonnegative square, E has no elements. Such a set is said to be *empty*. The following theorem shows, among other things, that the set of all unicorns is equal to the set of all letters in the English alphabet that come after the letter Z.

THEOREM 1. Any two empty sets are equal.

Proof. Let X and Y be empty sets. Since X is empty, we cannot find any element x of X, and so we certainly cannot produce an x in X that is not in Y. Similarly, we cannot find an element of Y that is not in X because Y has no elements. Thus we are forced to conclude that the statement $X \neq Y$ is false, so X and Y must be equal. □

By Theorem 1 we may speak of *the* empty set, since there is only one. It will be denoted by the symbol $\emptyset$. Whenever we define a property that a particular set may or may not have, it is a useful exercise to determine whether the empty set has the property.

A set A is a *subset* of a set B if every element of A is also an element of B. In this case, we also say A is *contained in* B or that B *contains* A. A subset A of B is called a *proper subset* if $A \neq B$, that is, if there is some element of B that is not in A. We will write $A \subseteq B$ when A is a subset of B and $A \subset B$ when A is a proper subset of B. Some authors prefer to write $A \subset B$ where we write $A \subseteq B$. The notation used here has been chosen because it parallels the use of $<$ and $\leq$ to denote inequality of real numbers. The statements $B \supseteq A$ and $B \supset A$ mean $A \subseteq B$ and $A \subset B$, respectively. We have $\emptyset \subseteq A$ and $A \subseteq A$ for any set A.

If the APL vectors *A* and *B* list the elements of two subsets A and B of **R**, then the assertion $A \subseteq B$ corresponds to the APL proposition ∧/*A*∊*B*. (An APL *proposition* is an APL expression with one entry, which is either 1 or 0. The APL membership operation ∊ is described in Section A1.5 and the operation ∧/, called "and reduction", is discussed in Section A1.6.) For example,

```
      A←1 3 5                         ∧/A∊B
      B←1 2 3 4 5             1
      C←2 3 5 7                       ∧/A∊C
                              0
```

Here *A* is a subset of *B* but not of *C*.

For the time being, we will rely on our intuition concerning the term "finite set", which will be defined in Section 3 of this chapter. If X is a finite set, then $|X|$ is the *cardinality* of X, that is, the number of elements of X.

Suppose $A_1, \ldots, A_k$ are sets of real numbers such that $|A_i| = m$ for $1 \leq i \leq k$. We can represent $\{A_1, \ldots, A_k\}$ by a k-by-m matrix *A* such that the *I*th row *A*[*I*;] of *A* lists the elements of A_I. For example, *CLASSLIB* contains a procedure *SSUB* such that *A*←*K SSUB N* defines *A* to be a matrix whose rows list the *K*-element subsets of ⍳*N*.

```
       ⎕IO←1
       2 SSUB 4
 1 2
 1 3
 2 3
 1 4
 2 4
 3 4
```

```
       ⎕IO←0
       4 SSUB 5
 0 1 2 3
 0 1 2 4
 0 1 3 4
 0 2 3 4
 1 2 3 4
       ⎕IO←1
```

The number of rows of `K SSUB N` is the binomial coefficient `K ! N`.

Given one or more sets, we can construct new sets in several ways. For example, if X is a set, we can form $P = \{A | A \subseteq X\}$, the set of all subsets of X. When X is finite, $|P| = 2^{|X|}$. Because of this, we will use 2^X to denote the set of all subsets of X for any set X, finite or infinite.

If X and Y are sets, then $X \cup Y$, the *union* of X and Y, is the set of elements that are members of at least one of the sets X and Y. Thus

$$X \cup Y = \{x | x \in X \quad \text{or} \quad x \in Y\},$$

where the "or" is inclusive "or".

The *intersection* $X \cap Y$ of X and Y is the set of all elements that are in both X and Y. Hence

$$X \cap Y = \{x | x \in X \quad \text{and} \quad x \in Y\}.$$

Two sets are *disjoint* if their intersection is empty. The set $\{x | P(x)\} \cap A$ is often written $\{x \in A | P(x)\}$.

The *difference* $X - Y$ is the set of elements of X that are not elements of Y. If Y is a subset of X, then $X - Y$ is also called the *complement* of Y in X.

Suppose the vectors `X` and `Y` list the elements of two subsets X and Y of **R**. Then it is quite easy to construct vectors that list the elements of the sets $X \cup Y$, $X \cap Y$, and $X - Y$. The vector `X,Y` lists the elements of $X \cup Y$. The `I`th component of the logical vector `X∊Y` is 1 if and only if `X[I]` is equal to some component of `Y`. Thus the compression `(X∊Y)/X` gives a list of the elements of $X \cap Y$. The 1's in `~X∊Y` correspond to the components of `X` that are not equal to any component of `Y`, and so

```
(~X∊Y)/X
```

lists the elements of $X - Y$.

```
      X←1 2 3 4
      Y←2 4 5
      X,Y
1 2 3 4 2 4 5
```

```
      (X∊Y)/X
2 4
      (~X∊Y)/X
1 3
```

The operations $\cup$, $\cap$, and $-$ on sets satisfy many useful properties. The most important are listed in the following theorem:

THEOREM 2. Let A, B, and C be sets. Then

(a) $A = A \cup A = A \cap A = A \cup \emptyset$.
(b) $A \cap \emptyset = \emptyset$.
(c) $A \cup B = B \cup A$ and $A \cap B = B \cap A$.
(d) The following are equivalent:
  (1) $A \subseteq B$.
  (2) $A \cup B = B$.
  (3) $A \cap B = A$.
  (4) $A - B = \emptyset$.
(e) $(A \cup B) \cup C = A \cup (B \cup C)$.
(f) $(A \cap B) \cap C = A \cap (B \cap C)$.
(g) $A \cap (B \cup C) = (A \cap B) \cup (A \cap C)$.
(h) $A \cup (B \cap C) = (A \cup B) \cap (A \cup C)$.
(i) $(C - A) \cup (C - B) = C - (A \cap B)$.
(j) $(C - A) \cap (C - B) = C - (A \cup B)$.
(k) If $A \subseteq B$ and $B \subseteq C$, then $A \subseteq C$.

Proof. We will leave most of the proof as an exercise, proving only part (i) as an illustration. First, let x be an element of $(C - A) \cup (C - B)$. Then x is in $C - A$ or in $C - B$. If x is in $C - A$, then $x \in C$ and $x \notin A \cap B$ and so x is in $C - (A \cap B)$. Similarly, if x is in $C - B$, then x is in $C - (A \cap B)$. Thus, in either case, x is in $C - (A \cap B)$, and so

$$(C - A) \cup (C - B) \subset C - (A \cap B).$$

Now suppose x is in $C - (A \cap B)$. Then x is in C and x is not in $A \cap B$. Thus x is either not in A or not in B. But this says x is in either $C - A$ or $C - B$, and so x is in $(C - A) \cup (C - B)$. Hence

$$(C - A) \cup (C - B) \supseteq C - (A \cap B).$$

Combining this with the previous result, we conclude that

$$(C - A) \cup (C - B) = C - (A \cap B). \square$$

Let X and Y be sets. The *Cartesian product* $X \times Y$ of X and Y is the set of all ordered pairs (x, y) with $x \in X$ and $y \in Y$. [The term "Cartesian" is derived from the name of the French mathematician and philosopher René Descartes (1596-1650).] The *diagonal* of $X \times X$ is the set $\{(x, x) | x \in X\}$.

A *partition* of a set X is a set Π of subsets of X or, equivalently, a subset of 2^X, such that

1. If $A \in \Pi$, then $A \neq \emptyset$.
2. If $A \in \Pi$ and $B \in \Pi$, then either $A = B$ or $A \cap B = \emptyset$.
3. Every element of X is in some element of Π.

Rephrasing, we can say that a partition of X is a family of nonempty subsets of X such that each element of X belongs to exactly one member of the family. The elements of the partition are called *blocks.* For example,

$$\{\{1\}, \{2, 3\}, \{4, 5, 6\}, \{7, 8, 9, 10\}\}$$

is a partition of $\{1, 2, \ldots, 10\}$ with four blocks. For any proper nonempty subset A of a set X, the set $\{A, X - A\}$ is a partition of X. A subset R of X is called a *set of representatives* for a partition Π of X if R contains exactly one element from each block of Π.

We close this section with an illustration of the ways our intuition can get us into trouble when we try to use sets that are "too big". It is possible to imagine sets that are members of themselves, such as the set of all sets or the set of all abstract concepts. Thus we can reasonably form

$$K = \{X | X \text{ is a set and } X \notin X\},$$

the set of all sets that are not members of themselves. Now we ask the question whether or not K is a member of itself. If $K \notin K$, then K is one of the sets that our definition says must be in K, and so $K \in K$. But if $K \in K$, then K cannot be one of the members of K, and so $K \notin K$. Thus it seems that the statements $K \notin K$ and $K \in K$ must both be false, since they each lead to a contradiction. This antinomy or paradox is due to the English mathematician Bertrand Russell (1872-1970). It and others like it had a great influence on the development of axiomatic set theory.

EXERCISES

1 What is a reasonable interpretation for $\{1, 2, \ldots, n\}$ when $n = 0$?

2 Let X be a set. Explain the difference between X and $\{X\}$. What are $|\emptyset|$ and $|\{\emptyset\}|$?

3 Let X and Y be finite sets. Show

(a) $|2^X| = 2^{|X|}$.

(b) $|X \cup Y| + |X \cap Y| = |X| + |Y|$.

(c) $|X \times Y| = |X| \times |Y|$.

4 What is $2^\emptyset$?

5. Complete the proof of Theorem 2.

6 Determine the number of partitions of a set X when $1 \le |X| \le 4$.

7 Suppose Π_1 and Π_2 are partitions of a set X. We say Π_1 is a *refinement* of Π_2 if, whenever $A \in \Pi_1$ and $B \in \Pi_2$, then either $A \subseteq B$ or $A \cap B = \emptyset$. Let $X = \{1, 2, \ldots, 10\}$ and let Π be the partition

$$\{\{1\}, \{2, 3\}, \{4, 5, 6\}, \{7, 8, 9, 10\}\}$$

of X. How many partitions of X are refinements of Π? For how many partitions Π_1 of X is Π a refinement of Π_1?

8 How many partitions of the empty set are there?

9 We stated that all algebraic objects studied in this text would be sets, but we did not define the ordered pair (x, y) as a set. Show that if we define (x, y) to be $\{\{x\}, \{x, y\}\}$, then $(x, y) = (u, v)$ if and only if $x = u$ and $y = v$.

10 Let X, Y, and Z be sets. Are $X \times (Y \times Z)$ and $(X \times Y) \times Z$ the same set?

In Exercises 11 to 15 *A*, *B*, and *C* are APL vectors listing the elements of subsets A, B, and C of $\mathbf{R}$, respectively.

11 Write APL propositions corresponding to each of the following assertions. Check your answers at a terminal with specific examples.

(a) $B \subseteq C$.
(b) $A \subset B$.
(c) $A = C$.
(d) $C = A \cup B$.
(e) $B = A \cap C$.
(f) $B \subseteq A - C$.
(g) $A \neq B$.
(h) $7 \in A$.
(i) $B \subseteq \mathbf{Z}$.
(j) $\{3, 5, 9\} \subseteq C$.
(k) $|A| = |C|$.
(l) $|B| < 5$.

12 Write APL expressions defining vectors that list the elements of the following sets. Check your answers at a terminal.

(a) $A \cap C$.
(b) $B \cup C$.
(c) $C - (A \cap B)$.
(d) $A \cup (B \cap C)$.
(e) $\{a \in A \mid a \geq 6\}$.
(f) $A \cap \mathbf{Z}$.

13 Why might it be a good practice to use *SSORT A,B* to represent $A \cup B$ instead of using *A,B*?

14 Write APL procedures *SETEQ*, *SETUN*, *SETINT*, and *SETDIFF* such that *A SETEQ B* is 1 if $A = B$ and 0 otherwise and *A SETUN B*, *A SETINT B*, and *A SETDIFF B* are vectors listing the elements of $A \cup B$, $A \cap B$, and $A - B$, respectively.

15 Verify assertions (a) to (c) and (e) to (j) of Theorem 2 at a terminal for several choices of the sets A, B, and C. The procedures defined in Exercise 14 may be used to construct representations for the sets involved and to test equality.

16 Let n be a positive integer. What would be a good way to represent a partition of $\{1, \ldots, n\}$ by an APL array? Given representations

of two partitions Π_1 and Π_2, how could you decide whether $\Pi_1 = \Pi_2$ or whether Π_1 is a refinement of Π_2?

2. RELATIONS

A *relation* from a set X to a set Y is a subset R of the Cartesian product $X \times Y$. A relation on X is a relation from X to X. In this context it is conventional to replace the statement $(x,y) \in R$ by xRy. For example, when we speak of the relation $>$, "is greater than", on **R**, we mean

$$\{(x,y) \mid x,y \in \mathbf{R} \quad \text{and} \quad x > y\},$$

and by the relation $\subseteq$, "is contained in", on 2^X, we mean

$$\{(A,B) \mid A \subseteq B \subseteq X\}.$$

On any set X we have the *identity relation*, the diagonal of $X \times X$, in which two elements of X are related if and only if they are the same. The relation $X \times X$ itself is the *trivial relation* on X and $\emptyset$ is the *empty relation*. Relations occur in everyday speech. Thus "is a child of", "has a common ancestor with", and "lives on the same street as" can be considered relations in our sense on the set of all people living today when they are interpreted as describing certain sets of ordered pairs of people.

Let R and S be relations from X to Y. Then $R \cup S$, $R \cap S$, and $(X \times Y) - R$ are also relations from X to Y. If $R \subseteq S$, we say R *implies* S. Thus $<$ implies $\leq$, and "is a son of" implies "is a child of". The set

$$R^{-1} = \{(y,x) \mid (x,y) \in R\}$$

is a relation from Y to X and is called the *inverse* of R. Clearly, $(R^{-1})^{-1} = R$ and R implies S if and only if R^{-1} implies S^{-1}. The inverse of $>$ is $<$, and the inverse of "is a child of" is "is a parent of".

If X is the set ⍳*M* and Y is the set ⍳*N*, then a relation R from X to Y can be conveniently represented by the *M*-by-*N* matrix *R* such that *R*[*I*;*J*] is 1 if the pair (*I*,*J*) is in R and 0 otherwise. Such matrices can often be constructed using APL outer products. Recall that if f is any APL dyadic scalar operation and if *U* and *V* are vectors, then *U*∘.*fV* is the matrix whose *I*,*J*-th entry is *U*[*I*]*f V*[*J*]. In the examples

```
      ⎕←S←(⍳5)∘.<⍳5
0 1 1 1 1
0 0 1 1 1
0 0 0 1 1
0 0 0 0 1
0 0 0 0 0
```

```
      ⎕←T←1≥|(⍳4)∘.-⍳6
1 1 0 0 0 0
1 1 1 0 0 0
0 1 1 1 0 0
0 0 1 1 1 0
```

S represents the relation < on `⍳5` and T represents "differs in absolute value by at most 1 from", considered as a relation from `⍳4` to `⍳6`.

If R is the logical matrix representing a relation R from `⍳M` to `⍳N`, then R will be called the *characteristic matrix* for R. The reason for this terminology will be discussed in Section 3.

If R is a relation from X to Y and S is a relation from Y to Z, then the *composition* of R and S is the relation from X to Z consisting of those pairs (x, z) for which there is an element y in Y such that $(x, y) \in R$ and $(y, z) \in S$. For example, person x is a grandchild of person z if and only if there is a person y such that x is a child of y and y is a child of z. Thus the relation "is a grandchild of" is the composition of the relation "is a child of" with itself. The composition of the relations "is at least 2 more than" and "is at least 3 more than" on **R** is the relation "is at least 5 more than", since for real number x and z the assertion $x \geq z + 5$ is equivalent to the assertion that there is a real number y such that $x \geq y + 2$ and $y \geq z + 3$. The composition of R and S is denoted $R \circ S$.

Suppose now that X is `⍳L`, Y is `⍳M`, and Z is `⍳N` and that R is described by the matrix R and S by the matrix S. How can we construct the matrix T representing $R \circ S$? The entry `T[I;K]` is 1 if and only if some `J` in `⍳M` the entries `R[I;J]` and `S[J;K]` are both 1. This is equivalent to saying that the row `R[I;]` and the column `S[;J]` have a 1 in the same position. Thus `T[I;K]` is

```
∨/R[I;]∧S[;K].
```

But this means that T is the inner product `R∨.∧S`. (APL inner products are discussed in Section A1.7.) Consider the following example.

```
      ⎕←R←(⍳6)∘.>⍳6                 ⎕←T←R∨.∧R
0 0 0 0 0 0                    0 0 0 0 0 0
1 0 0 0 0 0                    0 0 0 0 0 0
1 1 0 0 0 0                    1 0 0 0 0 0
1 1 1 0 0 0                    1 1 0 0 0 0
1 1 1 1 0 0                    1 1 1 0 0 0
1 1 1 1 1 0                    1 1 1 1 0 0
```

Here R represents > on `⍳6` and T represents the composition of > with itself, in other words, the relation "is at least two greater than".

THEOREM 1. Let R be a relation from X to Y, S a relation from Y to Z, and T a relation from Z to W. Then

(a) $(R \circ S) \circ T = R \circ (S \circ T)$.

(b) $(R \circ S)^{-1} = S^{-1} \circ R^{-1}$.

Proof. We note first that $R \circ S$ is a relation from X to Z, and so $(R \circ S) \circ T$ is defined and is a relation from X to W. Similarly, $R \circ (S \circ T)$ is defined and is also a relation from X to W. Suppose now that (x, w) is in $(R \circ S) \circ T$. Thus there exists z in Z with (x, z) in $R \circ S$ and (z, w) in T. But therefore there exists y in Y with (x, y) in R and (y, z) in S. This implies that (y, w) is in $S \circ T$, and so (x, w) is in $R \circ (S \circ T)$. We have now shown

$$(R \circ S) \circ T \subseteq R \circ (S \circ T).$$

The proof of the reverse containment and the proof of part (b) are left as exercises. □

Part (a) of Theorem 1 states that composition of relations is *associative*, and part (b) states that the inverse of the composition of two relations is the composition of the inverses in the opposite order.

There are three important properties that a relation R on a set X may have. We say R is *reflexive* if xRx for all x in X. This is the same as requiring that R contain the diagonal of $X \times X$ or that the identity relation imply R. The relation R is *symmetric* if whenever xRy, then yRx. Equivalently, R is symmetric if R implies R^{-1}. But if $R \subseteq R^{-1}$, then

$$R^{-1} \subseteq (R^{-1})^{-1} = R,$$

and so $R = R^{-1}$. Thus R is symmetric if and only if R is equal to its inverse. We say R is *transitive* if whenever xRy and yRz, then xRz. Since there exists a y in X with xRy and yRz if and only if (x, z) is in $R \circ R$, we see that R is transitive if and only if $R \circ R$ implies R. A relation that is reflexive, symmetric, and transitive is called an *equivalence relation*. We will see equivalence relations quite often in our study of algebra.

On any set X the identity relation and the trivial relation are equivalence relations. On **R** the relation $<$ is transitive but not symmetric or reflexive, while $\leq$ is reflexive and transitive but not symmetric. The relation "has a common parent with" is reflexive and symmetric but not transitive.

Suppose `R` is an `N`-by-`N` matrix describing a relation R on `⍳N`. The main diagonal of `R` is the vector `1 1⍉R`. (The dyadic transpose operation `⍉` is explained in Section A1.5. In origin 0 the diagonal of `R` is `0 0⍉R`.) Since R is reflexive if and only if the diagonal entries of `R` are all 1, the proposition `∧/1 1⍉R` corresponds to the assertion that R is reflexive. We leave it to the reader (see Exercise 18) to show that the propositions `∧/,R=⍉R` and `∧/,R>R∨.∧R` correspond to the assertions that R is symmetric and transitive, respectively. Let us look at a simple example.

```
      ⎕←E←4 4⍴1 0 0 1 0 1 1 0 0 1 1 0 1 0 0 1
 1 0 0 1
 0 1 1 0
 0 1 1 0
 1 0 0 1
      ∧/1 1⍉E
1
      ∧/,E=⍉E
1
      ∧/,E≥E∨.∧E
1
```

Here we see that the relation on ⍳4 represented by *E* is an equivalence relation.

The procedure *SEQREL* in *CLASSLIB* checks whether a logical matrix defines an equivalence relation.

```
      SEQREL E
1
```

The method used in *SEQREL* is more efficient than just checking in turn for reflexivity, symmetry, and transitivity. It is based on the ideas of Theorem 2, which follows. (See also Exercise 21.)

We will now show that there is a close connection between equivalence relations on a set X and partitions of X. For any relation R from X to Y and any subset A of X, let

$$AR = \{y \in Y \mid aRy \quad \text{for some} \quad a \in A\}.$$

Thus AR is the set of all elements of Y to which R relates at least one element of A. For example, if R is the relation "is a child of" and A is a set of people, then AR is the set of all people y such that some person a in A is a child of y. In other words, AR is the set of parents of the people in A.

THEOREM 2. Let E be an equivalence relation on X. Then

$$\Pi = \{\{x\}E \mid x \in X\}$$

is a partition of X.

Proof. Since E is reflexive, for all x in X we know that x is in $\{x\}E$. Thus every element of X is contained in some member of Π and all the members of Π are nonempty. Suppose that an element x of X is contained in both $\{y\}E$ and $\{z\}E$. We must show $\{y\}E = \{z\}E$. Suppose w is in $\{y\}E$. Then we have yEx and zEx and also yEw. By the symmetry of E this implies xEy. Using the transitivity of E twice, we get zEy and hence

zEw. Therefore w is in $\{z\}E$. This shows that $\{y\}E \subseteq \{z\}E$. By a similar argument we obtain $\{z\}E \subseteq \{y\}E$ and so $\{y\}E = \{z\}E$. Hence every element of X is contained in a unique member of Π. □

The partition Π of Theorem 2 is sometimes written X/E. Elements of Π are called *equivalence classes* of E.

It is also true that given a partition Π of a set X we can construct an equivalence relation on X. Define the relation E_Π on X so that $xE_\Pi y$ if and only if x and y are in the same block of Π.

THEOREM 3. Let Π be a partition of the set X and let R be an equivalence relation on X. Then

(a) E_Π is an equivalence relation on X and $\Pi = X/E_\Pi$.
(b) $R = E_{X/R}$.

Proof. The proof is left as an exercise. □

EXERCISES

1 Complete the proof of Theorem 1.

2 Suppose $|X| = 2$. How many relations on X are there? How many of these are reflexive? How many are symmetric? How many are transitive?

3 Find the error in the following "proof" that a symmetric, transitive relation R on a set X is automatically reflexive. For any x in X we can conclude by symmetry from xRy that yRx and then by transitivity that xRx.

4. Is the empty relation on a set reflexive? Is it symmetric or transitive?

5 Let R be a relation from X to Y and let A and B be subsets of X. Show

(a) If $A \subseteq B$, then $AR \subseteq BR$.
(b) $(A \cup B)R = (AR) \cup (BR)$.
(c) $(A \cap B)R \subseteq (AR) \cap (BR)$.

Give an example showing that equality does not always hold in part (c).

6. Let R be the relation $<$ on $\mathbf{R}$ and let $S = R \cap (\mathbf{Z} \times \mathbf{Z})$ be the corresponding relation on $\mathbf{Z}$. Show that $R \circ R = R$ but $S \circ S \neq S$.

7 Let R be a reflexive relation on a set X. Show that $R \subseteq R \circ R$.

8 Prove that if S and T are symmetric relations on a set X, then so are $S \cup T$, $S \cap T$, and $S \circ S$.

9 A relation on a set may or may not be reflexive, it may or may not be symmetric, and it may or may not be transitive. Show that all eight possibilities can occur by giving examples of everyday relations illustrating each type.

10 Let E and F be equivalence relations on a set X. Show that $E \cap F$ is an equivalence relation but that, in general, $E \cup F$ is not.

11 Is the converse of Theorem 2 true? That is, if E is a relation on X such that $\{\{x\}E | x \in X\}$ is a partition of X, does it necessarily follow that E is an equivalence relation?

12 Prove Theorem 3.

13 Suppose X is the set ⍳*M* and Y is the set ⍳*N*. Given that *R* and *S* are *M*-by-*N* matrices describing relations R and S from X to Y, write APL expressions for the matrices representing the following relations.

(a) $R \cup S$.
(b) $R \cap S$.
(c) $(X \times Y) - R$.
(d) R^{-1}.

14 Let R, S, *R*, *S* be as in Exercise 13. Write APL propositions equivalent to the following statements.

(a) R implies S.
(b) If R does not hold, then S holds. [By this we mean that if $x \in X, y \in Y$, and $(x, y) \notin R$, then $(x, y) \in S$.]
(c) R and S never hold simultaneously. [That is, there is no pair (x, y) with $(x, y) \in R$ and $(x, y) \in S$.]

15 Write APL expressions defining the matrices representing the following relations on ⍳10.

(a) =
(b) ≠
(c) ≤
(d) >
(e) "is at least 2 more than"
(f) "equals half of"

16 Explain how one could verify Theorem 1 in particular cases at a terminal.

17 Show that the *N*-by-*N* logical matrix *R* defines a reflexive relation on ⍳*N* if and only if `∧/,R≥(⍳N)∘.=⍳N`.

18 Prove the validity of the APL formulations of symmetry and transitivity given in the text.

19 Describe how the random number generator in an APL terminal system may be used to construct "random" logical matrices, that is, matrices *R* such that the entries `R[I;J]` are chosen independ-

ently from the set {0, 1} with each value having a probability of ½ of being chosen.

20 Using the technique of Exercise 19, construct 10 random, logical, 10-by-10 matrices *R*. How many times is *R*∨.∧*R* equal to 10 10⍴1? Explain your answer.

*21 Let *R* be an *N*-by-*N* logical matrix. To compute *R*∨.∧*R* in the natural way requires a time proportional to *N*⋆3. Show that it is possible to decide whether or not *R* defines an equivalence relation on ⍳*N* in a time proportional to *N*⋆2.

22 Let X and Y be finite subsets of **R**. How could we represent a relation from X to Y by one or more APL arrays?

23 The workspace *EXAMPLES* contains a 25-by-25 logical matrix *E*25. If one entry of *E*25 is changed, then *E*25 becomes the characteristic matrix of an equivalence relation on ⍳25. Which entry should be changed?

24 Let *E* be the characteristic matrix for an equivalence relation E on ⍳*N* and let *U* be a vector with distinct components listing the elements of a subset U of ⍳*N*. Write an APL proposition corresponding to the assertion that U is a set of representatives for the set of equivalence classes of E.

3. FUNCTIONS

A *function* from a set X to a set Y is a relation f from X to Y such that for each x in X the set $\{x\}f$ has exactly one element. The statement "f is a function from X to Y" is written symbolically $f{:}X \longrightarrow Y$ or $X \xrightarrow{f} Y$. The set X is called the *domain* of f, and Y is called the *codomain* or *range* of f. If $\{x\}f = \{y\}$, then y is called the *image* of x under f, and we write $y = xf$ or $f{:}x \longmapsto y$. The words "map" and "mapping" are synonyms for "function". Words such as "operator" and "transformation" are also used to refer to certain kinds of functions. The statement $f{:}x \longmapsto y$ can be read "f maps x to y." The set of all functions from X to Y is denoted Y^X. Exercise 3 explains why this notation was chosen.

Mathematicians have developed a great many different notations for representing the image of an element x under a function f. We have introduced the notation xf, but fx, $f(\mathrm{x})$, f_x, and x^f are also widely used. Other notations are possible when the symbol representing the function is something other than a letter. For example, $\hat{x}$ and x' could be the images of x under two functions ^ and ′, respectively. Analysts seem to favor writing the name of the function on the left, as in the familiar calculus expression $y = f(x)$. Algebraists, on the other hand, tend to prefer writing the name of the function on the right. Each convention has advantages as well as disadvantages. Faced with this abundance of notational possibilities, students of mathematics have little choice but to learn to be flexible and to adopt the conventions of the particular branch of mathematics they are studying. In this book we are torn between the preference of algebraists for writing the symbol for a function on the right and the rule in APL that requires the symbol for a monadic function to be placed to the left of its argument. The decision has been made to let these two notational conventions coexist, peacefully it is hoped. The wisdom of this decision will have to be judged by the reader.

Let us consider now some examples of functions.

Example 1. On any set X the identity relation e is a function, the *identity function.* For any x in X we have $xe = x$.

Example 2. If X and Y are sets, then the *projection* of $X \times Y$ onto X is the map taking (x, y) to x. The projection onto Y is similarly defined.

Example 3. A vector of length N is a function whose domain is ⍳N. Whether we use traditional notation, as in $v = (2, 4, 6)$, or APL notation, `V←2 4 6`, the ith component of a vector is the image of i under the vector.

Example 4. The most general definition of a matrix is a function whose domain is a Cartesian product $X \times Y$. Normally, we consider only the case in which X is ⍳M and Y is ⍳N for some positive integers M and N. If A is a matrix, then the image of the pair (i, j) under A is written A_{ij} or, in APL notation, $A[I;J]$. Some authors use the statement $A = [a_{ij}]$ to assert that A is a matrix whose IJth entry is a_{ij}. We will not use statements of this type here.

Example 5. For any set X we may define a function g mapping X to 2^X by $xg = \{x\}$.

Example 6. Let Π be a partition of the set X. For each x in X, let $\pi(x)$ be the block of Π containing x. Then π is called the *natural map* from X to Π.

Example 7. A *binary operation* on a set X is a function from $X \times X$ to X. The ordinary arithmetic operations $+$, $-$, and $\times$ are binary operations on $\mathbf{R}$. The operations $\cup$ and $\cap$ may be considered to be binary operations on 2^X for any set X. If f is a binary operation on X, then the image $(x, y)f$ of the ordered pair (x, y) is often written xfy. For example, $x + y$ and $x \times y$ are the images of (x, y) under the binary operations $+$ and $\times$, respectively. In fact, when the operation is clearly understood, we may even write xy, as is often done in the case of multiplication. (Of course, the omission of the symbol for any function is strictly prohibited in APL.)

It is convenient to use the symbol • to denote a typical binary operation on a set X, with $x \bullet y$ denoting the image of (x, y) under •.

Using the extended definition of a matrix given in Example 4, we may say that a binary operation on X is a matrix whose rows and columns are indexed by X and whose entries lie in X. Thus a binary operation on ⍳*N* is simply an *N*-by-*N* matrix with entries in ⍳*N*. For example, if

```
      ⎕←MAX←(⍳4)∘.⌈⍳4
1 2 3 4
2 2 3 4
3 3 3 4
4 4 4 4
```

then *MAX* is the binary operation on ⍳4 defined by the maximum operation ⌈.

There are two important properties a binary operation • on a set X may possess. If x and y are in X and $x \bullet y = y \bullet x$, we say x and y *commute.* If every pair of elements in X commutes, • is called a *commutative* binary operation. If $x \bullet (y \bullet z) = (x \bullet y) \bullet z$, then x, y, and z *associate*, and if every triple of elements in X associates, we say • is an *associative* operation. On **R** the operations + and × are commutative and associative, but − is neither.

Example 8. Let X be a fixed set. Each subset A of X determines a function f_A from X to $\{0, 1\}$. If x is in X, then xf_A is defined to be 1 if $x \in A$ and 0 if $x \notin A$. The function f_A is called the *characteristic function* of A. If $B \subseteq X$, then $A = B$ if and only if $f_A = f_B$. It is in this sense that f_A characterizes A.

Characteristic functions provide another method of representing subsets of ⍳*N*. If A is a subset of ⍳*N*, then the characteristic function of A is the logical vector *X* such that *X*[*I*] is 1 if and only if *I* is in A. We will refer to *X* as the *characteristic vector* of A. The basic set theoretic operations can easily be performed using characteristic vectors. For example, if *Y* is the characteristic vector for another subset B of ⍳*N*, then *X*∨*Y* is the characteristic vector for $A \cup B$. The formulation of the characteristic vectors for $A \cap B$, $A - B$, and the complement of A in ⍳*N* is left as an exercise.

In Section 1 we described how to represent a set of sets of real numbers by a matrix whose *i*th row lists the elements of the *i*th set. For example, the matrix

```
      ⎕←A←2 SSUB 4
1 2
1 3
2 3
1 4
2 4
3 4
```

represents the six two-element subsets of {1, 2, 3, 4}. This method of representation is best suited to the case in which all of the sets have the same number of elements. Characteristic vectors can be used as an alternative for representing sets of sets, particularly subsets of ι*N*, which easily handles sets of different cardinalities.

Suppose $A_1, \ldots, A_k$ are subsets of ι*N*. We can represent $\{A_1, \ldots, A_k\}$ by the matrix *C* whose *I*th row is the characteristic vector of A_I. The procedure *SCHV* computes this representation for the family of sets listed in the rows of a matrix.

```
      ⎕←C←4 SCHV A
1 1 0 0
1 0 1 0
0 1 1 0
1 0 0 1
0 1 0 1
0 0 1 1
```

The first argument of *SCHV* is an integer *N* such that all the sets are subsets of ι*N*.

```
      ⎕IO←0
      ⎕←D←5 SCHV A
0 1 1 0 0
0 1 0 1 0
0 0 1 1 0
0 1 0 0 1
0 0 1 0 1
0 0 0 1 1
```

Let us now take another look at our definition of the term "function". We have said that a function *f* from *X* to *Y* is a particular kind of relation from *X* to *Y*, that is, a subset of *X* × *Y*. Suppose we are given *f* in just this form, as a set of ordered pairs. Is it possible to determine *X* and *Y* from this information? Our definition says that for every *x* in *X* there is exactly one element *y* of *Y* such that (*x*, *y*) is in *f*. Thus *X* is the set of first components of the elements of *f*. Therefore the domain *X* is uniquely determined. However, the codomain Y is not unique. The set of second components of the elements of f is the image Xf of f, which is a subset of Y, but Y may be any set containing Xf. For example, if

$$f = \{(m, m^2 + 1) \mid m \in \mathbf{Z}\},$$

then, according to our definition, $f:\mathbf{Z} \longrightarrow \mathbf{Z}$, $f:\mathbf{Z} \longrightarrow \mathbf{Q}$, and $f:\mathbf{Z} \longrightarrow \mathbf{R}$.

In some areas of mathematics it is convenient to change the definition so that a function F from X to Y determines both X and Y uniquely. This is done by defining the function F to be a triple (X, Y, f), where X and Y are sets and f is a function in our sense from X to Y. In this context f is called the *graph* of F. We will not adopt this somewhat more cumbersome, although technically superior, definition of a function. The reader is warned, however, that a few of the definitions we will give concerning functions assume that the codomains as well as the domains are specified.

If $f:X \longrightarrow Y$ and $A \subseteq X$, we can construct a function h from A to Y by restricting the domain in the following sense. We set $h = f \cap (A \times Y)$ and refer to h as the *restriction* of f to A, writing $h = f|_A$. The elements of h are the ordered pairs in f whose first components lie in A.

THEOREM 1. Let $f:X \longrightarrow Y$ and $g:Y \longrightarrow Z$ and suppose $A \subseteq X$. Then
(a) $f \circ g$ is a function from X to Z.
(b) $f|_A$ is a function from A to Y.

Proof. (a) The composition $f \circ g$ is certainly a relation from X to Z. We must show $f \circ g$ is, in fact, a function. Suppose (x, z_1) and (x, z_2) are both in $f \circ g$. Then there exist y_1 and y_2 in Y such that $(x, y_i) \in f$ and $(y_i, z_i) \in g$ for $i = 1, 2$. However, since f and g are functions, $y_1 = xf = y_2$ and, therefore, $z_1 = y_1 g = y_2 g = z_2$. Thus x is related to at most one element of Z by $f \circ g$. For any x in X we have (x, xf) in f and $(xf, (xf)g)$ in g, and so $(x, (xf)g)$ is in $f \circ g$. Hence x is related to exactly one element of Z and $f \circ g$ is a function from X to Z.

The proof of part (b) is left as an exercise. □

A function from ⍳*L* to ⍳*M* is simply a vector *F* of length *L* whose components lie in ⍳*M*. If *G* is a vector that is a function from ⍳*M* to ⍳*N*, then the composition of *F* and *G* is the vector *H*←*G*[*F*].

```
      ⎕IO←1                              G[F]
      F←1 3 5                      2 6 10
      G←2 4 6 8 10
```

If $f:X \longrightarrow Y$ and $A \subseteq X$, then the set Af defined in Section 2 can be written $\{af \mid a \in A\}$. We say that f is *surjective* or that f maps X *onto* Y if $Xf = Y$. This means that for each y in Y there is at least one x in X such that $y = xf$. Note that the definition of the term "surjective" requires that the codomain Y be specified. We say that f is *injective* or *one to one* (abbreviated 1–1) if each element of Y is the image of at most one element of X. In general, the inverse f^{-1} of f is a relation from Y to X, but not a function. If B is a subset of Y, then Bf^{-1} is called the *inverse image* of B under f. It is clear that f is injective if and only if $|\{y\}f^{-1}| \leq 1$ for all y in Y. If f is both surjective and injective, then f is said to be *bijective.* A *surjection, injection,* or *bijection* is a function that is surjective, injective, or bijective, respectively. Bijections are also called 1–1 *correspondences.* The inverse of a bijection from X to Y is not only a function but a bijection from Y to X. A bijection from X to itself is called a *permutation* of X. The set of all permutations of the set X will be denoted $\Sigma(X)$. In the special case that X is ιN we write Σ_N for $\Sigma(X)$.

THEOREM 2. Suppose $f{:}X \longrightarrow Y$ and $g{:}Y \longrightarrow Z$. Then

(a) If f and g are surjective, then so is $f \circ g$.

(b) If f and g are injective, then so is $f \circ g$.

(c) If $f \circ g$ is surjective, then g is surjective.

(d) If $f \circ g$ is injective, then f is injective.

Proof. We will prove only part (a), leaving the rest to the reader. Let $z \in Z$. Then, since g maps Y onto Z, there is an element y in Y with $yg = z$. Because f is surjective, there is an element x in X with $xf = y$. Then $x(f \circ g) = z$, so $f \circ g$ is surjective. □

COROLLARY 3. If f and g are permutations of the set X, then so are $f \circ g$ and f^{-1}. □

COROLLARY 4. Suppose $f{:}X \longrightarrow Y$ and $g{:}Y \longrightarrow X$. Assume that $f \circ g$ is the identity function on X and $g \circ f$ is the identity function on Y. Then f and g are bijections and $g = f^{-1}$.

Proof. Since the identity function on a set is a bijection, we have in particular that $f \circ g$ is injective and $g \circ f$ is surjective. By parts (c) and (d) of Theorem 2 we see that f is both injective and surjective and hence is a bijection. If $xf = y$, then $yg = (xf)g = x(f \circ g) = x = yf^{-1}$. Thus $g = f^{-1}$. □

As an application of Corollary 4, we complete our discussion of the connection between equivalence relations and partitions begun in Section 2. For any set X, let Eq(X) denote the set of equivalence relations on X and let Part(X) be the set of partitions of X.

THEOREM 5. For any set X there exists a bijection from Eq(X) to Part(X).

Proof. Define $f{:}\mathrm{Eq}(X) \longrightarrow \mathrm{Part}(X)$ and $g{:}\mathrm{Part}(X) \longrightarrow \mathrm{Eq}(X)$ by $f{:}E \longmapsto X/E$ and $g{:}\Pi \longmapsto E_\Pi$. By Theorem 2.3, $f \circ g$ is the identity on Eq(X) and $g \circ f$ is the identity on Part(X). By Corollary 4, f is a bijection. □

Let X, Y, and Z be sets An element of $(X \times Y) \times Z$ has the form

$((x, y), z)$, and an element of $X \times (Y \times Z)$ has the form $(x, (y, z))$. The function that maps $((x, y), z)$ to $(x, (y, z))$ is clearly a bijection from $(X \times Y) \times Z$ to $X \times (Y \times Z)$. When such an obvious bijection exists from one set to another, we will often *identify* the two sets, considering them to be the same even though, strictly speaking, they are different. Thus we will normally write $X \times Y \times Z$ without parentheses and denote a typical element by (x, y, z).

We have been working informally with the notion of a finite set. It is now time to make this concept precise. A set X is *finite* if, whenever $f: X \longrightarrow X$ is injective, then f is surjective. The map $x \longmapsto x+1$ of $\mathbf{N}$ into itself is injective but not surjective. Thus, by our definition, $\mathbf{N}$ is not finite, that is, it is *infinite.* The connection between this definition of finiteness and the idea of a set with a finite number of elements is given by the following theorem.

THEOREM 6. A set X is finite if and only if there exists a nonnegative integer n and a bijection from $\{1, 2, \ldots, n\}$ to X. The integer n is unique.

Although this theorem is intuitively very natural and perhaps even "obvious", its formal proof requires a fairly large number of steps that do not significantly increase our basic understanding of the underlying concepts. For this reason we omit the proof. □

THEOREM 7. Let X be a finite set and suppose $f: X \longrightarrow X$ is surjective. Then f is injective.

Proof. For each x in X choose one element y of X such that $x = yf$ and define $g: X \longrightarrow X$ by $g: x \longmapsto y$. If $x_1 g = x_2 g$, then $x_1 = (x_1 g)f = (x_2 g)f = x_2$ and so g is injective. Since X is finite, g must also be surjective. Now suppose f is not injective. Then, for some $x_1 \neq x_2$ in X, we have $x_1 f = x_2 f$. Since g is surjective, there exist u_1 and u_2 in X such that $u_1 g = x_1$ and $u_2 g = x_2$. Thus

$$u_1 = (u_1 g)f = x_1 f = x_2 f = (u_2 g)f = u_2.$$

Hence $u_1 = u_2$, which is impossible, since $u_1 g \neq u_2 g$. □

Theorem 7 shows that the finite sets are precisely those sets X for which the notions of injectivity and surjectivity coincide for maps from X to X. This is the most important property of finite sets.

Suppose for each element i of a nonempty set I we have a set A_i. We can think of A as a function mapping the element i in I to the set A_i. Such a function is often referred to as a *family of sets indexed by I.* Given such a family, we define

$$\bigcup_{i \in I} A_i$$

to be the set of elements contained in at least one A_i and

$$\bigcap_{i \in I} A_i$$

to be the set of elements contained in every one of the A_i. We say the A_i are *pairwise disjoint* if $A_i \cap A_j = \emptyset$ whenever $i \neq j$. If, for each real number x, we let

$$A_x = \{y \in \mathbf{R} | x < y\},$$

then

$$\bigcup_{x \in \mathbf{R}} A_x = \mathbf{R}, \quad \bigcap_{x \in \mathbf{R}} A_x = \emptyset.$$

EXERCISES

1 According to our definition, a function f from X to Y can have only one argument: an element of its domain. What do we mean when we speak of a function with two or more arguments?

2 For each of the following subsets of $\mathbf{Z} \times \mathbf{Z}$ tell whether or not the set is a function from $\mathbf{Z}$ to $\mathbf{Z}$.

(a) $\{(m, m^2) | m \in \mathbf{Z}\}$.

(b) $\{(2m, m) | m \in \mathbf{Z}\}$.

(c) $\{(m, n) | m, n \in \mathbf{Z}, |m| = |n|\}$.

(d) $\{(m, 2m + 1) | m \in \mathbf{Z}\} \cup \{(m - 1, 2m - 1) | m \in \mathbf{Z}\}$.

3 Show that if X and Y are finite sets, then $|Y^X| = |Y|^{|X|}$. (What convention must be made for 0^0?)

4 Prove part (b) of Theorem 1.

5 Complete the proof of Theorem 2.

6 Give examples of pairs of functions f and g with $f \circ g$ defined such that

(a) $f \circ g$ is surjective but f is not surjective.

(b) $f \circ g$ is injective but g is not injective.

7 How many permutations of the empty set are there?

8 Show that the empty set is finite by our definition.

9 Prove that if X is a finite set, then the number of permutations of X is $|X|!$.

10 Using our definition of finiteness, prove

(a) If X is finite and $f: X \longrightarrow Y$ is a bijection, then Y is finite.

(b) Any subset of a finite set is finite.

11 Let X be a finite set and let f, g be maps of X into itself. Suppose

$f \circ g$ is the identity permutation of X. Show that f and g are permutations of X and $g = f^{-1}$. Give an example that shows that the finiteness of X is necessary.

12 Let $f: X \longrightarrow Y$ and suppose that A and B are subsets of Y. Show that

$$(A \cap B)f^{-1} = (Af^{-1}) \cap (Bf^{-1}).$$

(Compare Exercise 2.5c.)

13 Let X and Y be sets, let $f: X \longrightarrow Y$, and let E be the set of elements (x_1, x_2) in $X \times X$ such that $x_1 f = x_2 f$. Show that E is an equivalence relation on X.

14 What would be a reasonable definition of a ternary operation on a set X?

15 Show that the intersection of any nonempty family of equivalence relations on a set X is again an equivalence relation on X.

16 Let E be a family of equivalence relations on a set X indexed by **N**. Suppose $E_n \subseteq E_{n+1}$ for all n in **N**. Show that

$$\bigcup_{n \in \mathbf{N}} E_n$$

is an equivalence relation on X.

17 Prove De Morgan's Laws. That is, show that if A is a family of sets indexed by I and B is a set, then

(a) $B \cap \bigcup_{i \in I} A_i = \bigcup_{i \in I} (B \cap A_i)$.

(b) $B \cup \bigcap_{i \in I} A_i = \bigcap_{i \in I} (B \cup A_i)$.

(c) $B - \bigcup_{i \in I} A_i = \bigcap_{i \in I} (B - A_i)$.

(d) $B - \bigcap_{i \in I} A_i = \bigcup_{i \in I} (B - A_i)$.

18 Sketch a proof of Theorem 6, making clear what facts about the integers are used.

19 Let X and Y be the characteristic vectors for two subsets A and B of ιN. What are the characteristic vectors for $A \cap B$, $A - B$, and the complement of A in ιN?

20 Let A be a vector listing the elements of a subset A of ιN. Write an APL expression for the characteristic vector X of A. Suppose X is given. How can we obtain a list of the elements of A?

21 Show that the characteristic matrix R of a relation R from ιM to ιN is the characteristic function of R considered as a subset of the Cartesian product of ιM and ιN.

22 Let R be the characteristic matrix of a relation R from ιM to ιN.

Write an APL proposition equivalent to the assertion that R is a function from ⍳M to ⍳N.

23 We have two methods available for representing a function f from ⍳M to ⍳N by an APL array. We can use either the vector F such that F[I] is the image of I under f or the characteristic matrix R of f. Write APL expressions that construct R from F and F from R.

24 Let F be a vector describing a function f from ⍳M to ⍳N. Write APL propositions corresponding to the assertions that f is 1–1 and that f is onto. Let A be a vector listing the elements of a subset A of ⍳N. What is the characteristic vector of Af^{-1}?

25 Let F be a vector. Write an APL proposition equivalent to the assertion that F is a permutation of ⍳⍴F.

26 Assume that F is a permutation of ⍳⍴F. Write an APL expression for the inverse of F.

27 Let B be an N-by-N matrix. Write an APL proposition corresponding to the statement that B is a binary operation on ⍳N.

For Exercises 28 and 29, assume that B is a binary operation on ⍳N.

28 Write APL propositions for the following assertions.

(a) B is commutative.

(b) B is associative.

29 A *left identity element* for a binary operation • on a set X is an element e of X such that $e \bullet x = x$ for all x in X. Write an APL proposition asserting that I is a left identity element for B. Write an APL expression for the characteristic vector of the set of left identity elements for B.

30 Let $A_1, \ldots, A_m$ be subsets of ⍳N and let A be the matrix such that A[I;] is the characteristic vector for A_I. What are the characteristic vectors of $A_1 \cup \ldots \cup A_m$ and $A_1 \cap \ldots \cap A_m$?

31 Let R be the characteristic matrix for a relation R from ⍳M to ⍳N. Let A be a vector listing the elements of a subset A of ⍳M. Write an APL expression for the characteristic vector of the set AR.

32 Let $A_1, \ldots, A_r$ and $B_1, \ldots, B_s$ be two sequences of subsets of ⍳N and let X and Y be the matrices of characteristic vectors for these two sequences. Write APL expressions for the matrices whose i, jth entries are

(a) $|A_i \cap B_j|$.

(b) $|A_i \cup B_j|$.

(c) 1 if $A_i \subseteq B_j$ and 0 otherwise.

(d) 1 if $A_i = B_j$ and 0 otherwise.

(e) 1 if $A_i \cap A_j \neq \emptyset$ and 0 otherwise.

33 Let *X* be the characteristic vector for the subset of ⍳*N* listed in the vector *A*. Suppose *P* is a permutation of ⍳*N*. Write an APL expression involving only *X* and *P* for the characteristic vector of the set listed in *P*[*A*].

34 Write an APL expression for a (2⋆*N*)-by-*N* matrix whose rows list the characteristic vectors of all subsets of ⍳*N*.

35 Let • be an associative binary operation on a set X. If $x \in X$ and n is a positive integer, then x^n is defined to be the product $x \bullet x \bullet \ldots \bullet x$ with n factors. In order to compute x^n, it is not always necessary to calculate $n - 1$ products. For example, x^8 can be computed by forming $x^2 = x \bullet x$, $x^4 = x^2 \bullet x^2$, and $x^8 = x^4 \bullet x^4$, which requires only three products. Suppose that, in order to evaluate a product $x \bullet y$, a computer program must be run at a cost of \$.50. What is the minimum expenditure required to compute x^{127} for a given element x of X?

*4. SETS OF SETS USING APL

In the previous sections of this chapter we have discussed ways of representing subsets of ⍳*N* by APL arrays and techniques for performing set-theoretic operations with these representations. Working with a small number of sets presents few challenges, either theoretical or computational. However, the study of certain types of large families of sets suggests many interesting problems, some of which can lead to extensive machine computation. Although more properly a part of the branch of mathematics known as combinatorial theory, several problems of this type will be described briefly in the next section. In this section we will investigate some methods for answering various types of questions about moderately large families of subsets of ⍳*N* using APL. These techniques will be useful not only in Section 5 but also in Section 3.9. The reader is assumed to be familiar with inner products of vectors and matrices as described in Section A1.7.

As noted in Section 3, characteristic vectors are usually more convenient than lists of elements for set-theoretic calculations with subsets of ⍳*N* when *N* is not too large. The characteristic vector *X* for the set listed in the vector *A* is (⍳*N*)∊*A*.

```
      A←3 8 2 6 3
      ⎕←X←(⍳9)∊A
0 1 1 0 0 1 0 1 0
```

Given X, we can easily reconstruct a list of the elements.

```
      ⎕←B←X/⍳9
2 3 6 8
```

Note that A and B are not the same vector. We can also use the procedure *SCHV* to construct characteristic vectors.

```
      9 SCHV A
0 1 1 0 0 1 0 1 0
```

Let us now consider some simple questions about two families $\{A_1, \ldots, A_5\}$ and $\{B_1, \ldots, B_4\}$ of subsets of ⍳9, where A_i is the set listed in the ith row of

```
      ⎕←A←5 3⍴7 5 2 4 3 5 6 9 6 1 6 8 8 2 8
 7 5 2
 4 3 5
 6 9 6
 1 6 8
 8 2 8
```

and B_j is the set listed in the jth row of

```
      ⎕←B←4 4⍴7 2 8 6 9 1 8 9 6 3 9 3 7 3 4 2
 7 2 8 6
 9 1 8 9
 6 3 9 3
 7 3 4 2
```

How can we construct APL expressions that answer the following?

1. Is there an A_i that is a subset of some B_j?
2. Are the sets A_1, A_2, A_3, A_4, and A_5 distinct?
3. What are the numbers $|A_i \cap B_j|$?

To get an idea about how to form the right expressions, let us look at the case of just two sets. Suppose P and Q are the characteristic vectors for two subsets P and Q of ⍳N. Then $P \subseteq Q$ if and only if the APL proposition ∧/P≤Q is true. This proposition can be rewritten as P∧.≤Q. The example

```
      P←1 0 0 1 0 1                      P∧.≤Q
      Q←1 1 0 0 1 1          0
      ∧/P≤Q
0
```

verifies that {1, 4, 6} is not a subset of {1, 2, 5, 6}. The sets P and Q are equal if and only if ∧/P=Q or, equivalently, if and only if P∧.=Q is true.

```
      P∧.=Q
0
```

Finally, the characteristic vector for $P \cap Q$ is P∧Q, and so $|P \cap Q|$ is +/P∧Q or P+.∧Q.

```
      P+.∧Q
2
```

Now let

```
      □←X←9 SCHV A                      □←Y←9 SCHV B
0 1 0 0 1 0 1 0 0                 0 1 0 0 0 1 1 1 0
0 0 1 1 1 0 0 0 0                 1 0 0 0 0 0 0 1 1
0 0 0 0 0 1 0 0 1                 0 0 1 0 0 1 0 0 1
1 0 0 0 0 1 0 1 0                 0 1 1 1 0 0 1 0 0
0 1 0 0 0 0 0 1 0
```

To answer question 1 we form the matrix U such that U[I;J] is

X[I;]∧.≤Y[J;],

which is 1 if and only if $A_I \subseteq B_J$. By the definition of the inner product, U is X∧.≤⍉Y.

```
      U←X∧.≤⍉Y
```

Our answer to question 1 will be yes provided some entry of U is 1. From

```
      +/,U
2
```

we see that there are two pairs (i, j) such that $A_i \subseteq B_j$.

To answer question 2, we want the matrix V such that V[I;J] is X[I;]∧.=X[J;]. This is the matrix X∧.=⍉X.

```
      V←X∧.=⍉X
```

Since $A_i = A_i$ for each i, the five diagonal entries of V are all 1. The sets A_i are distinct provided V has no other entries equal to 1.

```
      +/,V
5
```

Since V has just 5 nonzero entries, the A_i are distinct.

Finally, to answer question 3, we must determine the matrix W such that W[I;J] is X[I;]+.∧Y[J;]. This means that W is X+.∧⍉Y.

```
      ⎕←W←X+.∧⍉Y
2 0 0 2
0 0 1 2
1 1 2 0
2 2 1 0
2 1 0 1
```

Here we see, for example, that $|A_4 \cap B_3| =$ W[4;3] $= 1$.

Sometimes we want to work with the set of all subsets of ⍳N. The characteristic vector of a subset of ⍳N is a logical vector of length N that we can consider as the vector of digits of an integer in binary or base 2 notation. (See Section A1.8.) The characteristic vectors of all 2^N subsets of ⍳N are simply the N-digit binary representations of the numbers

$$0, 1, \ldots, 2^N - 1.$$

The matrix Z←⍉(Nρ2)⊤⍳2*N lists these vectors. The order in which the vectors are listed depends on the origin. The order is more natural in origin 0.

```
      ⎕IO←0                        ⎕←Z←⍉(Nρ2)⊤⍳2*N
      N←3                     0 0 0
                              0 0 1
                              0 1 0
                              0 1 1
                              1 0 0
                              1 0 1
                              1 1 0
                              1 1 1
```

Given a logical vector R of length N, we can find the integer I such that R is Z[I;] by I←2⊥R.

```
      R←1 0 1                          Z[I;]
      ⎕←I←2⊥R                     1 0 1
5
```

It is often useful to be able to select one representative from each set in a given family of sets. If U is a matrix whose rows are characteristic vectors for nonempty subsets of ⍳*N*, then *V*←*SFEL U* is the vector whose *I*th component is the first element in *U*[*I*;]/⍳*N*. Using the matrix *X* previously defined, we get

```
      X                              ⎕IO←1
0 1 0 0 1 0 1 0 0                    SFEL X
0 0 1 1 1 0 0 0 0              2 3 6 1 2
0 0 0 0 0 1 0 0 1
1 0 0 0 0 1 0 1 0
0 1 0 0 0 0 0 1 0
```

It is possible to describe *SFEL U* by a single APL expression involving an inner product.

EXERCISES

1 Let *Z* be a logical matrix with *N* columns. Write an APL proposition corresponding to the assertion that the sets whose characteristic vectors are the rows of *Z* form a partition of ⍳*N*.

2 Let *E* be the characteristic matrix for an equivalence relation *E* on ⍳*N*. Write an APL expression for a vector that gives one representative from each equivalence class of *E*. Procedures in *CLASSLIB* may be used in the expression.

3 This problem deals with the sets $\{A_1, \ldots, A_5\}$ and $\{B_1, \ldots, B_4\}$ defined in the text. Write APL expressions that answer the following questions.

(a) For how many pairs (i, j) is $A_i \cap B_J = \emptyset$?

(b) What is the matrix of integers $|A_i \cup B_j|$?

(c) Which two-element set is contained in the largest number of B_j?

4 Let *Z* be a matrix all of whose entries are nonnegative integers less than *N* and let *F*←*N*⊥⍉*Z*. Show that the matrices *Z*∧.=⍉*Z* and *F*∘.=*F* are the same. Which construction takes less central processing unit (CPU) time?

5 Let *Z* be a logical matrix such that ∧/∨/*Z* is 1. Write an expression for *SFEL Z* using only primitive operations. (*Hint*. One way to solve the problem involves the inner product ⌈.×.)

6 Suppose the logical matrices *P1* and *P2* list the characteristic vectors of the blocks of two partitions Π_1 and Π_2 of ⍳*N*. Write an APL proposition corresponding to the assertion that Π_1 is a refinement of Π_2. (See Exercise 1.7.)

*5. BLOCK DESIGNS AND GRAPHS

In this section we will apply some of the techniques discussed in Section 4 to the study of two types of families of sets that are of particular interest. The first type is illustrated by the array *DESIGN*1 in *EXAMPLES*, which lists seven three-element subsets of ι7 with origin 1. For simplicity, let us rename this array *D*.

```
      ⎕IO←1
      ⎕←D←DESIGN1
1 2 3
1 4 5
1 6 7
2 4 6
2 5 7
3 4 7
3 5 6
```

It is not hard to verify that the sets listed in *D* satisfy the following properties.

1. Each set has three elements.
2. Each element of ι7 is contained in exactly three sets.
3. Each two-element subset of ι7 is contained in exactly one set.

We say that the rows of *D* form a block design with parameters 7, 7, 3, 3, 1.

In general, a *block design* with *parameters* v, b, k, r, λ is a pair $(X, \mathcal{B})$, where X is a finite set and $\mathcal{B}$ is a family of subsets of X such that

1. $|X| = v$.
2. $|\mathcal{B}| = b$.
3. If B is in $\mathcal{B}$, then $|B| = k$.
4. If x is in X, then x is contained in exactly r elements of $\mathcal{B}$.
5. Each two-element subset of X is contained in exactly λ elements of $\mathcal{B}$.

The elements of $\mathcal{B}$ are called the *blocks* of the design.

Let $\mathcal{D}$ be the family of sets listed in *D*. Using the methods of Section 4, we can easily check that (ι7, $\mathcal{D}$) is a block design. To make sure that the elements of $\mathcal{D}$ are subsets of ι7, we compute

```
      ∧/,D∊ι7
1
```

Obviously, ι7 has seven elements, and so v = 7. The remaining calculations are done using the matrix of characteristic vectors.

```
      ⎕←X←7 SCHV D
1 1 1 0 0 0 0
1 0 0 1 1 0 0
1 0 0 0 0 1 1
0 1 0 1 0 1 0
0 1 0 0 1 0 1
0 0 1 1 0 0 1
0 0 1 0 1 1 0
```

To check that |D|= 7, we need to see that the rows of X are distinct.

```
      +/,X∧.=⍉X
7
```

Since X∧.=⍉X has just seven entries equal to 1, we do have $b = 7$. The Ith component of +/X is the cardinality of the Ith element of D. From

```
      ∧/3=+/X
1
```

we see that all elements of D have $k = 3$ elements. The Ith component of +⌿X is the number of elements of D containing I. Since

```
      ∧/3=+⌿X
1
```

it follows that every element of ⍳7 is in $r = 3$ elements of D.

The fifth axiom is a little more complicated to verify. We first form

```
      Y←7 SCHV 2 SSUB 7
      ⍴Y
21 7
```

The Ith row of Y is the characteristic vector for the Ith two-element subset of ⍳7. If

```
      M←Y∧.≤⍉X
```

then M[I;J] is 1 if the Ith two-element set is contained in the Jth element of D and 0 otherwise. Therefore the Ith component of

```
      L←+/M
```

gives the number of elements of D containing the Ith two-element set. From

```
      ∧/1=L
1
```

we see that every two-element set is contained in $\lambda = 1$ elements of $\mathcal{D}$. Thus (⍳7, $\mathcal{D}$) is a block design with parameters $v = 7$, $b = 7$, $k = 3$, $r = 3$, $\lambda = 1$.

One type of block design that has been studied extensively is the projective plane. A *projective plane* is a block design whose parameters are of the form $v = b = q^2 + q + 1$, $k = r = q + 1$, $\lambda = 1$, where q is an integer greater than 1. The integer q is called the *order* of the plane. The design (⍳7, $\mathcal{D}$) is a projective plane of order 2. The order of every known projective plane is a power of a prime. No projective plane of order 6 exists, but it is still an unsolved question whether or not there is a projective plane of order 10. A great deal of effort has been devoted to this problem, including a very large amount of machine computation.

The concept of a graph provides another class of examples of interesting families of sets. An *undirected graph* (or, simply, a graph) is a pair (X, E), where X is a set and E is a set of two-element subsets of X. The elements of X are called the *vertices* of the graph and the elements of E are called the *edges*. One method of describing a graph is to choose one point in the plane for each vertex and connect two of these points by a line segment if the corresponding pair of vertices is an edge. For example, if X is ⍳5 in origin 0 and

$$E = \{\{0, 1\}, \{0, 2\}, \{1, 2\}, \{1, 3\}\},$$

then one possible diagram for the graph $G = (X, E)$ would be

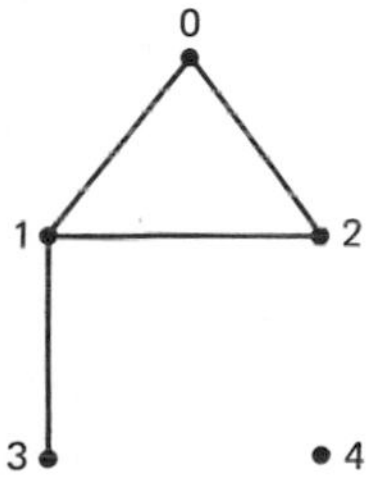

If $\{x, y\}$ is an edge of G, we say x is *connected* to y by an edge. Note in our example that 4 is not connected to any other vertex by an edge. One way to describe G by an APL array is to list the elements of E in a matrix such as

```
      ⎕←E←4 2⍴0 1 0 2 1 2 1 3
0 1
0 2
1 2
1 3
```

Of course, we must also remember that X is ⍳5 in order to determine G completely.

The number of graphs with the vertex set ⍳N is 2⋆2!N, since there are 2!N two-element subsets of ⍳N and the number of subsets of a set with M elements is 2⋆M. However, many of these graphs are so similar that they are not usually considered to be really distinct. For example, the graph H given by the diagram

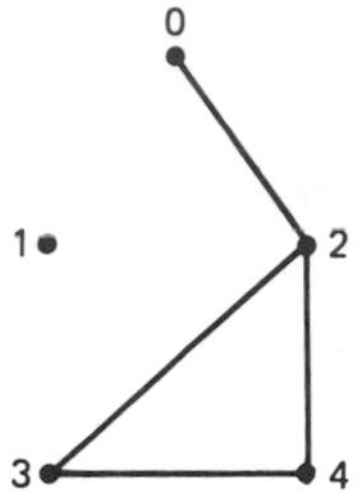

has the edge set {{0, 2}, {2, 3}, {2, 4}, {3, 4}} and thus is not the same graph as our example G. However, the diagrams for both consist of "a triangle with a tail and one extra point", and the second can be obtained from the first by renumbering the points in the diagram. We say that these two graphs are isomorphic.

More formally, we say that a graph (X_1, E_1) is *isomorphic* to a graph (X_2, E_2) if there is a bijection $f{:}X_1 \longrightarrow X_2$ such that the edges in E_2 are precisely the sets $\{xf, yf\} = \{x, y\}f$, where $\{x, y\}$ is an edge in E_1 or, equivalently, such that the map $\{x, y\} \longmapsto \{x, y\}f$ is a bijection of E_1 onto E_2. The map f is referred to as an *isomorphism* of (X_1, E_1) onto (X_2, E_2). If X_1 and X_2 are both ⍳N, then f is a permutation of ⍳N. If we set

```
      F←4 2 3 0 1                                ⎕IO←0
```

then the array

```
      F[E]
 4 2
 4 3
 2 3
 2 0
```

lists the edges of H. Since E lists the edges of G, we see that F is an isomorphism of G onto H.

Near the end of Chapter 3 we will return to the study of graphs and determine the number of essentially distinct, that is, nonisomorphic, graphs with n vertices, where $n \leq 5$. As an introduction to the techniques that will be used, let us consider several ways of representing a graph whose vertex set is ⍳N. We will use the preceding graph G = (⍳5,E) as an illustration.

Instead of listing the edges of G in the matrix E we could form the 5-by-5 logical matrix R in which `R[I;J]` is 1 if and only if I is connected to J by an edge of G. For our example we have

```
      ⎕←R←5 5⍴0 1 1 0 0 1 0 1 1 0 1 1 0 0 0 0 1 0 0 0 0 0 0 0 0
0 1 1 0 0
1 0 1 1 0
1 1 0 0 0
0 1 0 0 0
0 0 0 0 0
```

We call R the *adjacency matrix* for G.

Another approach is first to form

```
      ⎕←P←2 SSUB 5
 0 1
 0 2
 1 2
 0 3
 1 3
 2 3
 0 4
 1 4
 2 4
 3 4
```

and then to specify a graph with vertex set ⍳5 by describing which rows of P are edges. Thus the edges of G are the rows of P whose indices are listed in the vector

```
      V←0 1 2 4
```

If we form the corresponding characteristic vector,

```
      ⎕←Z←10 SCHV V
1 1 1 0 1 0 0 0 0 0
```

then we can list the edges of G in two ways.

```
      P[V;]                              Z/P
 0 1                                 0 1
 0 2                                 0 2
 1 2                                 1 2
 1 3                                 1 3
```

Finally, we can assign to G the integer

```
      ⎕←M←2⊥Z
928
```

which uniquely determines G. In this way we get a numbering of the graphs with vertex set ⍳5 from 0 to 1023. To see what the 364th graph in this numbering is, we compute

```
      ⎕←W←(10⍴2)⊤364
0 1 0 1 1 0 1 1 0 0
      W⌿P
 0 2
 0 3
 1 3
 0 4
 1 4
```

Thus the 364th graph is given by the diagram

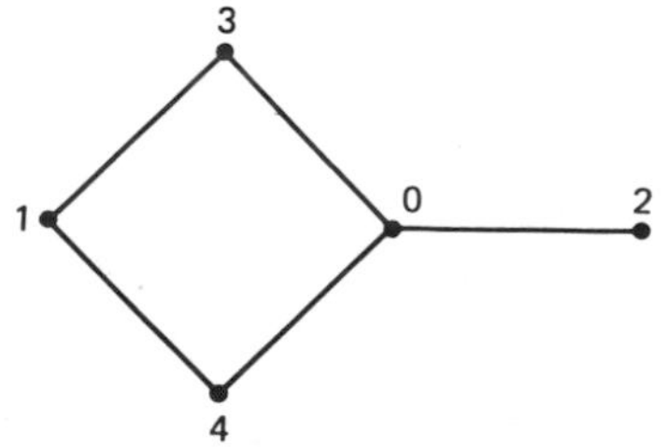

It should be emphasized that this numbering is relative to a fixed list of the two-element subsets of ⍳5, in this case given by P. A similar numbering of the graphs with vertex set ⍳N for any N can be constructed using the matrix 2 *SSUB* N.

EXERCISES

1 The workspace *EXAMPLES* contains arrays *DESIGN*2, *DESIGN*3, *DESIGN*4, and *DESIGN*5. Show that each of these arrays defines a block design and determine the parameters in each case.

2 Show that the parameters of a block design $(X, \mathcal{B})$ satisfy the equations $rv = kb$ and $\lambda(v - 1) = r(k - 1)$. [*Hint.* For the first equation, count the number of pairs (x, B), where x is in X and B is a block in $\mathcal{B}$ containing x.]

3 Sketch a diagram for each of the following graphs (X, E).

(a) $X = \{0, 1, 2, 3\}$, $E = \{\{0, 2\}, \{1, 3\}, \{2, 3\}\}$.

(b) $X = \{0, 1, 2, 3, 4\}$, $E = \{\{0, 2\}, \{0, 3\}, \{1, 3\}, \{1, 4\}, \{2, 4\}\}$.

4 Let G, H, and K be graphs. Show that
 (a) G is isomorphic to G.
 (b) If G is isomorphic to H, then H is isomorphic to G.
 (c) If G is isomorphic to H and H is isomorphic to K, then G is isomorphic to K.

5 Suppose the edges of a graph G with vertex set ⍳N are listed in the rows of a matrix E. Show how to construct the adjacency matrix R of G from N and E using one or more APL statements. (*Hint.* First form `C←N SCHV E`.)

6 Construct the adjacency matrices for the following graphs.
 (a)

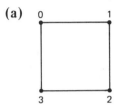

 (b)

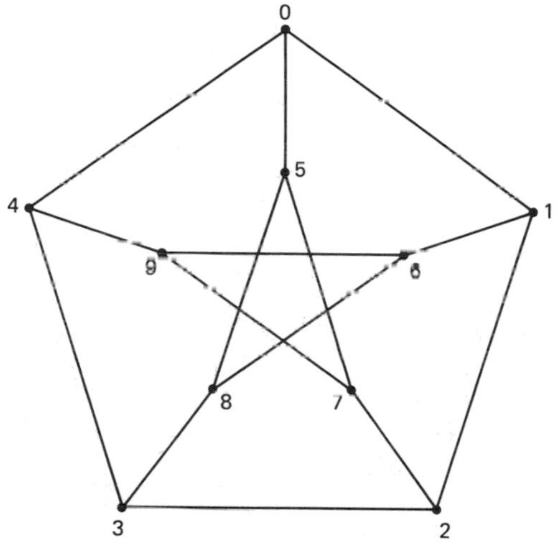

7 Let K be the graph in Exercise 6a. How many isomorphisms of K onto itself are there?

8 What is the number of the graph in Exercise 6a in the numbering of the graphs with vertex set ⍳4 based on the matrix `2 SSUB 4`? What is the 18th graph in this numbering?

9 Let (X, E) be a graph and let $f{:}X \longrightarrow Y$ be a bijection. Define $F = \{\{xf, yf\} \mid \{x, y\} \in E\}$. Show that (Y, F) is a graph isomorphic to (X, E).

10 Let X be a set and let Z be the set of two-element subsets of X.

Suppose E and E' are subsets of Z. Prove that the graphs (X, E) and (X, E') are isomorphic if and only if $(X, Z - E)$ and $(X, Z - E')$ are isomorphic.

11 Give an appropriate definition for one block design $(X_1, \mathcal{B}_1)$ to be isomorphic to another block design $(X_2, \mathcal{B}_2)$. Prove that isomorphic designs have the same parameters.

12 Show that any projective plane of order 2 is isomorphic to the design described by the matrix `DESIGN1`.

*13 Show that the arrays `DESIGN4` and `DESIGN5` describe block designs that are not isomorphic.

2

THE INTEGERS

In this chapter we will develop some important properties of the integers. Integers are encountered frequently in many branches of mathematics, but this is not the only reason they are of interest to us. As we pursue the study of algebra, we will repeatedly come across concepts whose significance was first noted in the context of the integers. Later these concepts were seen to be useful in much more general situations. Since readers are expected to have a basic familiarity with integers, we will proceed in an informal manner, much as we did in our review of sets in Chapter 1. Throughout this chapter the APL index origin will be assumed to be 1 unless there is an explicit statement to the contrary.

1. DIVISIBILITY

Space does not permit listing all the facts about the integers that we will assume without proof. However, there are two theorems that are so important that they need to be mentioned.

THEOREM 1. Every nonempty set of positive integers has a smallest element. □

THEOREM 2. Let m and n be integers with $n \neq 0$. There exist unique integers q and r such that $m = qn + r$ and $0 \leq r < |n|$. □

Let X be a nonempty set of positive integers. According to Theorem 1, there is an element x in X such that $x \leq y$ for all y in X. It is this property that makes it possible to prove theorems by mathematical induction.

In Theorem 2 the integers q and r are called the *integral quotient* and the *remainder*, respectively, when m is divided by n. Integral quotients and remainders are easily computed in APL. The remainder `R` when `M` is divided by `N` is `(|N)|M`. (Why not `N|M`?) The integral quotient `Q` is `(M-R)÷N`. For convenience, `CLASSLIB` contains procedures `ZQUOT` and `ZREM`, which can be used to calculate integral quotients and remainders, respectively.

```
      M←12                      N ZREM M
      N←¯5              2
      (|N)|M                    M ZQUOT N
2                      ¯2
      (M-2)÷N
¯2
```

Note that the order of the arguments for *ZQUOT* and *ZREM* is analogous to that for ÷ and |, respectively. The procedures *ZQUOT* and *ZREM* are extended to nonscalar arguments in the same entry-by-entry manner as are the primitive dyadic scalar operations.

```
      25 56 ZQUOT 7 11
3 5
      (⍳6) ZREM 31
0 1 1 3 1 1
```

An integer a is said to *divide* or be a *divisor* of another integer c if there is a third integer b such that $c = ab$. In this case we also say that c is a *multiple* of a. Every integer divides 0, but the only integer divisible by 0 is 0 itself. The APL proposition corresponding to the assertion that *A* divides *C* is 0=*A*|*C*. In traditional notation the statements "a divides c" and "a does not divide c" are abbreviated $a|c$ and $a\nmid c$, respectively. Because of the possible confusion with the APL remainder operation, we will not use the symbol | to denote "divides" in this book.

The following theorem summarizes some important facts about divisors.

THEOREM 3. The relation "is a divisor of" on **Z** is reflexive and transitive. If x and y are in **Z**, then

(a) x divides y and y divides x if and only if $|x| = |y|$.

(b) If x divides y and $y \neq 0$, then $|x| \leq |y|$.

Proof. The reader should already be familiar with these facts. Their proofs are left as exercises. □

EXERCISES

1 Let *N* be a positive integer. Write APL expressions for

- **(a)** The characteristic vector for the set of positive divisors of *N*, considered as a subset of ⍳*N*.
- **(b)** The list of positive divisors of *N*.
- **(c)** The number of positive divisors of *N*.

2 Write an APL expression for a vector *D* of length *N* such that *D*[*I*] is the number of positive divisors of *I*.

3 Prove Theorem 3.

4 Does every nonempty set of positive rational numbers have a smallest element?

5 Whenever a division is performed on an APL terminal system and the result is known to be an integer, it is a good practice to apply the floor operation ⌊ to the quotient. Explain why.

2. GREATEST COMMON DIVISORS

Let a and b be integers. A *common divisor* of a and b is an integer c such that c divides both a and b. For example, 4 is a common divisor of 12 and −20 and 1 is a common divisor of every pair of integers. A *greatest common divisor* of a and b is an integer d such that

1 $d \geq 0$.

2 d is a common divisor of a and b.

3 d is divisible by every common divisor of a and b.

The set of common divisors of 10 and 12 is $D = \{-2, -1, 1, 2\}$, and 2 is a nonnegative element of D that is divisible by every element in D. Thus 2 is a greatest common divisor of 10 and 12. The goal of this section is to prove that greatest common divisors exist and are unique and to present an efficient method for calculating greatest common divisors.

THEOREM 1. Any two integers a and b have at most one greatest common divisor.

Proof. Suppose d and e are both greatest common divisors of a and b. Then d is divisible by every common divisor of a and b and, in particular, d is divisible by e. By exactly the same argument, e is divisible by d. Since d and e are each nonnegative, we must have $d = e$. □

In order to show that a greatest common divisor of a and b always exists, we consider the set $S(a, b)$ of all integers of the form $ra + sb$, with r, s in **Z**. As the next theorem shows, $S(a, b)$ cannot be an arbitrary subset of **Z**.

THEOREM 2. If x and y are in $S(a, b)$, then $x + y$ and $x - y$ are also in $S(a, b)$.

Proof. This follows immediately from the observation that

$$(r_1 a + s_1 b) \pm (r_2 a + s_2 b) = (r_1 \pm r_2)a + (s_1 \pm s_2)b. \quad \square$$

A nonempty subset M of **Z** that contains the sum and difference of any two of its elements is called an *additive subgroup* of **Z**. (The general definition of the term "subgroup" will be given in Chapter 3.) Since $S(a, b)$

is obviously nonempty, Theorem 2 tells us that $S(a, b)$ is an additive subgroup of **Z**. If n is any integer, then the set $n\mathbf{Z}$ of all multiples of n is $S(n, 0)$ and so is an additive subgroup of **Z**. If n is contained in an additive subgroup M of **Z**, then M contains $n + n = 2n$, $2n + n = 3n, \ldots,$ and $n - n = 0$, $0 - n = -n$, $-n-n = -2n, \ldots.$ Thus M contains $n\mathbf{Z}$. We will now show that the sets $n\mathbf{Z}$ are the only additive subgroups of **Z**.

THEOREM 3. Let M be an additive subgroup of **Z**. Then there exists a unique integer $n \geq 0$ such that $M = n\mathbf{Z}$.

Proof. Since $M \neq \emptyset$, there is an element x in M and M contains $x - x = 0$. If $M = \{0\}$, then $M = 0\mathbf{Z}$. Thus we may assume $M \neq \{0\}$ and therefore M contains an element $y \neq 0$. Since both y and $-y$ are in M, M contains positive elements. By Theorem 1.1, M contains a smallest positive element n. Let m be any element of M. By Theorem 1.2, we can find integers q and r such that $m = qn + r$ and $0 \leq r < n$. But m and qn are both in M, so $r = m - qn$ is in M. By the choice of n, this forces r to be 0. Therefore $m = qn$ is in $n\mathbf{Z}$. This shows that $M \subseteq n\mathbf{Z}$. However, the inclusion $M \supseteq n\mathbf{Z}$ was noted previously, and so $M = n\mathbf{Z}$. We still have to establish the uniqueness of n. Suppose $m\mathbf{Z} = n\mathbf{Z}$ for some nonnegative integers m and n. Then m and n must each be a divisor of the other, and this implies that $m = n$. □

The integer n of Theorem 3 is called the *nonnegative generator* of M.

THEOREM 4. Let a and b be integers. Then a and b have a greatest common divisor d, d is unique, and d can be written in the form $ra + sb$, where r, s are in **Z**.

Proof. The uniqueness of d was shown in Theorem 1. By Theorems 2 and 3, there is a nonnegative integer d in $S(a, b)$ such that $S(a, b) = d\mathbf{Z}$. Since d is in $S(a, b)$, there exist r and s in **Z** such that $d = ra + sb$. Therefore any common divisor of a and b divides d. Now $a = 1a + 0b$ and $b = 0a + 1b$ are both in $S(a, b)$, and every element of $S(a, b)$ is divisible by d. Thus d is a common divisor of a and b. Since $d \geq 0$, we see that d is a greatest common divisor of a and b. □

Having proved Theorem 4, we may speak of *the* greatest common divisor of a and b, which we denote by $\gcd(a,b)$. Although we know $\gcd(a,b)$ exists, we do not yet have an efficient method for actually computing $\gcd(a, b)$. The following theorem and its corollary provide the basis for one such algorithm.

THEOREM 5. If a, b, and q are integers, then

$$S(a, b) = S(b, a) = S(a, -b) = S(a\ \ b + qa).$$

Also, $S(0, b) = |b|\mathbf{Z}$.

Proof. We will prove only that $S(a, b) = S(a, b + qa)$, leaving the rest as an exercise. For any r, s in **Z**, we have

$$ra + s(b + qa) = (r + sq)a + sb$$

and so $S(a, b + qa) \subseteq S(a, b)$. If we let $c = b + qa$, the same argument shows that $S(a, c - qa) \subseteq S(a, c)$. But $c - qa = b$ and, hence, $S(a, b) \subseteq S(a, b + qa)$. Therefore $S(a, b) = S(a, b + qa)$. □

COROLLARY 6. If a, b, and q are integers, then

$$\gcd(a, b) = \gcd(b, a) = \gcd(a, -b) = \gcd(a, b + qa).$$

Also, $\gcd(0, b) = |b|$.

Proof. Since $\gcd(a, b)$ is the nonnegative generator of $S(a, b)$, the corollary follows at once from Theorem 5. □

Suppose we are given integers a and b and we want to compute $\gcd(a, b)$. By Corollary 6, we may assume $0 \le a \le b$. If $a = 0$, then $\gcd(a, b) = b$. If $a \neq 0$, then we may write $b = qa + r$, where q is the integral quotient of b by a and r is the remainder. Since $r = b - qa$, Corollary 6 tells us that $\gcd(a, b) = \gcd(a, r) = \gcd(r, a)$. If $r = 0$, then $\gcd(a, b) = a$. Otherwise we may repeat the process, dividing r into a to get a new remainder. Since $0 \le r < a$, this procedure cannot continue indefinitely. Eventually, a remainder of 0 is reached, and $\gcd(a, b)$ is the last nonzero remainder. As an example, let us compute gcd(493, 533).

```
      493|533                          13|40
40                                1
      40|493                           1|13
13                                0
```

By Corollary 6,

$$\gcd(493, 533) = \gcd(40, 493) = \gcd(13, 40)$$
$$= \gcd(1, 13) = \gcd(0, 1) = 1.$$

The recursive procedure we have described for calculating $\gcd(a, b)$ is called the *Euclidean algorithm* after the Greek geometer Euclid who flourished around 300 B.C. The procedure *ZGCD* in *CLASSLIB* is based on this algorithm.

```
      493 ZGCD 533
1
```

Using *ZGCD* with nonscalar arguments gives the entry-by-entry greatest common divisor. One-entry arrays are expanded to match the other argument just as with the primitive scalar dyadic operations.

```
      12 14 ZGCD 15 27
3 1
      6 ZGCD ⍳6
1 2 3 2 1 6
```

Being able to compute greatest common divisors allows us to decide the existence of integer solutions of a single linear equation with integer coefficients.

THEOREM 7. Let a, b, and c be integers. There exist integers x and y satisfying the equation $ax + by = c$ if and only if $\gcd(a, b)$ divides c.

Proof. Suppose x and y are integers such that $ax + by = c$. Then c is in $S(a, b) = d\mathbf{Z}$, where $d = \gcd(a, b)$. Therefore d divides c. On the other hand, suppose d divides c, so that $c = md$ for some m in $\mathbf{Z}$. We know there exist integers r and s such that $d = ra + sb$. Multiplying by m gives $c = md = (mr)a + (ms)b$. Thus $x = rm$ and $y = sm$ satisfy $ax + by = c$. □

As an example, let us consider the equation $493x + 533y = 2$. We know already that $\gcd(493, 533) = 1$, which divides 2, and so integer solutions to this equation exist by Theorem 7. The proof of Theorem 7 even shows us how to find a solution provided we can determine integers r and s such that $1 = 493r + 533s$. The Euclidean algorithm can be extended to produce one pair r, s. We first compute the integral quotients corresponding to the remainders we found in calculating $\gcd(493, 533)$.

```
      533 ZQUOT 493                          40 ZQUOT 13
1                                3
      493 ZQUOT 40
12
```

Thus

$$\begin{aligned} 40 &= 533 - 1 \times 493, \\ 13 &= 493 - 12 \times 40, \\ 1 &= 40 - 3 \times 13. \end{aligned}$$

Working backward, we find

$$\begin{aligned} 1 &= 40 - 3 \times 13 \\ &= 40 - 3(493 - 12 \times 40) \\ &= 37 \times 40 - 3 \times 493 \\ &= 37(533 - 1 \times 493) - 3 \times 493 \\ &= (-40) \times 493 + 37 \times 533. \end{aligned}$$

Therefore $r = -40$ and $s = 37$ gives one solution to $493r + 533s = 1$, and $x = 2r = -80$ and $y = 2s = 74$ is a solution to the original equation $493x + 533y = 2$.

The procedure *ZGCD* computes r and s as the global variables $\underline{R}$ and $\underline{S}$.

```
      493 ZGCD 533                           S
1                                37
      R                                      (493×R)+533×S
¯40                              1
```

Thus, to find a solution to the equation $2497x - 3872y = 33$, we may proceed as follows.

```
      2497  ZGCD  ̄3872
1̄1
            R̲
107
            S̲
69
         33÷11
3
         □←X←3×R̲
321
         □←Y←3×S̲
207
         (2497×X)-3872×Y
33
```

Let a and b be integers. We say that a and b are *relatively prime* if $\gcd(a, b) = 1$. By Theorem 4, this is equivalent to requiring that there exist integers r and s such that $ra + sb = 1$.

THEOREM 8. Let a and b be integers and let $d = \gcd(a, b)$. Then a/d and b/d are relatively prime. (Here we adopt the APL convention that $0÷0$ is 1 for the case $a = b = 0$.)

Proof. There exist integers r and s such that $d = ra + sb$. If $d = 0$, then $a = b = 0$ and it is true that $\gcd(0/0, 0/0) = \gcd(1, 1) = 1$. If $d \neq 0$, then

$$1 = \frac{d}{d} = r\left(\frac{a}{d}\right) + s\left(\frac{b}{d}\right)$$

and so $\gcd(a/d, b/d) = 1$. □

THEOREM 9. Suppose a and b are relatively prime integers. If c is an integer and a divides bc, then a divides c.

Proof. Choose integers r and s such that $ra + sb = 1$. Then $c = cra + csb$. Since a divides cra and csb, a divides c. □

COROLLARY 10. Let a, b, and c be integers and let $d = \gcd(a, b)$. If a divides bc, then a/d divides c.

Proof. If a divides bc, then a/d divides $(b/d)c$. By Theorem 8, $\gcd(a/d, b/d) = 1$ and so, by Theorem 9, a/d divides c. □

So far we have considered only greatest common divisors of two integers. Given three integers a, b, and c, we have several choices for defining $\gcd(a, b, c)$. We could define it to be $\gcd(a, \gcd(b, c))$ or $\gcd(\gcd(a, b), c)$. However, we could also consider the set $S(a, b, c)$ of integers of the form $ra + sb + tc$, prove that $S(a, b, c)$ is an additive subgroup of $\mathbf{Z}$, and define gcd(a, b, c) to be the nonnegative generator of $S(a, b, c)$. In fact, all of these definitions are equivalent. (See Exercises 10 and 12.)

Related to the concept of the greatest common divisor is the notion of the least common multiple. A *common multiple* of two integers a and b is an integer m that is divisible by both a and b. We say m is a *least common multiple* of a and b if m is a nonnegative common multiple that divides every common multiple.

THEOREM 11. Every pair of integers a and b has a unique least common multiple m. If $d = \gcd(a, b)$, then $md = |ab|$.

Proof. If $a = 0$ or $b = 0$, then 0 is the only common multiple of a and b and so $m = 0$ is the unique least common multiple of a and b. Clearly, $md = |ab|$ in this case. Thus we may suppose neither a nor b is 0. This means that $d = \gcd(a, b)$ is also nonzero. Since

$$\frac{(ab)}{d} = a\left(\frac{b}{d}\right) = \left(\frac{a}{d}\right)b$$

and both b/d and a/d are integers, we see that $m = |ab|/d$ is a nonnegative common multiple of a and b. Suppose n is any common multiple of a and b. Thus $n = ra = sb$ for some integers r and s. Since a divides sb, by Corollary 10 we know that a/d divides s. Thus $s = t(a/d)$ for some t in $\mathbf{Z}$ and $n = sb = tab/d$. Thus m divides n and so m is a least common multiple of a and b. If m' is any other least common multiple of a and b, then m and m' are nonnegative integers that divide each other. Therefore $m = m'$ and so m is the unique least common multiple of a and b. From the definition of m, we have $md = |ab|$. □

The least common multiple of a and b is denoted lcm(a, b). The procedure *ZLCM* in *CLASSLIB* computes least common multiples.

```
      12 ZLCM 15
60
      120 ZLCM 105
840
```

The result when *ZLCM* is used with nonscalar arrays is the entry-by-entry least common multiple.

EXERCISES

1. Show directly from the definition that 0 is a greatest common divisor of 0 and 0.
2. Let M and N be additive subgroups of $\mathbf{Z}$. Prove that $M \cap N$ is also an additive subgroup of $\mathbf{Z}$.
3. Complete the proof of Theorem 5.
4. Without using *ZGCD*, compute $d = \gcd(a, b)$ by means of the Euclidean algorithm for each of the following pairs a, b.

(a) 12, 39. (c) 4953, 14697.

(b) 217, 413. (d) 737019, 2055168.

5 The Euclidean algorithm reduces the computation of $d = \gcd(a, b)$ to the computation of $\gcd(a, c)$, where $c = b - qa$ and $0 \le c < |a|$. Suppose we can express d as $r'a + s'c$. Show how to write d in the form $ra + sb$.

6 For each pair of integers a, b in Exercise 4, write $\gcd(a, b)$ in the form $ra + sb$, where r and s are integers.

7 For each of the following equations, find one integer solution or show that no integer solution exists.

(a) $7x + 5y = 9$. (c) $91x + 141y = -27$.

(b) $6x - 9y = 11$. (d) $577x - 828y = 1001$.

8 The function gcd is a binary operation on **Z**. Show that gcd is commutative and associative.

9 Let a and b be integers with $0 < a \le b$. Determine an upper bound in terms of a for the number of remainders that have to be calculated in order to compute $\gcd(a, b)$ by the Euclidean algorithm.

10 Let $a_1, \ldots, a_n$ be integers. A greatest common divisor of $a_1, \ldots, a_n$ is a nonnegative integer d such that d divides each a_i and whenever c is an integer dividing each a_i then c divides d. Show that d is unique. Let $S(a_1, \ldots, a_n)$ denote the set of integers of the form $r_1 a_1 + \ldots + r_n a_n$, where each r_i is an integer. Prove that $S(a_1, \ldots, a_n)$ is an additive subgroup of **Z** and that the nonnegative generator d of $S(a_1, \ldots, a_n)$ is the greatest common divisor of $a_1, \ldots, a_n$. [We will write $d = \gcd(a_1, \ldots, a_n)$.]

11 Let $a_1, \ldots, a_n, b$ be integers. Show that the equation $a_1 x_1 + \ldots + a_n x_n = b$ has an integer solution if and only if $\gcd(a_1, \ldots, a_n)$ divides b.

12 Show that the function gcd defined in Exercise 10 satisfies the condition $\gcd(a_1, \ldots, a_n) = \gcd(a_1, \gcd(a_2, \ldots, a_n))$.

13 Compute gcd(6409, 8177, 13949).

14 Find one integer solution to the equation $12x - 15y + 20z = 29$.

15 Let a and b be integers. Show that $M = (a\mathbf{Z}) \cap (b\mathbf{Z})$ is the set of common multiples of a and b. Prove that $M = m\mathbf{Z}$, where $m = \operatorname{lcm}(a, b)$.

16 Calculate the least common multiples of the pairs of integers in Exercise 4.

17 What should be the definition of $\operatorname{lcm}(a_1, \ldots, a_n)$? Compute lcm(6409, 8177, 13949).

18 Let A and B be positive integers. Write an APL expression defining a vector listing the positive common divisors of A and B.

19 Let A be a vector of integers and let M be an integer. Write APL propositions for the following statements.

(a) M is a common divisor of the components of A.

(b) M is a common multiple of the components of A.

20. For any integer vector A let $S(A)$ be the set of integers of the form $R+.\times A$, where R is an integer vector. By Exercise 10, the nonnegative generator of $S(A)$ is the greatest common divisor of the components of A. What is $S(\iota 0)$? Prove that

(a) $S(A) = S(|A)$.

(b) $S(A) = S((A\neq 0)/A)$.

(c) If M is any component of A, then $S(A) = S(M,M|A)$.

21 Let A be a nonempty vector with nonzero integer components. Suppose the statement

```
A←(A≠0)/A←M,(M←⌊/A)|A←|A
```

is repeatedly executed. Show that eventually A will have length 1 and, when that happens, the component of A is the greatest common divisor of the components in the original vector.

22 Write a procedure $GCDV$ based on Exercises 20 and 21 such that $D\leftarrow GCDV\ A$ makes D the greatest common divisor of the components of A.

23 Modify your procedure $GCDV$ so that it computes a global vector $\underline{R}$ such that $GCDV\ A$ is $\underline{R}+.\times A$.

24 Write a procedure $LCMV$ such that $M\leftarrow LCMV\ A$ defines M to be the least common multiple of the components of the vector A.

3. CONGRUENCE

Let n be a fixed positive integer. We say that an integer a is *congruent modulo n* to another integer b if n divides $a - b$. If this is the case, we write $a \equiv b \pmod{n}$. Thus $19 \equiv 33 \pmod{7}$ and $-8 \equiv 7 \pmod{5}$. The integer n is called the *modulus* of the congruence.

THEOREM 1. Congruence modulo n is an equivalence relation on $\mathbf{Z}$ with exactly n equivalence classes. The set $\{0, 1, \ldots, n-1\}$ is a set of representatives for the equivalence classes.

Proof. Let a, b, c be integers. Clearly, n divides $a - a$, and so $a \equiv a \pmod{n}$. If $a \equiv b \pmod{n}$, then n divides $a - b$, and so n divides $b - a$. Thus $b \equiv a \pmod{n}$. If $a \equiv b \pmod{n}$ and $b \equiv c \pmod{n}$, then n divides both $a - b$ and $b - c$. Therefore n divides $(a - b) + (b - c) = a - c$ and $a \equiv c \pmod{n}$. We have now shown that congruence modulo n is an equivalence relation. By Theorem 1.2, every integer m is congruent modulo n to an element of $X = \{0, 1, \ldots, n - 1\}$. Clearly, no two distinct elements of X are congruent modulo n, and so X is a set of representatives for the equivalence classes. □

An equivalence class of the relation congruence modulo n is called a *congruence class* modulo n. The set of all congruence classes modulo n is denoted $\mathbf{Z}_n$. The congruence class containing a particular integer a will be written $[a]_n$, or simply $[a]$ when the modulus n is clear.

We will now define two binary operations on $\mathbf{Z}_n$, that is, two functions from $\mathbf{Z}_n \times \mathbf{Z}_n$ to $\mathbf{Z}_n$. These binary operations will be denoted + and $\times$. We would like the equations

$$[a] + [b] = [a + b],$$
$$[a] \times [b] = [a \times b],$$

to hold for all a, b in $\mathbf{Z}$. This suggests that we should define + to be the subset

$$P = \{ ([a], [b]), [a + b] \ | a, b \in \mathbf{Z}\}$$

of $(\mathbf{Z}_n \times \mathbf{Z}_n) \times \mathbf{Z}_n$. But is P really a function? Certainly P is a relation from $\mathbf{Z}_n \times \mathbf{Z}_n$ to $\mathbf{Z}_n$. What we need to show is that each ordered pair (x, y) in $\mathbf{Z}_n \times \mathbf{Z}_n$ occurs as the first component in exactly one element $((x, y), z)$ of P. Suppose $x = [a]$ and $y = [b]$. If we take z to be $[a + b]$, then $((x, y), z)$ is in P. Suppose $((x, y), w)$ is also in P. Then we must have $x = [c]$, $y = [d]$, and $w = [c + d]$ for some c and d in $\mathbf{Z}$. Now $[a] = [c]$ and $[b] = [d]$, and so n divides $a - c$ and $b - d$. Therefore n divides $(a - c) + (b - d) = (a + b) - (c + d)$. Thus $a + b \equiv c + d \pmod{n}$ and $w = z$. Thus (x, y) is the first component of exactly one element of P, and P is a binary operation on $\mathbf{Z}_n$. By a similar argument, it can be shown that

$$T = \{\left(([a], [b]), [a \times b]\right) | a, b \in \mathbf{Z}\}$$

is a binary operation $\times$ on $\mathbf{Z}_n$. Modulo 7 we have

$$[2] + [6] = [1], \qquad [2] \times [6] = [5],$$
$$[4] + [5] = [2], \qquad [4] \times [5] = [6].$$

Showing, as we did previously, that a given relation is really a function is called proving that the function is *well defined*. Our arguments showing that the binary operations + and $\times$ on $\mathbf{Z}_n$ are well defined can be summarized in the following theorem.

THEOREM 2. If $a \equiv c \pmod{n}$ and $b \equiv d \pmod{n}$, then $a + b \equiv c + d \pmod{n}$ and $a \times b \equiv c \times d \pmod{n}$. □

Suppose we are given integers a and b. By a solution to the congruence

$$ax \equiv b \pmod{n} \qquad (*)$$

we mean an integer x for which the congruence is true. By Theorem 2, if x is a solution, then so is every element of the congruence class $[x]$ modulo n. Thus, solving the congruence $(*)$ amounts to finding those elements $[x]$ of $\mathbf{Z}_n$ such that $[a] \times [x] = [b]$. As the next theorem shows, the existence of solutions to $(*)$ depends on $\gcd(a, n)$.

THEOREM 3. The congruence $ax \equiv b \pmod{n}$ has solutions if and only if $d = \gcd(a, n)$ divides b. If a solution exists, then it is unique modulo n/d.

Proof. The integer x satisfies the congruence $ax \equiv b \pmod{n}$ if and only if there is an integer z such that $ax + nz = b$. By Theorem 2.7, the equation $ax + nz = b$ has solutions if and only if $d = \gcd(a, n)$ divides b. Suppose x satisfies $ax \equiv b \pmod{n}$. Let y be congruent to x modulo n/d. Thus $y = x + r(n/d)$ for some r in $\mathbf{Z}$. Then

$$ay = ax + ar\left(\frac{n}{d}\right) = ax + nr\left(\frac{a}{d}\right) \equiv ax \equiv b \pmod{n}.$$

Therefore $ay \equiv b \pmod{n}$. Conversely, suppose $ax \equiv ay \equiv b \pmod{n}$. Then

$$a(x - y) = ax - ay \equiv b - b = 0 \pmod{n}$$

and so n divides $a(x - y)$. By Corollary 2.10, n/d divides $x - y$ and hence $x \equiv y \pmod{n/d}$. Therefore, if solutions exist, then they are unique modulo n/d. □

COROLLARY 4. If a, b, and c are integers and $ab \equiv ac \pmod{n}$, then $b \equiv c \pmod{n/d}$, where $d = \gcd(a, n)$. In particular, if a and n are relatively prime, then b F c (mod n). □

Suppose we wish to solve the congruence $35x \equiv 77 \pmod{98}$. From

```
      35 ZGCD 98                                R
7                                         3
      7|77                                      S
0                                        ¯1
```

we see that $\gcd(35, 98) = 7$ and that 7 divides 77. By Theorem 3, the congruence has a solution that is unique modulo $98/7 = 14$. The global variables `R` and `S` satisfy `7=(35×R)+98×S`. Multiplying by $77/7 = 11$, we find that `11×R` is one solution. Reducing modulo 14,

```
      14|11×R
5
```

we obtain the solution $x \equiv 5 \pmod{14}$. Modulo 98, the solutions are 5, 19, 33, 47, 61, 75, and 89.

There is another way to solve the congruence $35x \equiv 77 \pmod{98}$ on an APL terminal system. If we let `X←⍳98`, then the `I`th component of the logical vector `77 = 98|35×X` is 1 if and only if `I` is a solution of the congruence. Thus we can obtain a list of the solutions by forming

```
      (77=98|35×X)/X←⍳98
5 19 33 47 61 75 89
```

This method of trying all possible values modulo 98 would not be feasible in hand computation nor would it be feasible even at a terminal for a congruence such as $876543x \equiv 123456 \pmod{10^6}$. However, it is useful for solving some congruences that are not of the form $ax \equiv b \pmod{n}$. For example, suppose we wish to find all integers x such that $x^3 + 5x^2 - 2x + 6 \equiv 0 \pmod{110}$. From

```
      (0=110|6+X×¯2+X×5+X)/X←⍳110
31 42 86 97
```

we see that the solutions are $x \equiv 31, 42, 86,$ and $97 \pmod{110}$. The APL expression used to evaluate the polynomial in this example is based on the identity

$$x^3 + 5x^2 - 2x + 6 = 6 + x(-2 + x(5 + x)).$$

This approach to polynomial evaluation, which reduces the number of arithmetic operations involved, is referred to as *Horner's method*.

EXERCISES

1 Our definition of congruence modulo n makes sense even when $n = 0$. What parts of Theorem 1 remain true when $n = 0$?

2 Let m and n be positive integers. The sets $\mathbf{Z}_m$ and $\mathbf{Z}_n$ are both partitions of $\mathbf{Z}$. Show that $\mathbf{Z}_m$ is a refinement of $\mathbf{Z}_n$ if and only if n divides m. (See Exercise 1.7.)

3 Show that the set T defined in this section is a function from $\mathbf{Z}_n \times \mathbf{Z}_n$ to $\mathbf{Z}_n$.

4 Find all solutions, if any, to the following congruences.
(a) $12x \equiv 11 \pmod{17}$. (c) $853x \equiv 472 \pmod{1999}$.
(b) $30x \equiv 44 \pmod{125}$. (d) $20623x \equiv 30143 \pmod{40877}$.

5 Solve the congruence $x^4 + x^3 - 3x^2 + 2x + 4 \equiv 0 \pmod{455}$.

6 Let A, B, and N be integers with $N > 0$. Write an APL proposition corresponding to the statement that A is congruent to B modulo N.

7 Let A, M, and N be integers with $N > 0$. Write an APL expression

for the element of the set `M+⍳N` that is congruent to A modulo N. (Assume `⎕IO←1`.)

8 Any solution of the congruence $ax \equiv b \pmod{n}$ is also a solution of $ax \equiv b \pmod{m}$ for any positive divisor m of n. We may use this fact to solve the congruence $876543x \equiv 123456 \pmod{10^6}$ by a modified trial-and-error process. Try all elements of `⍳100` to find the solution modulo 10^2. Each solution modulo 10^2 can yield up to 100 solutions modulo 10^4. Construct a vector of reasonable length that must contain every solution modulo 10^4 and test to find the solution modulo 10^4. Repeat the process to find all solutions modulo 10^6.

9 Use the ideas of Exercise 8 to solve the congruence $x^2 + 786448x + 128767 \equiv 0 \pmod{10^6}$.

4. PRIMES

A *prime number* or *prime* is a positive integer with exactly two positive divisors. Thus 1 is not a prime because it has only one positive divisor and 4, 6, 8, 9, and 10 are not primes because they each have at least three positive divisors. The first five primes are 2, 3, 5, 7, and 11. Since 1 and p are positive divisors of any positive integer p, we could also define a prime to be an integer greater than 1 with no positive divisors other than 1 and itself. An integer greater than 1 that is not a prime is said to be *composite*.

THEOREM 1. Let p be a prime and let a and b be integers. If p divides ab, then either p divides a or p divides b.

Proof. Let $d = \gcd(a, p)$. Then d is a positive divisor of p and so is 1 or p. If $d = p$, then p divides a. If $d = 1$, then p divides b by Theorem 2.9. □

It follows from Theorem 1 that if a prime p divides a product $a_1 a_2 \ldots a_r$ of integers, then p divides a_i from some i.

We will now prove a very important result.

THEOREM 2 ***(Fundamental Theorem of Arithmetic).*** Each positive integer n can be factored uniquely as $p_1 p_2 \ldots p_r$, where each p_i is a prime and $p_1 \leq p_2 \leq \ldots \leq p_r$.

Proof. We prove first by induction that n can be factored as a product of primes. If n is 1, then n is the product of the empty sequence of primes, so we can get our induction started. Suppose $n > 1$ and every positive integer less than n can be factored into primes. If n is a prime, then n is the product of the single prime n. Suppose n is not a prime. Then we can write n as uv, where u and v are integers greater than 1 and less than n. By induction, we can factor u and v into primes. Arranging the prime factors of u and v into one sequence gives the required factorization of n.

Now we prove uniqueness, again by induction on n. Suppose

$$n = p_1 p_2 \ldots p_r = q_1 q_2 \ldots q_s$$

with each p_i and q_j a prime and $p_1 \leq p_2 \leq \ldots \leq p_r$ and $q_1 \leq q_2 \leq \ldots \leq q_s$. We want to show $r = s$ and $p_i = q_i$, $1 \leq i \leq r$. If r or s is 0, then $n = 1$ and the result is trivial. Thus we may assume r and s are both at least 1. Since p_1 divides $q_1 \ldots q_s$, it follows that p_1 divides some q_j. But this implies that $p_1 = q_j$. Similarly, $q_1 = p_i$ for some i. Then

$$p_1 \leq p_i = q_1 \leq q_j = p_1,$$

so p_1 and q_1 must be equal. Let $m = n/p_1$. We have

$$m = p_2 \ldots p_r = q_2 \ldots q_s$$

and, by induction, $r = s$ and $p_i = q_i$, $2 \leq i \leq r$. □

It follows immediately from Theorem 2 that any positive integer n can be written uniquely in the form

$$\prod_{i=1}^{r} p_i^{e_i},$$

where $p_1 < p_2 < \ldots < p_r$, each p_i is a prime, and each e_i is a positive integer.

It is possible to write a single APL expression defining the vector of primes less than or equal to a given integer N. However, the space and CPU requirements for executing this expression are very large. A more efficient method of listing primes is the so-called *sieve procedure*, which we will illustrate with a small example. Let us assume that we wish to construct a vector P consisting of all primes not exceeding 25. We begin by setting P equal to the empty vector and letting Q be the vector of integers from 2 to 25.

```
      P←⍳0
      ⎕←Q←1↓⍳25
2 3 4 5 6 7 8 9 10 11 12 13 14 15 16 17 18 19 20 21 22 23 24 25
```

The first prime is the first component of Q, or 2. We add 2 to P and delete all multiples of 2 from Q.

```
      ⎕←P←P,2
2
      ⎕←Q←(0≠2|Q)/Q
3 5 7 9 11 13 15 17 19 21 23 25
```

The next prime is the new first component of Q. Again, we add this prime to P and delete its multiples from Q.

```
      ⎕←P←P,3
2 3
      ⎕←Q←(0≠3|Q)/Q
5 7 11 13 17 19 23 25
```

Repeating this process once more, we get

```
      ⎕←P←P,5
2 3 5
      ⎕←Q←(0≠5|Q)/Q
7 11 13 17 19 23
```

All of the remaining components of Q have no prime factors less than 7. If some component $Q[I]$ is not a prime, then $Q[I]$ is a product of two or more primes each greater than 6. Since none of the components of Q exceeds 25, this is impossible. Therefore all of the remaining components in Q are primes, and we complete our computation with

```
      ⎕←P←P,Q
2 3 5 7 11 13 17 19 23
```

The procedure *ZPRIMES* is *CLASSLIB* is based on this technique of sieving out the composite numbers from 1↓⍳N.

```
      ZPRIMES 25
2 3 5 7 11 13 17 19 23
```

It is very important to be able to factor a given positive integer into a product of primes, and a great deal of effort has been spent on trying to develop efficient factoring algorithms. We can not go very deeply into this subject here. For a more complete discussion, consult Knuth [Vol. 2]. See also Section 3.3.

Suppose we wish to factor 4693. Since no positive integer n can have more than one prime factor greater than $\sqrt{n}$, we try first to determine the prime factors of 4693 not exceeding $\sqrt{4693} = 68.5\ldots$ One way is as follows.

```
      P←ZPRIMES 68
      (0=P|4693)/P
13 19
```

We see that 4693 is divisible by 13 and 19 and by no other primes less than 68. Computing the quotient

```
      4693÷13×19
19
```

we find that $4693 = 13 \times 19 \times 19$. If we wish to factor an integer n and a list of primes is not readily available, the easiest thing to do is to divide n by 2 and by every odd number not exceeding $\sqrt{n}$.

It turns out to be easier to decide whether or not a large integer is prime than it is to factor a large integer known to be composite. One test for primality is discussed in Section 3.3. The procedure *ZFACTOR* in *CLASSLIB* looks for factors less than 50000.

```
      ⎕←P←ZFACTOR 33263
29 31 37
      ×/P
33263
      ZFACTOR 10*7
2 2 2 2 2 2 2 5 5 5 5 5 5 5
```

With *ZFACTOR*, the complete factorization of any integer up to 2.5×10^9 can be obtained, and often the factorizations of much larger numbers can be found. Using the results of Section 3.3, it is possible to formulate more powerful factoring algorithms.

We close this section with an important result concerning the solutions of simultaneous congruences. Let $m_1, \ldots, m_r$ be positive integers and let $a_1, \ldots, a_r$ be integers. A solution of the system

$$\begin{aligned} x &\equiv a_1 \pmod{m_1}, \\ x &\equiv a_2 \pmod{m_2}, \\ &\;\;\vdots \\ x &\equiv a_r \pmod{m_r}, \end{aligned}$$

of simultaneous congruences is one integer x that is a solution of each congruence. The existence and uniqueness of solutions to this system depend on the numbers $\gcd(m_i, m_j)$. We say that the integers $m_1, \ldots, m_r$ are *pairwise relatively prime* if $\gcd(m_i, m_j) = 1$ for $i \neq j$.

LEMMA 3. Let $m_1, \ldots, m_r$, $r \geq 1$, be a sequence of pairwise relatively prime nonzero integers. Set $n = m_1 m_2 \ldots m_r$ and $v_i = n/m_i$. Then $\gcd(v_1, \ldots, v_r) = 1$ and $\operatorname{lcm}(m_1, \ldots, m_r) = n$.

Proof. Let $d = \gcd(v_1, \ldots, v_r)$. If $d \neq 1$, then d is divisible by a prime p. Since d divides v_1 and v_1 divides n, it follows that d divides n. Thus p divides some m_i. Since $m_1, \ldots, m_r$ are pairwise relatively prime, p does not divide m_j for $j \neq i$. But

$$v_i = m_1 m_2 \ldots m_{i-1} m_{i+1} \ldots m_r$$

and so p does not divide v_i, contradicting our choice of p. We leave the proof that $\operatorname{lcm}(m_1, \ldots, m_r) = n$ as an exercise. □

THEOREM 4 ***(Chinese Remainder Theorem).*** Let $m_1, \ldots, m_r$, $r \geq 1$, be a sequence of pairwise relatively prime positive integers and let $a_1, \ldots, a_r$ be integers. There exists an integer x such that $x \equiv a_i \pmod{m_i}$, $1 \leq i \leq r$, and x is unique modulo $n = m_1 m_2 \ldots m_r$.

Proof. First, we prove uniqueness. Suppose x and y are both solutions of the simultaneous congruences. Then $x \equiv y \pmod{m_i}$, $1 \leq i \leq r$, and so $x - y$ is a multiple of m_i for each i. Therefore $x - y$ is divisible by $\text{lcm}(m_1, \ldots, m_r)$, which by Lemma 3 is n. Therefore $x \equiv y \pmod{n}$. It is also clear that if x is a solution and $y \equiv x \pmod{n}$, then y is a solution, too.

Now we prove that a solution of the simultaneous congruences exists. Let $v_i = n/m_i$, $1 \leq i \leq r$. By Lemma 3, $\gcd(v_1, \ldots, v_r) = 1$, and so there exist integers $c_1, \ldots, c_r$ such that

$$\sum_{j=1}^{r} c_j v_j = 1.$$

Define x by

$$x = \sum_{j=1}^{r} a_j c_j v_j.$$

We have $v_j \equiv 0 \pmod{m_i}$ for $j \neq i$, and so $c_i v_i \equiv 1 \pmod{m_i}$, while $c_j v_j \equiv 0 \pmod{m_i}$ for $j \neq i$. Therefore

$$x \equiv a_i c_i v_i \equiv a_i \pmod{m_i}, \qquad 1 \leq i \leq r,$$

and thus x is a solution. □

As an example, let us solve the system

$$x \equiv 19 \pmod{27},$$
$$x \equiv 7 \pmod{32}.$$

In the notation of Theorem 4 and its proof we have $m_1 = 27$, $m_2 = 32$, $a_1 = 19$, and $a_2 = 7$.

```
      M←27 32                              A←19 7
```

First, we compute n, v_1, and v_2.

```
      ⎕←N←×/M                              ⎕←V←N÷M
864                             32 27
```

Then we express $1 = \gcd(v_1, v_2)$ as $c_1 v_1 + c_2 v_2$.

```
      V[1] ZGCD V[2]                        C+.×V
1                               1
      ⎕←C←R,S
11 ¯13
```

We obtain one solution by

```
      ⎕←X←+/A×C×V
4231
```

Reducing modulo *N*,

```
      N|X
775
```

we see that $x \equiv 775 \pmod{864}$ is our solution. As a check, we can compute

```
      M|775
19 7
```

The procedure *ZCHREM* in *CLASSLIB* can be used to solve simultaneous congruences.

```
      19 7 ZCHREM 27 32
775
```

This procedure can solve several systems of congruences with the same vector of moduli. If *M* is a vector of nonzero integers and *A* is a matrix of integers, then `X←A ZCHREM M` is a vector of integers such that `X[I]` is congruent to `A[I;J]` modulo `M[J]` for all *I* and *J*, provided such a vector of integers exists. (See Exercise 9.) In addition, *ZCHREM* computes the global variable <u>M</u>, which is the least common multiple of the components of *M*.

```
      M←4 6 9
      ⎕←A←2 3⍴1 5 8 3 1 4
 1 5 8
 3 1 4
      ⎕←X←A ZCHREM M
17 36
      ⍉M∘.|X
 1 5 8
 3 1 4
      M̲
36
```

EXERCISES

1. Complete the proof of Lemma 3.
2. Construct the vector *P* of primes not exceeding 100 using the sieve procedure described in the text.
3. Write an APL expression for a vector listing the positive integers less than 100 that are relatively prime to 100.
4. At a terminal, compute the factorizations of the following integers into a product of primes without using the procedure *ZFACTOR*.

   (a) 6726. (c) 452521.

   (b) 2003. (d) 987027.
5. Find the first prime larger than n for $n = 10^6$, 10^7, 10^8, and 10^9.

6 Let N be a positive integer and assume `Q←SSORT P←ZFACTOR N`. Write an expression for the vector E such that N is `Q×.*E`.

7 Let P be a vector whose components are distinct primes and let E and F be vectors of length `ρP` with nonnegative integer components. Set `M←P×.*E` and `N←P×.*F`. Show that `M ZGCD N` is `P×.*E⌊F` and `M ZLCM N` is `P×.*E⌈F`.

8 Solve the following systems of simultaneous congruences.

(a) $x \equiv 3 \pmod{8}$,
$x \equiv 7 \pmod{9}$.

(b) $y \equiv 47 \pmod{81}$,
$y \equiv 101 \pmod{125}$.

(c) $z \equiv 19 \pmod{64}$,
$z \equiv 20 \pmod{81}$,
$z \equiv 21 \pmod{125}$.

9 Prove the generalization of the Chinese Remainder Theorem that states that there exists an integer x satisfying $x \equiv a_i \pmod{m_i}$, $1 \le i \le r$, if and only if $\gcd(m_i, m_j)$ divides $a_i - a_j$ for all i and j and, if a solution exists, then it is unique modulo $\operatorname{lcm}(m_1, \ldots, m_r)$.

10 Describe all solutions of the following simultaneous congruences.

$$7x \equiv 3 \pmod{16},$$
$$10x \equiv 6 \pmod{36},$$
$$55x \equiv 30 \pmod{75}.$$

11 Suppose $2^p - 1$ is a prime. Show that p is a prime.

12 Prove that $2^p - 1$ is a prime for $p = 2, 3, 5, 7, 13$ but not for $p = 11$.

13 Show that the set of primes is infinite.

14 Familiarize yourself with the procedure `ZCHREM` in `CLASSLIB`.

*5. MULTIPLE-PRECISION ARITHMETIC

The largest prime known, at the time this book was written, was $2^{44497} - 1$, a number with 13395 decimal digits. Primes of the form $2^p - 1$ are called *Mersenne primes* after the French mathematician Marin Mersenne (1588-1648). If m and n are positive integers, then

$$2^{mn} - 1 = (2^m - 1)(2^{(n-1)m} + 2^{(n-2)m} + \ldots + 2^m + 1).$$

Thus $N = 2^p - 1$ can be a prime only when p is a prime. We remarked in the last section that it is easier to decide whether or not a number is prime than it is to factor a number known to be composite. For *Mersenne numbers*, that is, numbers of the form $2^p - 1$ with p a prime, a very powerful test for primality has been discovered. This test involves the sequence of integers defined as follows. Let $S_1 = 4$ and set $S_{n+1} = (S_n)^2 - 2$ for $n \geq 1$. Thus $S_2 = 14$, $S_3 = 194$, and $S_4 = 37634$.

THEOREM 1 ***(Lucas and Lehmer).*** Let $N = 2^p - 1$, with p an odd prime. Then N is a prime if and only if N divides S_{p-1}.

Proof. A proof of this result may be found in Knuth [Vol. 2]. □

Let us apply the test in Theorem 1 to the first few Mersenne numbers. If $p = 3$, then $N = 2^3 - 1 = 7$, which does divide $S_{p-1} = S_2 = 14$. If $p = 5$, then $N = 2^5 - 1 = 31$ and $S_4 = 37634 = 31 \times 1214$. To check the case $p = 7$, it seems necessary to compute S_6, which is a number with 19 decimal digits. However, all we are really interested in is whether or not $S_6 \equiv 0 \pmod{N}$, where $N = 2^7 - 1 = 127$. To decide this, we need only compute the terms S_k modulo N for $1 \leq k \leq 6$. This can be done as follows.

```
      ⎕←N←¯1+2*7
127
      S←4
      ⎕←S←N|¯2+S×S
14
      ⎕←S←N|¯2+S×S
67
      ⎕←S←N|¯2+S×S
42
      ⎕←S←N|¯2+S×S
111
      ⎕←S←N|¯2+S×S
0
```

Thus S_6 is divisible by 127. The same procedure may be used to show that S_{10} is congruent to 1736 modulo $2^{11} - 1 = 2047$, so Theorem 1 tells us that 2047 is not a prime. Factoring,

```
      ZFACTOR 2047
23 89
```

we see that $2047 = 23 \times 89$.

Let us now jump to the case $p = 37$. On one terminal system the following results were obtained for the first few terms in the sequence S_n modulo $2^{37} - 1$.

```
      ⎕←N←¯1+2*37
137438953471
      S←4
      ⎕←S←N|¯2+S×S
14
      ⎕←S←N|¯2+S×S
194
      ⎕←S←N|¯2+S×S
37634
      ⎕←S←N|¯2+S×S
1416317954
      ⎕←S←N|¯2+S×S
111419319478
```

From this output it seems that S_6 is congruent to 111419319478 modulo $2^{37} - 1$. However, the correct congruence is

$$S_6 \equiv 111419319482 \pmod{2^{37} - 1}.$$

What went wrong? The answer is that we have exceeded the precision of this particular terminal system. The square of 1416317954 is

$$2005956546822746116,$$

which has 19 digits. Unfortunately, our terminal system performs arithmetic operations with an accuracy of only about 16 decimal digits, and the square was not correctly computed.

Clearly, we cannot compute $S_7, S_8, \ldots, S_{36}$ modulo $2^{37} - 1$ using the previous approach. The intermediate results require greater precision than is possible using a single APL scalar to store our integers. In *CLASSLIB* there are several procedures for performing arithmetic operations using multiple precision, in which very large integers are represented by vectors of smaller integers. The procedures in *CLASSLIB* require that a large integer be represented by a character vector listing the decimal digits of the number with an initial plus or minus sign optional. For example, the following character vectors are valid representations of multiple-precision integers.

```
        X←'375298010327654431'
        Y←'¯28971132410569821536'
        Z←'+9876543210123456789'
```

The procedures *MPZSUM* and *MPZPROD* compute sums and products of multiple-precision integers. (The prefix *MPZ* stands for "multiple-precision integer".)

```
        X MPZSUM Y
¯28595834400242167105
        X MPZPROD Z
3706647015674438332992604563952882059
```

There are also procedures for computing differences, powers, and remainders.

```
        Z MPZDIFF X
9501245199795802358
        X MPZPOWER 2
140848596555896211951322046153933761
        X MPZREM Y
302112394987224082
```

Note that the right argument of *MPZPOWER* is the exponent given by an ordinary APL integer. The method used by *MPZPOWER* to keep the number of multiplications to a minimum is discussed in Section 3.1. The result of *X MPZREM Y* is the remainder when *Y* is divided by *X*. The integer quotient in this division is saved in the global variable *Q*.

```
      '1234567' MPZREM '9876543210123'
1037288
      Q
8000005
      1234567 ZREM 9876543210123
1037288
      9876543210123 ZQUOT 1234567
8000005
```

The procedure *MPZMAG* computes the absolute value, while *MPZSGN* corresponds to the monadic signum operation ×.

```
      MPZMAG Y                        MPZSGN Y
28971132410569821536             ¯1
```

We can use the multiple-precision procedures in *CLASSLIB* to complete our testing of the Mersenne number $2^{37} - 1$ to see if it satisfies the Lucas-Lehmer criterion of Theorem 1. The reader should be aware, however, that a significant amount of CPU time will be required. If *S* is one term of the sequence S_i modulo *N*, then the next term modulo *N* is

```
      S←N MPZREM (S MPZPROD S) MPZDIFF '2'
```

Starting with

```
      ⎕←N←('2' MPZPOWER 37) MPZDIFF '1'
137438953471
      S←'4'
```

we find that S_{36} is congruent to 117093979072 modulo $2^{37} - 1$ and so $2^{37} - 1$ is not a prime. There is another way to see this

```
      ZFACTOR 137438953471
223 616318177
```

Since $2^{37} - 1$ turns out to have a rather small prime factor, *ZFACTOR* is able to handle it easily. For large primes p, however, the approach used in *ZFACTOR* is inferior to the Lucas-Lehmer criterion for deciding whether $2^p - 1$ is prime.

While it would divert us too much to explore the workings of the multiple-precision procedures in complete detail, it will be useful to examine some of the basic ideas involved, since they also arise in the polynomial manipulation procedures discussed in Section 4.4.

We normally represent integers using decimal notation. When we write the number 2573, we mean $2 \times 10^3 + 5 \times 10^2 + 7 \times 10 + 3$. The number 10 is called the *base* or *radix* of the notational system. The choice of 10 as a base came about mainly for anatomical, not mathematical, reasons. Our methods of measuring time and angles are derived from a Babylonian notational system with base 60. The internal representation of numbers in computers usually involves bases of 2, 8, or 16. Normal input and output of APL terminal systems is in decimal notation, but the encode and decode operations allow us to work with any base.

```
      10 10 10 10⊤2573
2 5 7 3
      (5⍴8)⊤2573
0 5 0 1 5
      (13⍴2)⊤2573
0 1 0 1 0 0 0 0 0 1 1 0 1
      8⊥1 7 7 6
1022
```

When different bases are used in the same discussion, the base is written as a subscript. The preceding computations show that

$$2573_{10} = 5015_8 = 101000001101_2 \qquad \text{and} \qquad 1776_8 = 1022_{10}.$$

Suppose we are given the vectors X and Y of the decimal digits of two positive integers x and y. How can we describe the vectors of digits for $x + y$ and xy? To answer this question, let us consider an example with $x = 9318$ and $y = 3728$.

```
      ⎕←X←(4⍴10)⊤9318
9 3 1 8
      ⎕←Y←(4⍴10)⊤3728
3 7 2 8
```

If we form

```
      ⎕←Z←X+Y
12 10 3 16
```

we get a vector Z such that `10⊥Z` is $x + y$

```
      10⊥Z                        9318+3728
13046                       13046
```

but Z is not a valid vector of digits in the base 10 system. We have not allowed for any carry from one digit position to the next position on the left. The vector of digits to be carried is

```
      Z ZQUOT 10
1 1 0 1
```

and the carry may be performed as follows.

```
      ⎕←Z←(0,10|Z)+(Z ZQUOT 10),0
1 3 0 4 6
```

To compute the product of x and y by hand, we would probably write

$$\begin{array}{r} 9318 \\ \underline{3728} \\ 74544 \\ 18636 \\ 65226 \\ \underline{27954} \\ 34737504 \end{array}$$

In carrying out this calculation, we had to multiply each digit of x by each digit of y. This suggests that to formulate a multiplication algorithm in APL we should first form the outer product

```
        ⎕←U←Y∘.×X
 27   9   3  24
 63  21   7  56
 18   6   2  16
 72  24   8  64
```

The rows of U correspond to the four intermediate rows in the hand computation, but in reverse order. However, in the hand computation the intermediate rows are shifted horizontally. We can perform this shifting as follows.

```
        ⎕←V←0 ¯1 ¯2 ¯3⌽U,4 3⍴0
 27   9   3  24   0   0   0
  0  63  21   7  56   0   0
  0   0  18   6   2  16   0
  0   0   0  72  24   8  64
```

Adding the columns of V and making the necessary carries,

```
        ⎕←W←+⌿V
27 72 42 109 82 24 64
        ⎕←W←(0,10|W)+(W ZQUOT 10),0
2 14 6 12 17 4 10 4
        ⎕←W←(0,10|W)+(W ZQUOT 10),0
0 3 4 7 3 7 5 0 4
        9318×3728
34737504
```

we obtain the vector of digits of xy. Note that adding the carry digits can cause further carries and that our method of handling carries sometimes introduces leading zeros in the digit vector.

It is possible to formulate algorithms for subtraction and division in decimal notation in much the same way as we have described decimal addition and multiplication. We could write APL procedures for doing multiple-precision arithmetic using decimal notation, but this would be a waste of both time and space. (Why?) The multiple precision procedures in *CLASSLIB* use a base of 10^6. The character vector arguments are converted to vectors of "digits" in base 10^6 notation, the appropriate arithmetic operation is performed, and the result is converted back to a character vector.

There is another method for performing arithmetic operations with large integers. Let $m_1, \ldots, m_r$ be a sequence of relatively prime integers and let $n = m_1 m_2 \ldots m_r$. By the Chinese Remainder Theorem, an integer x with $-n/2 < x \leq n/2$ is uniquely determined by the vector $(x_1, \ldots, x_r)$, where x_i is the remainder when x is divided by m_i. If y is another integer and $(y_1, \ldots, y_r)$ is the corresponding vector of remainders, then the vector of remainders for $x + y$ and xy are easily computed. (See Exercise 8.) This method of *modular arithmetic* is very efficient for carrying out addition, subtraction, and multiplication, provided the results are known in advance to lie in the interval $(-n/2, n/2)$. However, it is much harder to perform division and to decide which of two numbers is larger using this modular representation. A more detailed discussion of the advantages and disadvantages of modular arithmetic can be found in Knuth [Vol. 1].

EXERCISES

1 Arrange the following integers in increasing order: 1111101010_2, 1750_8, 1001_{10}, $6B2_{12}$, $3E7_{16}$. (For bases 12 and 16 the "digits" A, B, C, D, E, and F have decimal values 10, 11, 12, 13, 14, and 15, respectively.)

2 Write APL procedures *DECSUM* and *DECPROD* for performing addition and multiplication of positive integers in decimal notation. That is, if *X* and *Y* are the vectors of decimal digits for two positive integers x and y, then *X DECSUM Y* and *X DECPROD Y* should be the vectors of decimal digits for $x + y$ and xy, respectively. The procedures should be in the spirit of the examples in the text.

3 Describe how to perform long division in decimal notation using APL operations on the vectors of decimal digits.

4 Write a procedure *LUCASLEHMER* such that if *P* is an odd prime, then *LUCASLEHMER P* is 1 if `N←¯1+2*P` is a prime and 0 if *N* is composite. Show that the first nine Mersenne primes are the numbers $2^p - 1$, with $p = 2, 3, 5, 7, 13, 17, 19, 31, 61$. (Warning! This could be expensive if carried out in the most obvious manner.)

5 Use the procedure *MPZREM* and the Euclidean algorithm to compute the greatest common divisor of

533379186147253323339050654336402 1456194114

and

617534528419290220014708148588944 6083450374.

(Use the result of Exercise 2.9 to estimate the amount of CPU time that will be required.)

6 Compute the exact value of the term S_7 in the sequence S_n of the Lucas-Lehmer test.

7 Although the multiple-precision procedures in *CLASSLIB* are designed to manipulate integers, they can be used to compute with decimal fractions as well. Explain how this can be done. As an example, compute $\sqrt{2}$ to 30 decimal places.

8 Describe how to perform addition and multiplication using modular representations. That is, if M is a vector of positive integers and A and B are integers, show how to construct $M | A+B$ and $M | A \times B$ from $X \leftarrow M | A$ and $Y \leftarrow M | B$.

3
GROUPS

In this chapter we encounter the concept of a group, our first example of an abstract algebraic system. Groups occur in almost every branch of mathematics and are frequently used to describe the symmetry present in some mathematical structure. We will investigate many examples of specific or "concrete" groups. However, in order to isolate the common features that all of our examples share, we will also study "abstract" groups, sets with a particular type of binary operation. Throughout this chapter the APL index origin is assumed to be 0 unless there is an explicit statement to the contrary.

1. BINARY OPERATIONS

In Section 1.3 we defined a binary operation on a set X to be a function from $X \times X$ to X. We also adopted the convention of using the symbol $\bullet$ to denote a typical binary operation and of writing $x \bullet y$ for the image of the pair (x, y) under $\bullet$. We usually refer to $x \bullet y$ as the *product* of x and y, even though $\bullet$ may not have anything to do with ordinary multiplication of numbers. We say that $\bullet$ is commutative on X if $x \bullet y = y \bullet x$ for all x and y in X and that $\bullet$ is associative if $x \bullet (y \bullet z) = (x \bullet y) \bullet z$ for all x, y, and z in X.

A binary operation on `⍳N` is an `N`-by-`N` matrix `B` whose entries lie in `⍳N`. Such an operation `B` is commutative if and only if `B` is equal to `⍉B` and `B` is associative if and only if `B[I;B[J;K]]` and `B[B[I;J];K]` are the same for all `I`, `J`, and `K`, that is, if and only if the arrays `B[;B]` and `B[B;]` are equal. Thus the APL propositions corresponding to the assertions that `B` is commutative and that `B` is associative are `∧/,B=⍉B` and

```
∧/,B[;B]=B[B;],
```

respectively. In *EXAMPLES* the matrix `G6` is a binary operation on `⍳6`.

```
      G6                                          ⎕IO←0
0 1 2 3 4 5                                       ∧/,G6∊⍳6
1 0 4 5 2 3                     1
2 3 0 1 5 4
3 2 5 4 0 1
4 5 1 0 3 2
5 4 3 2 1 0
```

Substituting $G6$ for B in these two propositions,

```
      ∧/,G6=⍉G6                                   ∧/,G6[;G6]=G6[G6;]
0                               1
```

we find that $G6$ is not commutative but is associative.

If X is a small finite set, it is often convenient to describe a binary operation on X by a table. For example, if $X = \{x_0, x_1, x_2\}$, then the table

●	x_0	x_1	x_2
x_0	x_1	x_0	x_1
x_1	x_2	x_1	x_0
x_2	x_1	x_2	x_2

defines the binary operation ● on X in which $x_0 \bullet x_0 = x_1$, $x_2 \bullet x_1 = x_2$, and so on. We call the matrix

```
      ⎕←T←3 3⍴1 0 1 2 1 0 1 2 2
1 0 1
2 1 0
1 2 2
```

the binary operation table for ● with respect to this particular numbering of the elements of X. In general, if ● is a binary operation on the N-element set X and we have numbered the elements of X from 0 to N-1 (or 1 to N in origin 1), then the *binary operation table* for ● with respect to the given numbering of X is the N-by-N matrix T such that T[I;J] is the number of the product of the Ith and Jth elements of X. Clearly, T is itself a binary operation on ⍳N.

Let n be a positive integer. In Section 2.3 we defined two binary operations + and × on $\mathbf{Z}_n$, the set of congruence classes modulo n. It is easy to compute binary operation tables for these operations. The most natural numbering of the elements of $\mathbf{Z}_n$ is to let the ith congruence class be $[i]$, the class containing i, $0 \leq i < n$. Using this numbering, the binary operation table for + on $\mathbf{Z}_6$ is

```
      ⎕←Z6←6|(⍳6)∘.+⍳6
0 1 2 3 4 5
1 2 3 4 5 0
2 3 4 5 0 1
3 4 5 0 1 2
4 5 0 1 2 3
5 0 1 2 3 4
```

THEOREM 1. The binary operations + and × on $\mathbf{Z}_n$ are commutative and associative.

Proof. Let a and b be in **Z**. Then, in $\mathbf{Z}_n$, we have

$$[a] + [b] = [a+b] = [b+a] = [b] + [a],$$
$$[a] \times [b] = [a \times b] = [b \times a] = [b] \times [a].$$

Thus the commutativity of + and × on $\mathbf{Z}_n$ follows from the commutativity of the corresponding operations on **Z**. Associativity is proved in the same way and is left to the reader. □

Let X be any set. Composition of functions is an associative binary operation on X^X. The set $\Sigma(X)$ of all permutations of X is a subset of X^X and, by Corollary 1.3.3, the composition of two elements of $\Sigma(X)$ is again in $\Sigma(X)$. Thus we may restrict the operation of composition to $\Sigma(X)$ and get an associative binary operation on $\Sigma(X)$.

If $\bullet$ is a binary operation on a set X, we can define a new binary operation $\circledcirc$ on 2^X by

$$A \circledcirc B = \{a \bullet b \mid a \in A, b \in B\}.$$

If $B = \{b\}$, we often write $A \circledcirc b$ instead of $A \circledcirc \{b\}$. As noted earlier, the product $x \bullet y$ is sometimes abbreviated xy when the operation $\bullet$ is clear from context. In this case $A \circledcirc B$ is written AB. Even when the symbol $\bullet$ is not omitted, it is standard practice to write $\bullet$ for $\circledcirc$, although these are really different binary operations. Thus, if A and B are subsets of **Z**, then $A + B$ and $A + 1$ denote $\{a + b \mid a \in A, b \in B\}$ and $\{a + 1 \mid a \in A\}$, respectively.

THEOREM 2. If $\bullet$ is an associative binary operation on X, then $\circledcirc$ is associative on 2^X.

Proof. The proof is straightforward and left as an exercise. □

A *semigroup* is defined to be a pair $(X, \bullet)$ of a set X and an associative binary operation $\bullet$ on X. As noted in Section 1.3, the function $\bullet$ determines its domain $X \times X$, which in turn determines X. Thus X is determined by $\bullet$, and so we could just as well define the semigroup to be the binary operation $\bullet$, X being redundant. This is, in fact, almost never done. Actually, it is standard practice to refer to the semigroup $(X, \bullet)$ by the symbol X, even though a given set may have many associative binary operations on it. Of course, this should only be done when the binary operation is clear from context.

If • is an arbitrary binary operation on X, then an expression such as $x \bullet y \bullet z \bullet w$ is ambiguous, since we need to specify in which order the operations are to be carried out. (Traditional mathematical notation has no Right-to-Left Rule!) Parentheses must be added to indicate whether we mean $x \bullet (y \bullet (z \bullet w))$, $(x \bullet (y \bullet z)) \bullet w$, or one of the other three possibilities. However, in a semigroup parentheses are unnecessary. If • is associative, then

$$x \bullet (y \bullet (z \bullet w)) = x \bullet ((y \bullet z) \bullet w) = (x \bullet (y \bullet z)) \bullet w = ((x \bullet y) \bullet z) \bullet w = (x \bullet y) \bullet (z \bullet w).$$

In fact, an induction argument shows that any two ways of adding parentheses to $a_1 \bullet a_2 \bullet \ldots \bullet a_n$ yield expressions with the same value. In particular, we can define x^n to be the product $x \bullet x \bullet \ldots \bullet x$, with n factors whenever n is a positive integer. It is easy to check that the *laws of exponents*,

$$x^m \bullet x^n = x^{m+n},$$
$$(x^m)^n = x^{mn},$$

hold for any positive integers m and n.

An element e of X is a *left identity element* for the binary operation • on X if $e \bullet x = x$ for all x in X. Similarly, f is a *right identity element* if $x \bullet f = x$ for all x in X. An element that is both a left and a right identity element is called a *two-sided identity element*. Clearly, if the operation • is commutative, then the notions of left, right, and two-sided identity elements coincide.

If `B` is a binary operation on `⍳N`, then an element `E` of `⍳N` is a left identity if and only if `∧/B[E;]=⍳N` and a right identity if and only if `∧/B[;E]=⍳N`. For `G6`,

```
      G6
0 1 2 3 4 5
1 0 4 5 2 3
2 3 0 1 5 4
3 2 5 4 0 1
4 5 1 0 3 2
5 4 3 2 1 0
```

```
      ∧/G6[0;]=⍳6
1
      ∧/G6[;0]=⍳6
1
```

0 is a two-sided identity.

The integers 0 and 1 are two-sided identities for + and × on **Z**, respectively. The operation − on **Z** has a right identity, 0, but no left identity. The identity function on a set X is a two-sided identity for composition of functions on X. The congruence classes [0] and [1] modulo n are two-sided identities for + and $\times$ on $\mathbf{Z}_n$, respectively.

A binary operation may have many left identities or many right identities but not both.

THEOREM 3. Let $\bullet$ be a binary operation on X and let e be a left identity and f a right identity for $\bullet$. Then $e = f$, and so e is a two-sided identity. Moreover, e is the only identity element for $\bullet$, right, left, or two-sided.

Proof. Consider the element $e \bullet f$. Since e is a left identity, $e \bullet f = f$. However, since f is a right identity, $e \bullet f = e$. Thus $e = f$ and e is a two-sided identity. The same argument shows that any right identity is equal to e and, since e is also a right identity, that any left identity is also equal to e. □

A *monoid* is a semigroup $(X, \bullet)$ that has a two-sided identity element e. By Theorem 3, e is unique. The semigroups $(\mathbf{Z}, +)$, $(\mathbf{Z}, \times)$, $(\mathbf{Z}_n, +)$, $(\mathbf{Z}_n, \times)$, $(X^X, \circ)$, and $(\Sigma(X), \circ)$ are all monoids, where $\circ$ denotes composition of functions on the set X.

Let $(X, \bullet)$ be a monoid with identity element e and let x be in X. An element y of X is called a *left inverse* for x if $y \bullet x = e$. Similarly, a *right inverse* for x is an element z such that $x \bullet z = e$. A *two-sided inverse* for x is an element of X that is both a left and a right inverse for x. If $\bullet$ is commutative, then left, right, and two-sided inverses are the same. Every integer x has an inverse in the monoid $(\mathbf{Z}, +)$, but only 1 and -1 have inverses in $(\mathbf{Z}, \times)$. However, in $(\mathbf{Q}, \times)$ every nonzero rational number has an inverse. The inverse of the congruence class $[i]$ in $(\mathbf{Z}_n, +)$ is $[-i]$. If f is a permutation of the set X, then f^{-1} is a two-sided inverse for f in $(X^X, \circ)$.

Suppose the matrix *B* defines a monoid on ⍳*N* with identity element *E*. Then *I* in ⍳*N* has a right inverse if and only if some component of *B*[*I*;] is equal to *E*, that is, if and only if ∨/*B*[*I*;]=*E*. In fact, *B*[*I*;]=*E* is the characteristic vector for the set of right inverses of *I*. In the monoid (⍳6,*G*6) every element has a two-sided inverse. The inverses are listed in the vector *INV*6 in *EXAMPLES*.

```
      INV6
0 1 2 4 3 5
```

The inverse of *I* is *INV*6[*I*]. (See Exercise 16.)

Theorem 3 tells us that a semigroup cannot have many right identities and also many left identities. Similarly, we can show that an element of a monoid cannot have many right inverses and also many left inverses.

THEOREM 4. Let $(X, \bullet)$ be a monoid and let x be in X. Suppose y is a left inverse of x and z is a right inverse of x. Then $y = z$. If u and v are in X and each has a two-sided inverse, then so does $u \bullet v$.

Proof. Let e be the identity element of $(X, \bullet)$ and suppose $y \bullet x = x \bullet z = e$. By the associativity of $\bullet$, we have

$$y = y \bullet e = y \bullet (x \bullet z) = (y \bullet x) \bullet z = e \bullet z = z.$$

If s and t are two-sided inverses for u and v, respectively, then

$$(u \bullet v) \bullet (t \bullet s) = (u \bullet (v \bullet t)) \bullet s = (u \bullet e) \bullet s = u \bullet s = e.$$

Similarly, $(t \bullet s) \bullet (u \bullet v)$ is e, and so $t \bullet s$ is a two-sided inverse for $u \bullet v$. □

The first part of Theorem 4 implies that in a monoid, if x has a two-sided inverse y, then y is the only inverse of x, left, right, or two-sided.

In proving the second part of Theorem 4 we found that the inverse of $u \bullet v$ was the product of the inverse of v and the inverse of u. This fact is often stated as "the inverse of a product is the product of the inverses in the opposite order." It should also be pointed out that if y is a right inverse for x, then x is a left inverse for y. Thus if y is the two-sided inverse of x, then x is the two-sided inverse of y.

Let us now determine which elements in the monoid $(\mathbf{Z}_n, \times)$ have inverses.

THEOREM 5. Let n be a positive integer. The congruence class $[u]$ has an inverse in $(\mathbf{Z}_n, \times)$ if and only if $\gcd(u, n) = 1$.

Proof. The class $[x]$ is an inverse for $[u]$ if and only if $[u] \times [x] = [1]$ or, equivalently, $ux \equiv 1 \pmod{n}$. By Theorem 2.3.3, this congruence has a solution x if and only if $\gcd(u, n)$ divides 1, that is, $\gcd(u, n) = 1$. □

For example, suppose we wish to find the inverse of $[8]$ in $(\mathbf{Z}_{37}, \times)$ or, in other words, we want to solve the congruence $8x \equiv 1 \pmod{37}$. Using the method of Section 2.3, we express $1 = \gcd(8, 37)$ as $8x + 37y$.

```
      8 ZGCD 37                    R,S
1                       14 ¯3
```

We find that $x = 14$ is a solution of the congruence, and so $[14]$ is the inverse of $[8]$. As a check, we can compute

```
      37|8×14
1
```

Let U_n be the set of elements in $\mathbf{Z}_n$ with multiplicative inverses. By Theorem 5,

$$U_n = \{ [x] \mid \gcd(x, n) = 1 \}.$$

The product of any two elements of U_n is again in U_n. This can be seen either by applying the second part of Theorem 4 or by observing that the product of two integers that are relatively prime to n is itself relatively prime to n. Also, if $[x]$ is in U_n and $[y]$ is the multiplicative inverse of $[x]$, then

$[y]$ is in U_n too. Since multiplication in $\mathbf{Z}_n$ is associative and U_n clearly contains $[1]$, it follows that $(U_n, \times)$ is a monoid in which every element has a two-sided inverse.

As the final topic of this section, we will discuss the computation of powers in a semigroup $(X, \bullet)$. Given an element x of X and a positive integer n, we have defined x^n to be the product $x \bullet \ldots \bullet x$, with n factors. We can also define x^n recursively by setting $x^1 = x$ and $x^{n+1} = x \bullet x^n$ for $n \geq 1$. If n is large and the amount of work necessary to compute the product of two elements of X is also large, it is quite inefficient to compute x^n using this definition. For example, to compute x^{57} requires calculating 56 products. However, if we compute $x^2 = x \bullet x$, $x^4 = x^2 \bullet x^2$, $x^8 = x^4 \bullet x^4$, $x^{16} = x^8 \bullet x^8$, and $x^{32} = x^{16} \bullet x^{16}$, then we can get x^{57} as $x \bullet x^8 \bullet x^{16} \bullet x^{32}$, having calculated only 8 products. In general, let $a_r \ldots a_0$ be the binary (base 2) representation for n. We may compute $x, x^2, \ldots, x^{2^r}$ with r products. Then x^n is the product of the terms x^{2^i} for which $a_i = 1$. An alternative recursive formulation of this same approach is to say that

$$x^{2m} = (x^2)^m,$$
$$x^{2m+1} = x \bullet (x^2)^m,$$

for all $m > 0$. This method of calculating powers will be referred to as the *binary power algorithm.* Exercise 23 describes an APL procedure that computes powers in $(\mathbf{R}, \times)$ using the binary power algorithm. The binary power algorithm is particularly useful for calculating powers in the monoid $(\mathbf{Z}_n, \times)$.

Even the binary power algorithm does not always compute powers using the smallest number of multiplications. The first exponent for which an improvement is possible is 15. The binary power algorithm calculates x^{15} as $x \bullet x^2 \bullet x^4 \bullet x^8$, which requires 6 products. However, forming $y = x^3 = x \bullet x \bullet x$ and then $x^{15} = y^5 = y \bullet (y^2)^2$ needs only 5 products. For a more thorough discussion of the computation of powers in a monoid, consult Knuth [Vol. 2].

EXERCISES

1 Show that $\wedge$ and $\vee$ are commutative and associative binary operations on $\{0, 1\}$.

2 Complete the proof of Theorem 1.

3 Prove Theorem 2.

4 Prove that in any semigroup the identities $x^m \bullet x^n = x^{m+n}$ and $(x^m)^n = x^{mn}$ hold for all positive integers m and n.

5 There is exactly one binary operation $\bullet$ on the empty set. Is $(\emptyset, \bullet)$ a semigroup? Is it a monoid?

6 Let X be a set and let $W(X)$ denote the set of all vectors, including the empty vector, with components in X. We will use APL notation for all elements of $W(X)$. Suppose that *U* and *V* are in $W(X)$; then *U*,*V* is in $W(X)$. Show that $(W(X),,)$ is a monoid. [The elements of $W(X)$ are often called *words* in the elements of X.]

7 List the elements of U_n for $2 \leq n \leq 6$.

8 Prove that in a semigroup $(X, \bullet)$ it is not necessary to put parentheses in expressions such as $a_1 \bullet a_2 \bullet \ldots \bullet a_n$.

9 Let X be a set and let f be an element of the monoid X^X. Show that f is injective if and only if f has a right inverse in X^X and that f is surjective if and only if f has a left inverse.

10 Find the inverse of [27] in $(\mathbf{Z}_{73}, \times)$.

11 Show that the number of multiplications needed to compute x^n using the binary power algorithm does not exceed $2 \log_2 n$.

12 The workspace *EXAMPLES* contains matrices *B*1, *B*2, and *B*3, which are binary operations on ⍳6. Determine which of these operations are commutative and which are associative.

13 Let *N* be an integer greater than 1. Show that each of the following matrices is a binary operation on ⍳*N* (in origin 0). Which of these operations are associative?

(a) (⍳*N*)∘.⌈⍳*N*
(b) |(⍳*N*)∘.-⍳*N*
(c) *N*|(⍳*N*)∘.+⍳*N*
(d) *N*|(⍳*N*)∘.×⍳*N*
(e) ⌊(⍳*N*)∘.÷1+⍳*N*

14 Construct the binary operation table for $(\mathbf{Z}_6, \times)$ using the numbering of the elements of $\mathbf{Z}_6$ given in the text.

15 Let *W*←5 4 3 2 1 0. We can number the elements of $\mathbf{Z}_6$ so that the *I*th element is the congruence class containing *W*[*I*]. Construct the binary operation tables for $(\mathbf{Z}_6, +)$ and $(\mathbf{Z}_6, \times)$ using this numbering.

16 Suppose the matrix *B* defines a binary operation on ⍳*N* with two-sided identity *E* and suppose *V* is a vector of length *N* with components in ⍳*N*. Write an APL proposition corresponding to the assertion that *V*[*I*] is a right inverse for *I*. Do the same for left inverses and two-sided inverses. Use this approach to show that the vector *INV*6 in *EXAMPLES* does, in fact, list the two-sided inverses of the elements in (⍳6, *G*6).

17 Write an APL proposition corresponding to the assertion that the

inverse of the product of two elements of (ι6, *G*6) is the product of the inverses in the opposite order.

18 Let (ι*N*, *B*) be a monoid with identity element *E*. Write APL expressions for the characteristic vectors of the following subsets of ι*N*.

(a) The set of elements with right inverses.
(b) The set of elements with left inverses.
(c) The set of elements with two-sided inverses.

19 The columns of the matrix *F*←3 3 3⊤ι27 list all of the maps of $X = \{0, 1, 2\}$ into itself. Construct the binary operation table *T* for X^X using this numbering of the elements of X^X.

20 Let *F* be as in Exercise 19 and let *P* be the vector of length 6 such that the columns of *F*[;*P*] are the six permutations of $X = \{0, 1, 2\}$. Using *P* and the table *T* constructed in Exercise 19, write an expression for a binary operation table for the monoid $(\Sigma(X), \circ)$.

21 Let *N* be an integer greater than 1 and let *P* be a vector listing the set of primes dividing *N*. Write an APL expression defining a vector that lists all positive integers less than *N* and relatively prime to *N*.

22 Construct a binary operation table for $(U_{100}, \times)$.

23 Enter the following procedure definition at a terminal.

```
      ∇Y←X POWER N
[1]   Y←1
[2]   LOOP:→(N=0)/0
[3]   →(0=2|N)/EVEN
[4]   Y←X×Y
[5]   EVEN:X←X×X
[6]   N←⌊N÷2
[7]   →LOOP∇
```

Compute 2 *POWER* 23 and compare it with 2⋆23. Show that *POWER* is an implementation of the binary power algorithm for $(\mathbf{R}, \times)$.

24 The procedure *MPZPOWER* described in Section 2.5 uses the binary power algorithm to compute powers of integers using multiple precision. Calculate 2^{100} exactly with *MPZPOWER*.

25 The procedure *ZNPOWER* uses the binary power algorithm to compute powers in $(\mathbf{Z}_n, \times)$ where n is given by the global variable *N̲*. If *X* and *M* are integers with *M* nonnegative, then *X ZNPOWER M* is congruent to *X*⋆*M* modulo *N̲*. Use *ZNPOWER* to show that

$$2^{990000} \equiv 1 (\mathrm{mod}\ 1000001).$$

26 Write a procedure using the binary power algorithm to calculate powers of maps of ιM into itself. That is, given a vector F with components lying in $\iota\rho F$ and a nonnegative integer N, the procedure should calculate the vector G that would be obtained by composing F with itself $N-1$ times. (If $N = 0$, then G should be $\iota\rho F$.)

27 Write a recursive procedure for calculating powers in $(\mathbf{R}, \times)$ based on the observation that $x^{2m} = (x^m)^2$ and $x^{2m+1} = x(x^m)^2$ for $m > 0$.

28 Let f:$\mathbf{N} \longrightarrow \mathbf{N}$ map i to $i + 1$, $i = 1, 2, \ldots$. Show that f has infinitely many right inverses and no left inverses in the monoid $\mathbf{N}^{\mathbf{N}}$.

2. GROUPS

In our investigation of sets with binary operations in Section 1 we came across several monoids in which every element has a two-sided inverse. The monoids $(\mathbf{Z}, +)$, $(\mathbf{Z}_n, +)$, $(\Sigma(X), \circ)$, $(U_n, \times)$, and $(\iota 6, G6)$ all have this property. Such monoids are called *groups.* The rest of this chapter will be devoted to the study of groups.

Since groups play such an important role, not only in algebra but in all of mathematics, it is worthwhile restating the definition. A group is a pair $(G, \bullet)$ consisting of a set G and a binary operation $\bullet$ on G such that:

1. $\bullet$ is associative.
2. There exists an element e in G such that $e \bullet x = x \bullet e = x$ for all x in G.
3. For each x in G there is an element y of G such that $x \bullet y = y \bullet x = e$.

By Theorem 1.3 the identity elements e is unique and by Theorem 1.4 the inverse y of an element x of G is also unique. As we have remarked before, it is conventional to refer to G as the group as long as the operation $\bullet$ is clear.

An N-by-N matrix G is called a *group table* if $(\iota N, G)$ is a group. Suppose that $(G, \bullet)$ is a group and that G is a finite set. It is easy to show that the binary operation table G for $(G, \bullet)$ relative to any numbering of the elements of G is a group table. We say that G is a group table for $(G, \bullet)$. In *CLASSLIB* there are several procedures for working with groups defined by group tables. The names of these procedures all begin with the prefix *GT*. Most of these group table procedures perform computations in the group described by the three global variables *GTABLE*, *GTIO*, and *GTINV*. The matrix *GTABLE* is the group table defining the binary operation. Since

GTABLE cannot be a group table both in origin 0 and in origin 1, we need to know which origin to use with *GTABLE*. The variable *GTIO* gives the correct origin. In addition, all of the procedures assume that *GTIO* is the identity element for *GTABLE*. The *I*th component of the vector *GTINV* is the inverse of *I* in the group defined by *GTABLE*. We will refer to the group described by *GTABLE*, *GTIO*, and *GTINV* as the *current abstract group*.

In order to make (⍳6,*G*6) the current abstract group, we use the procedure *GTINIT* to initialize *GTABLE*, *GTIO*, and *GTINV*.

```
      GTINIT G6
      ∧/,GTABLE=G6
1
```

```
      GTIO
0
      GTINV
0 1 2 4 3 5
```

The single argument to *GTINIT* is the group table, which is copied into *GTABLE*. The variables *GTIO* and *GTINV* are then computed. In order to save time, *GTINIT* does not check whether *GTABLE* really does define a group. In order to test whether a given matrix is a group table for the appropriate choice of index origin, it is necessary to use *GTCHECK*.

```
      GTCHECK G6
1
```

The value returned by *GTCHECK* is 1 or 0, depending on whether the argument is or is not a group table, with the identity element being the smallest entry in the table. Executing *GTCHECK* does not affect the definition of the current abstract group; that is, *GTABLE*, *GTIO*, and *GTINV* are not changed.

If *A* and *B* are arrays whose entries are elements of the current abstract group, then *A GTPROD B* is the entry-by-entry product of *A* and *B* in the group.

```
      2 GTPROD 4
5
      GTABLE[2;4]
5
      3 GTPROD 1
2
      2 3 GTPROD 4 1
5 2
```

```
      0 GTPROD ⍳6
0 1 2 3 4 5
      (⍳6) GTPROD 0
0 1 2 3 4 5
      GTINV GTPROD ⍳6
0 0 0 0 0 0
      (⍳6) GTPROD GTINV
0 0 0 0 0 0
```

Note that arguments for *GTPROD* that have only one entry are expanded to match the other argument.

The workspace *EXAMPLES* contains two other origin 0 group tables, *G*24 and *G*60, which are 24-by-24 and 60-by-60 matrices, respectively.

```
      ⍴G24
24 24
      GTCHECK G24
1
```

```
      ⍴G60
60 60
      GTCHECK G60
1
```

The matrix *G*60 is too big to permit the associative law to be verified directly by comparing *G*60[;*G*60] and *G*60[*G*60;]. The approach used in *GTCHECK* is much more efficient in both time and space. (See Exercise 8.21.) The vectors *INV*24 and *INV*60 list the inverses of the elements in (⍳24, *G*24) and (⍳60, *G*60), respectively.

```
      ⍴INV24
24
      ⍴INV60
60
```

```
      GTINIT G24
      13 GTPROD INV24[13]
0
```

Although group tables are an important aid as one is beginning the study of groups, in practice one does not perform computations in a particular group G by first constructing a group table for G and then working with that table. The group G may have infinitely many elements and, even if G has only finitely many elements, a group table for G may be too large to fit into any computer. For example, to show that [471562] + [823495] = [307403] in the group $(\mathbf{Z}_{987654}, +)$ hardly requires the construction of a group table.

Let $(G, \bullet)$ be a group. If the operation $\bullet$ is commutative, we say the group is *commutative* or *abelian.* [The term "abelian" is derived from the name of Norwegian mathematician Niels Abel (1802-1829).] A *nonabelian* group is a group that is not abelian. If G is a finite set, $(G, \bullet)$ is said to be a *finite group* and $|G|$ is called the *order* of the group. Thus $(\mathbf{Z}_n, +)$ is abelian, finite, and of order n. The group $(U_n, \times)$ is also finite and abelian. Its order is usually denoted by $\phi(n)$, and ϕ is referred to as the *Euler phi-function.* [Leonhard Euler (1707-1783) was a very important Swiss methematician.] The function ϕ is important in number theory. We will exhibit a nice formula for $\phi(n)$ in Section 6.

The group $\Sigma(X)$ of all permutations of the set X is called the *symmetric group* on X. Recall that if X is the set ⍳*N*, we write Σ_N instead of $\Sigma(X)$.

THEOREM 1. The group Σ_n has order $n!$ and is nonabelian if $n \geq 3$.

Proof. The order of Σ_n was determined in Exercise 1.3.9. Suppose $n \geq 3$. Let f be the element of Σ_n that interchanges 0 and 1 and fixes 2, . . . ,

$n - 1$, and let g be the element interchanging 1 and 2 and fixing the remaining points. Then $f \circ g$ maps 0 to 2 while $g \circ f$ maps 0 to 1. Thus $f \circ g \neq g \circ f$, and so Σ_n is nonabelian. □

THEOREM 2. Suppose $(G, \bullet)$ is a group and u, x, y are in G. Then

(a) If $u \bullet x = u \bullet y$, then $x = y$.

(b) If $x \bullet u = y \bullet u$, then $x = y$.

Proof. Let e be the identity element of G and let v be the inverse of u. If $u \bullet x = u \bullet y$, then

$$x = e \bullet x = (v \bullet u) \bullet x = v \bullet (u \bullet x) = v \bullet (u \bullet y) = (v \bullet u) \bullet y = e \bullet y = y.$$

Part (b) is proved in the same way. □.

Parts (a) and (b) of Theorem 2 are called the *left* and *right cancellation laws*, respectively.

COROLLARY 3. Let $(G, \bullet)$ be a group, let u be a fixed element of G, and let v be the inverse of u. The maps $R_u : G \longrightarrow G$ and $L_u : G \longrightarrow G$ defined by $R_u : x \longmapsto x \bullet u$ and $L_u : x \longmapsto v \bullet x$ are permutations of G.

Proof. By Theorem 2, the maps R_u and L_u are injective. Let y be an element of G. Then $(y \bullet v)R_u = (y \bullet v) \bullet u = y \bullet (v \bullet u) = y$. Therefore R_u is surjective also. Similarly, L_u maps $u \bullet y$ to y, and so both R_u and L_u are permutations of G. □

The reason for defining xL_u to be $v \bullet x$ instead of $u \bullet x$ is explained in Exercise 3. It follows from Corollary 3 that if G is an N-by-N group table, then each row and each column of G is a permutation of ιN.

There are two standard notational conventions used in connection with groups. In *multiplicative notation* the symbol for the group operation is omitted so that $x \bullet y$ is written as xy and called the product of x and y. The identity element is denoted by e or, more commonly, by 1. The inverse of x is written x^{-1}. By Theorem 1.4, we have the formula $(xy)^{-1} = y^{-1}x^{-1}$. If n is a positive integer, then x^n is the product $xx \ldots x$, with n factors, and x^{-n} is defined to be $(x^{-1})^n$. Finally, x^0 is defined to be 1.

THEOREM 4. Let G be a group written multiplicatively, let m and n be integers, and let x and y be elements of G. Then

$$x^m x^n = x^{m+n},$$

$$(x^m)^n = x^{mn}.$$

If x and y commute, then $(xy)^n = x^n y^n$.

Proof. The proof is left to the reader. The case in which m and n are both positive has already been discussed in Section 1. □

As a consequence of Theorem 4, we have the fact that the inverse of x^n is x^{-n}.

Groups may also be written additively. In *additive notation* the group operation is denoted by a plus sign, even though the operation may have little to do with ordinary addition. The value of $x + y$ is called the *sum* of x and y, and zero is used to denote the identity element. The inverse of x is written $-x$ and the sum $x + (-y)$ is abbreviated $x - y$. If n is a positive integer and x is a group element, then the sum $x + x + \ldots + x$, with n summands, is written nx and called the nth *multiple* of x. The multiple $(-n)x$ is defined to be $n(-x)$, and $0x$ is defined to be the identity element 0.

Additive notation is used almost exclusively for abelian groups such as $(\mathbf{Z}, +)$ and $(\mathbf{Z}_n, +)$. Multiplicative notation is used for both abelian and nonabelian groups. We will use multiplicative notation for the symmetric groups $\Sigma(X)$. In making definitions and stating theorems about groups, we will normally write the groups multiplicatively.

There is an entertaining pastime indulged in by many students of group theory; it consists of trying to find the weakest assumptions that can be made about a binary operation $\bullet$ on a set X that imply that $(X, \bullet)$ is a group. The next result illustrates one attempt of this kind.

THEOREM 5. Let $(G, \bullet)$ be a semigroup with a left identity e such that relative to e every element of G has at least one left inverse. Then $(G, \bullet)$ is a group.

Proof. Since e is a left identity, $e \bullet e = e$. Let x be any element of G such that $x \bullet x = x$. If y is a left inverse of x relative to e, then

$$e = y \bullet x = y \bullet (x \bullet x) = (y \bullet x) \bullet x = e \bullet x = x.$$

Thus e is the only element of G that is equal to its square. Now let u be any element of G and let v be a left inverse for u. Then, since $v \bullet u = e$, we have

$$(u \bullet v) \bullet (u \bullet v) = u \bullet ((v \bullet u) \bullet v) = u \bullet (e \bullet v) = u \bullet v.$$

Therefore $u \bullet v$ must be e, and so v is a right inverse for u too. Also,

$$u \bullet e = u \bullet (v \bullet u) = (u \bullet v) \bullet u = e \bullet u = u,$$

and so e is a right identity. Thus we have a two-sided identity relative to which two-sided inverses exist, and so $(G, \bullet)$ is a group. □

It seems that it is very difficult to weaken the associative law in any significant way and still have a group. See Exercise 12 in this regard.

EXERCISES

1 Show that each of the following is a group.

(a) $(\{1, -1\}, \times)$.

(b) $(2\mathbf{Z}, +)$.

(c) $(\mathbf{Q} - \{0\}, \times)$.

(d) $(\mathbf{Z} \times \mathbf{Z}, \bullet)$, where $(x_1, x_2) \bullet (y_1, y_2) = (x_1 + y_1, x_2 + y_2)$.

2 Complete the proof of Theorem 2.

3 Let G be a group and let R_u and L_u be the permutations of G defined in Corollary 3. Show that for u and w in G we have $R_u \circ R_w = R_{uw}$ and $L_u \circ L_w = L_{uw}$. Would the second condition hold if we had defined xL_u to be $u \bullet x$?

4 Prove Theorem 4.

5 Write Theorem 4 using additive notation.

6 Suppose $(G, \bullet)$ is a group. Define a new binary operation $*$ on G by $x * y = y \bullet x$. Show that $(G, *)$ is a group.

7 Suppose $(G, \bullet)$ is a group and $f: G \longrightarrow X$ is a bijection. Define a binary operation $*$ on X by

$$x * y = ((xf^{-1}) \bullet (yf^{-1}))f.$$

Prove that $(X, *)$ is a group.

8 Let G be a group in which $x^2 = 1$ for all x in G. Show that G is abelian.

9 Let p be a prime and r a positive integer. Prove that

$$\phi(p^r) = p^{r-1}(p - 1).$$

10 Show that an integer n greater than 1 is a prime if and only if $\phi(n) = n - 1$.

11 Let $(X, \bullet)$ be a semigroup. Assume that X is not empty and that for all u and v in X there exist x and y in X such that $u \bullet x = v$ and $y \bullet u = v$. Show that $(X, \bullet)$ is a group.

*12 Let G be a nonempty set, let D be a subset of $G \times G$, and let $\bullet$ be a function from D to G. Suppose:

(a) For all u and v in G there exist x and y in G such that (u, x) and (y, u) are in D and $u \bullet x = v = y \bullet u$.

(b) If x, y, and z are in G, then $x \bullet (y \bullet z) = (x \bullet y) \bullet z$ provided all four products are defined, that is, if (x, y), (y, z), $(x, y \bullet z)$, and $(x \bullet y, z)$ are all in D.

Prove that $D = G \times G$ and $(G, \bullet)$ is a group.

13 Verify that `⍉G6` is a group table. (Compare Exercise 6.)

14 Let `P←2⍉⍳6` and `G←P[G6[⍋P;⍋P]]`. Verify that `G` is a group table. (Compare Exercise 7.)

15 Compute $\phi(n)$ for $n = 2, \ldots, 10$.

16 Prove that if G is a binary operation table for the finite group G, then G is a group table.

17 Given that the matrix G is a group table, write a simple APL expression for the identity element of G.

18 How many of the elements in the group (⍳24, G24) satisfy the condition $x^{-1} = x$? How many elements in (⍳60, G60) satisfy this condition?

19 Determine the number of n-by-n group tables for $1 \leq n \leq 3$.

3. SUBGROUPS

In Section 2.2 we defined an additive subgroup of **Z** to be a nonempty subset of **Z** that contains the sum and difference of any two of its elements. It is now time to place this definition in its proper context. Let G be a group. A *subgroup* of G is a nonempty subset H of G such that whenever x and y are in H, then xy and x^{-1} are also in H. We often say that a subgroup H is *closed* under products and inverses. (If we are using additive notation, we say H is closed under sums and negatives.) A subset M of **Z** is clearly a subgroup of (**Z**, +) if and only if M is an additive subgroup of **Z** in the sense of Section 2.2. Thus the subgroups of **Z** are the sets of the form $n\mathbf{Z}$, $n \geq 0$. The set {[0], [2], [4]} of congruence classes modulo 6 is a subgroup of ($\mathbf{Z}_6$, +).

Let x be an element of the set X and let H be the set of all permutations of X that fix x. If f and g are in H, then so are $f \circ g$ and f^{-1} and, therefore, H is a subgroup of $\Sigma(X)$. We call H the *stabilizer* of x in $\Sigma(X)$. For any group G the sets $\{1\}$ and G are always subgroups of G. A *proper subgroup* of G is a subgroup different from G, and a *nontrivial subgroup* is a subgroup different from $\{1\}$.

THEOREM 1. Let H be a subgroup of the group G. Then H is a group in its own right under the restriction to H of the binary operation on G. A subset K of H is a subgroup of G if and only if K is a subgroup of H.

Proof. Since H is closed under products, the restriction of the binary operation of G to H is a binary operation on H. Since the associative law holds in all of G, it holds in H. Now H is nonempty and so contains an element h. Therefore H contains $1 = hh^{-1}$, and so H has a two-sided identity. Since H is closed under inverses, H is a group. The remainder of the proof is left as an exercise. □

Suppose H and K are subgroups of a group G. In general, the set $H \cup K$ will not be a subgroup of G. For example, if we take $G = \mathbf{Z}$, $H = 2\mathbf{Z}$, and $K =$

$3\mathbf{Z}$, then 2 and 3 are in $H \cup K$ but $5 = 2 + 3$ is not. For intersections, however, things are much nicer.

THEOREM 2. Let G be a group. For each i in the nonempty set I let H_i be a subgroup of G. Then

$$K = \bigcap_{i \in I} H_i$$

is a subgroup of G.

Proof. Each of the subgroups H_i contains the identity element 1 of G. Therefore 1 is in K and K is nonempty. Let x and y be elements of K. Then, for each i in I, the subgroup H_i contains x and y and therefore also xy and x^{-1}. Thus xy and x^{-1} are in K, and so K is a subgroup of G. □

Let X be a subset of the group G. The set of subgroups H of G that contain X is not empty, since $H = G$ is one such subgroup. By Theorem 2, the intersection K of all subgroups of G containing X is a subgroup of G. Clearly, K contains X and is the smallest subgroup of G that contains X. We denote K by $<X>$ and refer to K as the subgroup of G *generated* by X. An interesting example is given by the case $X = \emptyset$. If $K = <\emptyset>$, then K is a subgroup, and so 1 is in K. But $\{1\}$ is a subgroup of G which contains $\emptyset$ and thus $K \subseteq \{1\}$. Therefore $K = \{1\}$, and so $\emptyset$ generates the trivial subgroup of G. The next theorem gives a more concrete description of $<X>$ when $X \neq \emptyset$.

THEOREM 3. Let X be a nonempty subset of the group G. The subgroup $<X>$ is the set of all elements of G that can be expressed in the form $x_1 \dots x_n$, where $n \geq 1$ and each x_i is either in X or is the inverse of an element in X.

Proof. Since subgroups are closed under products and inverses, any subgroup of G that contains X also contains every product $x_1 \dots x_n$ such that either x_i or x_i^{-1} is in X. Therefore the set H of all products of this form is in $<X>$. However, if $u = x_1 \dots x_n$ and $v = y_1 \dots y_m$ are in H, then $uv = x_1 \dots x_n y_1 \dots y_m$ and $u^{-1} = x_n^{-1} \dots x_1^{-1}$ are also in H. Thus H is closed under products and inverses. Since $H \neq \emptyset$, H is a subgroup of G. Clearly, $X \subseteq H$, and so $H \supseteq <X>$. Therefore $H = <X>$. □

If X and Y are both subsets of G, then $<X \cup Y>$ is normally written $<X, Y>$. We say that G is *finitely generated* if there is a finite subset X of G such that $G = <X>$. If x is an element of G, we write $<x>$ for $<\{x\}>$. If $<x>$ is finite, then $|<x>|$ is called the *order* of x. A group generated by a single element is said to be *cyclic*. The group $\mathbf{Z}$ is cyclic since it is generated by 1. The groups $\mathbf{Z}_n$ are generated by the congruence class $[1]$ and so are cyclic too. If $G = <x>$, then $G = \{x^m \mid m \in \mathbf{Z}\}$ by Theorem 3. Since $x^m x^n = x^{m+n} = x^n x^m$, G is abelian.

THEOREM 4. Let G be a finite cyclic group of order n generated by the element x. Then $x^n = 1$ and n is the smallest positive integer with this property. Moreover, $G = \{1, x, \ldots, x^{n-1}\}$. If $x^m = 1$, then n divides m.

Proof. Since G is finite, the elements $1 = x^0, x = x^1, x^2, x^3, \ldots$ cannot be distinct. Thus there exist integers r and s with $r > s \geq 0$ and $x^r = x^s$. Choose such a pair with r as small as possible. Multiplying both sides of the equation $x^r = x^s$ by $(x^s)^{-1} = x^{-s}$, we obtain $x^{r-s} = 1 = x^0$. By the minimality of r, this means $s = 0$, and so $x^r = 1$. Also, the elements $x^0, \ldots, x^{r-1}$ are distinct. Since $G = \langle x \rangle$, every element of G has the form x^i for some integer i. Let q be the integral quotient when i is divided by r. Then $i = j + qr$ and $0 \leq j < r$. Thus

$$x^i = x^{j+qr} = x^j x^{qr} = x^j (x^r)^q = x^j.$$

Therefore every element of G is in $\{1, x, \ldots, x^{r-1}\}$, and so $r = |G| = n$. If $x^m = 1$, then m must be divisible by n. □

COROLLARY 5. Let x be an element of a finite group G. Then there exists a positive integer m such that $x^{-1} = x^m$.

Proof. Let n be the order of x, that is, the order of $\langle x \rangle$. Then $n \geq 1$ and $x^n = 1$ by Theorem 5. Setting $m = 2n - 1$, we have $m > 0$ and

$$x^m = x^{2n-1} = (x^n)^2 x^{-1} = x^{-1}. \qquad \square$$

Now we are able to show that for finite groups the condition that a subgroup be closed under inverses is superfluous.

COROLLARY 6. Let H be a nonempty subset of the finite group G. Then H is a subgroup of G if and only if H is closed under products.

Proof. If H is a subgroup of G, then H is closed under products. Suppose H is a nonempty subset of G that is closed under products. By Corollary 5, if x is in H, then there is an integer $m > 0$ such that $x^{-1} = x^m$. Since H is closed under products, x^m is in H, and so H is a subgroup of G. □

Let `G` be an `N`-by-`N` group table and let `A` be a vector listing the elements of a nonempty subset A of `⍳N`. By Corollary 6, A is a subgroup of (`⍳N`, `G`) if and only if `G[A[I];A[J]]` is a component of `A` for all `I` and `J` in `⍳⍴A`. This is equivalent to the proposition `∧/,G[A;A]∊A`.

Let us test some subsets of `⍳6` to see if they are subgroups of (`⍳6`, `G6`).

```
      G6
0 1 2 3 4 5
1 0 4 5 2 3
2 3 0 1 5 4
3 2 5 4 0 1
4 5 1 0 3 2
5 4 3 2 1 0
```

```
      ∧/,G6[A;A]∊A←0 2
1
      ∧/,G6[B;B]∊B←0 3 4
1
      ∧/,G6[C;C]∊C←A,B
0
```

Here we see that {0, 2} and {0, 3, 4} are subgroups while, as we should expect, their union is not.

As a further example, we will determine the subgroup of (⍳24, *G*24) generated by the set $A = \{3, 8\}$. The calculation

```
      ∧/,G24[A;A]∊A←3 8
0
```

shows that A is not closed under products, and so we form the union of A with the set of products of pairs of elements in A.

```
      ⎕←A←SSORT A,,G24[A;A]
3 4 7 8 12 16
      ∧/,G24[A;A]∊A
0
```

We still do not have a subgroup, so we repeat the process.

```
      ⎕←A←SSORT A,,G24[A;A]
0 3 4 7 8 11 12 15 16 19 20 23
      ∧/,G24[A;A]∊A
1
```

The set listed in *A* is now a subgroup, and so <3, 8> = {0, 3, 4, 7, 8, 11, 12, 15, 16, 19, 20, 23}.

The procedure used in the preceding example is inefficient, since it computes many unnecessary products. A better algorithm is described in Exercise 13. This algorithm is used in the procedure *GTSGP* in *CLASSLIB*.

```
      GTINIT G6
      GTSGP 3
0 3 4
      GTINIT G24
      GTSGP 3 8
0 3 4 7 8 11 12 15 16 19 20 23
```

The procedure *GTSGP* takes a single argument that lists a set of elements in the current abstract group. The result is a list of the elements in the subgroup generated by the set.

It would be interesting to know how many subgroups there are in the group (⍳24, *G*24). However, as ⍳24 has 2^{24} or 16777216 subsets, it is clear that we will have to have a better way of finding out than just trying each subset in turn. To do this, we will need some more theory. It will turn out that a randomly chosen subset of a group is very unlikely to be a subgroup.

Throughout the rest of this section we will assume that H is a subgroup of the group G. If x and y are in G, we say x is *right congruent* to y modulo H if xy^{-1} is in H.

THEOREM 7. Right congruence modulo H is an equivalence relation on G.

Proof. For any x in G we know that $xx^{-1} = 1$ is in H, and so right congruence is reflexive. If xy^{-1} is in H, then $yx^{-1} = (xy^{-1})^{-1}$ is also in H, since H is closed under inverses. Therefore, if x is right congruent to y, then y is right congruent to x. Finally, suppose x is right congruent to y and y is right congruent to z. Then xy^{-1} and yz^{-1} are both in H. Hence $xz^{-1} = (xy^{-1})(yz^{-1})$ is in H as H is closed under products. Thus x is right congruent to z, and we have an equivalence relation. □

The equivalence classes of right congruence modulo H are called the *right cosets* of H in G. The next theorem gives an alternate description of the right cosets of H.

THEOREM 8. The right coset of H containing x is Hx, and there is a bijection from H to Hx.

Proof. The set Hx is defined to be $\{hx \mid h \in H\}$. Since $(hx)x^{-1} = h$, any element of Hx is right congruent to x modulo H. Suppose y is an element of G that is right congruent to x modulo H. Then $h = yx^{-1}$ is in H, and $y = (yx^{-1})x = hx$ is in Hx. Thus Hx is the right coset of H containing x. Now let f be the map from H to Hx that takes h to hx. The definition of Hx implies immediately that f is surjective, and the right cancellation law shows that f is injective. Therefore f is bijective. □

If the set of right cosets of H in G is finite, then the number of right cosets is called the *index* of H in G and is written $|G:H|$.

THEOREM 9 *(Lagrange).* If G is finite, then $|G| = |G:H| \times |H|$. In particular, $|H|$ divides $|G|$.

Proof. We have $|G:H|$ right cosets of H, each with $|H|$ elements. Since the right cosets of H form a partition of G, the number of elements in G is just the sum of the number of elements in each right coset. □

Theorem 9 is named for the Italian-French mathematician Joseph-Louis Lagrange (1736-1813).

COROLLARY 10. A group of prime order is cyclic.

Proof. Let G be a group of prime order p. Choose an element x in G different from the identity. The subgroup $\langle x \rangle$ generated by x contains x and the identity element and has order dividing p. By the definition of a prime, $|\langle x \rangle| = p$, and so $\langle x \rangle = G$. Thus G is generated by any one of its nonidentity elements. □

COROLLARY 11. If x is an element of the finite group G, then $x^{|G|} = 1$.

Proof. Let n be the order of x and let $m = |G:\langle x \rangle|$. By Lagrange's Theorem, $|G| = mn$, and so

$$x^{|G|} = x^{mn} = (x^n)^m = 1^m = 1,$$

since $x^n = 1$ by Theorem 4. □

If we apply Corollary 11 to the group U_n, we get an important result in number theory.

COROLLARY 12. Let a and n be integers with $n > 0$. If $\gcd(a, n) = 1$, then $a^{\phi(n)} \equiv 1 \pmod{n}$.

Proof. Suppose $\gcd(a, n) = 1$. Then the congruence class $[a]$ is in U_n, which has order $\phi(n)$. By Corollary 11, we have $[a]^{\phi(n)} = [1]$ or $a^{\phi(n)} \equiv 1 \pmod{n}$. □

Corollary 12 gives us a very powerful test for deciding whether a given integer is a prime.

COROLLARY 13. Let p be a prime and let a be an integer not divisible by p. Then $a^{p-1} \equiv 1 \pmod{p}$.

Proof. Since p is a prime, $\phi(p) = p - 1$. □

Suppose we fix a small integer a greater than 1, say $a = 2$ or $a = 3$. Given a large integer n, it is easy to see whether or not $\gcd(a, n) = 1$ using the Euclidean algorithm. If $\gcd(a, n) \neq 1$, then n is certainly not a prime. If $\gcd(a, n) = 1$, then we can use the binary power algorithm described in Section 1 to compute the remainder of a^{n-1} modulo n. If a^{n-1} is not congruent to 1 modulo n, then n is not a prime. If $a^{n-1} \equiv 1 \pmod{n}$, then we say that n is a *pseudoprime* relative to a. In this case the chances are very good that n is a prime. One way to prove that n is a prime is to show that U_n has order $n - 1$. The exercises describe an approach to this problem that assumes that the prime factors of $n - 1$ are known.

Having seen the definitions of right congruence and right cosets, one can easily construct the definitions of left congruence and left cosets. Two elements x and y of G are *left congruent* modulo H if $x^{-1}y$ is in H. Left congruence is readily seen to be an equivalence relation on G, and the equivalence classes are the *left cosets* of H. The analogue of Theorem 8 holds so that the left coset containing x is xH. If G is finite, then the number of elements in a right coset of H is the same as the number in a left coset, that is, $|H|$. Thus the number of right and left cosets must be the same. Even if G is infinite, there is still the "same number" of right and left cosets in the sense that there is a bijection from the set of right cosets to the set of left cosets. (See Exercise 22.)

Let us consider an example in which G is the group (`⍳24`, `.G24`). The cyclic subgroup H generated by 9 has order 4.

```
      GTINIT G24
      ⎕←H←GTSGP 9
0 9 16 18
```

Right and left congruence modulo H are equivalence relations on `⍳24`. How can we construct the characteristic matrices for these equivalence relations? The inverse of `I` in G is `INV24[I]`. Two elements `I` and `J` of `⍳24` are right congruent modulo H if and only if `G24[I;INV24[J]]` is in H. Thus the characteristic matrix for right congruence modulo H is

```
      R←G24[;INV24]∊H
```

Similarly, the characteristic matrix for left congruence is

```
      L←G24[INV24;]∊H
```

We can check directly that right and left congruence are equivalence relations.

```
      SEQREL R                              SEQREL L
1                                     1
```

We can also verify that `R` and `L` represent different relations.

```
      ∧/,R=L
0
```

The procedures `GTRCON` and `GTLCON` construct the characteristic matrices for right and left congruence, respectively, modulo a subgroup of the current abstract group. We could have obtained `R` and `L` by

```
      R←GTRCON H
      L←GTLCON H
```

EXERCISES

1 Give an example of a nonempty subset of **Z** that is closed under sums but is not a subgroup of **Z**.

2 Suppose H is a nonempty subset of a group G such that for all x and y in H the product xy^{-1} is in H. Show that H is a subgroup of G.

3 Let K be the set of rational numbers m/n with m and n integers and n a power of 3. Prove that K is a subgroup of (**Q**, +).

4 Which of the following sets are subgroups of $(\mathbf{Q} - \{0\}, \times)$?
 (a) The set of positive rational numbers.
 (b) The set of negative rational numbers.
 (c) $\{1, -1\}$.
 (d) The set of rational numbers x with $x \geq 1$.

5 Complete the proof of Theorem 1.

6 Let x be an element of a group and let $M = \{m \epsilon \mathbf{Z} | x^m = 1\}$. Show that M is a subgroup of $\mathbf{Z}$.

7 Let L be the set of all elements in Σ_7 that map the set $\{0, 1, 2\}$ into itself. Prove that L is a subgroup of Σ_7 and determine $|L|$.

8 Let X be a subset of the group G and let C be the set of elements g in G such that $xg = gx$ for all x in X. Show that C is a subgroup of G. (We call C the *centralizer* of X in G.)

9 Let `G` be an `N`-by-`N` group table and let `X` be a vector listing a subset of `ιN`. Write an APL expression for the characteristic vector of the centralizer of `X` in `(ιN, G)`. Determine the centralizer of 27 in `(ι60, G60)`.

10 Let H and K be subgroups of the abelian group G. Show that HK is a subgroup of G.

11 Let r and s be integers. Prove that $(r\mathbf{Z}) + (s\mathbf{Z}) = d\mathbf{Z}$ and $(r\mathbf{Z}) \cap (s\mathbf{Z}) = m\mathbf{Z}$, where $d = \gcd(r, s)$ and $m = \operatorname{lcm}(r, s)$.

12 Let X be a nonempty subset of the finite group G. Show that $<X>$ is the smallest subset H of G such that $X \subseteq H$ and $HX \subseteq H$.

13 Let G and X be as in Exercise 12. Show that the following algorithm terminates with $H = <X>$.
 (a) Set H and Y equal to $\{1\}$.
 (b) Set Y equal to $YX - H$.
 (c) If $Y = \emptyset$, stop.
 (d) Set H equal to $H \cup Y$ and go to step (b).

14 Suppose x is an element of a group and x has order 40. What are the orders of x^2, x^5, and x^{28}?

15 Find all subgroups of `(ι6, G6)`.

16. Find the cyclic subgroups of the groups `(ι24, G24)` and `(ι60, G60)`. How many elements of each order do these groups have?

17 Show that `(ι24, G24)` and `(ι60, G60)` can each be generated by two elements.

18 In the text we found a subgroup A of `(ι24, G24)` with order 12. Construct a group table for A.

19 Show that the group A in Exercise 18 has no subgroup of order 6.

20 Show that the set of elements x in the group A of Exercise 18 that satisfy $x^2 = 1$ forms a subgroup of A.

21 Let n be a positive integer. Show that the cosets of $n\mathbf{Z}$ in $\mathbf{Z}$ are the congruence classes modulo n and thus that $|\mathbf{Z}:n\mathbf{Z}| = n$.

22 Let H be a subgroup of the group G. Show that two elements x and y of G are right congruent modulo H if and only if x^{-1} and y^{-1} are left congruent modulo H. Use this to show that there is a bijection from the set of right cosets of H onto the set of left cosets of H.

23 Show that any group of order 4 is abelian. (*Hint.* Either there is an element of order 4 or Exercise 2.8 applies.)

24 Let p be a prime and let a be any integer. Prove that $a^p \equiv a \pmod{p}$.

25 Let x be an element of a group and let n be a positive integer. Suppose $x^n = 1$, but $x^{n/p} \neq 1$ for all primes p dividing n. Show that x has order n.

26 Let x be an element of a group and suppose an integer n is known such that $x^n = 1$. Assuming the prime factors of n are given, describe an efficient procedure for determining the order of x.

27 Find the order of $[2]$ in U_{29}.

28 Given that $2^{990000} \equiv 1 \pmod{1000001}$, determine the order of $[2]$ in $U_{1000001}$. (The procedure *ZNPOWER* described in Exercise 1.25 will be useful here.)

29 Let n be an integer greater than 1. Assume we have a sequence $a_1, \ldots, a_r$ of integers relatively prime to n and we know the order m_i of $[a_i]$ in U_n. Suppose $\operatorname{lcm}(m_1, \ldots, m_r) = n - 1$. Prove that n is a prime.

30 Construct an approach based on Exercises 26 and 29 and the remarks following Corollary 13 for deciding whether or not an integer n is a prime, assuming that the prime factors of $n - 1$ can be obtained.

*31 On most APL terminal systems, computation in U_n with $n > 10^8$ requires multiple-precision calculations. The procedure *ZNPOWER* uses double precision when necessary to produce correct results, even when $\underline{N}*2$ exceeds the usual precision of the system. Use *ZNPOWER* to determine which of the following integers are pseudoprimes relative to 2.

(a) 999999937.
(b) 159890287921.
(c) 1099511627689.

(These computations may require significant amounts of CPU time.)

*32 Show that $p = 1000037$ is a prime. From this it follows that U_p has order $p - 1$. Show that the class [2] has order $p - 1$ in U_p and therefore is a generator of U_p. This means that there is an integer m such that $[2]^m = [3]$ or, equivalently, $2^m \equiv 3 \pmod{p}$. Find one such m.

33 Let G be a finite group. Prove that there is a subset X of G such that $G = \langle X \rangle$ and $|X| \leq \log_2 |G|$.

34 Is $(\mathbf{Q}, +)$ finitely generated?

In Exercises 35 to 37 assume that the following statements have been executed as in the last example in the text.

```
GTINIT G24
H←GTSGP 9
R←GTRCON H
L←GTLCON H
```

35 Let `REP←SSORT SFEL R`. Explain why `REP` lists a set of representatives for the right cosets of H in (`⍳24, G24`). Verify that `(⍴REP)×⍴H` is 24, in agreement with Lagrange's Theorem.

36 Write an APL expression for `L` in terms of `R` and `INV24`. (*Hint.* See Exercise 22.)

37 Construct the characteristic vector for the set N of integers `I` in `⍳24` such that `R[I;] = L[I;]`. Check that N is a subgroup of (`⍳24, G24`). Is this an accident?

4. HOMOMORPHISMS

In this section we will study maps from one group to another. We will not be interested in just any maps but only in those that are "compatible" with the binary operations on the two groups. Let $(G, \bullet)$ and $(H, *)$ be groups and let h be a function from G to H. What does it mean for h to be compatible with $\bullet$ and $*$? We say that h is a *homomorphism* from G to H if, for all x and y in G, we have $(x \bullet y)h = (xh) * (yh)$. Informally, we say that under a homomorphism the image of a product is the product of images. (Of course, with additive notation we would say the image of a sum is the sum of the images.)

Let us look at some examples of group homomorphisms. Let G be any group and let x be an element of G. The first law of exponents in Theorem 2.4 states that the map that takes an integer m to x^m is a homomorphism from $\mathbf{Z}$ to G. Here we are writing the binary operation in $\mathbf{Z}$ additively and the operation in G multiplicatively. The image of $\mathbf{Z}$ under this homomorphism is the cyclic subgroup of G generated by x. A particular case of this homomorphism is the map from $\mathbf{Z}$ onto $\mathbf{Z}_n$ taking m to the congruence class $[m]$.

Let G and H be any two groups. The *trivial homomorphism* from G to H is the map that sends every element of G to the identity element of H. The identity function on G is a homomorphism from G to itself. If G is abelian, then for any x and y in G we have $(xy)^{-1} = y^{-1}x^{-1} = x^{-1}y^{-1}$, so the inverse map is a homomorphism from G to G.

Let *G* and *H* be *M*-by-*M* and *N*-by-*N* group tables, respectively. A homomorphism from (⍳*M*, *G*) to (⍳*N*, *H*) is a vector *F* of length *M* with components in ⍳*N* such that for all *I* and *J* the value of *F*[*G*[*I*;*J*]] is equal to *H*[*F*[*I*];*F*[*J*]] or, equivalently, such that the matrices *F*[*G*] and *H*[*F*;*F*] are the same. The workspace *EXAMPLES* contains a vector *H*24*TO*6. Let us rename this vector *F* and show that *F* is a homomorphism of (⍳24, *G*24) onto (⍳6, *G*6).

```
      ⍴F←H24TO6
24
      ∧/F∊⍳6
1
      ∧/,F[G24]=G6[F;F]
1
      SSORT F
0 1 2 3 4 5
```

Not only does *F* map products to products, it also takes the identity element of *G*24 to the identity element of *G*6.

```
      F[0]
0
```

and it maps inverses to inverses.

```
      ∧/F[INV24]=INV6[F]
1
```

This is true for any homomorphism.

THEOREM 1. Let h be a homomorphism from the group $(G, \bullet)$ to the group $(H, *)$. Then h takes the identity element of G to the identity element of H and for all x in G we have $(xh)^{-1} = x^{-1}h$. If K is a subgroup of G, then Kh is a subgroup of H, and if L is a subgroup of H, then Lh^{-1} is a subgroup of G.

Proof. Let e be the identity of G and let f be the identity of H. Then

$$(eh) * f = eh = (e \bullet e)h = (eh) * (eh),$$

and so, by the left cancellation law, $f = eh$. Suppose x is any element of G. Then

$$f = eh = (x \bullet x^{-1})h = (xh) \star (x^{-1}h)$$

and, by the uniqueness of inverses in H, it follows that $(xh)^{-1} = x^{-1}h$. Now let K be a subgroup of G. Then Kh is nonempty, since K is nonempty. Suppose u and v are in Kh. There exist elements x and y of G such that $u = xh$ and $v = yh$. Therefore

$$u \star v = (xh) \star (yh) = (x \bullet y)h$$

and

$$u^{-1} = (xh)^{-1} = x^{-1}h.$$

Since $x \bullet y$ and x^{-1} are both in K, we see that Kh is closed under products and inverses and so is a subgroup of H. We leave as an exercise the proof that the inverse image of a subgroup of H is a subgroup of G. □

Homomorphisms can be composed to form new homomorphisms.

THEOREM 2. Let $f:G \longrightarrow H$ and $g:H \longrightarrow K$ be homomorphisms of groups. Then $f \circ g$ is a homomorphism from G to K.

Proof. See Exercise 4. □

The following table defines a group with $\{0,1\}$ as its set of elements.

+	0	1
0	0	1
1	1	0

The group operation is addition modulo 2. Similarly, the table

×	1	−1
1	1	−1
−1	−1	1

describes the group $(\{1,-1\}, \times)$. Although these groups are distinct, they are very closely related. If each 0 in the table for the first group is replaced by a 1 and each 1 replaced by a -1, we get the table for the second group. That is, the second group can be obtained by "renaming" the elements of the first group. Two groups related in this way are said to be isomorphic. The concept of isomorphic algebraic structures is a very important idea.

We formalize the notion of isomorphic groups in the following way. Let G and H be groups. An *isomorphism* of G onto H is a bijective homomorphism from G to H. We say G is *isomorphic* to H if there exists an isomorphism of G onto H. For example, the function that takes each real number x to 2^x is an isomorphism of $(\mathbf{R},+)$ onto $(\mathbf{R}^+, \times)$, where $\mathbf{R}^+$ denotes the set of positive real numbers. Also, if n is a nonzero integer, then the map $m \longmapsto nm$ is an isomorphism of $\mathbf{Z}$ onto its subgroup $n\mathbf{Z}$. If G is isomorphic to H, we write $G \cong H$.

THEOREM 3. Let $f:G \longrightarrow H$ and $g:H \longrightarrow K$ be isomorphisms of groups. Then f^{-1} and $f \circ g$ are isomorphisms.

Proof. Theorem 2, along with Theorem 1.3.2, implies immediately that $f \circ g$ is an isomorphism. To show that f^{-1} is an isomorphism, we need only show that it is a homomorphism, since the inverse of a bijection is always a bijection. Let u and v be in H and set $x = uf^{-1}$ and $y = vf^{-1}$. Then $(xy)f = uv$, and so $(uv)f^{-1} = xy = (uf^{-1})(vf^{-1})$. □

COROLLARY 4. Let $G, H,$ and K be groups. Then

(a) $G \cong G$.
(b) If $G \cong H$, then $H \cong G$.
(c) If $G \cong H$ and $H \cong K$, then $G \cong K$.

Proof. The identity function on G is an isomorphism of G onto G. Thus G is isomorphic to itself. Suppose $f:G \longrightarrow H$ is an isomorphism. Then, by Theorem 3, f^{-1} is an isomorphism from H to G, so H is isomorphic to G. Part (c) is equally trivial. □

In view of Corollary 4 it is tempting to call $\cong$ an equivalence relation on the set of all groups. However, the set of all groups is one of those sets that is too big in the sense that its consideration can lead to logical contradictions. Nevertheless, the analogy with equivalence relations is useful.

Two groups that are isomorphic are often considered to be the same. In the 1–1 correspondence preserving products between the elements of the groups, subgroups correspond to subgroups and cosets correspond to cosets. In fact, their entire structures as groups are identical. For example, let $X = \{x\}$ be a set with one element. There is exactly one binary operation $\bullet$ on X, the one in which $x \bullet x$ is x. The pair $(X, \bullet)$ is a group. Clearly, any two such groups are isomorphic. Another way of saying this is that up to isomorphism there is only one group of order 1. Very often the phrase "up to isomorphism" is omitted in such statements.

Let G be a finite group of order N and let T be the group table for G relative to some numbering $x_0, \ldots, x_{N-1}$ of the elements of G. We have already remarked that $(\iota N, T)$ is a group. The map taking I to x_I is an isomorphism of $(\iota N, T)$ onto G.

EXERCISES

1 Let G be an abelian group and let n be an integer. Show that the map taking x to x^n is a homomorphism of G into itself.

2 Listed here are pairs consisting of a group G and a map f of G into G. In each case determine whether f is a homomorphism.

(a) $G = \mathbf{Z}, \; f:x \longmapsto 3x$.

(b) $G = (\mathbf{Q},+)$, $f{:}x \longmapsto x^2$.

(c) $G = (\mathbf{Q}-\{0\}, \times)$, $f{:}x \longmapsto x^2$.

(d) $G = (\mathbf{R}^+,\times)$, where $\mathbf{R}^+ = \{x \in \mathbf{R} | x > 0\}$, $f{:}x \longmapsto \sqrt{x}$.

3 Complete the proof of Theorem 1.

4 Prove Theorem 2.

5 Let $(G,\bullet)$ be a group and let $\star$ be a binary operation on a set H. Suppose there is a map f of G onto H such that $(x \bullet y)f = (xf) \star (yf)$ for all x and y in G. Prove that $(H,\star)$ is a group.

6 Let $f{:}G \longrightarrow H$ be a surjective homomorphism of groups. Show that:

(a) If G is abelian, then H is abelian.

(b) If $G = \langle x_1, \ldots, x_r \rangle$, then $H = \langle x_1 f, \ldots, x_r f \rangle$.

7 Let $f{:}X \longrightarrow Y$ be a bijection. Prove that $\Sigma(X)$ and $\Sigma(Y)$ are isomorphic.

8 Exhibit an injective homomorphism of $\mathbf{Z}_4$ into $\mathbf{Z}_8$ and a surjective homomorphism of $\mathbf{Z}_8$ onto $\mathbf{Z}_4$.

9 Let T be a group table for a finite group G of order N. Show that (⍳N,T) is isomorphic to G. Prove that a matrix S is a group table for G if and only if there is a permutation P of ⍳N such that S is `P[T[⍋P;⍋P]]`.

10 Show that up to isomorphism there is only one group of order 2 and one group of order 3.

11 Exhibit two nonisomorphic groups of order 4.

12 Give an upper bound for the number of groups of order n up to isomorphism.

13 Let G be a finite group and let $f{:}G \longrightarrow H$ be a group homomorphism. Assume that K and L are subgroups of G with $K \subseteq L$. Show:

(a) $Kf \subseteq Lf$ and $|Lf{:}Kf| \leq |L{:}K|$.

(b) If x is in G, then the order of xf is less than or equal to the order of x.

14 Show that (⍳6, `G6`) is isomorphic to Σ_3. (A group table for Σ_3 was constructed in Exercise 1.20.)

15 Let `G2←2|(⍳2)∘.+⍳2`. Construct homomorphisms of (⍳6, `G6`) and (⍳24, `G24`) onto (⍳2, `G2`).

16 In the text we noted that the group of real numbers under addition is isomorphic to the group of positive real numbers under multiplication. Can the same thing be said about the rationals? That is, are $(\mathbf{Q},+)$ and $(\mathbf{Q}^+,\times)$ isomorphic, where $\mathbf{Q}^+ = \{x \in \mathbf{Q} \,|\, x > 0\}$?

17 The workspace *EXAMPLES* contains vectors *F*1, *F*2, and *F*3 of length 24, each of which describes a map from ⍳24 to ⍳60. Which of these maps are homomorphisms from (⍳24,*G*24) to (⍳60,*G*60)?

18 Execute the following APL statements.

```
GTINIT G24
K←GTSGP 3 8
F←H24TO6
```

We know that *K* is a subgroup of (⍳24,*G*24) of order 12 and that *F* is a homomorphism of (⍳24,*G*24) onto (⍳6,*G*6). Compute a vector *KF* listing the elements of the image of *K* under *F*. Show directly that *KF* is a subgroup of (⍳6,*G*6). Now execute

```
GTINIT G6
L←GTSGP 1
```

Construct a vector *LFI* listing the elements of the inverse image of *L* under *F*. Show that *LFI* is a subgroup of (⍳24,*G*24).

5. NORMAL SUBGROUPS

In the last section we observed that the vector *F*←*H*23*TO*6 is a homomorphism from (⍳24,*G*24). onto (⍳6,*G*6). Let us find the set *K* of elements in ⍳24 mapped to the identity element 0 of (⍳6,*G*6) by *F*.

```
      ⎕←K←(F=0)/⍳24
0 7 16 23
```

Since {0} is a subgroup of (⍳6,*G*6), Theorem 4.1 tells us that *K* is a subgroup of (⍳24,*G*24). We can also check this directly.

```
      ∧/,G24[K;K]∊K
1
```

As we will see, *K* has an important property not possessed by all subgroups of (⍳24,*G*24). Let us define three logical matrices.

```
GTINIT G24                 L←GTLCON K
R←GTRCON K                 P←F∘.=F
```

The matrices *R* and *L* are the characteristic matrices for right and left congruence, respectively, modulo *K* and *P*[*I*;*J*] is 1 if and only if *F* maps

I and J to the same element of ι6. It turns out that all three of these matrices are the same.

```
      ∧/,R=L                        ∧/,R=P
1                             1
```

This section is devoted to showing why this is not a coincidence.

Let H be a subgroup of a group G. We say that H is *normal* in G and write $H \lhd G$ if for all x in G we have $x^{-1}Hx \subseteq H$, where, as usual, $x^{-1}Hx$ means $\{x^{-1}yx \mid y \in H\}$. It follows immediately from the definition that $\{1\}$ and G are normal subgroups of G. There are many conditions that are equivalent to normality. Some of these conditions are given in the next theorem.

THEOREM 1. Let H be a subgroup of a group G. Then the following are equivalent.

(a) H is normal in G.
(b) If x is in G, then $x^{-1}Hx = H$.
(c) Right and left congruence modulo H are the same.
(d) Every right coset of H is also a left coset of H.
(e) The product of two right cosets of H is a right coset of H.

Proof. As is usual in proofs of theorems of this type, we will show that each of the first four conditions implies the next and that (e) implies (a).

(a) implies (b). Suppose $H \lhd G$ and x is in G. Then

$$xHx^{-1} = (x^{-1})^{-1}Hx^{-1}$$

is contained in H, so

$$H = (x^{-1}x)H(x^{-1}x) = x^{-1}(xHx^{-1})x \subseteq x^{-1}Hx \subseteq H.$$

Thus $x^{-1}Hx = H$.

(b) implies (c). Assume (b) and let x and y be in G. If yx^{-1} is in H, then $x^{-1}(yx^{-1})x = x^{-1}y$ is in $x^{-1}Hx = H$. Therefore right congruence modulo H implies left congruence modulo H. The reverse implication is equally trivial.

(c) implies (d). Let x be in G. Then the right coset Hx is the set of elements of G right congruent to x modulo H, while the left coset xH is the set of elements left congruent to x. If right and left congruence are the same, then $Hx = xH$.

(d) implies (e). Assume (d) and consider a product $(Hx)(Hy)$ of two right cosets of H. By (d), Hx is a left coset of H and, since x is in Hx, we must have $Hx = xH$. Therefore

$$(Hx)(Hy) = xHHy = xHy = Hxy.$$

(e) implies (a). Assume (e) and let x be in G. Then $(Hx^{-1})(Hx)$ is a right coset and, since it contains $1x^{-1}1x = 1$, we have $(Hx^{-1})(Hx) = H$. But $Hx^{-1}Hx$ contains $x^{-1}Hx$, and so $x^{-1}Hx \subseteq H$. Therefore $H \lhd G$. □

COROLLARY 2. If H is a subgroup of G and $|G:H| = 2$, then $H \lhd G$.

Proof. Since $|G:H| = 2$, there are exactly two right cosets of H in G. By Exercise 3.22, there are exactly two left cosets of H in G. Let x be in $G - H$. Then the right cosets of H are H and Hx and the left cosets are H and xH. Thus $Hx = G - H = xH$, and every right coset of H is also a left coset. Thus $H \lhd G$ by Theorem 1. □

In the example at the beginning of this section we verified that the subgroup `K` of (`ι24`, `G24`) satisfies condition (c) of Theorem 1. Therefore `K` is a normal subgroup. In Section 3 we found that right and left congruence in (`ι24`, `G24`) modulo the cyclic subgroup generated by 9 are not the same. Thus not every subgroup of (`ι24`, `G24`) is normal.

Suppose N is a normal subgroup of a group G. By Theorem 1, we do not need to distinguish between left and right cosets of N. We will let G/N denote the set of cosets of N in G. If H is a group and $f: G \longrightarrow H$ is a homomorphism, then $\{1\}f^{-1}$, the inverse image of the trivial subgroup of H, is called the *kernel* of f. In the preceding example, `K` is the kernel of `F`.

THEOREM 3. Let $f: G \longrightarrow H$ be a homomorphism of groups. Then the kernel N of f is a normal subgroup of G, and two elements of G are mapped to the same element of H if and only if they are in the same coset of N. If $L \lhd H$, then $Lf^{-1} \lhd G$. If $K \lhd G$, then $Kf \lhd Gf$.

Proof. By Theorem 4.2, we know that $N = \{1\}f^{-1}$ is a subgroup of G. Suppose that x is in N and u is in G. Then

$$(u^{-1}xu)f = (u^{-1}f)(xf)(uf) = (uf)^{-1}1(uf) = 1,$$

so $u^{-1}xu$ is in N. Therefore $N \lhd G$.

Next let x and y be any two elements of G. Then $xf = yf$ if and only if $(xf)^{-1}(yf) = 1$. However, $(xf)^{-1}(yf) = (x^{-1}y)f$ and $(x^{-1}y)f = 1$ if and only if $x^{-1}y$ is in N. Therefore $xf = yf$ if and only if x and y are in the same coset of N.

Now suppose $L \lhd H$ and let x be in Lf^{-1} and y in G. Then

$$(y^{-1}xy)f = (yf)^{-1}(xf)(xf) \in (yf)^{-1}L(yf) = L,$$

so $y^{-1}xy$ is in Lf^{-1}. Therefore $Lf^{-1} \lhd G$. Finally, suppose $K \lhd G$ and let u be in K and v in G. Then

$$(vf)^{-1}(uf)(vf) = (v^{-1}uv)f \in Kf,$$

and, since uf is a typical element of Kf and vf is a typical element of Gf, it follows that $Kf \lhd Gf$. □

COROLLARY 4. In Theorem 3 the homomorphism f is injective if and only if the kernel N is trivial.

Proof. By Theorem 3 the map f is injective if and only if all the cosets of N have exactly one element and, by Theorem 3.8, this holds if and only if $N = \{1\}$. □

It is important to note that if $f: G \longrightarrow H$ is a group homomorphism and $K \lhd G$, then Kf is not generally normal in H but only normal in Gf.

Let N be a normal subgroup of the group G. In view of Theorem 3 it is natural to ask whether there exists a group H and a homomorphism $f: G \longrightarrow H$ such that N is the kernel of f. The answer is that we can find such a pair H and f. In fact, we can take the elements of H to be the cosets of N. By Theorem 1, the product of any two cosets of N is again a coset of N. Thus the set G/N has a binary operation defined on it.

THEOREM 5. Let $N \lhd G$. The set G/N of cosets of N is a group under the usual multiplication of subsets of G. The natural map for G to G/N is a homomorphism of G onto G/N with kernel N.

Proof. By Theorem 1.2, multiplication of subsets of G is associative on all of 2^G and therefore is associative on G/N. In the proof of Theorem 1 we showed that for any x and y in G we have $(Nx)(Ny) = Nxy$. Taking x to be 1, we see that $N1 = N$ is a left identity element in G/N, and taking x to be y^{-1}, we see that Ny^{-1} is a left inverse for Ny relative to N. By Theorem 2.5, G/N is a group. The natural map from G to G/N is the map f that takes an element x of G to the coset Nx. Clearly, f is surjective. The formula $(Nx)(Ny) = Nxy$ shows that f is a homomorphism. Suppose x is in the kernel of f. Then $N = xf = Nx$. This implies x is in N. Every element of N is in the kernel, and so the kernel of f is N. □

The group G/N is called the *quotient group* or the *factor group* of G by N.

THEOREM 6. The quotient group of $\mathbf{Z}$ by $n\mathbf{Z}$ is $\mathbf{Z}_n$, $n \geq 1$.

Proof. Any subgroup of an abelian group is normal. (See Exercise 1.) We defined the binary operation + on the set $\mathbf{Z}_n$ of congruence class modulo n, which is the same as the set of cosets of $n\mathbf{Z}$ in $\mathbf{Z}$, such that $[x] + [y] = [x + y]$. In the notation of cosets this is

$$(x + n\mathbf{Z}) + (y + n\mathbf{Z}) = (x + y) + n\mathbf{Z},$$

and this is precisely the binary operation in the quotient group $\mathbf{Z}/n\mathbf{Z}$. □

THEOREM 7 *(First Isomorphism Theorem).* Let $f: G \longrightarrow H$ be a surjective homomorphism of groups and let N be the kernel of f. Then H is isomorphic to the quotient group G/N.

Proof. Let $g:G \longrightarrow G/N$ be the natural map. We would like to define a map h from G/N to H such that the diagram *commutes* in the sense that the

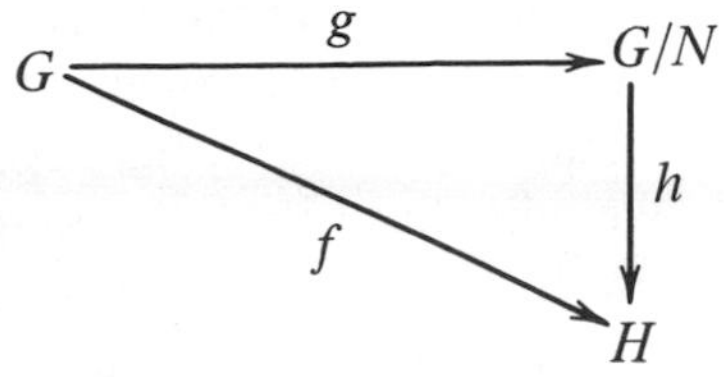

composition $g \circ h$ is equal to f. There is at most one such map h since, if x is in G, then

$$xf = x(g \circ h) = (xg)h = (Nx)h,$$

so h must map Nx to xf. But is the h so described well defined? That is, is the subset

$$h = \{(Nx, xf) \mid x \in G\}$$

a function from G/N to H? Certainly h is a relation from G/N to H. Suppose $Nx = Ny$. Then, by Theorem 3, $xf = yf$ and the coset Nx is the first component of exactly one ordered pair in h. Therefore h is a function and $f = g \circ h$. All that remains is to show that h is an isomorphism. If x and y are in G, then the product of $(Nx)h$ and $(Ny)h$ is $(xf)(yf) = (xy)f = (Nxy)h$ and Nxy is $(Nx)(Ny)$. Therefore h is a homomorphism. Since $g \circ h$ is surjective, we know h is surjective by Theorem 1.3.2. Suppose Nx is in the kernel of h. Then $xf = (Nx)h = 1$, so x is in N. Hence the kernel of H is trivial and, by Corollary 4, h is injective. Thus h is a bijective homomorphism, that is, an isomorphism. □

Theorem 7 allows us to describe all cyclic groups up to isomorphism.

THEOREM 8. Any infinite cyclic group is isomorphic to **Z**. Any finite cyclic group is isomorphic to $\mathbf{Z}_n$ for a unique $n > 0$.

Proof. Let $G = \langle x \rangle$ be a cyclic group. The map f taking m in **Z** to x^m is a homomorphism from **Z** onto G, and so by Theorem 7, G is isomorphic to the quotient group of **Z** by the kernel of f. The kernel of f is $n\mathbf{Z}$ for a unique nonnegative integer n. If $n > 0$, then $n\mathbf{Z}$ has n cosets. Therefore, if G is infinite, then $n = 0$ and $G \cong \mathbf{Z}/\{0\}$. We leave as an exercise the verification that $\mathbf{Z}/\{0\}$ is isomorphic to **Z**. (See Exercise 2.) If G is finite, then $n > 0$ and $G \cong \mathbf{Z}/n\mathbf{Z} = \mathbf{Z}_n$ by Theorem 6. □

The next theorem may be viewed as an addition to Theorem 4.1.

THEOREM 9. Let G and H be groups and let $f:G \longrightarrow H$ be a surjective homomorphism with kernel N. Then there is a bijection from the set of subgroups of G containing N onto the set of subgroups of H. If a subgroup

K of G containing N is mapped to a subgroup L of H, then $K \lhd G$ if and only if $L \lhd H$.

Proof. Let S be the set of subgroups of G containing N and let T be the set of subgroups of H. Given K in S, the image Kf of K under f is in T. Also, given L in T, the inverse image Lf^{-1} is a subgroup of G containing $\{1\}f^{-1} = N$, and so Lf^{-1} is in S. We will show that the maps $K \longmapsto Kf$ and $L \longmapsto Lf^{-1}$ are inverses of each other. If K is in S, then $(Kf)f^{-1} \supset K$. But if x is in $(Kf)f^{-1}$, then xf is in Kf and $xf = yf$ for some y in K. By Theorem 3, x is in Ny, which is contained in K, and so x is in K. Therefore $(Kf)f^{-1} = K$. Now suppose L is in T. By the definition of Lf^{-1}, $(Lf^{-1})f \subseteq L$. Given u in L, there exists x in G with $xf = u$, since f is surjective. Thus x is in Lf^{-1} and so $(Lf^{-1})f = L$. By Corollary 1.3.4, the maps $K \longmapsto Kf$ and $L \longmapsto Lf^{-1}$ are bijections and are inverses of each other. The final part of the theorem follows from Theorem 3. □

The bijection in Theorem 9 preserves indices in the following sense.

COROLLARY 10. In Theorem 9, if K is a subgroup of G containing N, then there is a 1–1 correspondence between the right cosets of K in G and the right cosets of Kf in H. In particular, if G is finite, then $|G:K| = |H:Kf|$.

Proof. For any x in G we have $(Kx)f = (Kf)(xf)$ and, therefore, the image under f of the coset Kx of K is a coset of Kf. Thus $Kx \longmapsto (Kf)(xf)$ defines a map g from the set of right cosets of K in G to the set of right cosets of Kf in H. Clearly, g is surjective. Suppose $(Kf)(xf) = (Kf)(yf)$. Then $(xf)(yf)^{-1} = (xy^{-1})f$ is in Kf and, since we showed in the proof of Theorem 9 that $(Kf)f^{-1} = K$, this means that $xy^{-1} \in K$ or $Kx = Ky$. Therefore g is injective too. □

We have already determined all of the subgroups of $\mathbf{Z}$. Using Theorem 9, we can find the subgroups of $\mathbf{Z}_n$ for $n > 0$.

COROLLARY 11. Let G be a finite cyclic group of order n. Then every subgroup of G is cyclic and G contains exactly one subgroup of order m for each positive integer dividing n.

Proof. By Theorem 8 we may assume $G = \mathbf{Z}_n$. Let f be the natural map of $\mathbf{Z}$ onto $\mathbf{Z}_n$. By Theorem 9 there is a 1–1 correspondence between the subgroups of $\mathbf{Z}_n$ and the subgroups of $\mathbf{Z}$ containing the kernel $n\mathbf{Z}$ of f. Any subgroup of $\mathbf{Z}$ has the form $d\mathbf{Z}$ for some $d \geq 0$ and $n\mathbf{Z} \subseteq d\mathbf{Z}$ if and only if d divides n. Suppose d does divide n. Then the image of $d\mathbf{Z}$ in $\mathbf{Z}_n$ is the cyclic subgroup generated by $df = [d]$. Let $m = n/d$. Then $m[d] = [md] = [n] = [0]$. If $0 < r < m$, then $0 < rd < n$ and $r[d] \neq [0]$. Therefore $[d]$ has order m in $\mathbf{Z}_n$. For each divisor m of n the value $d = n/m$ is uniquely determined. Thus $\mathbf{Z}_n$ has exactly one subgroup of order m. □

Theorem 7 is called the First Isomorphism Theorem. There are two additional isomorphism theorems, which we state for completeness.

THEOREM 12 ***(Second Isomorphism Theorem).*** Let N be a normal subgroup of a group G and let K be a subgroup of G. Then

(a) NK is a subgroup of G and N is a normal subgroup of NK.
(b) $N \cap K$ is a normal subgroup of K.
(c) The quotient groups $(NK)/N$ and $K/(N \cap K)$ are isomorphic.

Proof. See Exercise 15. □

THEOREM 13 ***(Third Isomorphism Theorem).*** Let H and K be normal subgroups of a group G with $H \subseteq K$. Then

$$G/K \cong (G/H) \,/\, (K/H).$$

Proof. See Exercise 16. □

EXERCISES

1 Prove that every subgroup of an abelian group is normal.

2 Show that for any group G the subgroups $\{1\}$ and G are normal in G and $G/\{1\}$ is isomorphic to G.

3 Prove that the intersection of any nonempty collection of normal subgroups of a group G is normal in G.

4 Let H be a subgroup of the group G and let $N = \{x \in G | Hx = xH\}$. Show that N is a subgroup of G containing H and $H \lhd N$. We call N the *normalizer* of H in G.

5 Suppose that H is a subgroup of G, N is the normalizer of H in G, and C is the centralizer of H in G. (See Exercise 3.8). Show that C is a subgroup of N and $C \lhd N$.

6 Let $f{:}G \longrightarrow H$ be a homomorphism of groups and let $x \in G$. Show that if x has finite order, then the order of xf divides the order of x.

7 Give an example of a group G, a normal subgroup K of G, and a homomorphism $f{:}G \longrightarrow H$ such that Kf is not normal in H.

8 Let n be a positive integer. Show that the order of the congruence class $[m]$ in $\mathbf{Z}_n$ is $n/\gcd(m,n)$.

9 Let x and y be commuting elements of a group G and suppose that x, y, and xy have orders m, n, and r, respectively. Set $d = \gcd(m,n)$. Show that r divides mn and that mn/d^2 divides r. Thus if $d = 1$, then $r = mn$.

10 Let x be an element of a group such that x^{10} has order 6. What is the order of x? Is the answer unique?

*11 Let G be an N-by-N group table and let H be a vector listing the elements of a subgroup of (ιN, G). Write an APL proposition corresponding to each of the conditions (a) to (e) of Theorem 1. You may assume that the vector INV gives the inverses of the group elements.

12 With G and H as in Exercise 11, write an APL expression defining a vector that lists the elements of the normalizer of H in (ιN, G).

13 Construct several subgroups of ($\iota 24$, $G24$) and ($\iota 60$, $G60$) as follows. Begin with a cyclic subgroup and construct the centralizer and normalizer of this subgroup. Then compute the centralizers and normalizers of these groups, continuing until no new groups are found. Repeat the process with another cyclic subgroup. (See Exercises 12 and 3.9.)

14 Let K be a vector listing the elements of a subgroup of the current abstract group G and let $R \leftarrow GTRCON\ K$. In Exercise 3.35 we saw that $REP \leftarrow SSORT\ SFEL\ R$ is a vector listing right coset representatives for K in G. Show how to construct a vector $COSET$ such that $COSET[I]$ is the number of the coset of K containing I, that is, I is right congruent to $REP[COSET[I]]$ modulo K. Assuming K is normal in G, write an APL expression defining a group table for the quotient group of G by K.

15 Let N, G, and K be as in Theorem 12 and let $f{:}G \longrightarrow G/N$ be the natural map. Show that $NK = (Kf)f^{-1}$, and so NK is a subgroup of G. Show that the kernels of the restrictions of f to NK and K are N and $N \cap K$, respectively. Complete the proof of Theorem 12 by using Theorem 7 and the observation that $(NK)f = Kf$.

16 Let H, K, and G be as in Theorem 13 and let $f{:}G \longrightarrow G/H$ be the natural map. Prove that $Kf = K/H$ and $K/H \lhd G/H$. Let $g{:}G/H \longrightarrow (G/H)/(K/H)$ be the natural map. Prove that $f \circ g$ maps G onto $(G/H)/(K/H)$ with kernel K. Complete the proof of Theorem 13.

6. DIRECT PRODUCTS

Given a group G, we can construct new groups by forming subgroups and quotient groups of G. In this section we will study a method for obtaining a new group from two given groups.

Let $(G, \bullet)$ and $(H, *)$ be groups. We can define a binary operation $\circledast$ on the Cartesian product $G \times H$ by setting

$$(x_1, y_1) \circledast (x_2, y_2) = (x_1 \bullet x_2, y_1 * y_2)$$

for all x_1, x_2 in G and all y_1, y_2 in H. Since both

$$(x_1, y_1) \circledast [(x_2, y_2) \circledast (x_3, y_3)]$$

and

$$[(x_1, y_1) \circledast (x_2, y_2)] \circledast (x_3, y_3)$$

are equal to $(x_1 \bullet x_2 \bullet x_3, y_1 * y_2 * y_3)$, the operation $\circledast$ is associative. The pair $(1, 1)$ is a two-sided identity element and (x^{-1}, y^{-1}) is a two-sided inverse for (x, y). Thus $(G \times H, \circledast)$ is a group, which we call the *external direct product* of $(G, \bullet)$ and $(H, *)$. If we are using multiplicative notation for G and H, we use multiplicative notation for $G \times H$ and write

$$(x_1, y_1)(x_2, y_2) = (x_1x_2, y_1y_2);$$

if we are using additive notation, we write $(x_1, y_1) + (x_2, y_2) = (x_1 + x_2, y_1 + y_2)$. When additive notation is being used, the group $G \times H$ is sometimes written $G \oplus H$ and called the *external direct sum* of G and H. However, in certain contexts the terms "direct sum" and "direct product" have different meanings. (See Exercise 12.)

If we have three groups R, S, and T, we can form the direct products $R \times (S \times T)$ and $(R \times S) \times T$. The reader should verify that our identification of these two objects as sets is a group isomorphism. We will generally not distinguish between these two groups and write $R \times S \times T$. The following is another simple but important fact about direct products. If $G_1 \cong G_2$ and $H_1 \cong H_2$, then $G_1 \times H_1 \cong G_2 \times H_2$. (See Exercise 2.)

The group $G = \mathbf{Z} \times \mathbf{Z}$ is easy to work with on an APL system. The elements of G are simply integer vectors of length 2. If `X` and `Y` are in G, then the sum of `X` and `Y` is `X+Y` and the negative of `X` is `-X`.

```
      X←7 ¯3
      Y←¯1 5
      X+Y
6 2
```

```
      ⎕←Z←-X
¯7 3
      X+Z
0 0
```

We can represent the elements of $\mathbf{Z}_7 \times \mathbf{Z}_7$ by vectors of length 2 with components in `⍳7`. The sum of `X` and `Y` is now `7|X+Y`, while the negative of `X` is `7|-X`.

```
      X←5 2
      Y←3 4
      7|X+Y
1 6
```

```
      ⎕←Z←7|-X
2 5
      7|X+Z
0 0
```

For $\mathbf{Z}_2 \times \mathbf{Z}_3$ we use vectors whose first components are in ⍳2 and whose second components are in ⍳3. The sum of *X* and *Y* is N|X+Y where N←2 3.

```
      N←2 3
      X←1 2
```

```
      N|X+X
0 1
```

Let us find the order of 1 1 in $\mathbf{Z}_2 \times \mathbf{Z}_3$.

```
      A←1 1
      ⎕←B←N|A+A
0 2
      ⎕←C←N|A+B
1 0
      ⎕←D←N|A+C
0 1
```

```
      ⎕←E←N|A+D
1 2
      ⎕←F←N|A+E
0 0
```

Here we see that the cyclic group generated by *A* contains all the elements of $\mathbf{Z}_2 \times \mathbf{Z}_3$. By Theorem 5.8, we have $\mathbf{Z}_2 \times \mathbf{Z}_3 \cong \mathbf{Z}_6$. Corollary 2, which follows, generalizes this observation.

Let G and H be groups. The projection map of $G \times H$ onto H is the map taking (x, y) in $G \times H$ to y. It is easy to check that this map is a group homomorphism with kernel $K = \{(x, 1) | x \in G\}$. Therefore K is a normal subgroup of $G \times H$ and $(G \times H)/K$ is isomorphic to H. Similarly, the map $(x, y) \longmapsto x$ is a homomorphism of $G \times H$ onto G with kernel $L = \{(1, y) | y \in H\}$. The maps $x \longmapsto (x, 1)$ and $y \longmapsto (1, y)$ are isomorphisms of G and H onto K and L, respectively. Since

$$(x, 1)(1, y) = (x, y) = (1, y)(x, 1),$$

elements of K commute with elements of L, and every element of $G \times H$ can be expressed uniquely as a product of an element in K and an element in L.

Let G be a group and let K and L be subgroups of G. It is impossible for G to be equal to the external direct product $K \times L$, since K and L are not subsets of $K \times L$. However, it is possible for G to be isomorphic to $K \times L$ in a particularly nice way. We say that G is the *internal direct product* of K and L if the map taking (x, y) in $K \times L$ to xy in G is an isomorphism of $K \times L$ onto G.

THEOREM 1. A group G is the internal direct product of its subgroups K and L if and only if each of the following holds.

(a) $K \lhd G$ and $L \lhd G$.
(b) $K \cap L = \{1\}$.
(c) $G = KL$.

Proof. In $K \times L$ we have the subgroups $K_1 = \{(x, 1) | x \in K\}$ and $L_1 =$

$\{(1, y) | y \in L\}$. By the previous discussion, both K_1 and L_1 are normal in $K \times L$, $K_1 \cap L_1$ is trivial, and $K \times L = K_1 L_1$. Let $f: K \times L \longrightarrow G$ map (x, y) to xy. Then $K_1 f = K$ and $L_1 f = L$. If f is an isomorphism, then conditions (a), (b), and (c) must hold. Now suppose (a), (b), and (c) hold. Let x be in K and y in L. Our first task is to show that x and y commute. Consider $u = x^{-1} y^{-1} xy$. Since x is in K and $K \lhd G$, both x^{-1} and $y^{-1} xy$ are in K, and so u is in K. But y is in L and $L \lhd G$ and hence $x^{-1} y^{-1} x$ is in L. Therefore u is in L too. Thus u is in $K \cap L = \{1\}$ and $u = 1$. This implies that

$$yx = yxu = yxx^{-1} y^{-1} xy = xy,$$

and so x and y commute. For all x_1, x_2 in K and all y_1, y_2 in L we have

$$[(x_1, y_1)(x_2, y_2)]f = (x_1 x_2, y_1 y_2)f = x_1 x_2 y_1 y_2 =$$
$$x_1 y_1 x_2 y_2 = [(x_1, y_1)f]\,[(x_2, y_2)f].$$

Therefore f is a homomorphism. By (c) we know that f is surjective. Suppose (x, y) is in the kernel of f. Then $xy = 1$ or $x = y^{-1}$. But x is in K and y^{-1} is in L, so x and y are in $K \cap L$. Thus $x = y = 1$ and the kernel of f is trivial. Therefore f is an isomorphism. □

If G is the internal direct product of its subgroups K and L, it is customary to write $G = K \times L$. It should always be clear from the context whether external or internal direct products are meant.

COROLLARY 2. Suppose m and n are relatively prime integers. Then $\mathbf{Z}_{mn} \cong \mathbf{Z}_m \times \mathbf{Z}_n$.

First Proof. In $\mathbf{Z}_{mn}$ the cyclic subgroup generated by $[n]$ is $K = \{[nx] | x \in \mathbf{Z}\}$. Now $[nx] = [ny]$ if and only if $x \equiv y \pmod{m}$, and so K has order m. Thus $K \cong \mathbf{Z}_m$. Similarly, $L = \{[mx] | x \in \mathbf{Z}\}$ is a subgroup of $\mathbf{Z}_{mn}$ isomorphic to $\mathbf{Z}_n$. Since $\mathbf{Z}_{mn}$ is abelian, both K and L are normal in $\mathbf{Z}_{mn}$. If $[u]$ is in $K \cap L$, then m and n divide u. Since $\gcd(m, n) = 1$, this means that mn divides u, so $[u] = [0]$. Therefore $K \cap L$ is trivial. Finally, given v in $\mathbf{Z}$, there are integers x and y such that $nx + my = v$. Thus $[nx] + [my] = [v]$ and $K + L = \mathbf{Z}_{mn}$. By Theorem 1, $\mathbf{Z}_{mn}$ is isomorphic to $K \times L$, which is isomorphic to $\mathbf{Z}_m \times \mathbf{Z}_n$.

Second Proof. If $x \equiv y \pmod{mn}$, then $x \equiv y \pmod{m}$ and $x \equiv y \pmod{n}$. Therefore we may define a function g from $\mathbf{Z}_{mn}$ to $\mathbf{Z}_m \times \mathbf{Z}_n$ by $[x]g = ([x]_m, [x]_n)$, where $[x]_m$ and $[x]_n$ denote the residue classes modulo m and n containing x, respectively. It is easily checked that g is a homomorphism. By the Chinese Remainder Theorem, g is surjective. If $[x]_m = [0]_m$ and $[x]_n = [0]_n$, then $[x] = [0]$ in $\mathbf{Z}_{mn}$, so g is injective. Therefore g is an isomorphism. □

A slight modification of the second proof of Corollary 2 yields the next result.

THEOREM 3. Let m and n be relatively prime positive integers. Then $U_{mn} \cong U_m \times U_n$.

Proof. For any x in $\mathbf{Z}$ it is easy to see that $\gcd(x, mn) = 1$ if and only if $\gcd(x, m) = 1$ and $\gcd(x, n) = 1$. Thus $[x]$ is in U_{mn} if and only if $[x]_m$ is in U_m and $[x]_n$ is in U_n. Therefore g in the second proof of Corollary 2 induces a bijection of U_{mn} onto $U_m \times U_n$. Not only does g preserve addition, it also preserves multiplication. Therefore the restriction of g to U_{mn} is an isomorphism of U_{mn} onto $U_m \times U_n$. □

Recall that $\phi(n)$ is defined to be $|U_n|$.

COROLLARY 4. If $\gcd(m, n) = 1$, then $\phi(mn) = \phi(m)\phi(n)$. If $p_1, \ldots, p_r$ are distinct primes and $n = p_1^{e_1} \ldots p_r^{e_r}$, $e_i \geq 1$, then

$$\phi(n) = \prod_{i=1}^{r} p_i^{e_i - 1} (p_i - 1).$$

Proof. By Theorem 3, if $\gcd(m, n) = 1$, then $|U_{mn}| = |U_m \times U_n| = |U_m| \times |U_n|$ and so $\phi(mn) = \phi(m)\phi(n)$. Therefore, if $p_1, \ldots, p_r$ are distinct primes, and $n = p_1^{e_1} \ldots p_r^{e_r}$, then

$$\phi(n) = \prod_{i=1}^{r} \phi(p_i^{e_i}).$$

It remains to determine $\phi(p^e)$ for p a prime and $e \geq 1$. Now $\gcd(x, p^e) = 1$ if and only if $\gcd(x, p) = 1$, and so the only elements of $\mathbf{Z}_{p^e}$ not in U_{p^e} are the classes $[py]$, $y \in \mathbf{Z}$. There are p^{e-1} such classes, so $\phi(p^e) = p^e - p^{e-1} = p^{e-1}(p - 1)$. □

Let n be a positive integer. The direct product $\mathbf{Z} \times \mathbf{Z} \times \ldots \times \mathbf{Z}$ with n factors is usually denoted $\mathbf{Z}^n$ and is a very important group. The elements of $\mathbf{Z}^n$ are integer vectors of length n with componentwise addition as the binary operation. In Chapter 6 we will study the subgroups and quotient groups of $\mathbf{Z}^n$ extensively. As a result of this investigation, we will be able to give a complete description of all finitely generated abelian groups. It will turn out that any such group is isomorphic to exactly one group of the form

$$\mathbf{Z}_{d_1} \times \mathbf{Z}_{d_2} \times \ldots \times \mathbf{Z}_{d_r} \times \mathbf{Z}^m,$$

where r and m are nonnegative integers, each d_i is an integer greater than 1, and d_i divides d_{i+1} for $1 \leq i < r$. This result, referred to as the Fundamental Theorem of Finitely Generated Abelian Groups, is Theorem 6.6.7.

EXERCISES

1 Let G and H be groups. Show that $G \times H \cong H \times G$.

2 Suppose G_1, G_2, H_1, and H_2 are groups with $G_1 \cong G_2$ and $H_1 \cong H_2$. Prove that $G_1 \times H_1 \cong G_2 \times H_2$.

3 Suppose that G and H are groups and that U is a subgroup of G and V is a subgroup of H. Show that $U \times V$ is a subgroup of $G \times H$.

4 Give an example of a group G with subgroups K and L such that $G \cong K \times L$ but G is not the internal direct product of K and L.

5 Let $A = < a_1, \ldots, a_n >$ be a finitely generated abelian group written additively. Show that the map taking $(m_1, \ldots, m_n)$ in $\mathbf{Z}^n$ to $m_1 a_1 + \ldots + m_n a_n$ is a homomorphism of $\mathbf{Z}^n$ onto A. Conclude that any finitely generated abelian group is isomorphic to a quotient group of $\mathbf{Z}^n$ for some n.

6 Show that any group of order 4 is isomorphic to $\mathbf{Z}_2 \times \mathbf{Z}_2$ or to $\mathbf{Z}_4$.

7 Show that any abelian group of order 6 is cyclic.

8 Let m and n be positive integers and set $d = \gcd(m, n)$ and $r = \text{lcm}(m, n)$. Generalize Corollary 2 by proving that $\mathbf{Z}_m \times \mathbf{Z}_n$ is isomorphic to $\mathbf{Z}_d \times \mathbf{Z}_r$.

9 Let $m_1, \ldots, m_r$ be positive integers. Use Exercise 8 to show that there exist positive integers $d_1, \ldots, d_r$ such that

$$\mathbf{Z}_{m_1} \times \ldots \times \mathbf{Z}_{m_r} \cong \mathbf{Z}_{d_1} \times \ldots \times \mathbf{Z}_{d_r}$$

and d_i divides d_{i+1}, $1 < i < r$.

10 Which of the following abelian groups of order 72 are isomorphic?

(a) $\mathbf{Z}_{72}$.
(b) $\mathbf{Z}_2 \times \mathbf{Z}_3 \times \mathbf{Z}_3 \times \mathbf{Z}_4$.
(c) $\mathbf{Z}_8 \times \mathbf{Z}_9$.
(d) $\mathbf{Z}_2 \times \mathbf{Z}_4 \times \mathbf{Z}_9$.
(e) $\mathbf{Z}_6 \times \mathbf{Z}_{12}$.
(f) $\mathbf{Z}_2 \times \mathbf{Z}_{36}$.

11 Show $U_8 \cong \mathbf{Z}_2 \times \mathbf{Z}_2$ and $U_{16} \cong \mathbf{Z}_2 \times \mathbf{Z}_4$.

12 For each i in the nonempty index set I, let G_i be a group. The Cartesian product

$$H = \prod_{i \in I} G_i$$

is defined to be the set of functions f from I to the union of the G_i such that for each i the value of $f(i)$ is in G_i. We can define a binary operation $*$ on H by $(f * g)(i) = f(i)g(i)$, where the product is computed in G_i. Show that $(H, *)$ is a group. This group is called the *direct product* of the groups G_i. Let K be the set of elements f of H such that $f(i)$ is the identity element of G_i for all but a finite number of elements i in I. Show that K is a subgroup of H. The group K is called the *direct sum* of the G_i. If I is finite, then $H = K$ but, if I is infinite, then K is a proper subgroup of H.

13 The goal of this exercise is to construct a group table for $G = \mathbf{Z}_3 \times \mathbf{Z}_3$. The nine elements of G are given by the columns of the matrix

```
      ⎕←V←3 3⊤⍳9
0 0 0 1 1 1 2 2 2
0 1 2 0 1 2 0 1 2
```

If `X←V[;I]`, we can compute `I` from `X` by `I←3⊥X`.

```
      ⎕←X←V[;5]                    3⊥X
1 2                            5
```

Write an APL expression for a rank 3 array `SUM` such that the vector `SUM[;I;J]` is the sum of `V[;I]` and `V[;J]` in G, that is, `3|V[;I]+V[;J]`. (*Hint.* First form `3|V∘.+V` and then take a suitable generalized transpose.) Our group table for G is the matrix `T` such that `T[I;J]` is `3⊥SUM[;I;J]`, the number of the column of `V` that is equal to `SUM[;I;J]`. Construct `T`.

14 Construct a group table for $\mathbf{Z}_4 \times \mathbf{Z}_6$.

15 Let `S` and `T` be an `M`-by-`M` and an `N`-by-`N` group table, respectively. Describe a method for constructing a group table for the direct product of the groups (`⍳M`, `S`) and (`⍳N`, `T`). As an example, take `S` and `T` to be `G6`.

16 Compute $\phi(10^6)$.

17 Write a procedure `PHI` to compute $\phi($`N`$)$.

18 Let $H_1, \ldots, H_r$ be subgroups of a group G. We say that G is the internal direct product of the H_i if the map of $H_1 \times \ldots \times H_r$ to G taking $(h_1, \ldots, h_r)$ to $h_1 \ldots h_r$ is an isomorphism. Give a condition in the spirit of Corollary 2 that is equivalent to the assertion that G is the internal direct product of the H_i.

19 Let G be a group and let D be the diagonal of $G \times G$. Show that D is a subgroup of $G \times G$. Show also that there is a 1–1 correspondence between normal subgroups of G and subgroups of $G \times G$ that contain D.

7. PERMUTATIONS

In this section we investigate individual elements of the symmetric group

$\Sigma(X)$ on a set X. In the following section we will study subgroups of $\Sigma(X)$.

Let $x_1, \ldots, x_r$ be a sequence of distinct elements of X. Then there is a unique permutation f of X such that

1. f fixes every element of $X - \{x_1, \ldots, x_r\}$.
2. $x_i f = x_{i+1}$, $1 \leq i < r$.
3. $x_r f = x_1$.

We denote f by $(x_1, \ldots, x_r)$ and refer to f as an *r-cycle* or a *cycle of length r*. We also say that *f cyclically permutes* the x_i. The cycle notation $(x_1, \ldots, x_r)$ is not uniquely determined by f. For example, if x is an element of X, then the 1-cycle (x) is the identity permutation. If f is not the identity, then the set $\{x_1, \ldots, x_r\}$ is uniquely determined, but there are r choices for x_1. Thus

$$f = (x_1, \ldots, x_r) = (x_2, \ldots, x_r, x_1) = \ldots = (x_r, x_1, \ldots, x_{r-1}).$$

For example, in the symmetric group Σ_3, there are just two distinct 3-cycles: (0, 1, 2) and (0, 2, 1). We say the cyclic permutations $(x_1, \ldots, x_r)$ and $(y_1, \ldots, y_s)$ are *disjoint* if one of them is the identity permutation or if $r > 1$, $s > 1$, and

$$\{x_1, \ldots, x_r\} \cap \{y_1, \ldots, y_s\} = \emptyset.$$

THEOREM 1. Disjoint cycles commute.

Proof. Let $f = (x_1, \ldots, x_r)$ and $g = (y_1, \ldots, y_s)$ be disjoint cycles. If either f or g is the identity permutation, then clearly $fg = gf$. Thus we may assume $r > 1$, $s > 1$, and $\{x_1, \ldots, x_r\} \cap \{y_1, \ldots, y_s\} = \emptyset$. If x is any element of X not in $\{x_1, \ldots, x_r\} \cup \{y_1, \ldots, y_s\}$, then f and g both fix x and $x(fg) = x = x(gf)$. Also, for $1 \leq i \leq r$, the points x_i and $x_i f$ are fixed by g, and

$$x_i(fg) = (x_i f)g = x_i f = (x_i g)f = x_i(gf).$$

Similarly, $y_j(fg) = y_j(gf)$ for $1 \leq j \leq s$. Therefore fg and gf agree on every element of X so $fg = gf$. □

Now let us assume that X is finite and that f is any element of $\Sigma(X)$. Choose x in X and set $x_0 = x$, $x_1 = x_0 f$, $x_2 = x_1 f$, and so on. Clearly, $x_i = xf^i$. Since X is finite, the elements $x_0, x_1, \ldots$ cannot all be distinct. Let n be the smallest integer such that $x_n = x_m$ for some $m < n$. If $m > 0$, then

$$x_{m-1} f = x_m = x_n = x_{n-1} f.$$

But f is a permutation of X, and so in particular f is 1–1. This means that $x_{m-1} = x_{n-1}$, contradicting our choice of n. Thus $m = 0$. The elements $x_0, \ldots, x_{n-1}$ are distinct, and f agrees with the n-cycle $f_x = (x_0, \ldots,$

x_{n-1}) on the set $\{x_0, \ldots, x_{n-1}\}$. We call f_x a *cycle* of f. Let C_f denote $\{f_x \mid x \in X\}$.

THEOREM 2. Let X be a finite set and let f be in $\Sigma(X)$. Suppose x and y are in X. Then either $f_x = f_y$ or f_x and f_y are disjoint.

Proof. Let $f_x = (x_1, \ldots, x_r)$ and $f_y = (y_1, \ldots, y_x)$. If f_x and f_y are not disjoint, then neither f_x nor f_y is the identity and $x_i = y_j$ for some i and j. But, by our preceding remarks concerning the uniqueness of the cycle notation, we may assume $i = j = 1$. Then, since $x_{k+1} = x_k f$, $1 \le k < r$ and $y_{k+1} = y_k f$, $1 \le k < s$, it follows immediately that $r = s$ and $x_k = y_k$, $1 \le i \le r$. □

Let f be the permutation of ι 7 given by the vector

```
4 5 6 3 1 0 2.
```

Then C_f has three elements.

$$f_0 = f_1 = f_4 = f_5 = (0,4,1,5),$$
$$f_2 = f_6 = (2,6),$$
$$f_3 = (3),$$

and f is the product (0,4,1,5) (2,6) (3) of the elements of C_f. As the next theorem shows, this is always true.

THEOREM 3. If X is finite, then any element f of $\Sigma(X)$ is the product of the elements of C_f.

Proof. Since X is finite, the set C_f is finite. By Theorems 1 and 2, any two elements of C_f commute, so we do not need to specify the order in which the elements of C_f are multiplied together to form the product g. Let x be in X. We must show that $xf = xg$. If $xf = x$, then x is fixed by every element of C_f and $xg = x$. Suppose $xf \neq x$. Then f_x is the only element of C_f that moves x, and every other element of C_f fixes xf_x. Therefore $xg = xf_x$. However, by the definition of f_x, $xf_x = xf$. Thus $f = g$. □

According to Theorem 3, any permutation of X can be expressed as a product of disjoint cycles. This decomposition is essentially unique, the only lack of uniqueness coming from the fact that the order of the factors may be changed, 1-cycles may be omitted and, when $r > 1$, a given r-cycle can be written in r different ways. Suppose we have a product of cyclic permutations that are not disjoint, such as

$$f = (0,2,4,1,3,5)\,(1,7,2,8)\,(0,9,7,6,1)$$

in Σ_{10}. We can easily obtain its decomposition into disjoint cycles. In the example, we first compute the cycle f_0. The first cycle takes 0 to 2, the second cycle takes 2 to 8, and the third cycle leaves 8 fixed. Thus $0f = 8$. We find that $8f = 0$ and so $f_0 = (0,8)$. The first point not moved by f_0 is 1, so we compute f_1. It turns out that $f_1 = (1,3,5,9,7,2,4,6)$; this accounts for all the elements of ⍳10. Thus

$$f = f_0 f_1 = (0,8)\,(1,3,5,9,7,2,4,6).$$

The inverse of a product of cycles can be found using the rule $(f_1 \ldots f_r)^{-1} = f_r^{-1} \ldots f^{-1}$ and the observation that

$$(x_1, \ldots, x_r)^{-1} = (x_r, x_{r-1}, \ldots, x_1) = (x_1, x_r, \ldots, x_2).$$

An r-cycle has order r. If $f_1, \ldots, f_s$ are disjoint cycles and $f = f_1 \ldots f_s$, then $f^n = f_1{}^n \ldots f_s{}^n$ and $f^n = 1$ if and only if $f_i^n = 1$ for each i. Therefore the order of f is the least common multiple of the lengths of the f_i. For example, (0,2) (1,3,4) has order 6 while (0,2) (1,3,4,5) has order 4.

Cycle notation is used extensively for hand computation but almost never for calculations in a computer. It is convenient to be able to enter permutations at a terminal in cycle notation and have the results of a computation printed in cycle notation. The procedures *GPCYCIN* and *GPCYCOUT* can be used to convert between cycle notation and the vector notation used by most of the procedures that work with permutations.

```
      ⎕←X←7 GPCYCIN C←'(0,5,1,3)(2,6)'
5 3 6 0 4 1 2
      ⎕←Y←10 GPCYCIN C
5 3 6 0 4 1 2 7 8 9
      GPCYCOUT X
(0,5,1,3)(2,6)(4)
      GPCYCOUT Y
(0,5,1,3)(2,6)(4)(7)(8)(9)
```

The cycle representation of a permutation is a character string. Cycles of length 1 may be omitted in the second argument of *GPCYCIN*, so the first argument is used to determine the set ⍳*N* on which the permutation acts. Cycles of length 1 are always listed by *GPCYCOUT*. The current origin is used to decide whether the permutations act on {0, . . . , *N*-1} or {1, . . . , *N*}. Both *GPCYCIN* and *GPCYCOUT* work with just one permutation at a time.

There is one more notation for permutations that is occasionally used. If $X = \{x_1, \ldots, x_n\}$, then the permutation of X that takes x_1 to y_1, x_2 to y_2, and so on may be written

$$\begin{pmatrix} x_1 & x_2 & \ldots & x_n \\ y_1 & y_2 & \ldots & y_n \end{pmatrix}.$$

Thus the element (0,3,6,2) (1,4,5) of Σ_7 could be written

$$\begin{pmatrix} 0 & 1 & 2 & 3 & 4 & 5 & 6 \\ 3 & 4 & 0 & 6 & 5 & 1 & 2 \end{pmatrix} \quad \text{or} \quad \begin{pmatrix} 6 & 5 & 4 & 3 & 2 & 1 & 0 \\ 2 & 1 & 5 & 6 & 0 & 4 & 3 \end{pmatrix}.$$

Sometimes we will want to work with subsets of Σ_n for small values of n. We will represent such a subset A by a matrix whose ith row is the ith element of A. For example, the matrix *GP*8 in *EXAMPLES* describes a set of eight elements of Σ_4 in origin 0.

```
      GP8
0 1 2 3
0 3 2 1
1 2 3 0
1 0 3 2
2 3 0 1
3 2 1 0
2 1 0 3
3 0 1 2
```

The procedure *GPSYMG* can be used to produce a list of all elements of the symmetric group Σ_n. The result is origin dependent.

```
      ⎕←P←GPSYMG 3
0 1 2
1 0 2
0 2 1
1 2 0
2 0 1
2 1 0
      ⎕IO←1
```

```
      ⎕←Q←GPSYMG 3
1 2 3
2 1 3
1 3 2
2 3 1
3 1 2
3 2 1
      ⎕IO←0
```

The rows of *GP*8 actually form a subgroup of Σ_4. Exercises 13 and 14 describe how to verify this fact.

EXERCISES

1 Find the decomposition as a product of disjoint cycles for the following permutations.

- **(a)** (0,1) (1,2).
- **(b)** (0,1) (2,3) (0,2,4).
- **(c)** (1,7,2,8,4) (0,5,2,6,3) (1,5,3,9).

2 Show that $[x] \longmapsto [3x + 2]$ is a permutation of $\mathbf{Z}_8$ and determine its cycles.

3 What is the inverse of (0,2,5,3,4,8,1,7,6)?

4 Let *N* be a positive integer and set `F←1⌽⍳N`, `G←¯1⌽⍳N`, and `H←⌽⍳N`. Determine the cycles of *F*, *G*, and *H*. Show that *G* is the inverse of *F* and that `F[H]` is equal to `H[G]`.

5 Let *N* be a positive integer and *M* any integer. Show that all cycles of `M⌽⍳N` have length `N÷M ZGCD N`.

6 Let *P* be a vector that is known to be a permutation of `⍳N`. Show that the vectors *F*, *G*, and *H* defined by

```
F←⍋P
G←P⍳⍳N
H←Nρ2
H[P]←⍳N
```

are all equal to the inverse of *P*. Which method uses the least CPU time for large *N*?

7 What are the possible orders for an element of Σ_6?

8 What is the largest order an element of Σ_{10} can have?

9 Determine the cycles of all the permutations listed in the matrix `GP8`.

10 Let *A* and *B* be two matrices whose rows are permutations of `⍳N`. Write an APL proposition corresponding to the assertion that the set of permutations listed in *A* is the same as the set listed in *B*.

11 Show that a permutation *P* of `⍳N` is uniquely determined by the integer `M←N⊥P`. How can *P* be reconstructed from *M* and *N*? We call *M* the *packed representation* of *P*.

12 Let *A* be a matrix whose rows are permutations of `⍳N`. Write an expression for the vector *V* such that `V[I]` is the packed representation of `A[I;]`.

13 In order to show that the rows of `GP8` form a subgroup of Σ_4, we must prove that the product of any two rows is again a row. Let

```
Q←GP8[;GP8]
```

Then `Q[I;J;]` is the product of `GP8[J;]` and `GP8[I;]`. The most straightforward way to show that each vector `Q[I;J;]` is a row of `GP8` is to form

```
C←∨/Q∧.=⍉GP8
```

Check that every entry of C is 1 and explain why this proves that the rows of $GP8$ form a subgroup of Σ_4. (Clearly, this approach is appropriate only for very small subgroups of Σ_n.)

14 In order to show in Exercise 13 that the rows of $GP8$ form a subgroup of Σ_4, we computed the inner product `Q∧.=GP8`. This calculation requires a significant amount of CPU time. Use the ideas in Exercises 11 and 12 to obtain a more efficient test for showing that the rows of $GP8$ are closed under products.

15 Construct a group table for the subgroup of Σ_4 listed in $GP8$.

8. PERMUTATION GROUPS

A permutation group on a set X is a subgroup of $\Sigma(X)$. The importance of permutation groups is made clear by the following theorem, which states that up to isomorphism all groups are permutation groups. This theorem is named for English mathematician Arthur Cayley (1821-1895), who first gave the definition of an abstract group.

THEOREM 1 ***(Cayley).*** If G is a group, then G is isomorphic to a subgroup of $\Sigma(G)$.

Proof. In Corollary 2.3 we defined for each u in G a permutation R_u of G by $xR_u = xu$. We will show now that R is an injective homomorphism of G into $\Sigma(G)$. For all x in G we have

$$x(R_u R_v) = (xR_u)R_v = (xu)v = x(uv) = xR_{uv},$$

so $R_u R_v = R_{uv}$. This means that R is a homomorphism of G into $\Sigma(G)$. Suppose u is in the kernel of R. Then $xu = x$ for all x in G. But if $xu = x$ for even one x in G, then $u = 1$. Therefore R is injective. If we let H be the image of G under R, then R is an isomorphism of G onto H. □

The homomorphism R used in the proof of Theorem 1 is called the *right regular representation* of G. The word "representation" is often used to denote a homomorphism of a group H that we know little about into a group K that we know more about. In the case of Theorem 1, we represent elements of the "abstract" group G by permutations, which are considered to be more "concrete".

In Corollary 2.3 we also defined a permutation L_u of G by the formula $xL_u = u^{-1}x$. It is easily shown (see Exercise 1) that L is also an injective homomorphism of G into $\Sigma(G)$. We call L the *left regular representation* of G.

If G is an N-by-N group table, then the image of $(\iota N, G)$ under its right regular representation is the set of columns of G, considered as elements of Σ_N. The image of (⍳*N*, *G*) under its left regular representation is the set of rows of *G*.

In Section 3 we discussed the problem of finding the subgroup generated by a set of elements in a group defined by a group table *G*. Suppose we are given a matrix *S* whose rows are elements of Σ_N. How might we construct the subgroup of Σ_N generated by the rows of *S*? One could imagine constructing a group table *T* for Σ_N and using the methods of Section 3 but, if *N* > 4, then the storage requirements of *T* would be prohibitive. Moreover, we do not need a group table to compute products in Σ_N. Given the matrix *S* of generators, we can form *S*[;*S*], and this rank 3 array gives all possible products of two rows of *S*. If some of these products are new, we can add them to the rows of *S* and repeat the process until we get no new permutations. Stated in this form, the procedure still requires quite a bit of space and CPU time. However, using the ideas in Exercises 3.13, 7.11, and 7.12, a procedure can be constructed that works reasonably well for small values of *N*. The workspace *CLASSLIB* contains one such procedure, *GPSGP*.

```
      ⎕←S←GP8[1 2;]
 0 3 2 1
 1 2 3 0
      GPSGP S
 0 1 2 3
 0 3 2 1
 1 2 3 0
 1 0 3 2
 2 3 0 1
 3 2 1 0
 2 1 0 3
 3 0 1 2
      ⎕←T←2 4⍴1 2 3 0 1 0 2 3
 1 2 3 0
 1 0 2 3
      H←GPSGP T
      ⍴H
24 4
```

The procedure *GPSGP* returns a matrix listing all the elements of the subgroup generated by the rows of its single argument. In Section 7 we saw that the rows of the matrix *GP*8 form a subgroup of Σ_4. Here we see that this subgroup is generated by the two elements given in *S* and that the rows of *T* generate all of Σ_4.

The techniques used in *GPSGP* are very crude, and this procedure should be used only to investigate small subgroups of Σ_N, where N itself is small. Much more powerful algorithms exist for determining the order of the group generated by a given set of permutations. However, these algorithms are beyond the scope of this text, and their computer implementation, even in APL, is somewhat complicated. Further information can be found in my paper listed in the bibliography.

One way to look for subgroups of Σ_N when N is small is to choose small subsets of Σ_N at random and use *GPSGP* to find the subgroups generated by these subsets. Suppose two elements f and g of Σ_4 are chosen randomly and independently. The order of $< f, g >$ must be one of the numbers 1, 2, 3, 4, 6, 8, 12, or 24. Let $p(m)$ denote the probability that $< f, g >$ has order m. It can be shown that the values of $p(m)$ are given by the following table.

m	$p(m)$
1	1/576 = .0017. . .
2	3/64 = .0468. . .
3	1/18 = .0555. . .
4	5/48 = .1041. . .
6	1/8 = .125
8	1/8 = .125
12	1/6 = .1666. . .
24	3/8 = .375

Thus the most likely value of m is 24 and the next most likely is 12. If two random elements f and g are chosen in Σ_n, with $n > 4$, then the exact probability that $< f, g >$ will have a particular order is difficult to compute and not of much interest. However, the most likely order is $n!$ and the next most likely is $n!/2$. In fact, it has been shown [Dixon] that in the limit, as n goes to infinity, $< f, g > = \Sigma_n$ with probability 3/4 and $< f, g >$ has index 2 in Σ_n with probability 1/4. This does not mean that Σ_n has no subgroups of index greater that 2, only that the other subgroups are comparatively small and not very numerous.

Dixon's theorem is beyond the scope of this text. However, we will prove the important result that Σ_n always has a subgroup of index 2 when $n \geq 2$. To do this, we will exhibit a homomorphism of Σ_n onto a group of order 2. Assume $n \geq 2$. If f is in Σ_n, we define the *sign* of f to be the rational number

$$\operatorname{sgn}(f) = \prod_{i>j} \frac{(if) - (jf)}{i - j} .$$

Here the product is over all integers i and j, with $0 \leq j < i < n$. As an example, the sign of (0,2,1) in Σ_3 is

$$\left(\frac{0-2}{1-0}\right)\cdot\left(\frac{1-2}{2-0}\right)\cdot\left(\frac{1-0}{2-1}\right)=1$$

and the sign of (0) (1,2) is

$$\left(\frac{2-0}{1-0}\right)\cdot\left(\frac{1-0}{2-0}\right)\cdot\left(\frac{1-2}{2-1}\right)=-1.$$

The following theorem justifies the use of the term "sign".

THEOREM 2. Let f be an element of Σ_n, $n \geq 2$. Then sgn(f) is either 1 or -1.

Proof. By definition, we have

$$\operatorname{sgn}(f) = \prod_{i>j} \frac{(if)-(jf)}{i-j} = \frac{\prod\limits_{i>j} [(if)-(jf)]}{\prod\limits_{i>j} (i-j)} .$$

Fix i and j with $0 \leq j < i < n$. We wish to show that the factor $(i-j)$ in the denominator of sgn(f) also occurs, perhaps with its sign changed, in the numerator of sgn(f). Let $u = if^{-1}$ and $v = jf^{-1}$. Then $uf = i$ and $vf = j$. If $u > v$, then the factor $(uf) - (vf) = (i-j)$ occurs in the numerator of sgn(f); if $u < v$, then $(vf) - (uf) = j-i$ occurs in the numerator. Choosing a different pair (i, j) leads to a different factor in the numerator, so there is a 1–1 correspondence between the factors in the denominator and the factors in the numerator such that corresponding factors differ at most by a sign. Therefore sgn(f) is 1 or -1. □

There are other ways of computing sgn(f). Since $|\operatorname{sgn}(f)| = 1$, it is enough to count the number of terms in the product

$$\prod_{i>j} \frac{(if)-(jf)}{i-j}$$

that are negative. The factor corresponding to the pair (i, j) is negative if and only if $if < jf$. Thus $\operatorname{sgn}(f) = (-1)^m$, where m is the number of pairs (i, j), with $i > j$ and $if < jf$. Such a pair is called an *inversion* in f. The procedure *GPSGN* uses this method to compute the sign of one or more permutations.

```
      GP8
 0 1 2 3
 0 3 2 1
 1 2 3 0
 1 0 3 2
 2 3 0 1
 3 2 1 0
 2 1 0 3
 3 0 1 2
      ⎕←S←GPSGN GP8
1 ¯1 ¯1 1 1 1 ¯1 ¯1
      GPSGN 3 GPCYCIN '(0,2,1)'
1
      GPSGN 3 GPCYCIN '(0)(1,2)'
¯1
```

Here S[I] is the sign of GP8[I;].

THEOREM 3. If $n \geq 2$, then sgn is a homomorphism of Σ_n onto the group $(\{1, -1\}, \times)$.

Proof. By Theorem 2, sgn is a map from Σ_n to $\{1, -1\}$. It is easy to check (see Exercise 6) that the sign of the 2-cycle (0, 1) is -1. What remains is to show that sgn is a homomorphism. Let f and g be in Σ_n. We must prove $\text{sgn}(fg) = [\text{sgn}(f)]\ [\text{sgn}(g)]$. By definition,

$$\text{sgn}(fg) = \prod_{i>j} \frac{(ifg) - (jfg)}{i - j}.$$

Multiplying by

$$1 = \prod_{i>j} \frac{(if) - (jf)}{(if) - (jf)}$$

and rearranging the denominator, we obtain

$$\text{sgn}(fg) = \prod_{i>j} \frac{(ifg) - (jfg)}{(if) - (jf)} \prod_{i>j} \frac{(if) - (jf)}{i - j}$$

$$= \text{sgn}(f) \prod_{i>j} \frac{(ifg) - (jfg)}{(if) - (jf)}.$$

Fix i and j with $i > j$ and set $u = if$ and $v = jf$. If $u > v$, then

$$\frac{(ifg) - (jfg)}{(if) - (jf)} = \frac{(ug) - (vg)}{u - v}$$

is a factor of sgn(g). If $u < v$, then

$$\frac{(ifg)-(jfg)}{(if)-(jf)} = \frac{(ug)-(vg)}{u-v} = \frac{(vg)-(ug)}{v-u}$$

is again a factor of sgn(g). Thus

$$\prod_{i>j} \frac{(ifg)-(jfg)}{(if)-(jf)} = \text{sgn}(g)$$

and sgn(fg) = [sgn(f)] [sgn(g)] . □

The kernel of sgn is called the *alternating group* and is denoted A_n. The index of A_n in Σ_n is 2, and A_n is normal in Σ_n. Elements of A_n are said to be *even permutations,* while elements of $\Sigma_n - A_n$ are said to be *odd.* The reason for this terminology will be discussed in Section 10.

Now we will show how a permutation group G on a set X defines an equivalence relation on X. If x and y are in X, let us write $x \sim y$ if there exists an element g of G such that $y = xg$.

LEMMA 4. The relation $\sim$ is an equivalence relation on X.

Proof. If e is the identity permutation on X, then $e \in G$ and $x = xe$. Thus $\sim$ is reflexive. If $y = xg$, then $x = yg^{-1}$ and $\sim$ is symmetric. If $y = xg$ and $z = yh$, then $z = x(gh)$, proving that $\sim$ is transitive. □

The equivalence classes of $\sim$ are called the *orbits* of G, and the orbit containing a point x of X is $xG = \{xg | g \in G\}$. The set $G_x = \{g \in G | xg = x\}$ is a subgroup of G and is called the *stabilizer* of x in G.

THEOREM 5. Let G be a permutation group on X. If x is in X, then there is bijection from the orbit xG containing x to the set of right cosets of G_x in G.

Proof. Every element of xG has the form xg for some g in G. We would like to map xg to the coset $G_x g$, but is this well defined? If $xg = xh$, then $xgh^{-1} = x$ and gh^{-1} is in G_x. This implies $G_x g = G_x h$, so our map is well defined. The map is clearly surjective. All that remains is to verify injectivity. If $G_x g = G_x h$, then $h = ug$ for some u in G_x. Therefore $xh = x(ug) = xg$, and the map is injective too. □

COROLLARY 6. If, in Theorem 5, the set X is finite, then $|xG| = |G{:}G_x|$. In particular, $|xG|$ divides $|G|$. □

Suppose the rows of the matrix P give the elements of some subgroup G of Σ_N. Then the orbit of G containing a point I is given by the vector `SSORT P[;I]`. However, suppose we are given only a set of generators for G, say as the rows of a matrix S. One method for computing the orbit of G containing I would be to form `P←GPSGP S`. But P may be a very large matrix that will not fit in a workspace. It is possible to determine the orbit containing I without constructing P first. The method is sketched in Exercise 12. The procedure *GPORBIT* is based on this approach. The first argument of *GPORBIT* is a matrix *S* whose rows list a set of permutations. The second argument is a single integer *I*. The result of *GPORBIT* is a vector listing the points in the orbit containing *I* of the group generated by the rows of *S*. In addition, the characteristic vector of the orbit is produced as the global variable X̲.

```
      P20
 6 16 12  0  8 11 18 13 15 19  9  5  2  1 14  4  7 17  3 10
12  7 18  2 10 13  0  1  4 19 15 11  6 17  5  8 14 16  3  9
      GPCYCOUT P20[0;]
(0,6,18,3)(1,16,7,13)(2,12)(4,8,15)(5,11)(9,19,10)(14)(17)
      GPCYCOUT P20[1;]
(0,12,6)(1,7)(2,18,3)(4,10,15,8)(5,13,17,16,14)(9,19)(11)
      P20 GPORBIT 0
0 2 3 6 12 18
      P20 GPORBIT 1
1 5 7 11 13 14 16 17
      P20 GPORBIT 4
4 8 9 10 15 19
      X̲
0 0 0 0 1 0 0 0 1 1 1 0 0 0 0 1 0 0 0 1
```

In this example the rows of *P20* are elements of Σ_{20}. The orbits of the group generated by these permutations are {0,2,3,6,12,18}, {1,5,7,11, 13,14,16,17}, and {4,8,9,10,15,19}.

Given a matrix of permutations, we can use the procedure *GPALLORB* to obtain a summary of the orbits of the group *G* generated by these permutations.

```
      GPALLORB P20
 0 1 4
 6 8 6
      Q̲
0 1 0 0 4 1 0 1 4 4 4 1 0 1 1 4 1 1 0 4
      (Q̲=1)/⍳20
1 5 7 11 13 14 16 17
```

The procedure *GPALLORB* returns a matrix with two rows. The first row lists the first element of each orbit of *G* and the second row lists the number of elements in each orbit. In addition, *GPALLORB* constructs a global vector Q̲ such that Q̲[*I*] is the first point in the orbit of *G* containing *I*. Thus Q̲=Q̲[*I*] is the characteristic vector for the orbit containing *I*. The procedure *GPALLORB* plays an important part in the calculations done in Section 9.

At this point it will be helpful in our discussion of permutation groups to look at an example. Let the vertices of a regular polygon P with n sides be labeled from 0 to $n-1$ as follows.

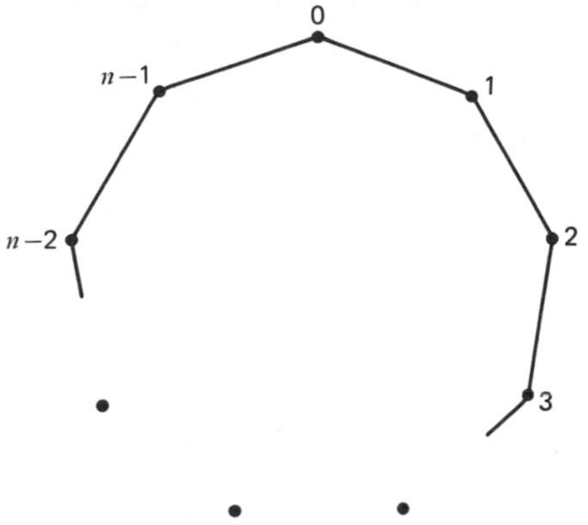

A symmetry of P is a permutation of the vertices that maps edges into edges. For example, the symmetry $a = (0,1,2, \ldots, n-1)$ corresponds to a clockwise rotation of P through an angle of $2\pi/n$ radians. The symmetry $b = (0)\,(1, n-1)\,(2, n-2) \ldots$ is the reflection in the line through vertex 0 and the center of P. The set G of all symmetries of P is a subgroup of Σ_n. Since a is in G and the powers of a map 0 to each of the other vertices, G has one orbit: $\{0,1, \ldots, n-1\}$. By Lagrange's Theorem (Theorem 3.9), we have $|G| = |G{:}G_0||G_0|$ and, by Corollary 6, we know that $|G{:}G_0| = n$. The only vertices joined to vertex 0 by an edge are 1 and $n-1$. Thus any element of G_0 either fixes 1 or maps 1 to $n-1$. Thus the orbit $1G_0$ of G_0 containing 1 has at most two elements. Since b is in G_0 and b interchanges 1 and $n-1$, the orbit $1G_0$ has exactly two elements. If we let $G_{01} = (G_0)_1$, the set of elements in G that fix both 0 and 1, then $|G_0| = |G_0{:}G_{01}||G_{01}| = 2|G_{01}|$ and so $|G| = 2n|G_{01}|$. If x is in G_{01}, then x fixes 0 and 1. Thus x fixes 2 because 2 is the only vertex other than 0 connected to 1. But then x fixes 3, since 3 is the only vertex besides 1 connected to 2. Continuing in this manner, we find that x fixes every vertex. Thus $|G_{01}| = 1$ and $|G| = 2n$. The group G is called the *dihedral group* of order $2n$ and is denoted D_{2n}. (Some authors write D_n.) Exercises 17 and 18 provide more information about D_{2n}.

Cycles of length 2 have a significant role in the study of symmetric groups. The term *transposition* is often used to refer to a 2-cycle.

THEOREM 7. For any integer $n > 1$, the symmetric group Σ_n is generated by the transpositions (0,1), (1,2), . . . , $(n-2, n-1)$.

Proof. Let $G = \langle(0,1), \ldots, (n-2, n-1)\rangle$. Clearly, $G = \Sigma_n$ if $n = 2$. We will assume $n > 2$ and proceed by induction. For any $i < n-1$, the image of $n-1$ under the product $(n-1, n-2)(n-2, n-3)\ldots(i+1, i)$ is i. Thus $\{0, \ldots, n-1\}$ is an orbit of G. Consider the stabilizer G_{n-1} of $n-1$ in G. The transpositions (0,1), . . . , $(n-3, n-2)$ are all in G_{n-1} so, by induction, the order of G_{n-1} is at least $(n-1)!$. However, by Theorem 5, the index $|G:G_{n-1}|$ is n. Thus

$$n! \geq |G| = |G:G_{n-1}||G_{n-1}| \geq n[(n-1)!] = n!.$$

Thus $|G| = |\Sigma_n|$ and $G = \Sigma_n$. □

The following slight generalization of Theorem 5 will be used in the next section.

THEOREM 8. Let G be a group, let X be a set, and let $f:G \rightarrow \Sigma(X)$ be a homomorphism. For any x in X there is a bijection from the orbit containing x of the image Gf to the set of right cosets of the stabilizer $G_x = \{u \in G | x(uf) = x\}$.

Proof. We know by Theorem 5 that there is a bijection from $x(Gf)$ to the set of right cosets of $(Gf)_x$ in Gf. But $G_x = [(Gf)_x]f^{-1}$ and so, by Corollary 5.11, there is a bijection from the set of right cosets of $(Gf)_x$ in Gf to the set of right cosets of G_x in G. □

EXERCISES

1 Show that the left regular representation of a group G is an injective homomorphism of G into $\Sigma(G)$.

2 Let G be a group and let H and K be the images of G under the right and left regular representations of G, respectively. Prove that elements of H commute with elements of K.

3 In Exercise 2 show that K is the centralizer of H in $\Sigma(G)$ and that H is the centralizer of K. (See Exercise 3.8.)

4 Use *GPSGP* to find the orders of the groups generated by

(a) (0) (1,2,3,4,5) and (0,1) (2) (3,4) (5).

(b) (0,1,3,2,5,6,4) and (0) (1) (2) (3,4) (5,6).

5 The procedure

```
        ∇M←TRIAL N;S
[1]     S←(2,N)ρ(N?N),N?N
[2]     M←1↑ρGPSGP S∇
```

chooses two elements of Σ_N at random and returns the order of the group generated by them. Execute *TRIAL* 5 at least 20 times and use the results to estimate the probability that two elements of Σ_5 chosen randomly generate Σ_5.

6 Let n be an integer greater than 1. Show that the sign of the 2-cycle (0,1) in Σ_n is -1.

7 Count the inversions in the following permutations.

(a) (0,3,4) (1,2).
(b) (0,7) (1,6) (2,5) (3,4).
(c) (0,3,5) (2) (1,4,6).
(d) (0,7,1,6,2,5,3,4).

8 Let *P* be a vector that is a permutation of ⍳⍴*P*. Write an APL expression for the number of inversions in *P*.

9 Show that the product of two even permutations is even, the product of an even and an odd permutation is odd, and the product of two odd permutations is even. Show also that $\text{sgn}(f^{-1}) = \text{sgn}(f)$ for all f in Σ_n.

10 We defined Σ_n to be the symmetric group on the set $\{0, \ldots, n-1\}$ and we showed that Σ_n has a subgroup of index 2 when $n \geq 2$. Suppose X is any finite set with $|X| \geq 2$. Prove that $\Sigma(X)$ has a subgroup of index 2.

11 Let $n \geq 2$. Show that the probability that two elements chosen randomly in Σ_n generate Σ_n is at most 3/4.

12 Let X be a finite set and let U be a subset of $\Sigma(X)$. Set $G = \langle U \rangle$. Prove that if x is in X, then the orbit xG is the smallest subset Y of X such that $x \in Y$ and for all y in Y and all u in U the image yu is in Y. Use this idea to find the orbits of the subgroup $\langle u, v \rangle$ of Σ_{10}, where

$u = (0,6,8)\,(1,5)\,(2,7)\,(3,9)\,(4)$,

$v = (0)\,(1)\,(2,6,7)\,(3)\,(4,9)\,(5)\,(8)$.

13 Let G be the subgroup of Σ_{50} generated by 5⌽⍳50, ⌽⍳50, and 1 0,2↓⍳50. What are the orbits of G? What is $|G:G_0|$? (Try to solve the problem by hand and then check your answer using *GPALLORB*.)

14 Show that Σ_n is generated by (0,1) and $(0,1,\ldots,n-1)$.

15 Show that Σ_n is generated by (0,1) and $(1,2,\ldots,n-1)$.

16 Let $G = (X, E)$ be an undirected graph. An *automorphism* of G is an isomorphism of G with itself. (See Section 1.5.) Show

that the set Aut(G) of all automorphisms of G is a subgroup of $\Sigma(X)$.

17 List the elements of the dihedral group D_8.

18 Let a and b be the elements of the dihedral group D_{2n} defined in the text. Show that $a^n = b^2 = 1$ and that $bab = a^{-1}$. Prove that every element of D_{2n} can be expressed uniquely as $b^i a^j$, with $0 \leq i \leq 1$ and $0 \leq j < n$. Show that every subgroup of $\langle a \rangle$ is normal in D_{2n}.

19 Let G and H be permutation groups on X and Y, respectively. We say G and H are *isomorphic as permutation groups* if there is a bijection $\theta: X \longrightarrow Y$ and an isomorphism $f: G \longrightarrow H$ such that $(xu)\theta = (x\theta)(uf)$ for all x in X and all u in G. Show that Σ_4 has two cyclic subgroups that are isomorphic as groups but not as permutation groups.

20 Let G be an N-by-N matrix defining a binary operation on ιN. Show that G is a group table if and only if there is a two-sided identity element and the set of rows of G is a subgroup of Σ_N.

*21 The time needed to verify that a given n-by-n matrix G is a group table directly from the definition of a group is proportional to n^3, since associativity must be checked for all possible triples of elements. Show that it is possible to decide whether G is a group table in a time proportional to $n^2 \log_2 n$ using an approach based on the following outline.

(a) Let $U = \{0,1, \ldots, n-1\}$ and for x, y in U let $xy = G[x;y]$.
(b) Find, if possible, a two-sided identity e.
(c) Find, if possible, a subset X of U such that $|X| \leq \log_2 n$, for each x in X the map $y \longmapsto yx$ is in Σ_n, and the group generated by these permutations has U as an orbit.
(d) For all x in X and y and z in U, check that $(yz)x = y(zx)$.

*22 Construct an algorithm for deciding whether an n-by-n matrix is a group table such that the execution time for the algorithm is proportional to n^2.

23 Let X be a finite set and let G be a subgroup of $\Sigma(X)$. For g in G let $\chi(g)$ be the number of elements fixed by g. Show that

$$\sum_{g \in G} \chi(g) = k|G|,$$

where k is the number of orbits of G on X. The function χ is called the *permutation character* of G. *Hint.* Let $U = \{(g, x) | g \in G, x \in X,$

$xg = x\}$. Thus $\chi(g)$ is the number of elements of U whose first component is g. Therefore

$$\sum_{g \in G} \chi(g) = |U|.$$

However, for a given x in X the pair (g, x) is in U if and only if g is in the stabilizer G_x. Thus

$$|U| = \sum_{x \in X} |G_x|.$$

Use Theorem 5 to show that if Y is an orbit of G, then

$$\sum_{x \in Y} |G_x| = |G|.$$

*9 GRAPHS WITH A SMALL NUMBER OF VERTICES

In this section we will show how the theory of groups developed so far can be used to determine, up to isomorphism, the undirected graphs with n vertices, where $n \leq 5$.

Let X be a set and let $Y = 2^X$, the set of all subsets of X. Suppose f is in $\Sigma(X)$. If A is in Y, that is, if $A \subseteq X$, then $\{af | a \in A\}$ is in Y also. Thus we may define a function $\bar{f}: Y \longrightarrow Y$ by $A\bar{f} = \{af | a \in A\}$.

LEMMA 1. The map $\bar{\ }: f \longmapsto \bar{f}$ is a homomorphism of $\Sigma(X)$ into $\Sigma(Y)$.

Proof. Suppose f and g are in $\Sigma(X)$ and A is in Y. Then

$$\begin{aligned} A(\bar{f} \circ \bar{g}) &= (A\bar{f})\bar{g} = (\{af | a \in A\})\bar{g} = \{(af)g | a \in A\} \\ &= \{a(f \circ g) \mid a \in A\} = A(\overline{f \circ g}). \end{aligned}$$

Thus $\bar{f} \circ \bar{g} = \overline{f \circ g}$. Let e be the identity function on X. Then $A\bar{e} = \{ae | a \in A\} = \{a | a \in A\} = A$, and $\bar{e}$ is the identity function on Y. Now, taking $g = f^{-1}$, we have

$$\bar{f} \circ \overline{f^{-1}} = \overline{f \circ f^{-1}} = \bar{e} = \overline{f^{-1} \circ f} = \overline{f^{-1}} \circ \bar{f}.$$

Therefore $\bar{f}$ is a permutation of Y by Corollary 1.3.4 and $\bar{\ }$ maps $\Sigma(Y)$ into $\Sigma(Y)$. Since $\bar{f} \circ \bar{g} = \overline{f \circ g}$, $\bar{\ }$ is a homomorphism. □

It is easy to see that if we had defined Y to be the set of all subsets of X with a given number, say k, of elements, then we would again have obtained a homomorphism of $\Sigma(X)$ into $\Sigma(Y)$.

Now let n be a positive integer and let $X = \{0,1, \ldots, n-1\}$. Define Y to be the set of two-element subsets of X and Z to be the set of all subsets of Y. If E is an element of Z, then E is a set of two-element subsets of X and (X, E) is an undirected graph. As before, we have homomorphisms of $\Sigma_n = \Sigma(X)$ into $\Sigma(Y)$ and of $\Sigma(Y)$ into $\Sigma(Z)$. For f in Σ_n, let $\bar{f}$ be the image of f under the composition of these homomorphisms. Thus, if E is in Z, then

$$E\bar{f} = \{\{xf, yf\} \mid \{x, y\} \in E\}.$$

Let H be the image of Σ_n under $\bar{\ }$, that is, $\{\bar{f} \mid f \in \Sigma_n\}$. Since H is a subgroup of $\Sigma(Z)$, we may define the orbits $Z_1, \ldots, Z_r$ of H on Z.

THEOREM 2. Let $G = (U, F)$ be an undirected graph with n vertices. Then G is isomorphic to a graph (X, E), where E is in Z. Moreover, two graphs (X, E) and (X, E'), with E and E' in Z, are isomorphic if and only if E and E' are in the same orbit of H. Up to isomorphism there are exactly r undirected graphs with n vertices.

Proof. Since $|U| = |X| = n$, there is a bijection $g: U \longrightarrow X$. Let $E = \{\{ug, vg\} \mid \{u, v\} \in F\}$. Then E is in Z and g is an isomorphism of G onto (X, E). Now let E and E' be any two elements of Z. The graphs (X, E) and (X, E') are isomorphic if and only if there is a permutation f of X such that

$$E' = \{\{xf, yf\} \mid \{x, y\} \in E\} = E\bar{f},$$

that is, if and only if E and E' are in the same orbit of H. □

Let us apply Theorem 2 for some small values of n. If $n = 1$, then $|X| = 1$, $|Y| = 0$, and $|Z| = 1$. Clearly, there is up to isomorphism only one undirected graph with one vertex. If $n = 2$, then $|X| = 2$, $|Y| = 1$, and $|Z| = 2$. The two elements of Z correspond to graphs that are obviously nonisomorphic. Thus, up to isomorphism, there are two graphs having two vertices.

If $n = 3$, then $X = \{0,1,2\}$ and $Y = \{a,b,c\}$, where

$$a = \{0,1\},$$
$$b = \{0,2\},$$
$$c = \{1,2\}.$$

There are eight subsets of Y, so Z consists of the sets

$$\begin{aligned} E_0 &= \emptyset, & E_4 &= \{a\}, \\ E_1 &= \{c\}, & E_5 &= \{a,c\}, \\ E_2 &= \{b\}, & E_6 &= \{a,b\}, \\ E_3 &= \{b,c\}, & E_7 &= \{a,b,c\}. \end{aligned}$$

(The numbering has been chosen to agree with our APL formulation that follows.) Now, by Exercise 8.14, the symmetric group Σ_3 is generated by $x = (0,1)$ and $y = (0,1,2)$. The group H is therefore generated by the images $\bar{x}$ and $\bar{y}$ of x and y under the homomorphism from Σ_3 to $\Sigma(Z)$. To compute $\bar{x}$, we first note that x induces $x' = (a)(b,c)$ on Y. To see what $\bar{x}$ does to a particular set E_i, we simply apply x' to the elements of E_i and check in the list to see which set E_j is obtained. A simple computation shows that

$$\bar{x} = (E_0)(E_1, E_2)(E_3)(E_4)(E_5, E_6)(E_7),$$

which we will abbreviate as

$$\bar{x} = (0)(1,2)(3)(4)(5,6)(7),$$

writing i instead of E_i. Similarly, we see that y induces $y' \doteq (a,b,c)$ on Y and

$$\bar{y} = (0)(1,2,4)(3,6,5)(7)$$

on Z. The orbits of $H = \langle \bar{x}, \bar{y} \rangle$ are

$$\{0\}, \{1,2,4\}, \{3,5,6\}, \{7\}.$$

Thus, up to isomorphism, there are four graphs with three vertices. Representatives for the isomorphism classes are given by (X,E_i) for $i = 0,1,3,7$. The corresponding diagrams follow.

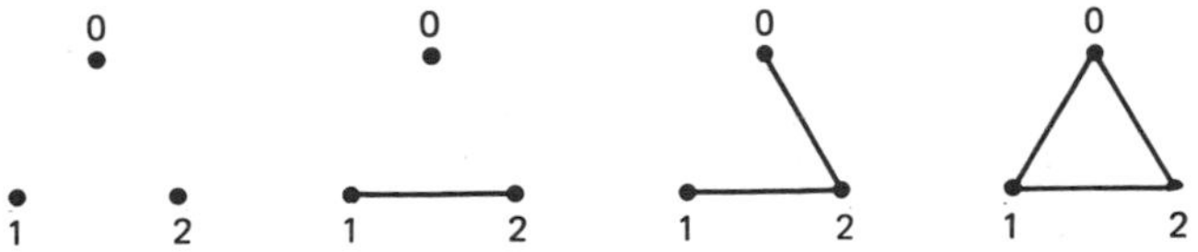

For $n = 4$ we have $|X| = 4$, $|Y| = 6$, and $|Z| = 64$. The computations that were easy for $n = 3$ now are somewhat tedious to carry out by hand. Let us try to reformulate, using APL, what we did for $n = 3$ in a way that can be used for larger n. The elements of Y are listed in the following matrix.

```
      ⎕←S←2 SSUB N←3
 0 1
 0 2
 1 2
```

In general Σ_n is generated by $x = (0,1)$ and $y = (0,1,2, \ldots, n-1)$, which are given by the vectors

```
      ⎕←X←1 0,2↓⍳N                    ⎕←Y←1⌽⍳N
 1 0 2                            1 2 0
```

Our first task is to determine x' and y', which give the action of x and y on Y. Now the images of the rows of S under X are given by the rows of

```
      X[S]
  1 0
  1 2
  0 2
```

Unfortunately, the matrix X[S] is not just S with its rows permuted. The entries of the first row have also been permuted. This illustrates the problems that can arise when we list the elements of sets instead of using characteristic vectors. The matrix

```
      ⎕←T←N SCHV S
 1 1 0
 1 0 1
 0 1 1
```

gives the characteristic vectors for the elements in Y. If we let

```
      ⎕←U←N SCHV X[S]
 1 1 0
 0 1 1
 1 0 1
```

then U gives the characteristic vectors for the images of the elements of Y under x. The matrix U has the form T[X1;] for a uniquely determined permutation X1. To obtain X1 explicitly, we use a trick. Let

```
      ⎕←B←2⊥⍉T
6 5 3
```

Then B[I] is the integer whose binary digits are given by T[I;], and B[I] uniquely determines I. Therefore, if

```
      ⎕←C←2⊥⍉U
6 3 5
```

then C[I] must be B[X1[I]]. But, by the definition of the dyadic operation ⍳, this means that

```
      ⎕←X1←B⍳C
0 2 1
```

To compute the vector Y1 corresponding to y' on Y, we calculate

```
      ⎕←V←N SCHV Y[S]
 0 1 1
 1 1 0
 1 0 1
```

```
      ⎕←D←2⊥⍉V
3 6 5
      ⎕←Y1←B⍳D
2 0 1
```

The elements of Z are subsets of the rows of S. We may identify these with the subsets of ⍳1↑⍴*S*, the index set for the rows of *S*. The characteristic vectors of the elements of *Z* are therefore given by the rows of

```
      ⎕←R←⍉(3⍴2)⊤⍳8
0 0 0
0 0 1
0 1 0
0 1 1
1 0 0
1 0 1
1 1 0
1 1 1
```

To describe $\bar{x}$, we need to determine the vector *XB* such that *R*[*XB*[*I*];] is the characteristic vector for the image under *X*1 of the set whose characteristic vector is *R*[*I*;].

LEMMA 3. Let *Q* be the characteristic vector for the subset of ⍳*M* listed in the vector *A*. If *P* is a permutation of ⍳*M*, then *Q*[⍋*P*] is the characteristic vector for *P*[*A*].

Proof. Let *QP*←*Q*[⍋*P*]. The *I*th component of *QP* is *Q*[*J*], where *J* is (⍋*P*)[*I*]. Now ⍋*P* is the inverse of *P*, so *I* is *P*[*J*]. Since *Q*[*J*] is 1 if and only if *J* is in *A*, it follows that *QP*[*I*] is one if and only if *I* is the image under *P* of an element *J* in *A*, that is, if and only if *I* is in *P*[*A*]. □

By Lemma 3, the matrix

```
      ⎕←RX←R[;⍋X1]
0 0 0
0 1 0
0 0 1
0 1 1
1 0 0
1 1 0
1 0 1
1 1 1
```

is such that *RX*[*I*;] is the characteristic vector for the image under *X*1 of the set whose characteristic vector is *R*[*I*;]. Since we constructed *R* so that *R*[*I*;] is the vector of binary digits for *I*, we can find out which row of *R* is equal to *RX*[*I*;] by considering *RX*[*I*;] to be a number in binary notation. Thus

```
      ⎕←XB←2⊥⍉RX
0 2 1 3 4 6 5 7
```

Note that XB does give the permutation $\bar{x}$ that we computed previously. We can also perform the check

```
      ∧/,RX=R[XB;]
1
```

To compute the vector YB corresponding to $\bar{y}$, we form

```
      ⎕←YB←2⊥⍉R[;⍋Y1]
0 2 4 6 1 3 5 7
```

Again the result checks with our earlier calculation.

All that remains is to compute the orbits of the group H generated by XB and YB. To do this, we construct a matrix whose rows are our generators and apply $GPALLORB$.

```
      ⎕←P←(2,ρXB)ρXB,YB
 0 2 1 3 4 6 5 7
 0 2 4 6 1 3 5 7
      GPALLORB P
 0 1 3 7
 1 3 3 1
```

The matrix returned by $GPALLORB$ indicates that H has four orbits with representatives 0, 1, 3, and 7 and that the lengths of these orbits are 1, 3, 3, and 1, respectively. All of this is in agreement with our previous computation.

This same calculation can be repeated with N=4 and with a moderately large workspace for N=5 as well. This and other similar investigations are left as exercises. Exercise 10.23 describes a method for computing the number of nonisomorphic graphs with n vertices when $n > 5$.

EXERCISES

1 Use the method presented in the text to determine up to isomorphism the undirected graphs with four vertices. Draw a diagram for one graph of each isomorphism type.

2 Do Exercise 1 for the graphs with five vertices.

3 Let n, X, Y, Z, and H be as in the discussion preceding Theorem 1. Let E be in an orbit of H on Z that contains m elements. Show that the order of the automorphism group of the graph (X, E) is

$n!/m$. Determine the orders of the automorphism groups of the graphs diagrammed as part of Exercises 1 and 2.

*4 Find generating elements for Aut(G) for the representative graphs G obtained in Exercises 1 and 2.

5 Show how the methods of this section can be modified to construct the nonisomorphic graphs with a specified number of vertices and a specified number of edges.

6 Suggest ways of reducing the amount of work required to determine the graphs with six vertices. For example, describe the graphs with six vertices in which each vertex is connected to at most two other vertices.

7 Let U be the set of all 3-element subsets of ⍳6 and let V be the set of 3-element subsets of U. We have homomorphisms $\Sigma_6 \longrightarrow \Sigma(U)$ and $\Sigma(U) \longrightarrow \Sigma(V)$. Let H be the image of Σ_6 under the composition of these maps. Determine representatives for the orbits of H.

10. CONJUGACY

The relations of right and left congruence modulo a subgroup are examples of equivalence relations defined on the elements of a group. We will now describe another important equivalence relation associated with any group.

Let G be a group and let x and y be elements of G. We say that x and y are *conjugate* in G if there is an element z of G such that $y = z^{-1}xz$ or, equivalently, such that $xz = zy$.

THEOREM 1. In any group G conjugacy is an equivalence relation.

Proof. For any x in G we have $x = 1^{-1}x1$, and so conjugacy is reflexive. If $y = z^{-1}xz$, then $x = zyz^{-1} = (z^{-1})^{-1}yz^{-1}$, so we also have symmetry. Finally, if $y = u^{-1}xu$ and $z = v^{-1}yv$, then

$$z = v^{-1}u^{-1}xuv = (uv)^{-1}x(uv),$$

proving that conjugacy is transitive. □

The equivalence classes of the conjugacy relation are called *conjugacy classes.* If x and z are in G, then $y = z^{-1}xz$ is called the *conjugate* of x by z. Note that $y = x$ if and only if x and z commute.

Suppose `G` is an `N`-by-`N` group table. Then two integers `I` and `J` are conjugate in (⍳`N`, `G`) if and only if `G[I;K]=G[K;J]` for some `K` in ⍳`N`. This corresponds to the proposition `∨/G[I;]=G[;J]`, which can also be written as `G[I;]∨.=G[;J]`. Thus the characteristic matrix for conjugacy in (⍳`N`, `G`) is `G∨.=G`.

```
      E24←G24∨.=G24
      SEQREL E24
1
```

Here we have verified that conjugacy is an equivalence relation in (⍳24, G24).

There is an important connection between the conjugacy classes of a group G and the isomorphisms of G onto itself. Such an isomorphism is called an *automorphism* of G, and the set of all automorphisms of G is denoted Aut(G).

THEOREM 2. For any group G the set Aut(G) is a subgroup of $\Sigma(G)$.

Proof. By the definition of isomorphism, any automorphism of G is a permutation of G and thus is an element of $\Sigma(G)$. The identity map on G is an automorphism of G, so Aut(G) is nonempty. Finally, by Theorem 4.3, we know that Aut(G) is closed under composition and inverses, so Aut(G) is therefore a subgroup of $\Sigma(G)$. □

If z is an element of the group G, we will denote by τ_z the map of G into itself, taking x in G to $z^{-1}xz$. Thus x and y are conjugate if and only if $y = x\tau_z$ for some z in G.

THEOREM 3. For each z in G the map τ_z is an automorphism of G. Moreover, τ is a homomorphism of G into Aut(G).

Proof. We must first show that τ_z is in Aut(G). Suppose x and y are in G. Then

$$(xy)\tau_z = z^{-1}xyz = z^{-1}xzz^{-1}yz = (x\tau_z)(y\tau_z)$$

and hence τ_z is a homomorphism of G into itself. If $1 = z^{-1}xz = x\tau_z$, then $x = zz^{-1} = 1$ and τ_z is therefore injective. Finally, for any x in G, we have $x = z^{-1}zxz^{-1}z = (zxz^{-1})\tau_z$. Thus τ_z is surjective and τ_z is in Aut(G). Now, for any x, z, and w in G, we have

$$x(\tau_z\tau_w) = (x\tau_z)\tau_w = w^{-1}z^{-1}xzw = (zw)^{-1}x(zw) = x\tau_{zw}.$$

Therefore $\tau_z\tau_w = \tau_{zw}$, and τ is a homomorphism of G into Aut(G). □

The kernel of τ is called the *center* of G and is denoted $Z(G)$. Clearly, Z(G) is the set of elements in G that commute with every element of G. The image in Aut(G) of G under τ is called the group of *inner automorphisms* of G and is written Inn(G). By the First Isomorphism Theorem, Inn(G) is isomorphic to $G/Z(G)$. It is not hard to show that Inn(G) is a normal subgroup of Aut(G). (See Exercise 33.) The quotient group Aut(G)/Inn(G) is called the *outer automorphism group* of G. The term "outer automorphism" has two common meanings and may refer either to an element of Aut(G)/Inn(G) or to an element of Aut(G)–Inn(G).

Two elements x and y of a group G are conjugate in G if and only if they are in the same orbit of Inn(G). We define the *centralizer* $C_G(x)$ of x in G to be the set of elements in G that commute with x. It is easily shown that $C_G(x)$ is a subgroup of G. (See Exercise 3.8.)

THEOREM 4. If x is an element of the finite group G, then the number of conjugates of x in G is $|G:C_G(x)|$.

Proof. The number of conjugates of x is the number of elements in the orbit of $\text{Inn}(G) = G\tau$ that contains x. Since $C_G(x) = \{z \in G | x = x\tau_z\}$, the theorem follows from Theorem 8.8. □

Let us turn now to the problem of determining the conjugacy classes of $\Sigma(X)$, where X is a finite set.

LEMMA 5. Let g and h be elements of $\Sigma(X)$ and suppose $(x_1, \ldots, x_r)$ is a cycle of h. Then $(x_1 g, \ldots, x_r g)$ is a cycle of $g^{-1}hg$.

Proof. This is equivalent to saying that if h takes x to y, then $g^{-1}hg$ takes xg to yg. But, if $xh = y$, then

$$(xg)(g^{-1}hg) = x(gg^{-1}hg) = x(hg) = (xh)g = yg. \quad \square$$

THEOREM 6. Let X be a finite set. Two elements of $\Sigma(X)$ are conjugate if and only if they have the same number of cycles of each length.

Proof. Let g and h be in $\Sigma(X)$. If the cycles of h are

$$(x_1, \ldots, x_r)(y_1, \ldots, y_s)(z_1, \ldots, z_t) \ldots,$$

then, by Lemma 5, the cycles of $g^{-1}hg$ are

$$(x_1 g, \ldots, x_r g)(y_1 g, \ldots, y_s g)(z_1 g, \ldots, z_t g) \ldots,$$

so h and $g^{-1}hg$ have the same number of cycles of each length. Now suppose h' is an element of $\Sigma(X)$ with cycles

$$(x'_1, \ldots, x'_r)(y'_1, \ldots, y'_s)(z'_1, \ldots, z'_t) \ldots.$$

Then the symbol

$$\begin{pmatrix} x_1\, x_2 \ldots x_r\, y_1 \ldots y_s\, z_1 \ldots z_t \ldots \\ x'_1\, y'_2 \ldots x'_r\, y'_1 \ldots y'_s\, z'_1 \ldots z'_t \ldots \end{pmatrix}$$

represents an element g of $\Sigma(X)$ such that $h' = g^{-1}hg$. □

A *partition* of the positive integer n is a sequence $m_1, \ldots, m_r$ of positive integers such that $m_1 \leq m_2 \leq \ldots \leq m_r$ and $n = m_1 + m_2 + \ldots + m_r$.

COROLLARY 7. The number of conjugacy classes of Σ_n is equal to the number of partitions of n.

Proof. The lengths $m_1, \ldots, m_r$ of the cycles of an element of Σ_n form a partition of n (when arranged in increasing order) and, by Theorem

6, two elements of Σ_n are conjugate in Σ_n if and only if the corresponding partitions of n are the same. □

We can now state a simple condition for a permutation in Σ_n to be even.

THEOREM 8. An element g of Σ_n, $n \geq 2$, is even if and only if g has an even number of cycles of even length.

Proof. Let U be the group $(\{1, -1\}, \times)$ and let $\text{sgn}: \Sigma_n \longrightarrow U$ be the surjective homomorphism defined in Section 8. If f and g are in Σ_n, then

$$\text{sgn}(f^{-1}gf) = \text{sgn}(f)^{-1}\,\text{sgn}(g)\text{sgn}(f) = \text{sgn}(g),$$

since U is abelian. Thus $\text{sgn}(g)$ depends only on the conjugacy class of g and hence only on the cycle structure of g. As we saw earlier, the transposition $(0,1)$ is odd and, therefore, so is every transposition. Since the transpositions generate Σ_n, an element of Σ_n is odd or even according to whether it can be written as a product of an odd or an even number of transpositions. Now

$$(x_1, \ldots, x_r) = (x_1, x_2)(x_1, x_3) \ldots (x_1, x_r).$$

Therefore an r-cycle is the product of $r-1$ transpositions, so cycles of even length are odd permutations and cycles of odd length are even permutations. Thus an element of Σ_n is even if and only if it has an even number of cycles of even length. □

Suppose N is a normal subgroup of a group G. If x is in N and g is in G, then the conjugate $g^{-1}xg$ of x is in $g^{-1}Ng = N$. Thus N is a union of conjugacy classes. We say G is *simple* if $G \neq \{1\}$ and the only normal subgroups of G are G and $\{1\}$. It is not hard to show (see Exercise 1) that G is simple if and only if G is nontrivial and $G = \langle C \rangle$ for every conjugacy class $C \neq \{1\}$. Since the kernel of a homomorphism is a normal subgroup, a homomorphism from a simple group into any other group is either injective or trivial.

Let G be a nontrivial finite group. Among the proper normal subgroups of G, let N be one of largest order. By Theorem 5.10, the quotient group G/N has no proper nontrivial normal subgroups, so G/N is simple. If N is nontrivial, then we can find a normal subgroup M of N such that N/M is simple. This process can be repeated until the trivial subgroup of G is reached. Thus G gives rise to one or more finite simple groups from which G is in some sense built up. Therefore we can consider the finite simple groups as the basic building blocks from which all finite groups are constructed. For this reason the determination of the finite simple groups has been one of the central problems in the theory of finite groups for nearly 100 years. This project has recently been completed.

Abelian simple groups are easy to describe. They are the cyclic groups of prime order. There are many infinite families of finite, nonabelian simple groups. The exercises contain an outline of a proof that the alternating groups A_n with $n \geq 5$ are nonabelian simple groups. In addition, there are 26 finite, simple groups that do not fit nicely into any infinite family. The existence of many of these groups is difficult to establish, and machine computation was used to prove that some of them exist. The paper by Gorenstein in the bibliography gives an overview of the classification of the finite simple groups. In particular, Chapter II of that paper describes all of the finite, nonabelian simple groups.

So far we have talked only about conjugacy of elements in a group. We can extend the notion to subgroups quite easily. Two subgroups H and K of a group G are *conjugate* if there is an element g of G such that $K = g^{-1}Hg$. It is not difficult to show that conjugacy is an equivalence relation on the set of subgroups of G and that if G is finite, then the number of conjugates of a given subgroup H is the index in G of the stabilizer $N_G(H) = \{g \in G | g^{-1}Hg = H\}$. We call $N_G(H)$ the *normalizer* of H in G, since $N_G(H)$ is the largest subgroup of G in which H is normal. Thus $N_G(H)$ consists of the elements g of G for which the right coset Hg is the same as the left coset gH. An element or a subgroup of $N_G(H)$ is said to *normalize H*.

Repeated computation of centralizers and normalizers is one method for discovering subgroups in a given group. For example, in (⍳60, G60), the element 1 has order 2.

```
      GTINIT G60
      GTSGP 1
0 1
```

The centralizer H of 1 has order 4.

```
      ⎕←H←(G60[1;]=G60[;1])/⍳60
0 1 56 59
```

To construct the normalizer K of H, we first find the characteristic matrices for right and left congruence modulo H

```
      R←GTRCON H
      L←GTLCON H
```

and then obtain K as the set of group elements I for which the right coset of H containing I is the same as the left coset of H containing I.

```
      ⎕←K←(∧/R=L)/⍳60
0 1 22 23 29 30 31 32 36 37 56 59
```

Since K has order 12, the number of conjugates of H in (ι60, G60) is 5.

The following theorem summarizes some elementary facts concerning automorphisms.

THEOREM 9. Let τ be an automorphism of the group G. If x is in G, then x and $x\tau$ have the same order. If H is a subgroup of G, then H and $H\tau$ are isomorphic. If $H \lhd G$, then $H\tau \lhd G$ and $G/(H\tau)$ is isomorphic to G/H.

Proof. Since τ fixes 1 and $(x^n)\tau = (x\tau)^n$, it follows that $x^n = 1$ if and only if $(x\tau)^n = 1$. Thus x and $x\tau$ have the same order. The restriction $\tau|_H$ of τ to H is an isomorphism of H onto $H\tau$. Finally, if $H \lhd G$, let $\pi: G \longrightarrow G/H$ be the natural map and let $\sigma = \tau^{-1} \circ \pi$. Then σ is a homomorphism of G onto G/H. The element g of G is in the kernel of π if and only if $g\tau^{-1}$ is in the kernel H of π or, equivalently, g is in $H\tau$. Thus the kernel of σ is $H\tau$. Therefore $H\tau \lhd G$ and, by the First Isomorphism Theorem,

$$G/(H\tau) \cong G/H. \quad \square$$

COROLLARY 10. Conjugate elements of a group have the same order and conjugate subgroups are isomorphic.

Proof. Since conjugation by a group element is an automorphism, the corollary is an immediate consequence of Theorem 9. □

We close this section with an application of Theorem 4. If p is a prime, then a *p-group* is a finite group whose order is a power of p.

THEOREM 11. If G is a nontrivial p-group, then $Z(G)$ is nontrivial.

Proof. Let $C_1, \ldots, C_r$ be the conjugacy classes of G and suppose the numbering has been chosen so that $C_1 = \{1\}$. Set $c_i = |C_i|$. Then $c_1 = 1$ and $c_1 + \ldots + c_r = |G| = p^m$ for some integer $m \geq 1$. Thus

$$p^m - 1 = c_2 + \ldots + c_r.$$

By Theorem 4, each c_i divides $|G|$ and thus each c_i is a power of p. Since $m \geq 1$, p does not divide $p^m - 1$. Therefore some c_i with $i \geq 2$ is not divisible by p. However, the only power of p that is not divisible by p is $p^0 = 1$. Thus $c_i = 1$ and $C_i = \{x\}$, where $x \neq 1$. But this means that $C_G(x) = G$, so x is in the center of G. Hence $Z(G) \neq 1$. □

EXERCISES

1 Show that a group G is simple if and only if $G \neq \{1\}$ and every conjugacy class other than $\{1\}$ generates G.

2 Prove that the center of Σ_n is $\{1\}$ for $n > 2$.

3 Let H be a subgroup of G and suppose $C = C_G(H)$ and $N = N_G(H)$.

Prove that N/C is isomorphic to a subgroup of Aut(H). (See Exercise 5.5.)

*4 Suppose G is a finite group with more than two elements. Show that Aut$(G) \neq 1$.

5 Determine the conjugacy classes in $\Sigma_n, n \leq 7$.

*6 Let G be a group and let $R: G \longrightarrow \Sigma(G)$ be the right regular representation of G. Let N be the normalizer in $\Sigma(G)$ of the image of G under R. Show that the stabilizer N_1 in N of the identity element 1 of G is Aut(G). (The group N is often called the *holomorph* of G.)

7 Determine the automorphism groups of the groups $\mathbf{Z}_3$, $\mathbf{Z}_2 \times \mathbf{Z}_2$, $\mathbf{Z}_4$, and $\mathbf{Z}_5$.

8 Show that every automorphism of Σ_3 is an inner automorphism.

9 Prove that Aut$(\mathbf{Z}_n)$ is isomorphic to U_n.

10 Let g be an element of Σ_n with k_1 cycles of length n_1, k_2 cycles of length n_2, ... Show that the number of conjugates of g in Σ_n is

$$\frac{n!}{n_1^{k_1} n_2^{k_2} \ldots k_1! k_2! \ldots}.$$

11 Let x by an n-cycle in Σ_n. Prove that the centralizer of x in Σ_n is $\langle x \rangle$.

12 Is it possible to find elements x and y in A_n such that x and y are conjugate in Σ_n but are not conjugate in A_n?

13 Suppose G is a group and $G/Z(G)$ is cyclic. Show that G is abelian.

14 Prove that if p is a prime, then any group of order p^2 is abelian.

15 Fill in the details of the proof of Theorem 9.

16 Let N be a normal subgroup of the group G. By Theorem 5.10, there is a 1–1 correspondence between the subgroups of G containing N and the subgroups of G/N. Suppose a subgroup K of G corresponds to a subgroup L of G/N. Show that $N_G(K)$ corresponds to $N_{G/H}(L)$.

17 Let H be a proper subgroup of the p-group G. Prove that $N_G(H)$ contains H properly. [*Hint.* Consider the two cases $Z(G)$ contained in H and $Z(G)$ not contained in H.]

18 Let H and K be groups and let $\sigma: H \longrightarrow$ Aut(K) be a homomorphism. Denote the image of h under σ by σ_h. Show that the set $H \times K$

with the product$(h_1, k_1)(h_2, k_2) = (h_1h_2, (k_1\sigma_{h_2})k_2)$ is a group. (Such a group is called a *semidirect product* of H and K.)

19 Determine all possible semidirect products of H and K where

(a) $H = \mathbf{Z}_3, K = \mathbf{Z}_2 \times \mathbf{Z}_2$.
(b) $H = \mathbf{Z}_2 \times \mathbf{Z}_2, K = \mathbf{Z}_3$.
(c) $H = \mathbf{Z}_4, K = \mathbf{Z}_3$.

(See Exercise 7.)

20 How many nonisomorphic groups of order 12 arise in Exercise 19?

21 Using the matrices `G24∨.=G24` and `G60∨.=G60` and the techniques of Exercise 3.35, determine sets of representatives for the conjugacy classes in the groups `(ι24, G24)` and `(ι60, G60)`. How many elements are in each class? Show that `(ι60, G60)` is a simple group.

22 Let X and G be as in Exercise 8.23 and let χ be the permutation character of G. Show that if g and h are conjugate elements in G, then $\chi(g) = \chi(h)$. Suppose $g_1, \ldots, g_r$ are representatives for the conjugacy classes of G and $c_i = |C_G(g_i)|$. Prove that the number of orbits of G is

$$\sum_{i=1}^{r} \frac{\chi(g_i)}{c_i} .$$

23 Let $X = \{0, \ldots, n-1\}$, let Y be the set of two-element subsets of X, and let Z be the set of all subsets of Y. For g in Σ_n, let $\bar{g}$ be the element of $\Sigma(Y)$ induced by g and let $\hat{g}$ be the element of $\Sigma(Z)$ induced by $\bar{g}$. For $n \leq 7$, choose representatives $g_1, \ldots, g_r$ for the conjugacy classes in Σ_n. (See Exercise 5.) Determine the cycle structure of $\bar{g}_i$ and, from this, the value of the permutation character $\chi(\hat{g}_i)$. Using Theorem 9.2 and Exercise 22, compute the number of nonisomorphic graphs with n vertices.

The next four exercises outline a proof that A_n is simple for $n \geq 3, n \neq 4$.

24 Show that A_n is generated by 3-cycles. (It suffices to show that a product of any two transpositions can be written as a product of 3-cycles. Why?)

25 Prove that if N is a normal subgroup of A_n and N contains a 3-cycle, then $N = A_n$.

26 Show that $A_3 \cong \mathbf{Z}_3$ and that A_4 is not simple.

27 Assume that $n \geq 5$ and that N is a nontrivial normal subgroup of A_n. Let g be a nonidentity element in N. By raising g to a power

if necessary, we may assume g has prime order p so that all cycles of g have length p or 1. Prove:

(a) If g fixes two points x and y and moves a third point z, then $u^{-1}g^{-1}ug$ is a 3-cycle in N, where $u = (x, y, z)$.

(b) If $p \geq 5$ and $(x_1, x_2, \ldots, x_p)$ is a cycle of g, then $g^{-1}u^{-1}gu$ is a 3-cycle in N, where $u = (x_1, x_2, x_3)$.

(c) If $p = 3$ and case (a) does not apply, then g has at least two 3-cycles (x_1, x_2, x_3) and (y_1, y_2, y_3) and $g^{-1}u^{-1}gu$ is a 5-cycle in N, where $u = (x_1, x_2, y_1)$.

(d) If $p = 2$, then g has at least two 2-cycles and either a fixed point or two more 2-cycles. If $g = (x)(y_1, y_2)(z_1, z_2)\ldots$, then $g^{-1}u^{-1}gu$ is a 5-cycle in N, where $u = (x, y_1, z_1)$. If $g = (x_1, x_2)(y_1, y_2)(z_1, z_2)\ldots$, then $u^{-1}g^{-1}ug$ is a nonidentity element of N fixing at least two points, where

$$u = (x_1, x_2, y_1).$$

Now conclude that N must contain a 3-cycle and $N = A_n$ by Exercise 25.

28 Let G be an N-by-N group table. Write an APL expression for an N-by-N matrix C such that $C[I;J]$ is the conjugate of I by J in $(\iota N, G)$. Assume that the vector INV lists the inverses of the group elements.

29 Show that an abelian group is simple if and only if it is cyclic of prime order.

*30 Prove that $(\iota 60, G60)$ has an outer automorphism.

31 Let G be a permutation group on the set X and let x and y be two elements of X in the same orbit. Show that the stabilizers G_x and G_y are conjugate in G.

32 Let H and K be conjugate subgroups of the group G. Show that $C_G(H)$ is conjugate to $C_G(K)$ and $N_G(H)$ is conjugate to $N_G(K)$.

33 Show that $\mathrm{Inn}(G) \lhd \mathrm{Aut}(G)$ for any group G.

34 Determine all subgroups of Σ_4. Into how many conjugacy classes do they fall?

11. THE SYLOW THEOREMS

Lagrange's Theorem (Theorem 3.9) states that if G is a finite group of order n and m is the order of some subgroup of G, then m divides n. The converse of this result is false. It is not generally true that if m is a positive divisor of n, then G has a subgroup of order m. However, the theorems that will be proved in this section show that if we assume in addition that m is a power of a prime p, then G does, in fact, have a subgroup of order m. Moreover, if m is the largest power of p dividing n, then all subgroups of G of order m are conjugate in G. A *p-subgroup* of G is a subgroup whose order is a power of p. Our existence theorem on p-subgroups depends on the following number-theoretic lemma.

LEMMA 1. Let n be a positive integer and let p be a prime. If p^a divides n, then the largest power of p dividing the binomial coefficient $\binom{n}{p^a}$ is the same as the highest power of p dividing n/p^a.

Proof. Since

$$\binom{n}{p^a} = \left(\frac{n}{p^a}\right)\left(\frac{n-1}{p^a-1}\right)\left(\frac{n-2}{p^a-2}\right)\cdots\left(\frac{n-p^a+1}{1}\right),$$

it suffices to prove that for any integer i, $1 \le i < p^a$, the power of p in $n - i$ is the same as the power of p in $p^a - i$. Suppose p^b divides $n - i$. If $b \ge a$, then p^a divides $n - i$ and, since p^a divides n, this means that p^a divides i, contradicting our assumption that $1 \le i < p^a$. Thus $b < a$. Hence p^b divides i. Therefore p^b divides $p^a - i$. Essentially the same argument shows that if p^b divides $p^a - i$, then p^b divides i and p^a and so divides $n - i$. □

THEOREM 2. Let G be a group of order n and let p be a prime. If p^a divides n, then G has a subgroup of order p^a.

Proof. Let $\mathcal{P}$ denote the set of subsets of G with p^a elements. If g is is in G and A is in $\mathcal{P}$, then Ag is in $\mathcal{P}$. We can define a homomorphism $\sigma: G \longrightarrow \Sigma(\mathcal{P})$ by $A(g\sigma) = Ag$. The number of elements in $\mathcal{P}$ is $\binom{n}{p^a}$. If p^b is the highest power of p in n, then by Lemma 1 we know that p^{b-a} is the highest power of p in $|\mathcal{P}|$. Now $\mathcal{P}$ is the union of the orbits of $G\sigma$, so there must be an orbit $\mathcal{O}$ such that $|\mathcal{O}|$ is not divisible by p^{b-a+1}. Let A be an element of $\mathcal{O}$ and let H be the stabilizer of A in G; that is,

$$H = \{g \in G | Ag = A\}.$$

Then $|G| = |\mathcal{O}|\,|H|$ and p^a divides $|H|$. Hence $|H| \ge p^a$. Fix x in A. If g is in H, then xg is in A and g is in $x^{-1}A$. Thus $|H| \le |x^{-1}A| = p^a$. Therefore $|H| = p^a$, and we have found a subgroup of G of order p^a. □

A *Sylow p-subgroup* of a finite group G is a p-subgroup of G whose order is the largest power of p dividing $|G|$. [Ludwig Sylow (1832-1918) was a Norwegian mathematician.]

COROLLARY 3 *(First Sylow Theorem)*. If G is a finite group and p is a prime, then G possesses a Sylow p-subgroup. □

In order to prove the remaining Sylow theorems (Theorems 6 and 7 that follow), we will need two lemmas. If H and K are subgroups of a group G, then the set HK may or may not be a subgroup of G. The following lemmas give a formula for $|HK|$ and state a sufficient condition for HK to be a group.

LEMMA 4. Let H and K be subgroups of the finite group G. Then

$$|HK| = \frac{|H| \times |K|}{|H \cap K|}.$$

Proof. The set HK is a union of right cosets of H of the form Hx with x in K. If x and y are in K, when does $Hx = Hy$? We know that $Hx = Hy$ if and only if xy^{-1} is in H, which is equivalent to xy^{-1} being in $H \cap K$, since xy^{-1} is always in K. Thus $Hx = Hy$ if and only if $(H \cap K)x = (H \cap K)y$. Therefore the number of right cosets of H in HK is the same as the number of right cosets of $H \cap K$ in K, which is $|K : H \cap K|$ or $|K|/|H \cap K|$. Every right coset of H has $|H|$ elements, and the formula now follows immediately. □

LEMMA 5. Let H and K be subgroups of a group G and suppose H is contained in $N_G(K)$. Then $HK = KH$ and HK is a subgroup of G.

Proof. Since $K \lhd N_G(K)$, KH is a subgroup by Theorem 5.12a. If h is in H, then $Kh = hK$, so $KH = HK$. □

THEOREM 6 ***(Second Sylow Theorem).*** The number of conjugates of a Sylow p-subgroup of a finite group G is congruent to 1 modulo p.

Proof. If p does not divide $|G|$, then G has exactly one Sylow p-subgroup, the trivial subgroup. Thus we may assume p divides $|G|$. Let H be a Sylow p-subgroup of G and let S denote the set of conjugates of H in G. All elements of S are Sylow p-subgroups of G. Let $\tau: H \longrightarrow \Sigma(S)$ be the homomorphism defined by $K(h\tau) = h^{-1}Kh$ for all K in S and all h in H. By Theorem 8.8, all orbits of $H\tau$ on S have lengths that are powers of p. Suppose $\{K\}$ is an orbit of $H\tau$ of length l. Then, for all h in H, we have $h^{-1}Kh = K$ and so, by Lemmas 5 and 6, the set HK is a subgroup of G of order p^{2a-b}, where $|H| = |K| = p^a$ and $|H \cap K| = p^b$. Since $|HK|$ must divide $|G|$, we have $2a - b \leq a$ or $a \leq b$ but, since $H \cap K$ is a subgroup of H, we have $b \leq a$. Thus $a = b$ and $H \cap K = H$, that is, $H = K$. Thus S consists of $\{H\}$ and a number of orbits of $H\tau$ all of whose lengths are divisible by p. Thus $|S|$ is congruent to 1 modulo p. □

THEOREM 7 ***(Third Sylow Theorem).*** Let H be a Sylow p-subgroup of the finite group G and let K be any p-subgroup of G. Then K is contained in a conjugate of H. In particular, all Sylow p-subgroups of G are conjugate.

Proof. Let S denote the set of conjugates of H in G. By Theorem 6 we know that $|S| \equiv 1 \pmod p$. Now K induces a group of permutations on S by conjugation. All orbits of this group have lengths that are powers of p. Since p does not divide $|S|$, there must be an orbit $\{L\}$ with one element. Then K normalizes L and, as in the proof of Theorem 6, the set KL is a p-subgroup of G that contains L. This means that $KL = L$ and $K \subseteq L$. If K happens to be a Sylow p-subgroup of G, then $|K| = |L|$, so $K = L$. Thus K is conjugate to H. □

We close this section with a proof that the congruence $(p - 1)! \equiv -1 \pmod p$ holds for any prime p. This result in number theory is known as Wilson's Theorem. [John Wilson (1741-1793) was an English mathematician.]

THEOREM 8. Let G be a finite abelian group written multiplicatively and let

$$z = \prod_{g \in G} g,$$

the product of the elements of G. Then $z = 1$ unless G has exactly one element of order 2, and, in this case, z is that element.

Proof. Since G is abelian, the set H of elements x in G with $x^2 = 1$ is a subgroup of G. By Theorem 2, the order of H is a power of 2. If y is in $G - H$, then $y \neq y^{-1}$, and y and y^{-1} occur as distinct factors in the product defining z and cancel each other. Thus z is the product of the elements of H. If $H = \{1\}$, then $z = 1$, and if $H = \{1, x\}$, then $z = x$. Thus we may assume 4 divides $|H|$. Let x be any nonidentity element of H and set $K = \langle x \rangle = \{1, x\}$. The product of the elements in any coset $Ky = \{y, xy\}$ of K in H is x, since $y^2 = 1$. Thus $z = x^m$, where $m = |H:K|$ is even. Therefore $z = 1$. □

COROLLARY 9 ***(Wilson's Theorem).*** If p is a prime, then $(p - 1)! \equiv -1 \pmod p$.

Proof. If $p = 2$, we have the assertion $1 \equiv -1 \pmod 2$, which is certainly true. Thus we may assume $p \geq 3$. The product of the elements of the abelian group U_p is $[1]\ [2]\ \ldots\ [p - 1] = [(p - 1)!]$. By Theorem 8, we need only show that $[-1]$ is the unique element of order 2 in U_p. Clearly, $[-1]$ has order 2. If $[x]$ has order 2, then $x^2 \equiv 1 \pmod p$ and p divides $x^2 - 1 = (x + 1)(x - 1)$. Therefore p divides $x + 1$ or $x - 1$ and hence $x \equiv \pm 1 \pmod p$. Since $[1]$ has order 1, $[-1]$ is the only element of order 2 in U_p. □

The following calculation illustrates Wilson's Theorem.

```
      11|!10
10
```

EXERCISES

1 Determine the number of Sylow p-subgroups of Σ_3 for $p = 2$ and 3.

2 Find a Sylow 2-subgroup and a Sylow 3-subgroup in Σ_4 and determine the number of conjugates of each.

3 Find one Sylow p-subgroup in (ι60, G60) for $p = 2, 3,$ and 5 and determine the number of conjugates of each group.

4 Let H and K be subgroups of a group G. Show HK is a group if and only if $HK = KH$.

5 Let G be a group of order 15. Prove that a Sylow 3-subgroup and a Sylow 5-subgroup of G are normal and conclude that $G \cong \mathbf{Z}_{15}$. Give examples of other composite integers n for which every group of order n is cyclic.

4
RINGS

In our study of groups we referred to the example $(\mathbf{Z},+)$ quite often. Besides the binary operation +, the set of integers has another binary operation, multiplication. The operations + and $\times$ on **Z** are closely related. For example, for all x, y, z, in **Z**, the familiar distributive law $x \times (y + z) = x \times y + x \times z$ holds. In this chapter we will study sets with two binary operations that possess many of the properties that hold for addition and multiplication in **Z**. The APL index origin will normally be 0.

1. DEFINITION AND EXAMPLES

A *ring* is a triple $(R, +, \times)$ consisting of a set R and two binary operations + and $\times$ on R satisfying the following conditions.

1. $(R, +)$ is an abelian group.
2. $(R, \times)$ is a semigroup.
3. The distributive laws

$$x \times (y + z) = x \times y + x \times z$$

$$(x + y) \times z = x \times z + y \times z$$

hold for all $x, y,$ and z in R.

We use additive notation for $(R, +)$ and multiplicative notation for $(R, \times)$, with the symbol $\times$ usually omitted. Thus we will normally write xy for $x \times y$. When the operations + and $\times$ are clear from context, we will refer to R as the ring. However, it must be remembered that a given set may be the set of elements of more than one ring.

The most familiar examples of rings are the sets **Z**, **Q**, **R**, and **C** of integers, rational numbers, real numbers, and complex numbers, respectively, with the usual operations. These rings all have a *multiplicative identity*, an element 1 such that $1x = x1 = x$ for all x in the ring. The set $2\mathbf{Z}$ of even integers is a ring but has no multiplicative identity. In this book we will be concerned almost entirely with rings having multiplicative identities.

We will therefore adopt the convention that the word "ring" will mean "ring with multiplicative identity" unless there is an explicit statement to the contrary.

Although the reader should have some familiarity with complex numbers, it is perhaps useful to review the definition of the ring **C**. An element z of **C** is represented as $a + bi$, where a and b are real numbers and i has the property that $i^2 = -1$. We call a the *real part* and b the *imaginary part* of z. If $w = c + di$, then

$$z + w = (a + c) + (b + d)i,$$
$$zw = (ac - bd) + (ad + bc)i.$$

Thus if $z = 2 - 3i$ and $w = 1 + 2i$, then $z + w = 3 - i$ and $zw = 8 + i$.

The preceding examples of rings are all infinite. Finite rings also exist.

THEOREM 1. For any positive integer n the set $\mathbf{Z}_n$ is a ring.

Proof. In Section 2.3 we defined binary operations + and $\times$ on $\mathbf{Z}_n$, and we have already shown that $(\mathbf{Z}_n, +)$ is an abelian group and that multiplication is associative. Clearly, the congruence class $[1]$ is a multiplicative identity. All that remains is to check distributivity. If x, y, z are in **Z**, then

$$[x] \times ([y] + [z]) = [x] \times [y + z] = [x(y + z)] = [xy + xz] =$$
$$[xy] + [xz] = [x] \times [y] + [x] \times [z].$$

The other distributive law now follows from the commutativity of multiplication in $\mathbf{Z}_n$. Thus $\mathbf{Z}_n$ is a ring. □

Addition in a ring R is commutative by definition. Thus if f is a function from the finite set X to R, then the expression

$$\sum_{x \in X} f(x)$$

is unambiguous, since the order in which the terms are summed does not matter. However, the expression

$$\prod_{x \in X} f(x)$$

may be ambiguous and is not used unless R is known to be a *commutative ring*, one in which multiplication is commutative. We will see some examples of noncommutative rings shortly.

As the next lemma shows, many elementary facts about **Z** and **Q** are true in any ring. However, a property such as $(xy)^n = x^n y^n$ for any positive integer n does not generally hold, since it depends on the commutativity of multiplication.

LEMMA 2. Let x and y be elements of a ring R. Then

(a) $0x = x0 = 0$.
(b) $x(-y) = (-x)y = -(xy)$.
(c) $(-x)(-y) = xy$.
(d) $(-1)x = -x$.

Proof. (a) Since 0 is the identity element of $(R, +)$, we have $0 = 0 + 0$. Thus

$$x0 = x(0 + 0) = x0 + x0.$$

By the cancellation laws in $(R, +)$ it follows that $x0 = 0$. Similarly, $0x = 0$.

(b) By distributivity,

$$xy + x(-y) = x(y - y) = x0 = 0.$$

Thus $x(-y)$ is the additive inverse of xy, that is, $-(xy)$. Similarly, $(-x)y = -(xy)$.

(c) By (b),

$$(-x)(-y) = -(x(-y)) = -(-(xy)) = xy.$$

(d) Again by (b),

$$(-1)x = -(1x) = -x. \quad \square$$

Let us consider some more examples of rings. For the most part, proofs of the assertions made will be left as exercises.

Example 1. If $R = \{x\}$, then defining $x + x$ and $x \times x$ both to be x makes R into a ring in which $1 = 0 = x$. We say R is *trivial.* The ring $\mathbf{Z}_1$ of integers modulo 1 is an example of a trivial ring.

Example 2. The set of polynomials $a_0 + a_1X + a_2X^2 + \ldots + a_nX^n$ with integer coefficients is a ring under the usual operations of polynomial addition and multiplication. This ring is denoted $\mathbf{Z}[X]$. Rings of polynomials will be studied in Section 3.

Example 3. Let $M_2(\mathbf{Z})$ denote the set of 2-by-2 matrices with integer entries. If A and B are in $M_2(\mathbf{Z})$, then $A + B$ is obtained by adding corresponding entries and AB is the usual matrix product. In APL notation the sum of A and B is `A+B` and their product is `A+.×B`. With these binary operations $M_2(\mathbf{Z})$ is a ring. If we define the elements `U` and `V` of $M_2(\mathbf{Z})$ by

```
      ⎕←U←2 2ρ1 ¯2 3 0                ⎕←V←2 2ρ¯2 1 0 2
 1 ¯2                              ¯2  1
 3  0                               0  2
```

then, computing the products `U+.×V` and `V+.×U`,

```
        U+.×V                          V+.×U
¯2 ¯3                            1 4
¯6  3                            6 0
```

we see that $M_2(\mathbf{Z})$ is a noncommutative ring. More general matrix rings will be studied in Section 4.

Example 4. Let $C[0,1]$ be the set of continuous real valued functions on the interval $[0,1]$. For f and g in $C[0,1]$, define $f+g$ and fg by

$$(f+g)(x) = f(x) + g(x),$$
$$(fg)(x) = f(x)g(x).$$

Then $C[0,1]$ is a commutative ring.

Example 5. Let $(R, +, \times)$ be any ring. Define a new binary operation $\star$ on R by $x \star y = y \times x$. Then $(R, +, \star)$ is a ring, called the *opposite ring* of R, and is denoted R^{op}.

Example 6. Small finite rings may be described by giving addition and multiplication tables for them. The workspace *EXAMPLES* contains two 8-by-8 matrices *PLUS* and *TIMES*, which in origin 0 define binary operations on ⍳8.

```
      PLUS                             TIMES
0 1 2 3 4 5 6 7                  0 0 0 0 0 0 0 0
1 0 3 2 5 4 7 6                  0 1 2 3 4 5 6 7
2 3 0 1 6 7 4 5                  0 2 0 2 0 2 0 2
3 2 1 0 7 6 5 4                  0 3 2 1 4 7 6 5
4 5 6 7 0 1 2 3                  0 4 2 6 4 0 6 2
5 4 7 6 1 0 3 2                  0 5 0 5 0 5 0 5
6 7 4 5 2 3 0 1                  0 6 2 4 4 2 6 0
7 6 5 4 3 2 1 0                  0 7 0 7 0 7 0 7
```

The triple (⍳8, *PLUS*, *TIMES*) is a ring.

Most of the computations necessary to check the ring axioms for (⍳8, *PLUS*, *TIMES*) are quite straightforward. For example,

```
      ∧/,PLUS=⍉PLUS                    ∧/PLUS[0;]=⍳8
1                                1
```

shows that addition is commutative and that 0 is an additive identity element and

```
      ∧/,TIMES[TIMES;]=TIMES[;TIMES]
1
      ∧/(TIMES[1;]=ι8),TIMES[;1]=ι8
1
```

shows that multiplication is associative and that 1 is a multiplicative identity.

The distributive laws are a little more complicated. The first distributive law states that $x(y + z) = xy + xz$. To check this in our example, we must verify that

```
TIMES[X;PLUS[Y;Z]] = PLUS[TIMES[X;Y];TIMES[X;Z]]
```

for all *X*, *Y*, and *Z* in ι8. Now

```
TIMES[X;PLUS[Y;Z]]
```

is a typical entry in the array *TIMES*[;*PLUS*] while

```
PLUS[TIMES[X;Y];TIMES[X;Z]]
```

is a typical entry in the generalized transpose

```
0 1 0 2⍉PLUS[TIMES;TIMES]
```

Thus the calculation

```
      ∧/,TIMES[;PLUS]=0 1 0 2⍉PLUS[TIMES;TIMES]
1
```

verifies the first distributive law in (ι8, *PLUS*, *TIMES*). The verification of the remaining axioms is left to the reader.

Example 7. Let R_1 and R_2 be rings. We have defined what is meant by the direct sum $R_1 \oplus R_2$ as an abelian group. If we define $(x_1, y_1) \times (x_2, y_2)$ to be $(x_1 \times x_2, y_1 \times y_2)$, then $R_1 \oplus R_2$ becomes a ring.

Let R be a nontrivial ring, one with $1 \neq 0$. An element u of R is called a *unit* if u has a two-sided inverse v in the monoid $(R, \times)$. By Theorem 3.1.4, v is unique and may be denoted u^{-1}. By the same theorem, the set U of units of R is a group under multiplication. The group of units of $\mathbf{Z}$ is $\{1, -1\}$. In $\mathbf{Q}$ every nonzero element is a unit. The group U_n, $n \geq 2$, defined in Section 3.1, is the group of units in the ring $\mathbf{Z}_n$.

A *division ring* is a ring with $1 \neq 0$ in which every nonzero element is a unit. Noncommutative division rings exist, but we will not see any until Section 5.4. A commutative division ring is called a *field*. The rings $\mathbf{Q}$, $\mathbf{R}$, and $\mathbf{C}$ are fields. Recall that the inverse of the nonzero complex number $a + bi$ is

$$\frac{a}{a^2 + b^2} - \frac{bi}{a^2 + b^2} .$$

If p is a prime, then by Theorem 3.1.5 every nonzero element of $\mathbf{Z}_p$ is a unit. Thus $\mathbf{Z}_p$ is a field. Fields make up the most important class of rings. If a and b are elements of a field and $b \neq 0$, then ab^{-1} is often written $a \div b$ or a/b.

It is possible for the product of two nonzero elements of a ring to be 0. For example, in $\mathbf{Z}_6$ the product of [2] and [3] is $[6] = [0] = 0$. If R is a commutative ring, then an element a of R is called a *zero-divisor* if $a \neq 0$ and for some nonzero element b of R the product ab is 0. An *integral domain* is a commutative ring with $1 \neq 0$ that has no zero-divisors. Every field is an integral domain. The ring $\mathbf{Z}$ is an example of an integral domain that is not a field.

Theorem 3. A finite integral domain is a field.

Proof. Let D be a finite integral domain. For any a in D with $a \neq 0$ we must show $ab = 1$ for some b in D. Define a map $f: D \longrightarrow D$ by $f(x) = ax$. Suppose for some x and y in D we have $f(x) = f(y)$. Then $ax = ay$ or $a(x - y) = 0$. Since a is not a zero-divisor, we must have $x - y = 0$ or $x = y$. Thus f is injective. However, for maps of the finite set D into itself injectivity is equivalent to surjectivity. Thus f is surjective and so, in particular, there is an element b of D such that $f(b) = ab = 1$. Hence D is a field. □

It is also true that any finite division ring is a field. This fact, originally proved by J. H. M. Wedderburn (1882-1948), who was born in Scotland and worked in the United States, is much harder to prove than Theorem 3. An excellent discussion of Wedderburn's Theorem can be found in Herstein.

In the proofs of certain results about determinants that are presented in Section 6 we will need to expand products each factor of which is a sum. A simple example is $(a + b)(c + d)$. If a, b, c, and d are all elements of a ring R, the distributive laws imply that $(a + b)(c + d) = ac + ad + bc + bd$. Similarly, the product

$$(a_1 + a_2 + a_3)(b_1 + b_2 + b_3)(c_1 + c_2 + c_3)$$

can be written as the sum of the 27 terms $a_i b_j c_k$, where $1 \leq i, j, k \leq 3$. The following theorem describes the general situation.

Theorem 4. For $1 \leq i \leq m$ and $1 \leq j \leq n$ let b_{ij} be an element of the ring R and set $a_i = b_1 + \ldots + b_{in}$. Then

$$a_1 a_2 \ldots a_m = \sum_{\rho} b_{1,1\rho} b_{2,2\rho} \ldots b_{m,m\rho},$$

where ρ ranges over all n^m functions from $\{1, \ldots, m\}$ to $\{1, \ldots, n\}$.

Proof. The formula states that $a_1 a_2 \ldots a_m$ is the sum of all possible products $c_1 c_2 \ldots c_m$, where c_i is a summand of a_i. A rigorous proof of this result can be obtained using induction on m and is left as an exercise. □

One final remark is required concerning notation. It is possible to define vectors, matrices, and arrays of higher rank with entries lying in a fixed ring R. If A and B are arrays of this type, then $-A$, $A + B$, $A - B$, and $A \times B$ will denote the results of entry-by-entry calculations in R. In particular, the symbol $\times$ will never be used to denote matrix multiplication.

EXERCISES

1 In each of the Examples 1 to 7 show that all of the ring axioms are satisfied.

2 Determine the group of units for the rings $\mathbf{Z}[X]$ and $\mathbf{C}[0, 1]$.

3 Let R be a ring. Describe the units of the opposite ring R^{op}.

4 What are the units in the ring ($\iota 8$, *PLUS*, *TIMES*)?

5 Let R_1 and R_2 be rings whose groups of units are U_1 and U_2, respectively. Show that the group of units of $R_1 \oplus R_2$ is $U_1 \times U_2$.

6 Is the direct sum of two integral domains an integral domain?

7 Show that $\mathbf{Z}[X]$ is an integral domain.

8 Prove that $\mathbf{Z}_n$ is an integral domain if and only if n is a prime.

9 The matrix

$$A = \begin{bmatrix} 3 & 2 \\ 7 & 5 \end{bmatrix}$$

is a unit in $M_2(\mathbf{Z})$. Find A^{-1}.

10 Let X be a set and R be a ring. Let S be the set of all functions $f{:}X \longrightarrow R$ and define operations of addition and multiplication on S by

$$(f+g)(x) = f(x) + g(x),$$
$$(fg)(x) = f(x)g(x).$$

Show that S is a ring.

11 Prove that the Binomial Theorem holds in any commutative ring R. That is, for all a, b in R and all positive integers n,

$$(a+b)^n = \sum_{m=0}^{n} \binom{n}{m} a^{n-m} b^m.$$

12 Let X be a set and let P be the set of all subsets of X. For A, B in P, define

$$A + B = (A - B) \cup (B - A).$$
$$AB = A \cap B.$$

Show that P is a ring.

13 A *Boolean ring* is a ring R in which $x^2 = x$ for all x in R. Show that every Boolean ring is commutative. The ring in Exercise 12 is a Boolean ring. [George Boole (1815-1864) was an English logician.]

14 Let A be an abelian group. An *endomorphism* of A is a homomorphism from A to itself. Let End(A) be the set of all endomorphisms of A. If f and g are in End(A), define $f+g$ by $a(f+g) = af + ag$ and set $fg = f \circ g$ so that $a(fg) = (af)g$. Show that both $f+g$ and fg are in End(A) and that End(A) is a ring.

15 Let R be a ring and let G be a finite group. The *group ring* $R[G]$ of G over R is the set of all functions from G to R with the following operations. If f and g are in $R[G]$, then

$$(f+g)(x) = f(x) + g(x),$$

$$(fg)(x) = \sum_{yz=x} f(y)g(z).$$

Here the sum in the definition of fg is over all pairs (y, z) in $G \times G$ such that $yz = x$. Show that $R[G]$ is a ring. It is standard practice to identify an element x of G with the characteristic function of the subset $\{x\}$ of G, that is, the element f_x of $R[G]$ such that

$$f_x(y) = \begin{cases} 1, & y = x, \\ 0, & y \neq x. \end{cases}$$

Prove that under this identification G is a subgroup of the group of units of $R[G]$.

16 Let G be the group (⍳6, G6) and let S be the group ring $\mathbf{Z}[G]$. Elements of S may be represented by integer vectors of length 6. Suppose A and B are elements of S. Write an APL expression for the product of A and B in S. What is the product of

A← 2 ¯3 1 0 ¯1 2 and B← ¯1 0 3 4 ¯2 1

in S?

17 Show that in any ring the following generalization of the distributive law holds.

$$\left(\sum_{i=1}^{m} a_i\right)\left(\sum_{j=1}^{n} b_j\right) = \sum_{i,j} a_i b_j.$$

18 Prove Theorem 4.

2. SUBRINGS AND HOMOMORPHISMS

Early in our study of groups we introduced the concepts of a subgroup, a normal subgroup, a homomorphism, and a quotient group. In this section we will make the corresponding definitions for rings.

Let $(R, +, \times)$ be a ring. A *subring* of R is a subset S of R containing the identity element 1 of R such that S is a subgroup of $(R, +)$ and a submonoid of $(R, \times)$. Thus S is closed under sums, negatives, and products. For example, $\mathbf{Z}$ is a subring of $\mathbf{Q}$, which is a subring of $\mathbf{R}$, which is a subring of $\mathbf{C}$. If a subring happens to be a field, it is often referred to as a *subfield.*

In defining subgroups of a group G we did not have to make the explicit statement that the identity element of G must belong to every subgroup. This followed automatically. Since we are assuming that all of our rings have multiplicative identity elements, it is convenient always to have the identity of a subring be the same as the identity of the larger ring. The following example shows that we must make this a part of the definition. Let $R = \mathbf{Z} \oplus \mathbf{Z}$ and let $S = \{(x, 0) \mid x \in \mathbf{Z}\}$. Then S is a subset of R that is closed under addition, subtraction, and multiplication and S is a ring under these operations. However, the identity element $(1, 0)$ of S is not the same as the identity $(1, 1)$ of R, so we do not consider S to be a subring of R.

Here are some additional examples of subrings.

Example 1. The set of complex numbers $x + yi$ with x and y in $\mathbf{Z}$ is a subring of $\mathbf{C}$. This subring is denoted $\mathbf{Z}[i]$ and called the ring of *Gaussian integers.* [German mathematician Carl Friedrich Gauss (1777-1855) is considered to be one of the greatest mathematicians of all time.]

Example 2. We can generalize Example 1 slightly as follows. Let m be an integer and let $\sqrt{m}$ denote a fixed square root of m in $\mathbf{C}$. The set of complex numbers $x + y\sqrt{m}$ with x and y in $\mathbf{Z}$ is a subring of $\mathbf{C}$ and is denoted $\mathbf{Z}[\sqrt{m}]$. If $m \geq 0$, then $\mathbf{Z}[\sqrt{m}]$ is a subring of $\mathbf{R}$. Normally, we assume that $|m| > 1$ and that m is *square free*, that is, not divisible by the square of an integer greater than 1.

Example 3. In $M_2(\mathbf{Z})$ the set of matrices of the form

$$\begin{bmatrix} a & b \\ 0 & c \end{bmatrix}$$

is a subring.

Example 4. The set of polynomials $a_0 + a_1X + \ldots + a_nX^n$ in $\mathbf{Z}[X]$ with $a_1 = 0$ is a subring of $\mathbf{Z}[X]$.

Example 5. The set {0, 1, 2, 3} is a subring of (⍳8, *PLUS*, *TIMES*).

```
      A←⍳4                              ∧/,TIMES[A;A]∊A
      ∧/,PLUS[A;A]∊A          1
1
```

We defined a homomorphism of groups to be a map that is compatible with the group operations. For rings we require that a homomorphism be compatible with both addition and multiplication and also that it map the identity correctly. Let R and S be rings. A *homomorphism* from R to S is a map $f{:}R \longrightarrow S$ such that $1f = 1$ and for all x and y in R we have $(x + y)f = xf + yf$ and $(xy)f = (xf)\,(yf)$. The map taking m in $\mathbf{Z}$ to $(m, 0)$ in $\mathbf{Z} \oplus \mathbf{Z}$ is compatible with both addition and multiplication. However, it does not map 1 to the identity of $\mathbf{Z} \oplus \mathbf{Z}$. The map $m \longmapsto (m, m)$ is a ring homomorphism of $\mathbf{Z}$ into $\mathbf{Z} \oplus \mathbf{Z}$. Since a ring homomorphism f is in particular an additive homomorphism, it follows that $0f = 0$. As with groups, a bijective ring homomorphism is called a ring *isomorphism.* Two rings R and S are *isomorphic* if there is a ring isomorphism of R onto S.

The kernel of a group homomorphism is the set of elements in the domain that are mapped to the identity element. Let $f{:}R \longrightarrow S$ be a ring homomorphism. Since our rings have two binary operations, each with its own identity element, we seem to have to choices for the kernel of f, either $I = \{x \in R \,|\, xf = 0\}$ or $J = \{x = R \,|\, xf = 1\}$. A little investigation shows that both I and J are closed under multiplication, but only I is closed under addition. In fact, I has several other nicc properties, and we define the *kernel* of f to be I.

THEOREM 1. Let $f{:}R \longrightarrow S$ be a homomorphism of rings and let I be the kernel of f. Then I is a subgroup of $(R, +)$ and, for all x in I and r in R, both rx and xr are in I.

Proof. Since f is a ring homomorphism, f is a homomorphism of the abelian group $(R, +)$ into the group $(S, +)$ and I is the kernel of this group homomorphism. Thus I is a subgroup of $(R, +)$. If $x \in I$ and $r \in R$, then $(rx)f = (rf)\,(xf) = (rf)0 = 0$. Thus rx is in I. Similarly, xr is in I. □

A subset I of a ring R that is an additive subgroup and contains all products rx and xr with x in I and r in R is called an *ideal* of R. Theorem 1 states that the kernel of a ring homomorphism of R to S is an ideal of R. The sets {0} and R are always ideals of R.

What are the ideals of $\mathbf{Z}$? A subgroup of $(\mathbf{Z}, +)$ has the form $n\mathbf{Z}$. It is trivial to verify that $n\mathbf{Z}$ is an ideal of $\mathbf{Z}$ for any integer n and so, for $\mathbf{Z}$, the notions of ideal and additive subgroup coincide.

The following theorem states analogues of parts of Theorems 3.4.1 and 3.5.3.

THEOREM 2. Let $f:R \longrightarrow S$ be a ring homomorphism. Then

(a) If U is a subring of R, then Uf is a subring of S.
(b) If V is a subring of S, then Vf^{-1} is a subring of R.
(c) If I is an ideal of R, then If is an ideal of Rf.
(d) If J is an ideal of S, then Jf^{-1} is an ideal of R.

Proof. See Exercise 11. □

Note that in Theorem 2c we can be sure If is an ideal of S only when f is surjective.

THEOREM 3. Let $f:R \longrightarrow S$ be a surjective ring homomorphism with kernel I. There is a 1–1 correspondence between the set of ideals of R containing I and the set of ideals of S.

Proof. This is similar to Theorem 3.5.9 and follows easily from parts (c) and (d) of Theorem 2. □

Ideals seem to have a place in the theory of rings roughly equivalent to that of normal subgroups in the theory of groups. It is therefore natural to ask whether, given an ideal I of a ring R, we can define a *quotient ring* R/I. By the definition of an ideal, we know that I is a subgroup of the group $(R, +)$, which is abelian. Therefore I is a normal subgroup of $(R, +)$, and we can define the quotient group $(R/I, +)$. In order to make R/I into a ring, we need to define multiplication. Addition in R/I is defined by $(x + I) + (y + I) = (x + y) + I$. Thus it is reasonable to try to define the product of the cosets $x + I$ and $y + I$ to be $xy + I$. But is this well defined? If $x + I = x' + I$ and $y + I = y' + I$, then $x' = x + u$ and $y' = y + v$ with u and v in I. Hence

$$x'y' = (x + u)(y + v) = xy + uy + xv + uv$$

and, since I is an ideal, uy, xv, and uv are all in I. Therefore $x'y' + I = xy + I$ and multiplication of cosets is well defined. The coset $1 + I$ is clearly a multiplicative identity. The remaining ring axioms are easily verified. For example,

$$(x + I)[(y + I)(z + I)] = (x + I)(yz + I) = x(yz) + I$$

and

$$[(x + I)(y + I)](z + I) = (xy + I)(z + I) = (xy)z + I.$$

Since $x(yz) = (xy)z$ in R, multiplication in R/I is associative. The distributive laws are left for the reader to check.

The following theorem is an immediate consequence of our definition of the ring structure of R/I.

THEOREM 4. If I is an ideal of the ring R, then the natural map of R onto R/I is a ring homomorphism with kernel I. □

We have already seen one example of a quotient ring. For any positive integer n the ring $\mathbf{Z}_n$ is the quotient ring of $\mathbf{Z}$ modulo the ideal $n\mathbf{Z}$.

The three isomorphism theorems for groups, Theorems 3.5.7, 3.5.12, and 3.5.13, have natural analogues for rings.

THEOREM 5. Let $f:R \longrightarrow S$ be a surjective ring homomorphism with kernel I. Then S is isomorphic to R/I.

Proof. By Theorem 3.5.7, the map h taking $x + I$ in R/I to xf in S is an isomorphism of $(R/I, +)$ onto $(S, +)$. This map also preserves products and maps $1 + I$ in R/I to 1 in S. Thus h is a ring isomorphism. □

THEOREM 6. Let S be a subring of the ring R and let I be an ideal of R. Then $S \cap I$ is an ideal of S, $S + I$ is a subring of R, and

$$S/(S \cap I) \cong (S + I)/I.$$

Proof. Let $f: R \longrightarrow R/I$ be the natural map. Then $S + I = (Sf)f^{-1}$ and so $S + I$ is a subring of R by Theorem 2ab. In addition, f maps both S and $S + I$ onto Sf with kernels $S \cap I$ and I, respectively. Thus, by Theorem 5,

$$S/(S \cap I) \cong Sf = (S + I)/I. \quad \square$$

THEOREM 7. Let I and J be ideals of the ring R with $I \subseteq J$. Then J/I is an ideal of R/I and $(R/I)/(J/I)$ is isomorphic to R/J.

Proof. See the proof of Theorem 3.5.13. □

For any element a of an abelian group A, written additively, and any integer m we have defined an element ma in A so that the map $m \longmapsto ma$ is a homomorphism of abelian groups. If R is a ring and we take $A = (R, +)$ and $a = 1$, then we get even more.

THEOREM 8. If R is a ring, then the map from $\mathbf{Z}$ to R taking m to $m1$ is a ring homomorphism.

Proof. If m and n are nonnegative integers, then the formula $(m1)(n1) = (mn)1$ is proved either by a simple induction on m or by invoking Theorem 1.4. The cases in which m or n is negative are then handled using the identity $(-m)1 = -(m1)$. □

Let R be a ring and let $f: \mathbf{Z} \longrightarrow R$ be the homomorphism of Theorem 8. The kernel of f is $n\mathbf{Z}$ for a unique integer $n \geq 0$. We call n the *characteristic* of R. The image of $\mathbf{Z}$ under f is a subring of R isomorphic to $\mathbf{Z}_n$. Since $n1 = 0$ in R, we have $nx = (n1)x = 0$ for any x in R. Of course, for rings of characteristic 0 this does not say anything new.

THEOREM 9. Let R be an integral domain or a division ring. Then the characteristic of R is either 0 or a prime.

Proof. Let n be the characteristic of R. Any commutative subring of R must be an integral domain. (Why?) Now R contains a subring isomorphic to $\mathbf{Z}/n\mathbf{Z}$, which is an integral domain if and only if $n = 0$ or n is a prime. □

COROLLARY 10. Let F be a finite field. Then $|F| = p^m$, where p is the characteristic of F, a prime, and m is a positive integer.

Proof. Since F is finite, F cannot contain a subring isomorphic to $\mathbf{Z}$. Thus, by Theorem 9, the characteristic p of F is a prime. As remarked before, $px = 0$ for all x in F. Thus every element of the finite abelian group $(F, +)$ has order 1 or p. By the First Sylow Theorem, this implies that no prime other than p divides $|F|$. □

As the next result shows, fields have very few ideals.

THEOREM 11. Let R be a commutative ring with $1 \neq 0$. Then R is a field if and only if the only ideals of R are $\{0\}$ and R.

Proof. Suppose first that R is a field and that $I \neq \{0\}$ is an ideal of R. Then I contains an element $x \neq 0$. Since R is a field, x has an inverse in R and I contains $xx^{-1} = 1$. Thus if r is an element of R, then $r = r1$ is in I. Hence $I = R$.

Now suppose R has only the ideals $\{0\}$ and R. Let x be a nonzero element of R. It is easy to show that $I = Rx = \{rx \mid r \in R\}$ is an ideal of R. (See Exercise 9.) Since I contains x, we have $I \neq \{0\}$. Therefore $I = R$ and 1 is in I. This means that there is an element y of R such that $yx = xy = 1$. Hence x is a unit in R. Thus R is a commutative ring with $1 \neq 0$ in which every nonzero element is a unit. That is, R is a field. □

We close this section with a brief description of a generalization of the concept of an ideal. A *right ideal* U of a ring R is a subgroup of $(R, +)$ such that for all u in U and all r in R the product ur is in U. Similarly, a *left ideal* is an additive subgroup of R that is closed under multiplication on the left by elements of R. Ideals, as we have defined them, are both left ideals and right ideals. To emphasize this fact, an ideal is often referred to as a *two-sided ideal.* In a commutative ring the concepts of a left ideal, a right ideal, and a two-sided ideal are the same.

EXERCISES

In the following exercises R, S, and T are always rings.

1 Suppose R is a subring of S and S is a subring of T. Show that R is a subring of T.

2 Prove that the intersection of any nonempty collection of subrings of R is again a subring of R.

3 Let X be a subset of R. The subring of R *generated* by X is the intersection of all subrings of R containing X. Give another description of S analogous to Theorem 3.3.3.

4 Suppose $f{:}R \longrightarrow S$ and $g{:}S \longrightarrow T$ are ring homomorphisms. Show that $f \circ g$ is a ring homomorphism.

5 Let $f{:}R \longrightarrow S$ be a bijective ring homomorphism. Prove that f^{-1} is a ring homomorphism from S to R.

6 Determine the units in $\mathbf{Z}[i]$.

7 Find a unit other than ± 1 in $\mathbf{Z}[\sqrt{2}]$.

8 Show that $n\mathbf{Z}$ is an ideal of $\mathbf{Z}$ for any integer n.

9 Assume that R is commutative and that $x_1, \ldots, x_m$ are elements of R. Show that the set of elements of R of the form $r_1x_1 + r_2x_2 + \ldots + r_mx_m$ with each r_i in R is an ideal of R.

10 How many ideals are there in the ring $\mathbf{Z}_{12}$?

11 Prove Theorem 2.

12 Fill in the details of the proof of Theorem 3.

13 Describe the ideal of $\mathbf{Z}[X]$ generated by 4, $2X$, and X^2.

14 Prove Theorem 7.

15 An *automorphism* of R is an isomorphism of R with itself. Show that the set Aut(R) of all automorphisms of R is a subgroup of $\Sigma(R)$.

16 Let m be an integer that is not a square in $\mathbf{Z}$. Prove that the map taking $a + b\sqrt{m}$ to $a - b\sqrt{m}$ is well defined and is an automorphism of $\mathbf{Z}[\sqrt{m}]$.

17 Show that Aut(**Q**) is trivial but that Aut(**C**) is nontrivial.

18 Show that Aut(**R**) is trivial.

19 Let F be a field of characteristic 0. Prove that F contains a subring isomorphic to **Q**.

20 Describe the ideal of $\mathbf{Z}[i]$ generated by $1 + i$.

21 Let I be a nontrivial ideal of $\mathbf{Z}[i]$. Show that $\mathbf{Z}[i]/I$ is a finite ring.

*22 Let $z = a + bi$ be a nonzero Gaussian integer. Determine the order of $\mathbf{Z}[i]/I$, where I is the ideal of $\mathbf{Z}[i]$ generated by z.

23 Let `P` and `T` be `N`-by-`N` matrices such that R = (`⍳N, P, T`) is a ring. Assume that 1 is the identity of R. Suppose the vector `A` lists the elements of a subset A of R. Write APL propositions corresponding to the following assertions.

(a) A is a subring of R.
(b) A is an ideal.

24 For each integer `I` in `⍳8` determine the subring of R = (`⍳8, PLUS, TIMES`) generated by `I` and the ideal of R generated by `I`.

3. COMPUTING IN RINGS USING APL

Already in this chapter we have introduced many rings, and we will be describing more examples as we go along. Frequently, we will need to compute with elements of these rings and to manipulate vectors and matrices with entries in these rings. This section is devoted to a discussion of some of the ways APL can be used to work with arrays that have entries in one of the rings **Z**, $\mathbf{Z}_n$, **Q**, **R**, **C**, or **Z**[i], as well as arrays with entries in a small finite ring described by its addition and multiplication tables. As the APL language is currently defined, arrays in APL have real numbers as entries. Thus, for $\mathbf{Z}_n$, **C**, and **Z**[i], we will first have to discuss methods for representing arrays over these rings.

Integer arrays are the easiest to work with in APL. As long as the entries remain small, say less than 10^{16}, calculations with integer arrays are performed exactly on APL terminal systems. Since the entires are real numbers, the primitive APL operations may be used to perform most arithmetic computations.

A matrix A with entries in $\mathbf{Z}_n$ can be represented by a matrix `A` of integers such that `A[I;J]` is a representative for the corresponding entry in A. The modulus n will always be denoted by the global variable `N̲`. A given matrix A can be represented by infinitely many integer matrices. Two integer matrices `A` and `B` represent the same matrix over $\mathbf{Z}_n$ if and only if `∧/,(N̲|A)=N̲|B` or, equivalently, `∧/,0=N̲|A-B`. If `A` represents A, then `B←N̲|A` also represents A and `B` is the unique matrix representing A that has all of its entries in the set $\{0, \ldots, n-1\}$. We will call `B` the *standard representation* of A. All procedures in `CLASSLIB` with the prefix `ZN` return standard representations.

If the integer arrays `A` and `B` represent arrays A and B over $\mathbf{Z}_n$ with the same shape, then the standard representations for $-A, A+B, A-B$, and $A \times B$ are `N̲|-A`, `N̲|A+B`, `N̲|A-B`, and `N̲|A×B`, respectively. For convenience, the procedures `ZNNEG`, `ZNSUM`, `ZNDIFF`, and `ZNPROD` have been included in `CLASSLIB` to compute these results. When `N̲` is greater than 10^7, then `ZNPROD` uses multiple-precision multiplication to compute `A×B`, since the entries in `A×B` may exceed the size of the largest integer that can be represented by the terminal system using single precision. If every entry in A is a unit in $\mathbf{Z}_n$, then the array of inverses of the entries in A is represented by `ZNINV A`. The matrix of `M`th powers of the entries in A is represented by `A ZNPOWER M`.

```
      N̲←1000001
      A←237938 791247
      ⎕←B←ZNINV A
536136 701616
```

```
      A ZNPROD B
1 1
      A ZNPOWER 1000000
405718 918192
```

With the dyadic procedures *ZNSUM*, *ZNDIFF*, *ZNPROD*, and *ZNPOWER* arguments having one entry are expanded to match the other argument.

The symbols for the APL primitive arithmetic operations indicate exact computations with real numbers. However, on a terminal system only a finite set of rational numbers can be represented and the results of the arithmetic operations are only approximations. For arrays whose entries may be irrational numbers we have little choice but to make do with the APL primitives and to remember that results may not be exact. Round-off errors often produce results that are nonzero numbers with small absolute value where the correct result is 0. Because of this, all procedures in *CLASSLIB* with the prefix *R* normally set to 0 all entries in an array that have absolute values less than *EPSILON* times the largest absolute value of the entries in the array, where *EPSILON* is a global variable normally set to 10^{-13}. The following example shows how this can be done.

```
      A←1E4 1 1E¯4 1E¯10
      ⎕←A←A×(|A)≥EPSILON×⌈/,|A
10000 1 0.0001 0
```

With arrays over **Q** we have a choice. We can treat them simply as arrays over **R** and use the APL primitive operations. This is fast but often leads to incorrect results. An alternative is to use integer vectors of length 2 to represent rational numbers. For example, if we really want to specify the number 1/3 and not some approximation such as 0.3333333333, then we can use the integer vector `1 3`. In general, if *P* and *Q* are integers with *Q*≠0, then the rational number *P*÷*Q* can be represented by the vector *P*,*Q*. Addition and multiplication can be performed using the formulas

$$\frac{a}{b} + \frac{c}{d} = \frac{ad + bc}{bd},$$

$$\left(\frac{a}{b}\right)\left(\frac{c}{d}\right) = \frac{ac}{bd},$$

which require only integer computations.

A matrix A of rational numbers can be represented by a rank 3 integer array *A* such that *A*[*I*;*J*;] gives the numerator and denominator of the corresponding entry in A. Thus

$$\begin{bmatrix} 7/6 & -3/5 \\ 1/2 & 3 \end{bmatrix}$$

would be represented by

```
      ⎕←A←2 2 2⍴7 6 ¯3 5 1 2 3 1
 7  6
¯3  5

 1  2
 3  1
```

The array A is unique if we assume that numerators and denominators are relatively prime and that denominators are positive. To improve readability, the procedure *DAQ* (for "display array of rationals") may be used to print out scalars, vectors, and matrices with rational entries.

```
      DAQ A
 7/6    ¯3/5
 1/2     3/1
```

The procedures *QNEG*, *QINV*, *QSUM*, *QDIFF*, *QPROD*, *QQUOT*, and *QPOWER* are the analogues of the monadic operations - and ÷ and the dyadic operations +, -, ×, ÷, and *.

```
      DAQ B←2 2 2⍴1 7 2 1 ¯8 3 1 5
 1/7     2/1
¯8/3     1/5
      DAQ A QSUM B
 55/42     7/5
¯13/6     16/5
      DAQ A QPROD B
 1/6    ¯6/5
¯4/3     3/5
      DAQ A QPOWER 2
49/36    9/25
 1/4     9/1
```

Note that *QPROD* computes the entry-by-entry product and not the matrix product, which is discussed in Section 5. The five dyadic procedures allow arguments of different shapes, provided the arrays they represent, considered as arrays of rational numbers, are conformable for scalar arithmetic operations. For example, to add 1/2 to each entry in the matrix represented by A, we form

```
      DAQ 1 2 QSUM A
5/3        ¯1/10
1/1         7/12
```

The fractions produced by the procedures with the prefix *Q* are always reduced to lowest terms. This involves computing many greatest common divisors. As a result, these procedures can use a significant amount of CPU time.

To represent complex numbers in APL, we again use vectors. The complex number $3-2i$ can be represented by the vector `3 ¯2` of its real and imaginary parts. Matrices over **C** are represented by rank 3 APL arrays and are displayed using *DARV* (for "display array of real vectors"). Thus

$$\begin{bmatrix} 1.0-i & 2.1+3.5i \\ 0.5 & 7.0i \end{bmatrix}$$

is represented by

```
      ⎕←U←2 2 2⍴1 ¯1 2.1 3.5 0.5 0 0 7
 1                   ¯1
 2.1                  3.5

 0.5                  0
 0                    7
```

which can be displayed as a matrix as follows.

```
      1 DARV U
 1.0 ¯1.0        2.1  3.5
  .5   .0         .0  7.0
```

The first argument of *DARV* is the number of decimal places desired. The procedure *DAZV* (for "display array of integer vectors") produces the same result as *DARV* with a first argument of 0 and may be used to display arrays of Gaussian integers.

The procedures *CINV*, *CSUM*, *CDIFF*, *CPROD*, *CQUOT*, and *CPOWER* perform the same operations with complex arrays that the corresponding procedures with the prefix *Q* perform for arrays of rational numbers. For example,

```
      1 DARV V←2 2 2⍴¯1 1 2.3 0 0 ¯0.4 2.1 ¯3.2
¯1.0  1.0      2.3   .0
  .0   .4      2.1 ¯3.2
      1 DARV U CSUM V
  .0   .0      4.4  3.5
  .5 ¯ .4      2.1  3.8
      2 DARV U CPROD V
   .00    2.00       4.83     8.05
   .00    ¯.20      22.40    14.70
```

The five dyadic procedures allow arguments of different shapes. Thus, to multiply all entries in the array represented by U by $2 + i$, we form

```
      1 DARV 2 1 CPROD U
 3.0   ¯1.0       .7    9.1
 1.0     .5    ¯7.0   14.0
```

There is no need for a procedure *CNEG*, since negation in **C** can be performed using the primitive APL negation operation. The procedures with the prefix *C* set to zero entries that are smaller in absolute value than *EPSILON* times the largest absolute value of any entry.

The *norm* of a complex number $z = x + yi$, with x and y real, is $N(z) = x^2 + y^2$, and the *magnitude* or *absolute value* of z is $|z| = \sqrt{x^2 + y^2}$, the nonnegative square root of $N(z)$. The *conjugate* of z is $\overline{z} = x - yi$. The procedures *CNORM*, *CMAG*, and *CCONJ* compute the norms, magnitudes, and conjugates of the entries in a complex array.

```
      1 DARV U
 1.0  ¯1.0       2.1   3.5
  .5    .0        .0   7.0
      CNORM U
 2       16.66
 0.25 49
      CMAG U
1.414213562 4.081666326
0.5         7
      1 DARV CCONJ U
 1.0   1.0       2.1  ¯3.5
 0.5    .0        .0  ¯7.0
```

Procedures with the prefix *FR* perform computations with arrays whose entries are in a specified finite ring R. A complete description of R requires four global variables. The addition and multiplication tables for R are assumed to be given by the matrices *FRPLUS* and *FRTIMES*. Origin 0 is assumed, and 0 and 1 must be the additive and multiplicative identities, respectively. The negative or inverse of *I* in the group $(R, +)$ is given by *FRNEG*[*I*]. If *I* is a unit in R, then *FRINV*[*I*] is the inverse of u. If *I* is not a unit, then *FRINV*[*I*] must be 0. We will refer to R as the *current finite ring*.

The procedure *FRINIT* may be used to initialize *FRPLUS*, *FRTIMES*, *FRNEG*, and *FRINV*. The arguments are the addition and multiplication tables for the finite ring. The procedure copies these arrays into *FRPLUS* and *FRTIMES* and then computes *FRNEG* and *FRINV*. To work with the ring R of order 8 discussed in Section 1, we enter

```
      PLUS FRINIT TIMES
      FRNEG
0 1 2 3 4 5 6 7
      FRINV
0 1 0 3 0 0 0 0
```

To save time, *FRINIT* does not actually check to see whether its arguments really define a finite ring.

Let us consider the matrices

```
      ⎕←A←2 2ρ3 2 6 1                 ⎕←B←2 2ρ7 5 0 2
 3 2                             7 5
 6 1                             0 2
```

to have entries in R. Then the sum and product of *A* and *B* are given by

```
      A FRSUM B                       A FRPROD B
 4 7                             5 2
 6 3                             0 2
```

The procedures *FRDIFF* and *FRPOWER* compute differences and powers.

EXERCISES

1 Experiment with the procedures *ZNNEG*, *ZNSUM*, *ZNDIFF*, *ZNPROD*, *ZNINV*, and *ZNPOWER*.

2 Enter rank 3 integer arrays representing the matrices

$$A = \begin{bmatrix} 11/3 & -2 \\ 1/7 & 9/4 \end{bmatrix}, \quad B = \begin{bmatrix} 3 & 7/9 \\ 13/2 & 0 \end{bmatrix}.$$

Use *DAQ* to display these arrays. Experiment with *QNEG*, *QINV*, *QSUM*, *QDIFF*, *QPROD*, *QQUOT*, and *QPOWER*.

3 Enter rank 3 arrays representing the complex matrices

$$A = \begin{bmatrix} 3-2i & 1+0.5i \\ -4 & 3.2i \end{bmatrix}, \quad B = \begin{bmatrix} 7+3i & -1+2i \\ 2-5i & 6+i \end{bmatrix}.$$

Experiment with *DARV*, *DAZV*, *CINV*, *CSUM*, *CDIFF*, *CPROD*, *CQUOT*, and *CPOWER*.

4 Let *A* be an integer matrix. Write APL expressions for the rank 3 arrays that represent the same matrix of integers considered as a matrix of rational numbers and as a matrix of Gaussian integers.

5 Show for any complex numbers z and w that the following identities hold.

(a) $N(z) = z\bar{z} = |z|^2$.
(b) $N(zw) = N(z)N(w)$.
(c) $|z + w| \leq |z| + |w|$.
(d) $\overline{z + w} = \bar{z} + \bar{w}$.
(e) $\overline{zw} = \bar{z}\bar{w}$.

6 Experiment with the procedures *CNORM*, *CMAG*, and *CCONJ*.

7 Assume that *FRPLUS* and *FRTIMES* are the addition and multiplication tables for a finite ring R with 0 and 1 as the additive and multiplicative identities. Write APL expressions for the vector *FRNEG* giving additive inverses and for the vector *FRINV* giving multiplicative inverses in R.

8 Construct addition and multiplication tables for $\mathbf{Z}_6$. Set $\underline{N}$←6 and set the current finite ring to $\mathbf{Z}_6$ using *FRINIT*. Compare the speeds of execution of the procedures with the prefixes *ZN* and *FR* for some sample calculations with arrays over $\mathbf{Z}_6$.

4. POLYNOMIAL RINGS

In this section we will describe how, given a ring R, we can construct a new ring called the *ring of polynomials* with coefficients in R. In order to do this, we will need to give a formal definition of the term "polynomial". Since polynomials are introduced first in high school algebra and studied extensively in calculus, it may seem unnecessary to dwell very long on them here. The following dialogue between an instructor and an undergraduate student of algebra may help to explain why we are devoting a whole section to rings of polynomials.

I: What is a polynomial with real coefficients?

U: Why do you ask? I've been working with polynomials ever since high school, adding them, multiplying them, and even factoring them sometimes. I certainly know what a polynomial is!

I: Well, if they're that familiar, you should be able to give me a definition.

U: All right. A polynomial with real coefficients is an expression $a_0 + a_1X + \ldots + a_nX^n$, where the a_i are real numbers.

I: Thank you. Let me write down some examples to see if I understand your definition.

$$1 + 2X,$$
$$1 + X + X^2 + X^3,$$
$$1 + 2X + 0X^2.$$

U: That's right. But the first and third polynomials are really the same.

I: They're different expressions. Why aren't they different polynomials?

U: Well, when some a_i is 0, we usually don't bother to write down the term a_iX^i.

I: Oh. Can you tell me how to add polynomials?

U: Sure. You just add corresponding coefficients. For example, $1 + X + X^2$ plus $2 + 3X + 4X^2$ is $3 + 4X + 5X^2$.

I: Okay. Suppose I want to add the polynomials X and X^{100}. What's the coefficient of X^{100} in X?

U: Zero.

I: Really? What's the coefficient of X^{1000} in X?

U: Zero.

I: How about $X^{1000000}$?

U: Still zero!

I: You seem to be saying a polynomial has infinitely many coefficients, one for each power X^i of X, $i = 0, 1, 2, \ldots$.

U: Yes, I guess I am.

I: Now, can you tell me how to multiply polynomials?

U: Well, I'm not sure I can give a general formula, but I can certainly work out an example.

$$\begin{array}{rrrrrr}
1 + & X + & 2X^2 + & 4X^3 & & \\
1 - & 2X + & 3X^2 & & & \\
\hline
1 + & X + & 2X^2 + & 4X^3 & & \\
 - & 2X - & 2X^2 - & 4X^3 - & 8X^4 & \\
 & & 3X^2 + & 3X^3 + & 6X^4 + & 12X^5 \\
\hline
1 - & X + & 3X^2 + & 3X^3 - & 2X^4 + & 12X^5
\end{array}$$

I: It seems to me you could save a lot of time if you didn't bother to write down all those X's and just worked with the coefficients. Like this:

$$\begin{array}{rrrrrr}
1 & 1 & 2 & 4 & & \\
1 & -2 & 3 & & & \\
\hline
1 & 1 & 2 & 4 & & \\
 & -2 & -2 & -2 & -8 & \\
 & & 3 & 3 & 6 & 12 \\
\hline
1 & -1 & 3 & 3 & -2 & 12
\end{array}$$

U: That's true in this example, but if you tried to multiply $1 + X^{10} +$

X^{20} by $X^{15} + X^{40}$ your way, you would have an awful lot of zeros that I don't have to bother with. In fact, I can do this product in my head my way.

I: Good point. Let me summarize what you've said about polynomials. A polynomial is determined by the sequence $a_0, a_1, \ldots$ of its coefficients. This sequence is an infinite sequence of real numbers that happens to consist of all zeros from some point on. To compute the coefficients of the sum or the product of two polynomials f and g, one needs only to be given the coefficients of f and g.

U: Yes, I'll agree with that.

I: Well, if a polynomial is determined by its sequence of coefficients and if the procedures for adding and multiplying two polynomials require only a knowledge of the coefficients of those polynomials, then I, as an algebraist, would be tempted to say that a polynomial *is* the sequence of its coefficients. Thus I would define a polynomial with real coefficients to be a function f from $\{0, 1, 2, \ldots\}$ to $\mathbf{R}$ such that $f(i) = 0$ for all sufficiently large values of i.

U: I think I see that you're driving at, but I sure don't think of polynomials that way!

I: Frankly, most of the time neither do I.

Let us formalize what the instructor and the undergraduate have been saying about polynomials and, at the same time, broaden the concept of a coefficient to include things other than real numbers. Let R be a ring and let M be the set of nonnegative integers. A *polynomial* with coefficients in R is a function $f:M \longrightarrow R$ such that for some integer n, depending on f, $f(i) = 0$ for $i > n$. For the moment, let us denote the set of all such polynomials by P. If f and g are in P, we define two new polynomials $f + g$ and fg by

$$(f+g)(k) = f(k) + g(k),$$

$$(fg)(k) = \sum_{i=0}^{k} f(i)g(k-i).$$

THEOREM 1. The set P with the operations of addition and multiplication is a ring.

Proof. We will leave most of the details as an exercise. As an illustration, we will prove that multiplication is associative. Suppose f, g, and h are all in P. Then

$$[(fg)h](k) = \sum_{i=0}^{k} (fg)(i)\, h(k-i)$$
$$= \sum_{i=0}^{k} \left(\sum_{j=0}^{i} f(j)\, g(i-j) \right) h(k-i)$$
$$= \sum_{i=0}^{k} \sum_{j=0}^{i} f(j)\, g(i-j)\, h(k-i).$$

Also,

$$[f(gh)] = \sum_{i=0}^{k} f(i)\,(gh)\,(k-i)$$
$$= \sum_{i=0}^{k} f(i) \left(\sum_{j=0}^{k-i} g(j)\, h(k-i-j) \right)$$
$$= \sum_{i=0}^{k} \sum_{j=0}^{k-i} f(i)\, g(j)\, h(k-i-j).$$

It is now easy to see that both $[(fg)h](k)$ and $[f(gh)]\ (k)$ are equal to the sum of all possible products $f(u)g(v)h(w)$, where u, v, w are nonnegative integers and $u + v + w = k$. Thus $(fg)h = f(gh)$. Note that this proof, like the proofs of the other ring axioms, does not require that R be commutative. The additive identity for P is the sequence 0, 0, 0, . . . and the multiplicative identity is 1, 0, 0, . . . □

For r in R define f_r to be the element of P such that $f_r(0) = r$ and $f_r(i) = 0, i \geq 1$.

THEOREM 2. The map $r \longmapsto f_r$ is an injective ring homomorphism of R into P.

Proof. It follows immediately from the definition of addition and multiplication in P that $f_r + f_s = f_{r+s}$ and $f_r f_s = f_{rs}$. Since f_1 is the multiplicative identity of P, the map $r \longmapsto f_r$ is a ring homomorphism. Now $f_r = 0$ if and only if $r = 0$, so this map is injective. □

If f is a nonzero polynomial in P, then the *degree* of f is the largest integer n such that $f(n) \neq 0$. By the definition of a polynomial, $f(i) \neq 0$ for only finitely many values of i, and so n exists. We write $n = \deg(f)$. If f has degree n, then $f(n)$ is called the *leading coefficient* of f. The degree of the zero polynomial is not usually defined but sometimes it is said to be $-\infty$.

THEOREM 3. Suppose R is an integral domain. If f and g are nonzero elements of P, then $fg \neq 0$. Thus P is an integral domain. Moreover, deg(fg) = deg(f)+deg(g), and the leading coefficient of fg is the product of the leading coefficients of f and g.

Proof. Let m = deg(f) and n = deg(g) and let k be in M. Then

$$(fg)(k) = \sum_{i=0} f(i)g(k-i).$$

If $k > m + n$ and $0 \le i \le k$, then either $i > m$ or $k - i > n$. Thus $(fg)(k) = 0$. Similarly, we see that $(fg)(m + n) = f(m)g(n)$. Since R is an integral domain, $f(m)g(n) \neq 0$, and so deg(fg) = $m + n$ and $f(m)g(n)$ is the leading coefficient of fg. □

We have introduced polynomials as infinite sequences to avoid having to define what is meant by an "expression" $a_0 + a_1X + \ldots + a_nX^n$ and to say when two such expressions represent the same polynomial. Our definition also allows addition and multiplication in P to be described easily. It is now time to admit that we will normally use the more familiar notation $a_0 + a_1X + \ldots + a_nX^n$ and variants of it, such as $a_nX^n + \ldots + a_0$, to denote polynomials. But what is X? Usually X is referred to as a variable or indeterminant, with no attempt made to define what these terms mean. We can give another answer. The symbol X stands for a particular element of P: the function $f{:}M \longrightarrow R$ such that $f(1) = 1$ and $f(i) = 0$, $i \neq 1$. It is easily shown that in P the nth power X^n of X is the polynomial g such that $g(n) = 1$, $g(i) = 0$, $i \neq n$. If we identify r in R with the corresponding *constant polynomial* f_r defined previously, then we may consider R to be a subring of P. When this is done, we may write a polynomial f in P as $f(0) + f(1)X + \ldots + f(n)X^n$, where n is any integer such that $f(i) = 0$ for $i > n$. Such an expression is a sum of products of elements in P and, as such, is unambiguous.

When we denote the polynomial 0, 1, 0, . . . by X, we usually write P as $R[X]$. If, for some reason, we prefer to use Y to stand for the polynomial 0, 1, 0, . . . , then we refer to P as $R[Y]$. One situation where this is done is the case in which the coefficient ring is itself a ring of polynomials. Having started with R and constructed the ring P of polynomials with coefficients in R, we might wish to consider the ring Q of polynomials with coefficients in P. If we have chosen to let X be a particular element of P, we cannot use X to stand for the polynomial 0, 1, 0, . . . in Q. Thus we write $Q = P[Y] = R[X][Y] = R[X,Y]$, not $Q = R[X][X]$.

Let f be a polynomial in $R[X]$. According to our formal definition, the expression $f(i)$ represents the ith coefficient of f. This notation is in conflict with notation used in calculus. There we are taught that if $f = 2 - X + 3X^2$, then $f(2)$ means $2 - 2 + 3(2^2) = 12$, not the coefficient 3 of X^2. Since we have already agreed to adopt the familiar notation $f =$

$a_0 + a_1X + \ldots + a_nX^n$ for polynomials, it should not come as a surprise that from now on (except in certain exercises at the end of this section) *whenever b is in R we will mean by $f(b)$ the element $a_0 + a_1b + \ldots + a_nb^n$ in R.*

THEOREM 4. Let R be a commutative ring and let b be an element of R. The map $f \longmapsto f(b)$ is a ring homomorphism of $R[X]$ into R.

Proof. Let f and g be in $R[X]$. The equality $(f+g)(b) = f(b) + g(b)$ holds even without the assumption that R is commutative. However, to prove that $(fg)(b) = f(b)g(b)$, we need to know that $(cb^i)(db^j) = cdb^{i+j}$, and this requires that R be commutative, or at least that b commute with every element in R. Since the identity $1 + 0X + \ldots$ of $R[X]$ is mapped to 1 in R, the map $f \longmapsto f(b)$ is a ring homomorphism. □

The map $f \longmapsto f(b)$ in Theorem 4 is called *evaluation* at b. Here b is fixed and f ranges over $R[X]$. We can also assume that f is fixed and obtain a map $b \longmapsto f(b)$ of R into R. Such a map is called the *polynomial function* on R defined by f. In calculus, it is not necessary to distinguish between polynomials and polynomial functions, since two polynomials f and g in $\mathbf{R}[X]$ define the same polynomial function of $\mathbf{R}$ into $\mathbf{R}$ if and only if $f = g$. However, when R is a finite ring, different polynomials may define the same polynomial function. For example, $\mathbf{Z}_2$ has two elements, so there are exactly $2^2 = 4$ functions from $\mathbf{Z}_2$ to $\mathbf{Z}_2$. The ring $\mathbf{Z}_2[X]$ is infinite, and each of the functions from $\mathbf{Z}_2$ to $\mathbf{Z}_2$ is defined by infinitely many different polynomials. For instance, the identity function on $\mathbf{Z}_2$ is defined by each of the polynomials $X, X^2, X + X^2 + X^3$, and $X^2 + X^7 + X^{10}$.

The library `CLASSLIB` contains procedures for manipulating arrays of polynomials with coefficients in the rings $\mathbf{Z}, \mathbf{Z}_n$, or $\mathbf{R}$ as well as in small finite rings. A polynomial $a_0 + a_1X + \ldots + a_nX^n$ in $\mathbf{R}[X]$ is represented by the vector of its coefficients. A matrix A with entries in $\mathbf{R}[X]$ is represented by a rank 3 array `A` in which `A[I;J;]` lists the coefficients of the corresponding entry in A. Thus

$$A = \begin{bmatrix} 0.1 + 2X & 1 - X^2 \\ X^2 & 2.3 + X + X^2 \end{bmatrix}$$

is represented by

```
      ⎕←A←2 2 3⍴0.1 2 0 1 0 ¯1 0 0 1 2.3 1 1
 0.1  2    0
 1    0   ¯1

 0    0    1
 2.3  1    1
```

The procedure *DARV* may be used to display *A* in a more convenient manner.

```
      1 DARV A
 .1   2.0    .0     1.0     .0 ¯1.0
 .0    .0   1.0     2.3    1.0  1.0
```

The procedures *RXSUM*, *RXDIFF*, and *RXPROD* compute entry-by-entry sums, differences, and products of arrays of real polynomials. For example,

```
       2 DARV B←2 2 3⍴¯1 1 0 0 1 ¯1 0 1 0 0 .73 0
 ¯1.00  1.00   .00       .00  1.00 ¯1.00
   .00  1.00   .00       .00   .73   .00
```

represents the matrix

$$B = \begin{bmatrix} -1 + X & X - X^2 \\ X & 0.73X \end{bmatrix}.$$

We can compute $A + B$, $A - B$, and $A \times B$ as follows.

```
      2 DARV A RXSUM B
 ¯.90 3.00   .00    1.00  1.00 ¯2.00
  .00 1.00  1.00    2.30  1.73  1.00
      2 DARV A RXDIFF B
 1.10 ¯1.00   .00    1.00 ¯1.00   .00
  .00  1.00  1.00    2.30   .27  1.00
      2 DARV A RXPROD B
 ¯.10 ¯1.90  2.00    .00   .00    .00  1.00 ¯1.00 ¯1.00  1.00
  .00   .00   .00   1.00   .00    .00  1.68   .73   .73   .00
```

The procedure *RXDEGREE* computes the degrees of the polynomials represented by its argument.

```
      RXDEGREE A                    RXDEGREE 0 0 0
 1 2                          ¯1
 2 2
```

The value returned by *RXDEGREE* for the zero polynomial is ¯1. This is a nonstandard definition. The procedure *RXLEAD* calculates the leading coefficients of an array of real polynomials.

```
      ⎕←C←RXLEAD A                  RXLEAD 0 0 0
  2 ¯1                          1
  1  1
```

Here *C*[*I*;*J*] is the leading coefficient of *A*[*I*;*J*;]. Note that *RXLEAD* considers the leading coefficient of the zero polynomial to be 1, another nonstandard definition. Recall the statement in the previous section that entries in real arrays that are smaller in absolute value than *EPSILON* times the largest absolute value of an entry in the array are considered to be 0. This applies to *RXSUM*, *RXDIFF*, *RXPROD*, *RXDEGREE*, and *RXLEAD*.

Arrays of polynomials in $\mathbf{Z}[X]$ are best manipulated using the procedures with the prefix *ZX* but procedures such as *RXSUM* and *RXPROD* may be used as long as all entries remain less than ÷*EPSILON*.

Arrays of polynomials in $\mathbf{Z}_n[X]$ are described by arrays of integers that represent the coefficients. The modulus n is, as always, the global variable N̲. To multiply $2 + 5X + 3X^2 + 9X^3$ and $7 + 10X + 2X^3 + 7X^4$ in $\mathbf{Z}_{11}[X]$, we enter

```
      N←11
      2 5 3 9 ZNXPROD 7 10 0 2 7
3 0 5 9 4 8 6 8
```

The result is $3 + 5X^2 + 9X^3 + 4X^4 + 8X^5 + 6X^6 + 8X^7$. The procedures *ZNXSUM*, *ZNXDIFF*, *ZNXDEGREE*, and *ZNXLEAD* are used in a similar way.

If R is the current finite ring described by the four arrays *FRPLUS*, *FRTIMES*, *FRNEG*, and *FRINV*, then computations in $R[X]$ can be performed using the procedures prefixed by *FRX*. Here are some examples in which R is the ring with 8 elements described in Section 1.

```
      PLUS FRINIT TIMES
      DAZV C←2 2 2ρ1 1 5 2 3 1 7 4
 1 1    5 2
 3 1    7 4
      DAZV D←2 2 2ρ7 0 4 1 6 2 5 5
 7 0    4 1
 6 2    5 5
      DAZV C FRXSUM D
 6 1    1 3
 5 3    2 1
      DAZV C FRXPROD D
 7 7 0    0 5 2
 6 4 2    7 7 0
```

We can evaluate polynomials using the procedures *RXEVAL*, *ZNXEVAL*, and *FRXEVAL*. For example, to evaluate all of the polynomials in the preceding matrix A at 3.7, we calculate

```
        1 DARV A
 .1   2.0    .0      1.0        .0 ¯1.0
 .0    .0  1.0       2.3       1.0  1.0
      A RXEVAL 3.7
 7.5                   ¯12.69
13.69                   19.69
```

To evaluate each entry of A at a different point, we make the second argument of *RXEVAL* a matrix.

```
      ⎕←X←2 2⍴3.7 ¯1.2 1 0.6
3.7  1.2
1    0.6
      A RXEVAL X
7.5                  ¯0.44
1                     3.26
```

In general, *RXEVAL* evaluates each polynomial in its first argument at the corresponding entry of its second argument. Arrays with one entry are expanded in the usual manner. The procedures *ZNXEVAL* and *FRXEVAL* evaluate polynomials over $\mathbf{Z}_n$ and the current finite ring, respectively.

EXERCISES

1 Complete the proof of Theorem 1.

2 Let A and B be vectors listing the coefficients of two polynomials f and g in $\mathbf{R}[X]$. Write an APL expression for a vector listing the coefficients of $f + g$. Allow for the possibility that A and B have different lengths.

3 The vectors `A←1 ¯2 3` and `B←1 1 2 4` list the coefficients of two polynomials f and g in $\mathbf{R}[X]$. The coefficients of fg are the column sums of the matrix

```
      D
1  1  2  4  0  0
0 ¯2 ¯2 ¯4 ¯8  0
0  0  3  3  6 12
```

The matrix

```
      ⎕←E←A∘.×B
 1  1  1  4
¯2 ¯2 ¯4 ¯8
 3  3  6 12
```

is closely related to D. Write an APL expression constructing D from E. Use this to give an APL definition for the product of two polynomials.

4 We have defined the polynomial ring in two variables $R[X,Y]$ to be $R[X][Y]$. Let $M = \{0, 1, \ldots\}$. Show that we can construct $R[X, Y]$ as the set Q of all functions $f:M \times M \longrightarrow R$ such that $f(i, j) = 0$ whenever i or j is sufficiently large. Define carefully the operations of addition and multiplication in Q.

*5 A polynomial in $\mathbf{R}[X, Y]$ can be represented by a matrix whose I,Jth entry is the coefficient of $X^I Y^J$. If A and B represent polynomials f and g in $\mathbf{R}[X, Y]$, write expressions for matrices representing $f+g$ and fg.

6 Experiment with the procedures for polynomial manipulation in *CLASSLIB*.

7 Let I be an ideal of the ring R and let J be the ideal of $R[X]$ generated by I. Describe J and show that $(R/I)\,[X]$ is isomorphic to $R[X]/J$.

8 Find a quadratic polynomial f in $\mathbf{Z}_4[X]$ such that $f(a) = 0$ for all a in $\mathbf{Z}_4$.

9 Find a function from $\mathbf{Z}_4$ to $\mathbf{Z}_4$ that is not defined by any polynomial in $\mathbf{Z}_4[X]$.

10 Let $f:R \longrightarrow S$ be a ring homomorphism. Show that the map taking $a_0 + a_1X + \ldots + a_nX^n$ in $R[X]$ to $f(a_0) + f(a_1)X + \ldots + f(a_n)X^n$ in $S[X]$ is a homomorphism.

5. MATRIX RINGS

This book has been written with the assumption that most readers are already familiar with matrix multiplications for matrices with entries in $\mathbf{R}$. In this section we will review the definition of matrix multiplication and extend it to matrices with entries in an arbitrary ring.

Let R be a ring and let A and B be matrices over R, that is, matrices with entries in R. Assume that A is an ℓ-by-m matrix and B is an m-by-n matrix. The *matrix product* of A and B is the ℓ-by-n matrix C such that

$$C_{ij} = \sum_{k=1}^{m} A_{ik} B_{kj} .$$

The matrix product is only defined when the number of columns of the first factor is equal to the number of rows of the second. If A and B are real matrices, then in APL notation the matrix product C of A and B is defined by the condition `C[I;J]=+/A[I;]×B[;J]` so that C is the inner product `A+.×B`. For example,

```
      ⎕←A←2 3⍴2 ¯1 4 0 5 1
 2 ¯1  4
 0  5  1
      ⎕←B←3 4⍴1 8 ¯3 4 ¯2 6 1 3 2 7 1 5
 1  8 ¯3  4
¯2  6  1  3
 2  7  1  5
      A+.×B
 12  38  ¯3  25
 ¯8  37   6  20
```

The matrix product of A and B is usually denoted simply as AB. If it is necessary to use an explicit symbol for matrix multiplication, we will borrow the symbol `+.×` from APL. As noted earlier, the expression $A \times B$ will always denote the entry-by-entry product of two arrays with the same shape.

Matrix multiplication satisfies conditions that may be viewed as generalizations of the associative law for multiplication and the distributive law in rings.

THEOREM 1. Let A, B, and C be ℓ-by-m, m-by-n, and n-by-r matrices, respectively, over the ring R. Then the products $(AB)C$ and $A(BC)$ are defined and are equal.

Proof. The matrices AB and BC are ℓ-by-n and m-by-r matrices, respectively. Thus the products $(AB)C$ and $A(BC)$ are defined. Moreover,

$$\begin{aligned}
[A(BC)]_{ij} &= \sum_{s=1}^{m} A_{is}(BC)_{sj} \\
&= \sum_{s=1}^{m} A_{is} \sum_{t=1}^{n} B_{st}C_{tj} \\
&= \sum_{s=1}^{m} \sum_{t=1}^{n} A_{is}B_{st}C_{tj} \\
&= \sum_{t=1}^{n} \sum_{s=1}^{m} A_{is}B_{st}C_{tj} \\
&= \sum_{t=1}^{n} \left(\sum_{s=1}^{m} A_{is}B_{st} \right) C_{tj}
\end{aligned}$$

$$= \sum_{t=1}^{n} (AB)_{it}C_{tj} = [(AB)C]_{ij}. \quad \square$$

THEOREM 2. Let B and C be m-by-n matrices and let A and D be ℓ-by-m and n-by-r matrices, respectively, all over the ring R. Then

$$A(B + C) = AB + AC,$$

$$(B + C)D = BD + CD.$$

Proof. All of the indicated products are defined. The proofs of these equalities are similar to the proof of Theorem 1 and are left as exercises. □

It should be noted that the proofs of Theorems 1 and 2 do not require R to be commutative.

If A and B are both n-by-n matrices over R, then the sum $A + B$ and the product AB are also n-by-n matrices. It follows immediately from Theorems 1 and 2 that the set $M_n(R)$ of all n-by-n matrices over R is a ring under the operations + and +.×. The zero element is the n-by-n matrix of zeros and the multiplicative identity is the matrix I such that $I_{ij} = 0$, $i \neq j$ and $I_{ii} = 1$, $1 \leq i \leq n$. Here is an example in $M_2(\mathbf{Z})$.

```
      ⎕←I←(⍳2)∘.=⍳2                  I+.×A
1 0                              4 ¯7
0 1                              1  3
      ⎕←A←2 2⍴4 ¯7 1 3               A+.×I
 4 ¯7                            4 ¯7
 1  3                            1  3
```

The group of units of $M_n(R)$ is denoted $GL_n(R)$ or $GL(n, R)$ and is called the n-by-n *general linear group* over R. In later sections we will devote considerable attention to the problem of deciding which elements of $M_n(R)$ are in $GL_n(R)$, particularly when R is commutative.

It is convenient to extend the operation +.× to pairs of arrays other than matrices. This will be done in exactly the same way that +.× is defined for real arrays in APL. Commonly, one factor will be a matrix and the other will be a vector. If A is an m-by-n matrix and V and W are vectors of length m and n, respectively, then VA and AW are the vectors with components

$$(VA)_j = \sum_{i=1}^{m} V_i A_{ij},$$

$$(AW)_i = \sum_{j=1}^{n} A_{ij} W_j \,.$$

The *transpose* B of a matrix A is the matrix obtained from A by interchanging rows and columns. Thus $B_{ij} = A_{ji}$. We usually write $B = A^t$. In APL notation this would be *B*←⍉*A*.

THEOREM 3. Let A and B be ℓ-by-m and m-by-n matrices, respectively, over the commutative ring R. Then $(AB)^t = B^t A^t$.

Proof. We have

$$[(AB)^t]_{ij} = (AB)_{ji}$$

$$= \sum_{k=1}^{m} A_{jk} B_{ki}.$$

Also,

$$[B^t A^t]_{ij} = \sum_{k=1}^{m} (B^t)_{ik} (A^t)_{kj}$$

$$= \sum_{k=1}^{m} B_{ki} A_{jk} .$$

Since R is commutative, the ijth entries of $(AB)^t$ and $B^t A^t$ are the same. □

The primitive APL operations +.× and ⍉ can be used to compute products and transposes of matrices with entries in **Z** or **R**. In *CLASSLIB* the procedures with the suffix *MATPROD* perform matrix multiplication with matrices over $\mathbf{Z}_n$, **Q**, $\mathbf{Z}_n[X]$, $\mathbf{Z}[X]$, and $\mathbf{R}[X]$, as well as finite rings. For example, to compute the product

$$\begin{bmatrix} 2/5 & -7/3 \\ 4/9 & 1 \end{bmatrix} \begin{bmatrix} 3 & 5/7 \\ 2/3 & -1/8 \end{bmatrix}$$

in $M_2(\mathbf{Q})$ we can enter

```
      DAQ A←2 2 2⍴2 5 ¯7 3 4 9 1 1
 2/5     ¯7/3
 4/9      1/1
      DAQ B←2 2 2⍴3 1 5 7 2 3 ¯1 8
 3/1      5/7
 2/3     ¯1/8
      DAQ A QMATPROD B
¯16/45        97/168
  2/1         97/504
```

In $M_2(\mathbf{Z}_5[X])$ we can compute the product

$$\begin{bmatrix} 1 + X + 3X^2 & 1 + 2X \\ 3 + 4X & X \end{bmatrix} \begin{bmatrix} 2 + X & 3 \\ 2X + X^2 & 4 + X \end{bmatrix}$$

as follows.

```
      DAZV C←2 2 3ρ1 1 3 1 2 0 3 4 0 0 1 0
1 1 3    1 2 0
3 4 0    0 1 0
      DAZV D←2 2 3ρ2 3 0 3 0 0 0 2 1 4 1 0
2 3 0    3 0 0
0 2 1    4 1 0
      N←5
      DAZV C ZNXMATPROD D
2 2 4 1      2 2 1 0
1 2 4 1      4 1 1 0
```

Thus the answer is

$$\begin{bmatrix} 2+2X+4X^2+X^3 & 2+2X+X^2 \\ 1+2X+4X^2+X^3 & 4+\ X+X^2 \end{bmatrix}.$$

For *ZNMATPROD* and *ZNXMATPROD* the modulus *N* must not exceed 10^7.

The procedure *TRAV* (for "transpose array of vectors") can be used to compute transposes of arrays with entries in **Q**, $\mathbf{Z}_n[X]$, $\mathbf{Z}[X]$, or $\mathbf{R}[X]$. For example,

```
      DAQ A                            DAQ TRAV A
 2/5     ¯7/3                      2/5     4/9
 4/9      1/1                     ¯7/3     1/1
```

If R is the current finite ring described by the arrays *FRPLUS*, *FRTIMES*, *FRNEG*, and *FRINV*, then products of matrices over R can be calculated using the procedure *FRMATPROD*.

Suppose now that R is any ring. We can form the two rings $S = M_n(R[X])$ and $T = M_n(R)[X]$. Here elements of S are matrices with polynomial entries and elements of T are polynomials with matrix coefficients. What relationship, if any, holds between S and T? It is possible to define a map from S to T as follows. Let F be in S. The ijth entry F_{ij} of F is a polynomial in $R[X]$. Let F_{ijk} denote the coefficient of X^k in F_{ij}. For a fixed k, let A_k be the element of $M_n(R)$ whose ijth entry is F_{ijk}. The polynomial

$$f = A_0 + A_1X + A_2X^2 + \ldots$$

is an element of T. As an example, we may take $R = \mathbf{Z}$, $n = 2$, and

$$F = \begin{bmatrix} 3-X\ +X^2 & 2X-4X^2 \\ -1+3X+4X^2 & 4-X^2 \end{bmatrix}.$$

Then the corresponding element of $M_2(\mathbf{Z})[X]$ is

$$f = \begin{bmatrix} 3 & 0 \\ -1 & 4 \end{bmatrix} + \begin{bmatrix} -1 & 2 \\ 3 & 0 \end{bmatrix} X + \begin{bmatrix} 1 & -4 \\ 4 & 1 \end{bmatrix} X^2.$$

THEOREM 4. For any ring R and any positive integer n, the rings $M_n(R[X])$ and $M_n(R)[X]$ are isomorphic.

Proof. We leave it as an exercise to verify that the map $F \longmapsto f$ just defined is a ring isomorphism. □

EXERCISES

1 Prove Theorem 2.

2 Verify that the ring axioms hold for $M_n(R)$.

3 Compute the following products in $M_2(\mathbf{Z}_7)$. Use *ZNMATPROD* to check your answers.

(a) $\begin{bmatrix} 6 & 5 \\ 4 & 1 \end{bmatrix} \begin{bmatrix} 2 & 5 \\ 6 & 1 \end{bmatrix}$ (b) $\begin{bmatrix} 3 & 4 \\ 1 & 2 \end{bmatrix} \begin{bmatrix} 1 & 5 \\ 3 & 5 \end{bmatrix}$

4 Compute the following product in $M_2(\mathbf{Z}[X])$ and use *ZXMATPROD* to check your answer:

$$\begin{bmatrix} 3 - X^2 & 1 + X \\ 7X & X + 3X^2 \end{bmatrix} \begin{bmatrix} -2 & X^2 \\ 1 - X & 2 + 3X \end{bmatrix}$$

5 Prove Theorem 4.

6 Suppose R is a commutative ring and A and B are matrices over R such that AB is defined. Show that for all r in R we have $(rA)B = A(rB) = r(AB)$, where rA is obtained from A by multiplying all entries by r.

7 Let $f: R \longrightarrow S$ be a ring homomorphism and let $g: M_n(R) \longrightarrow M_n(S)$ be defined so that the image of a matrix A in $M_n(R)$ is obtained by applying f to the entries in A. Show that g is a homomorphism.

8 State Theorem 3 using APL notation.

9 The ring $R = M_2(\mathbf{Z}_2)$ has 16 elements. Construct addition and multiplication tables for R as follows. Let

```
A←16 2 2⍴⍉(4⍴2)⊤⍳16.
```

Then `A` has shape `16 2 2` and the 16 matrices `A[I;;]` represent the 16 elements of R. Construct 16-by-16 matrices `P` and `T` (for "plus" and "times") such that `2|A[I;;]+A[J;;]` is `A[P[I;J];;]` and `2|A[I;;]+.×A[J;;]` is `A[T[I;J];;]`.

10 Let A and B be matrices with entries in a ring such that the product AB is defined. Show that the ith row of AB is the product of the ith row of A with B and the jth column of AB is the product of A with the jth column of B.

11 Suppose A, B, C, and D are matrices over a ring R with shape m-by-r, m-by-s, n-by-r, and n-by-s, respectively. By the *block matrix*

$$M = \begin{bmatrix} A & B \\ C & D \end{bmatrix}$$

we mean the $(m + n)$-by-$(r + s)$ matrix obtained by arranging the entries of A, B, C and D in the indicated pattern.

(a) Give a formula for $M[i;j]$.
(b) Give an APL expression describing M.

12 Show that block matrices may be multiplied as if they were matrices with matrix entries. That is, suppose

$$M = \begin{bmatrix} A & B \\ C & D \end{bmatrix} \quad \text{and} \quad N = \begin{bmatrix} S & T \\ U & V \end{bmatrix}$$

are block matrices over a ring. Prove that

$$MN = \begin{bmatrix} AS + BU & AT + BV \\ CS + DU & CT + DV \end{bmatrix},$$

provided the indicated products are defined.

13 Let R be a ring and let m and n be positive integers. Prove that $M_m(M_n(R))$ is isomorphic to $M_{mn}(R)$.

6. DETERMINANTS

Let A be a square matrix with entries in the commutative ring R. The *determinant* of A is a certain element of R that we denote det A. For example, if A is the 2-by-2 matrix

$$\begin{bmatrix} a & b \\ c & d \end{bmatrix},$$

then det $A = ad - bc$. Sometimes we write

$$\det A = \begin{vmatrix} a & b \\ c & d \end{vmatrix},$$

replacing the brackets used in displaying matrices by vertical straight lines. For example,

$$\begin{vmatrix} 7 & -1 \\ 4 & 2 \end{vmatrix} = (7)(2) - (-1)(4) = 18.$$

In this section we will define the determinant of any square matrix with entries in R and establish some basic properties of the determinant function. Additional facts about determinants will be proved in the next section.

Determinants are usually encountered first in the solution of simultaneous linear equations. Suppose a, b, c, d, e, f are real numbers and we wish to solve the equations

$$ax + by = e,$$

$$cx + dy = f.$$

Multiplying the first equation by d and the second by b and subtracting, we obtain

$$(ad - bc)x = ed - bf.$$

If $ad - bc \neq 0$, then

$$x = \frac{ed - bf}{ad - bc} = \frac{\begin{vmatrix} e & b \\ f & d \end{vmatrix}}{\begin{vmatrix} a & b \\ c & d \end{vmatrix}}.$$

Similarly, we find that

$$y = \frac{\begin{vmatrix} a & e \\ c & f \end{vmatrix}}{\begin{vmatrix} a & b \\ c & d \end{vmatrix}},$$

again assuming the denominator is not zero.

The determinant of the 3-by-3 matrix

$$\begin{bmatrix} a & b & c \\ d & e & f \\ g & h & i \end{bmatrix}$$

is defined to be

$$aei + bfg + cdh - ceg - afh - bdi.$$

This formula can be remembered using the following diagram.

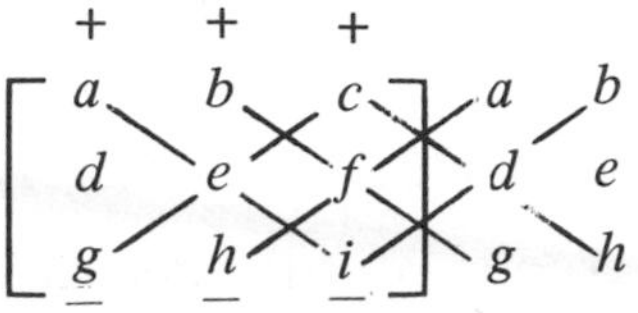

The terms with the plus signs are the products along the three diagonals sloping down and to the right. The terms with the minus signs are the products along the diagonals sloping up and to the right. Using the formula, we see that

$$\begin{vmatrix} 3 & 1 & -2 \\ 2 & -4 & 0 \\ -1 & 5 & -3 \end{vmatrix} = (3)(-4)(-3) + (1)(0)(-1) + (-2)(2)(5) - (-2)(-4)(-1) - (3)(0)(5) - (1)(2)(-3)$$

$$= 30.$$

Warning

> The "obvious" generalization of this scheme does not work for matrices larger than 3-by-3.

The determinant of an n-by-n matrix is a sum of $n!$ terms, half of which have a plus sign and half of which have a minus sign. To describe these terms, we use elements of the symmetric group Σ_n and the homomorphism sgn of Σ_n into $\{1, -1\}$, defined in Section 3.8. Permutations in Σ_n will be denoted by lower-case Greek letters. It will be convenient to use APL notation for indexing matrices so that A_{ij} will be written $A[i;j]$. The index origin will be assumed to be 1 throughout this section.

Let A be an n-by-n matrix over the commutative ring R. The determinant of A is the element of R given by the expression

$$\det A = \sum_{\sigma} (\operatorname{sgn} \sigma) \prod_{i=1}^{n} A[i;i\sigma],$$

where the sum is over all elements σ of Σ_n. The product

$$\prod_{i=1}^{n} A[i;i\sigma]$$

has exactly one factor from each row of A and, because σ is a permutation, exactly one factor from each column of A.

If $n = 2$, then Σ_n has two elements. If $\sigma = (1)(2)$, then the corresponding term in det A is $(+1)A[1;1]A[2;2]$. With $\sigma = (1, 2)$, we get $(-1)A[1;2]A[2;1]$. Thus

$$\det A = A[1;1]A[2;2] - A[1;2]A[2;1],$$

which agrees with the formula given for the determinant of a 2-by-2 matrix.

Let us consider a somewhat larger example. We will compute the determinant of the integer matrix

```
      ⎕←A←4 4⍴¯3 1 ¯3 1 ¯3 4 ¯1 0 ¯1 3 1 4 3 4 2 ¯1
¯3  1 ¯3  1
¯3  4 ¯1  0
¯1  3  1  4
 3  4  2 ¯1
```

directly from the definition. First, set

```
      ⎕IO←1
      S←GPSGN P←GPSYMG 4
```

The rows of `P` list the 24 elements in Σ_4, and `S[I]` is the sign of `P[I;]`. If σ is the `I`th row of `P`, then the `I`th term in det `A` is the product of `S[I]` and

```
A[1;1σ]×A[2;2σ]×...  = A[1;P[I;1]]×A[2;P[I;2]]×... .
```

Let

```
      B←2 1 2⍉A[;P]
```

Then `B[I;J]` is `A[J;P[I;J]]` and the `I`th term in the determinant of `A` is `S[I]××/B[I;]`. Thus the 24 terms of det `A` are given by the components of the vector `S××/B`. Therefore det `A` is

```
      +/S××/B
¯181
```

We can also write det `A` as `S+.××/B`.

Although our APL formulation of the definition of the determinant is valid for any square real matrix, typical workspace limits do not permit its use for matrices larger than 4-by-4.

As defined, the determinant of an n-by-n matrix has $n!$ terms, each with n factors. This would seem to make the determinant of a large matrix, say a 10-by-10 matrix, very difficult to compute. Actually, the determinant of an n-by-n matrix with entries in $\mathbf{R}$ or $\mathbf{Z}_p$, p a prime, can be computed in a time proportional to n^3, which grows much more slowly than $n!$. In *CLASSLIB* there are procedures for calculating determinants of matrices with entries in various commutative rings. The procedures *ZDET*, *ZNDET*, *RDET*, *RXDET*, *ZNXDET*, and *FRDET* compute determinants of matrices over $\mathbf{Z}$, $\mathbf{Z}_n$, $\mathbf{R}$, $\mathbf{R}[X]$, $\mathbf{Z}_n[X]$, and the current finite ring, respectively. The algorithms used by most of these procedures are discussed in Chapter 6. However, *FRDET* uses the method described in Exercise 7.19. When *ZNDET* is used, the modulus $\underline{N}$ may not exceed 10^7; when *ZNXDET* is used, $\underline{N}$ must, in addition, be prime. It should also be noted that *FRDET* does not check whether the current finite ring is commutative. The procedure *MPZDET* computes the determinant of an integer matrix as a multiple-precision integer. It can be used when the value of the determinant or the result of some intermediate calculation exceeds the precision of an APL scalar. See Exercise 15.

By way of an example, let us use *ZDET* to compute the determinants of the preceding matrix *A* and of its transpose.

```
      ZDET A                          ZDET ⍉A
¯181                             ¯181
```

As the next theorem proves, the fact that *A* and ⍉*A* have the same determinant is no accident.

THEOREM 1. If A is any square matrix over R, then $\det A^t = \det A$.

Proof. Let $B = A^t$. From the definition of the function det, we have

$$\det A = \sum_{\sigma} (\operatorname{sgn} \sigma) \prod_{i=1}^{n} A[i;i\sigma],$$

$$\det B = \sum_{\tau} (\operatorname{sgn} \tau) \prod_{j=1}^{n} B[j;j\tau]$$

$$= \sum_{\tau} (\operatorname{sgn} \tau) \prod_{j=1}^{n} A[j\tau;j],$$

where σ and τ range over Σ_n. Fix i and σ for a moment and define j to be $i\sigma$. Then $i = j\sigma^{-1}$ and $A[i;i\sigma] = A[j\sigma^{-1};j]$. If we keep σ fixed and let i vary from 1 to n, then j ranges over the set $\{1, \ldots, n\}$. Thus

$$\prod_{i=1}^{n} A[i;i\sigma] = \prod_{j=1}^{n} A[j\sigma^{-1};j].$$

Since $\operatorname{sgn} \sigma^{-1} = \operatorname{sgn} \sigma$ (see Exercise 3.8.9), we have

$$\det A = \sum_{\sigma} (\operatorname{sgn} \sigma^{-1}) \prod_{j=1}^{n} A[j\sigma^{-1};j].$$

If we write τ for σ^{-1} and note that τ runs over Σ_n as σ does, we obtain

$$\det A = \sum_{\tau} (\operatorname{sgn} \tau) \prod_{j=1}^{n} A[j\tau;j]$$

$$= \det B. \quad \square$$

A square matrix A is said to be *upper triangular* if $A[i;j] = 0$ for $i > j$. This is equivalent to saying that all entries below the main diagonal are 0. The matrix

```
      ⎕←U←3 3⍴2 ¯1 0 0 8 2 0 0 ¯3
 2 ¯1  0
 0  8  2
 0  0 ¯3
```

is upper triangular.

THEOREM 2. If A is an upper triangular matrix, then $\det A$ is the product of the entries of A on the main diagonal.

Proof. Let σ be in Σ_n. If σ is not the identity permutation, then for some i in $\{1, \ldots, n\}$ we have $i\sigma < i$. To see this, suppose $i\sigma \geq i$ for all i. Let $(i_1, \ldots, i_r)$ be a nontrivial cycle of σ. Then $i_1 \leq i_2 \leq \ldots \leq i_r \leq i_1$. This means $i_1 = i_2$, which cannot be the case. Thus if σ is not the identity, the product

$$\prod_{i=1}^{n} A[i;i\sigma]$$

contains a factor below the main diagonal of A and hence the product is 0. Therefore

$$\det A = \prod_{i=1}^{n} A[i;i]. \quad \square$$

The computation

```
      ZDET U                              ×/1 1⍉U
¯48                                 ¯48
```

verifies Theorem 2 in a particular case or provides a check on the correctness of the coding of *ZDET*, depending on one's point of view. As a special case of Theorem 2, we see that the determinant of the n-by-n identity matrix is 1. *Lower triangular* matrices are defined in the obvious way, and Theorem 2 holds with "upper" replaced by "lower".

The next result is one of the most important properties of the determinant.

THEOREM 3. Let A and B be square matrices of the same size with entries in R. Then $\det(A +.\times B) = (\det A)(\det B)$.

Proof. This theorem states that det is a homomorphism from the monoid $(M_n(R),+.\times)$ to the monoid $(R, \times)$. There are many different approaches that can be taken in proving this fact. The one chosen here is the most straightforward, although perhaps not the most elegant. Using the definition of the determinant to evaluate each side of the equality to be verified, we obtain two somewhat complicated expressions that must be proved equal. This is done by showing that a great deal of cancellation occurs in one of the expressions.

Let $C = A+.\times B$. Then

$$\det C = \sum_{\sigma} (\operatorname{sgn} \sigma) \prod_{i=1}^{n} C[i;i\sigma]$$

and

$$C[i;i\sigma] = \sum_{j=1}^{n} A[i;j]B[j;i\sigma].$$

If in Theorem 1.4 we let $a_i = C[i;i\sigma]$ and $b_{ij} = A[i;j]B[j;i\sigma]$, then we get

$$\prod_{i=1}^{n} C[i;i\sigma] = \sum_{\rho} \prod_{i=1}^{n} A[i;i\rho]B[i\rho;i\sigma],$$

where ρ ranges over the maps of $\{1, \ldots, n\}$ into itself. For a fixed σ and ρ, let

$$d(\sigma,\rho) = \prod_{i=1}^{n} A[i;i\rho]B[i\rho;i\sigma].$$

Then

$$\det C = \sum_{\sigma} (\operatorname{sgn} \sigma) \sum_{\rho} d(\sigma,\rho). \qquad (*)$$

Now let us expand $(\det A)(\det B)$. We have

$$\det A = \sum_{\sigma} (\operatorname{sgn} \sigma) \prod_{i=1}^{n} A[i;i\sigma],$$

$$\det B = \sum_{\tau} (\operatorname{sgn} \tau) \prod_{j=1}^{n} B[j;j\tau],$$

where σ and τ run over Σ_n. By the distributive laws (see Exercise 1.17), we have

$$(\det A)(\det B) = \sum_{\sigma,\tau} (\operatorname{sgn} \sigma)(\operatorname{sgn} \tau)\left(\prod_{i=1}^{n} A[i;i\sigma]\right)\left(\prod_{j=1}^{n} B[j;j\tau]\right).$$

Now $(\operatorname{sgn} \sigma)(\operatorname{sgn} \tau) = \operatorname{sgn} \sigma\tau$. Also, if we write $j = i\sigma$, then j runs over $\{1, \ldots, n\}$ as i does and

$$\prod_{j=1}^{n} B[j;j\tau] = \prod_{i=1}^{n} B[i\sigma;i\sigma\tau].$$

Thus

$$(\det A)(\det B) = \sum_{\sigma,\tau} (\operatorname{sgn} \sigma\tau) \prod_{i=1}^{n} A[i;j\sigma]B[i\sigma;i\sigma\tau]. \qquad (**)$$

The sum $(*)$ has $n!n^n$ terms, while the sum $(**)$ has $(n!)^2$ terms. To prove that these two sums are equal, we must show that a great deal of cancellation occurs in $(*)$. To do this, we need a lemma.

LEMMA 4. Let ρ be a map of $\{1, \ldots, n\}$ into itself that is not a permutation. There is a transposition θ in Σ_n such that $\theta\rho = \rho$. For any such θ we have $d(\sigma,\rho) = d(\theta\sigma,\rho)$ for all σ in Σ_n.

Proof. Since ρ is not a permutation, there exist i and j with $i \neq j$ and $i\rho = j\rho$. If we take θ to be the transposition (i,j), then $\theta\rho = \rho$. By definition,

$$d(\theta\sigma,\rho) = \prod_{i=1}^{n} A[i;i\rho]B[i\rho;i\theta\sigma]$$

$$= \left(\prod_{i=1}^{n} A[i;i\rho]\right)\left(\prod_{i=1}^{n} B[i\rho;i\theta\sigma]\right)$$

$$= \left(\prod_{i=1}^{n} A[i;i\rho]\right)\left(\prod_{i=1}^{n} B[i\theta\rho;i\theta\sigma]\right).$$

As i ranges over $\{1, \ldots, n\}$, so does $i\theta$. Thus

$$d(\theta\sigma,\rho) = \prod_{i=1}^{n} A[i;i\rho] \prod_{i=1}^{n} B[i\rho;i\sigma]$$

$$= \prod_{i=1}^{n} A[i;i\rho]B[i\rho;i\sigma] = d(\sigma,\rho). \quad \square$$

If ρ and θ are as in Lemma 4 and σ is in Σ_n, then sgn $\theta\sigma = -$sgn σ, since sgn $\theta = -1$. Thus $d(\sigma,\rho)$ and $d(\theta\sigma,\rho)$ occur with opposite signs in $(*)$ and therefore cancel. Since $< \theta >$ has order 2, we see that the sum of (sgn σ)$d(\sigma,\rho)$ as σ runs over a right coset of $< \theta >$ in Σ_n is 0. Hence

$$\sum_{\sigma} (\text{sgn } \sigma) d(\sigma,\rho) = 0$$

unless ρ is in Σ_n. Therefore

$$\det C = \sum_{\sigma} (\text{sgn } \sigma) \sum_{\rho} d(\sigma,\rho),$$

where now the sum on ρ is over Σ_n. If for a given σ and ρ we define τ in Σ_n by $\sigma = \rho\tau$, then

$$\det C = \sum_{\sigma} (\text{sgn } \sigma) \sum_{\rho} \prod_{i=1}^{n} A[i;i\rho]B[i\rho;i\sigma]$$

$$= \sum_{\rho,\tau} (\text{sgn } \rho\tau) \prod_{i=1}^{n} A[i;i\rho]B[i\rho;i\rho\tau]$$

and so, by $(**)$,

$$\det C = (\det A)(\det B). \quad \square$$

Theorem 3 has an important corollary that gives a necessary condition for a matrix to be a unit in $M_n(R)$.

COROLLARY 5. If A is a unit in $M_n(R)$, then det A is a unit in R and

$$(\det A^{-1}) = (\det A)^{-1}.$$

Proof. Since $AA^{-1} = I$, we have (det A) (det A^{-1}) = det I = 1, and so det A^{-1} is the inverse of det A in R. $\square$

We will continue our study of the units in $M_n(R)$ in the next section.

EXERCISES

1 Show that the formula

$$\det \begin{bmatrix} a & b & c \\ d & e & f \\ g & h & i \end{bmatrix} = aei + bfg + cdh - ceg - afh - bdi$$

is consistent with our general definition of the determinant.

2 Evaluate the following determinants by hand and check your answers using *ZDET*.

(a) $\begin{vmatrix} 7 & -2 \\ 4 & 3 \end{vmatrix}$ (b) $\begin{vmatrix} 3 & -1 & 2 \\ 0 & 4 & -3 \\ 5 & 1 & -2 \end{vmatrix}$

3 Compute the determinant of the matrix

$$\begin{bmatrix} 1 + 3X + 2X^2 & 4 + X + 3X^2 \\ 2 + X^2 & 3X + 2X^2 \end{bmatrix}$$

in $M_2(\mathbf{Z}_5[X])$ and check your result using *ZNXDET*.

4 At what points in the proofs of Theorems 1 and 3 was the commutativity of R used?

5 Verify the identity $\det(A+.\times B) = (\det A)(\det B)$ of Theorem 3 in a particular case by direct calculation using the matrices

$$A = \begin{bmatrix} 9 & -3 & 4 \\ 7 & 0 & -2 \\ 1 & -6 & 5 \end{bmatrix}, \quad B = \begin{bmatrix} -4 & 1 & 2 \\ 1 & 3 & -5 \\ -1 & 7 & 4 \end{bmatrix}$$

6 Suppose A is a square matrix with entries in R, and suppose A has a row of zeros. Show $\det A = 0$.

7 Let A be a square matrix with entries in R and let r be in R. Prove that multiplying all the entries in a row of A by r multiplies the determinant of A by r.

8 Let $f{:}R \longrightarrow S$ be a homomorphism of rings and let A be a square matrix with entries in R. Suppose B is the matrix with entries in S obtained by applying f to the entries of A. Show that $\det B = (\det A)f$.

9 Choose a random 4-by-4 integer matrix by executing

```
A←(?4 4⍴11)-6.
```

Select a value for the modulus $\underline{N}$ and check that $\underline{N}$ | *ZDET A* is the same as *ZNDET A*. Explain.

10 Let `A←(?6 6⍴11)-6`. Compare *D←RDET A* and *E←ZDET A*. Note the execution times for each computation. The procedure

ZDET takes longer because it uses only operations that yield integer results at each step.

11 Set ⎕*IO*←0, *N̲*←11, and *A*←?3 3 3⍴11. We will consider *A* to represent an element of $M_3(\mathbf{Z}_{11}[X])$. Let *F*←*ZNXDET A*. Select various values of *X* modulo 11 and verify that

ZNDET A ZNXEVAL X

is the same as *F ZNXEVAL X*. Explain.

12 Initialize the current finite ring to $\mathbf{Z}_6$ as in Exercise 3.8. Let ⎕*IO*←0, *N̲*←6, and *A*←?4 4⍴6. Compare the execution times for evaluating *ZNDET A* and *FRDET A*. Make further comparisons for larger random matrices over $\mathbf{Z}_6$.

13 Let *A*←(?10 10⍴201)−101. Compare the results and execution times of *RDET A* and *MPZDET A*.

14 Let A be an n-by-n real matrix and suppose every entry in A has absolute value not exceeding a. Show that

$$|\det A| \le n!a^n \le (na)^n.$$

[It is possible to prove the stronger result $|\det A| \le (a\sqrt{n})^n$.]

15 Suppose A is a square integer matrix and we know a real number c such that $|\det A| \le c$. Assume that $p_1, \ldots, p_r$ is a sequence of distinct primes such that $2c < p_1 p_2 \ldots p_r$. Show that $\det A$ can be determined if we know the determinant of the matrix over $\mathbf{Z}_{p_i}$ represented by A, $1 \le i \le r$. (The procedure *MPZDET* uses a bound similar to the one given in Exercise 14 to compute c. Then *ZNDET* is used to compute the determinant modulo

BIGPRIMES[*I*]

for sufficiently many values of *I*.)

7. UNITS IN MATRIX RINGS

Let R be a commutative ring. The primary goal of this section is to obtain a description of the group $GL_n(R)$ of units in the ring $M_n(R)$ of n-by-n matrices with entries in R. Along the way we will prove some additional facts about determinants and introduce the concept of a row operation. Besides being useful here, row operations will be needed in Chapter 5, and they will play a fundamental role in the algorithms developed in Chapter 6. As in the previous section, the index origin will be assumed to be 1.

A *row operation* over R is a particular kind of function that maps a matrix with entries in R to another matrix with the same shape. There are three types of row operations. In the following descriptions of the types of row operations, A will be assumed to be an m-by-n matrix over R.

Type 1. Choose an integer i between 1 and m and a unit u of R. The row operation $O_1(i,u)$ maps A to the matrix B, where B is identical with A except that the ith row $B[i;]$ of B is u times the ith row of A. For example, if $R = \mathbf{Z}$ and

```
      ⎕←B←A←3 3⍴⍳9                   B
1 2 3                          1   2   3
4 5 6                         ¯4  ¯5  ¯6
7 8 9                          7   8   9
      B[2;]←-B[2;]
```

then B is obtained from A by applying $O_1(2,-1)$.

Type 2. Choose two different integers i and j between 1 and m. The row operation $O_2(i,j)$ maps A to the matrix C obtained by interchanging the ith and jth rows of A. Continuing the previous example, we may apply $O_2(1,3)$ to A as follows.

```
      C←A                             C
      C[1 3;]←C[3 1;]           7 8 9
                                4 5 6
                                1 2 3
```

Type 3. Choose distinct integers i and j between 1 and m and an element r of R. The row operation $O_3(i,j,r)$ maps A to the matrix D obtained by adding r times $A[i;]$ to $A[j;]$. That is, $D[j;] = A[j;] + rA[i;]$ and $D[k;] = A[k;]$ for $k \neq j$. To apply $O_3(3,2,4)$ to A, we form

```
      D←A                              D
      D[2;]←D[2;]+4×D[3;]      1   2   3
                              32  37  42
                               7   8   9
```

We will write row operations to the left of their arguments. Thus if O is a row operation and A is a matrix, then the result applying O to A will be denoted OA. Row operations may be composed to form new maps. If O' is another row operation, then $(OO')A$ is $O(O'A)$.

THEOREM 1. The following identities hold for row operations.

(a) $O_1(i,u)O_1(i,v) = O_1(i,uv)$.

(b) $O_2(i,j)^2 = e$, the identity function.

(c) $O_3(i,j,r)O_3(i,j,s) = O_3(i,j,r+s)$.

Proof. We will prove part (c), leaving parts (a) and (b) as exercises. Suppose B is obtained from A by applying $O_3(i,j,s)$ and C is obtained from B by applying $O_3(i,j,r)$. If $k \neq j$, then $C[k;] = B[k;] = A[k;]$, and

$$C[j;] = B[j;] + rB[i;] = A[j;] + sA[i;] + rA[i;] = A[j;] + (r+s)A[i;].$$

Thus C is the image of A under $O_3(i,j,r+s)$. □

Not all row operations over R have the same domain. For example, $O_2(i,j)$ is defined only for matrices over R having at least m rows, where m is the larger of i and j.

THEOREM 2. Each row operation over R is a permutation of its domain. Moreover,

(a) $O_1(i,u)^{-1} = O_1(i,u^{-1})$.
(b) $O_2(i,j)^{-1} = O_2(i,j)$.
(c) $O_3(i,j,r)^{-1} = O_3(i,j,-r)$.

Proof. If u is a unit in R, then by Theorem 1a we have

$$O_1(i,u)O_1(i,u^{-1}) = O_1(i,1) = O_1(i,u^{-1})O_1(i,u).$$

Since $O_1(i,1)$ is the identity function on the set of matrices with at least i rows, it follows from Corollary 1.3.4 that $O_1(i,u)$ and $O_1(i,u^{-1})$ are inverse permutations. Parts (b) and (c) are derived in a similar manner from the corresponding parts of Theorem 1. □

The next theorem shows the relationship between row operations and matrix multiplication.

THEOREM 3. Let O be a row operation over R and let E be the result of applying O to the m-by-m identity matrix. If A is any matrix over R with m rows, then OA is the matrix product $E+.\times A$.

Proof. Set $B = E+.\times A$. We must consider three cases corresponding to the possible types for O. Suppose first that $O = O_1(i,u)$. Then $E[k;\ell] = 0$ for $k \neq \ell$. Thus

$$B[k;\ell] = \sum_t E[k;t]A[t;\ell] = E[k;k]A[k;\ell].$$

If $k \neq i$, then $E[k;k] = 1$ and $B[k;] = A[k;]$. Also, $B[i;] = E[i;i]A[i;] = uA[i;]$, and so $B = OA$.

Next suppose $O = O_2(i,j)$. Then $E[k;k] = 1$ if $k \neq i,j$, $E[i;j] = E[j;i] = 1$, and all other entries in E are 0. If $k \neq i,j$, then

$$B[k;\ell] = \sum_t E[k;t]A[t;\ell] = E[k;k]A[k;\ell] = A[k;\ell].$$

Also,

$$B[i;\ell] = \sum_t E[i;t]A[t;\ell] = E[i;j]A[j;\ell] = A[j;\ell].$$

Similarly, $B[j;] = A[i;]$ and, hence, $B = OA$.

Finally, suppose $O = O_3(i,j,r)$. Then, for all k, we have $E[k;k] = 1$ and $E[j;i] = r$. All other entries in E are 0. Thus, if $k \neq j$, then

$$B[k;\ell] = \sum_t E[k;t]A[t;\ell] = E[k;k]A[k;\ell] = A[k;\ell].$$

In addition,

$$B[j;\ell] = \sum_t E[j;t]A[t;\ell] = A[j;\ell] + rA[i;\ell].$$

Therefore $B[j;] = A[j;] + rA[i;]$ and $B = OA$. □

Any matrix E that is the result of applying a row operation to an identity matrix is called an *elementary matrix*. The *type* of elementary matrix is the type of the row operation used to construct it. Theorem 3 states that a row operation may be applied to a matrix by multiplying the matrix on the left by the appropriate elementary matrix. Let us verify Theorem 3 in a particular case. In the prior example we obtained D as the image of A under $O_3(3,2,4)$. The corresponding elementary matrix is constructed as follows.

```
      E←(⍳3)∘.=⍳3
      E[2;]←E[2;]+4×E[3;]
      E
1 0 0
0 1 4
0 0 1
```

The calculation

```
      ⎕←F←E+.×A                          ∧/,F=D
  1  2  3                          1
 32 37 42
  7  8  9
```

shows that indeed D is $E+.\times A$. If

```
      E1←E2←(⍳4)∘.=⍳4
      E1[2;]←¯3×E1[2;]
      E2[1 3;]←E2[3 1;]
      E1
 1  0  0  0
 0 ¯3  0  0
 0  0  1  0
 0  0  0  1
      E2
0 0 1 0
0 1 0 0
1 0 0 0
0 0 0 1
```

then $E1$ and $E2$ are the 4-by-4 elementary matrices corresponding to the row operations $O_1(2,-3)$ and $O_2(1,3)$, respectively, over **R**.

THEOREM 4. Every m-by-m elementary matrix E is a unit in $M_m(R)$. Moreover, E^{-1} is also an elementary matrix.

Proof. Let $E = OI$, where O is a row operation and I is the m-by-m identity matrix. Set O' equal to the row operation that is the inverse of O and set $E' = O'I$. Then, suppressing the symbol +.× for matrix multiplication, we have

$$I = (OO')I = O(O'I) = E(E'I) = EE'.$$

Similarly, $E'E = I$, and so E and E' are inverses of each other in $M_m(R)$. □

In general, the elementary matrices generate a proper subgroup of the group of units in $M_m(R)$. However, it will turn out that, for many familiar rings R, every unit in $M_m(R)$ is a product of elementary matrices.

Our next task is to compute the determinants of the various types of elementary matrices. First, we define a generalization of elementary matrices of type 2. Let σ be an element of Σ_m. The *permutation matrix* corresponding to σ is the m-by-m matrix $B = B(\sigma)$ such that $B[i;j] = 1$ if $j = i\sigma$ and $B[i;j] = 0$ otherwise. For example, if $m = 4$ and $\sigma = (1,3,2,4)$, then $B(\sigma)$ is the matrix

$$\begin{bmatrix} 0 & 0 & 1 & 0 \\ 0 & 0 & 0 & 1 \\ 0 & 1 & 0 & 0 \\ 1 & 0 & 0 & 0 \end{bmatrix}$$

If σ is the 2-cycle (i,j), then $B(\sigma)$ is just the m-by-m elementary matrix corresponding to $O_2(i,j)$.

THEOREM 5. Let σ be in Σ_m. Then $\det B(\sigma) = \operatorname{sgn} \sigma$.

Proof. Let $B = B(\sigma)$. Then, by definition,

$$\det B = \sum_{\tau} (\operatorname{sgn} \tau) \prod_{i=1}^{m} B[i;i\tau].$$

Since $B[i;i\tau] = 0$ unless $i\tau = i\sigma$, the only nonzero term in $\det B$ occurs when $\tau = \sigma$. Thus

$$\det B = (\operatorname{sgn} \sigma) \prod_{i=1}^{m} B[i;i\sigma] = \operatorname{sgn} \sigma. \quad \square$$

THEOREM 6. Let E be the elementary matrix obtained by applying the row operation O to the m-by-m identity matrix.

(a) If $O = O_1(i,u)$, then $\det E = u$.
(b) If $O = O_2(i,j)$, then $\det E = -1$.
(c) If $O = O_3(i,j,r)$, then $\det E = 1$.

Proof. (a) If $O = O_1(i,u)$, then E is upper (and also lower) triangular. Therefore $\det E$ is the product of the entries $E[k;k]$, which are all 1 except for $E[i;i] = u$. Thus $\det E = u$. (b) If $O = O_2(i,j)$, then $i \neq j$ and $E = B(\sigma)$, where $\sigma = (i,j)$. By Theorem 5, $\det E = \operatorname{sgn} \sigma = -1$. (c) If $O = O_3(i,j,r)$, then E is either upper triangular or lower triangular, according to whether $i > j$ or $i < j$. In either case, $\det E$ is the product of the diagonal entries in E, which are all 1. Hence $\det E = 1$. $\square$

THEOREM 7. Let B be the matrix obtained by applying the row operation O to the m-by-m matrix A.

(a) If $O = O_1(i,u)$, then $\det B = u(\det A)$.
(b) If $O = O_2(i,j)$, then $\det B = -\det A$.
(c) If $O = O_3(i,j,r)$, then $\det B = \det A$.

Proof. Let E be the m-by-m elementary matrix corresponding to O. Then B is $E+.\times A$ and so, by Theorem 6.3, we have $\det B = (\det E)(\det A)$. The theorem now follows from Theorem 6. $\square$

The following corollary is an important consequence of Theorem 7.

COROLLARY 8. If two rows of a square matrix A over R are the same, then $\det A = 0$.

Proof. Assume $A[i;] = A[j;]$ and $i \neq j$. Applying $O_3(i,j,-1)$ to A, we obtain a matrix B whose jth row consists entirely of zeros. By Theorem 7 and Exercise 6.6, $\det A = \det B = 0$. $\square$

So far we have dealt only with row operations. One can also define column operations in a completely analogous manner. We leave it as an exercise to formulate and prove the theorems about column operations that correspond to our results concerning row operations. We will make use of the following analogues of Theorem 7b and Corollary 8. If A is a square matrix, then interchanging two columns of A changes the sign of det A. If two columns of A are equal, det $A = 0$.

The 3-by-3 determinant

$$\begin{vmatrix} a & b & c \\ d & e & f \\ g & h & i \end{vmatrix} = aei + bfg + cdh - ceg - afh - bdi$$

can be written as

$$a(ei - fh) - b(di - fg) + c(dh - eg) = a \begin{vmatrix} e & f \\ i & h \end{vmatrix} - b \begin{vmatrix} d & f \\ g & i \end{vmatrix} + c \begin{vmatrix} d & e \\ g & h \end{vmatrix} .$$

It will be useful to obtain similar expansions for any determinant.

Let A be in $M_n(R)$. By definition, det A is the sum of the $n!$ terms

$$(\text{sgn } \sigma) \prod_{k=1}^{n} A[k;k\sigma].$$

Which of these terms contains a particular entry $A[i;j]$ as a factor? Clearly, the answer is the terms that correspond to permutations σ with $i\sigma = j$. Since each term in det A contains exactly one factor from the ith row of A, we have

$$\det A = \sum_{j=1}^{n} A[i;j] \sum_{i\sigma=j} (\text{sgn } \sigma) \prod_{k \neq i} A[k;k\sigma].$$

Define C to be the n-by-n matrix with

$$C[i;j] = \sum_{i\sigma=j} (\text{sgn } \sigma) \prod_{k \neq i} A[k;k\sigma].$$

Then, for $1 \leq i \leq n$, we have

$$\det A = \sum_{j=1}^{n} A[i;j]C[i;j].$$

We call $C[i;j]$ the ijth *cofactor* of A and C the matrix of cofactors of A. Since each term in det A has exactly one factor from the jth column of A, we also have

$$\det A = \sum_{i=1}^{n} A[i;j] \sum_{i\sigma=j} (\text{sgn } \sigma) \prod_{k \neq i} A[k;k\sigma] = \sum_{i=1} A[i;j]C[i;j].$$

Thus det A is the sum of the products of the entries in any one row or column of A with the corresponding cofactors.

By its definition, $C[i;j]$ does not depend on the entries in $A[i;]$ or $A[;j]$. Fix i and j and let A' be the matrix obtained from A by setting all entries in $A[i;]$ and $A[;j]$ to 0, except for $A[i;j]$, which is set equal to 1. For example, if

$$A = \begin{bmatrix} 3 & -1 & 4 & 2 \\ 1 & 2 & 5 & -3 \\ 7 & 2 & -2 & 1 \\ 8 & -6 & 3 & 2 \end{bmatrix}$$

and $i = 3, j = 2$, then

$$A' = \begin{bmatrix} 3 & 0 & 4 & 2 \\ 1 & 0 & 5 & -3 \\ 0 & 1 & 0 & 0 \\ 8 & 0 & 3 & 2 \end{bmatrix}.$$

Let C' be the matrix of cofactors of A'. Then

$$\det A' = \sum_{t=1}^{n} A'[i;t]C'[i;t] = C'[i;j].$$

However, $C'[i;j] = C[i;j]$, since $C[i;j]$ does not depend on the entries in the ith row or jth column of A. Thus $C[i;j] = \det A'$.

We have seen that if we multiply $A[i;j]$ by the cofactor $C[i;j]$ and sum over j, then the result is det A. Suppose, instead, we multiply $A[i;j]$ and $C[k;j]$ with $k \neq i$ and sum over j. What do we get then? To help answer this question, let us evaluate

$$\sum_{j=1}^{4} A[2;j]C[4;j],$$

where A is the preceding 4-by-4 matrix. Let

$$B = \begin{bmatrix} 3 & -1 & 4 & 2 \\ 1 & 2 & 5 & -3 \\ 7 & 2 & -2 & 1 \\ 1 & 2 & 5 & -3 \end{bmatrix}.$$

Here the fourth row of A has been changed so that it is equal to the second row. Let D be the matrix of cofactors of B. We wish to compare the entries

$C[4;j]$ and $D[4;j]$. Take $j = 2$ as an example. By our previous discussion, $C[4;2]$ is the determinant of the matrix

$$\begin{bmatrix} 3 & 0 & 4 & 2 \\ 1 & 0 & 5 & -3 \\ 7 & 0 & -2 & 1 \\ 0 & 1 & 0 & 0 \end{bmatrix}$$

But $D[4;2]$ is the determinant of the same matrix, and so $C[4;2] = D[4;2]$. In the same way we see that $C[4;j] = D[4;j]$, $1 \leq j \leq 4$. Since $B[4;j] = A[2;j]$, we have

$$\sum_{j=1}^{4} A[2;j]C[4;j] = \sum_{j=1}^{4} B[4;j]D[4;j] = \det B.$$

But $\det B = 0$ by Corollary 8. Thus

$$\sum_{j=1}^{4} A[2;j]C[4;j] = 0.$$

THEOREM 9. Let A be an n-by-n matrix and let C be the matrix of cofactors of A. If $k \neq i$ and $j \neq \ell$, then

$$\sum_{t=1}^{n} A[i;t]C[k;t] = \sum_{t=1}^{n} A[t;j]C[t;\ell] = 0.$$

Proof. Let B be the matrix obtained from A by replacing $A[k;]$ with $A[i;]$. Just as in our example, we see that $C[k;t]$ is the ktth cofactor of B, so

$$\sum_{t=1}^{n} A[i;t]C[k;t] = \sum_{t=1}^{n} B[k;t]C[k;t] = \det B = 0.$$

Similarly, we find that

$$\sum_{t=1}^{n} A[t;j]C[t;\ell] = 0. \quad \square$$

For any square matrix A, let A^* be the transpose of the matrix of cofactors of A. We call A^* the *adjoint* of A.

THEOREM 10. If A is a square matrix, then $AA^* = A^*A = dI$, where $d = \det A$ and I is the identity matrix.

Proof. Let n be the number of rows in A and set $P = AA^*$. Then

$$P[i;k] = \sum_{t=1}^{n} A[i;t]A^*[t;k] = \sum_{t=1}^{n} A[i;t]C[k;t],$$

where C is the matrix of cofactors of A. Thus $P[i;i] = \det A = d$ and, by Theorem 9, $P[i;k] = 0$ if $i \neq k$. Therefore $P = dI$.

Now set $Q = A^*A$. Then

$$Q[\ell;j] = \sum_{t=1}^{n} A^*[\ell;t]A[t;j] = \sum_{t=1}^{n} A[t;j]C[t;\ell].$$

Thus Q is also equal to dI. □

We can now give the promised description of the units in $M_n(R)$.

THEOREM 11. Let A be an n-by-n matrix over a commutative ring R. Then A is a unit in $M_n(R)$ if and only if $\det A$ is a unit in R.

Proof. If A is a unit in $M_n(R)$, then $d = \det A$ is a unit in R by Corollary 6.5. Now suppose d is a unit. Then, by Theorem 10,

$$A(d^{-1}A^*) = d^{-1}AA^* = d^{-1}dI = I = (d^{-1}A^*)A.$$

Therefore $d^{-1}A^*$ is a two-sided inverse for A in $M_n(R)$ and A is a unit. □

We now wish to show that each cofactor of an n-by-n matrix A is, up to sign, the determinant of an $(n-1)$-by-$(n-1)$ submatrix of A. We begin an example using the matrix

$$A = \begin{bmatrix} 3 & -1 & 4 & 2 \\ 1 & 2 & 5 & -3 \\ 7 & 2 & -2 & 1 \\ 8 & -6 & 3 & 2 \end{bmatrix}.$$

We know that the cofactor $C[3;2]$ of A is the determinant of the matrix

$$A' = \begin{bmatrix} 3 & 0 & 4 & 2 \\ 1 & 0 & 5 & -3 \\ 0 & 1 & 0 & 0 \\ 8 & 0 & 3 & 2 \end{bmatrix}.$$

Let B be the 3-by-3 matrix obtained by deleting the third row and second column of A. Thus

$$B = \begin{bmatrix} 3 & 4 & 2 \\ 1 & 5 & -3 \\ 8 & 3 & 2 \end{bmatrix}.$$

Then $\det A' = -\det B$. We could verify this fact by direct calculation of det A' and det B. However, we will use another approach that will give us an understanding of why the result holds.

If we interchange the second and third columns of A' and then interchange the third and fourth columns, we obtain the matrix

$$D = \begin{bmatrix} 3 & 4 & 2 & 0 \\ 1 & 5 & -3 & 0 \\ 0 & 0 & 0 & 1 \\ 8 & 3 & 2 & 0 \end{bmatrix}.$$

Each column interchange reversed the sign of the determinant, and so det D = det A'. Now, if we interchange the third and fourth rows of D, we get

$$E = \begin{bmatrix} 3 & 4 & 2 & 0 \\ 1 & 5 & -3 & 0 \\ 8 & 3 & 2 & 0 \\ 0 & 0 & 0 & 1 \end{bmatrix}$$

and det E = –det D. Note that the submatrix of E consisting of the first three rows and first three columns is B. Now

$$\det E = \sum_{\sigma} (\operatorname{sgn} \sigma) \prod_{k=1}^{4} E[k;k\sigma],$$

where σ ranges over Σ_4. If $4\sigma \neq 4$, then $E[4;4\sigma] = 0$. Thus we need only sum over those σ with $4\sigma = 4$. If $4\sigma = 4$, then the restriction τ of σ to $\{1,2,3\}$ is an element of Σ_3 and the map $\sigma \longmapsto \tau$ is a 1–1 correspondence. Moreover, since the sign of a permutation depends only on the number of cycles of even length, sgn τ = sgn σ. Since $E[4;4] = 1$, we have

$$\det E = \sum_{\tau} (\operatorname{sgn} \tau) \prod_{k=1}^{3} E[k;k\tau] = \sum_{\tau} (\operatorname{sgn} \tau) \prod_{k=1}^{3} B[k;k\tau] = \det B.$$

Thus $C[3;2]$ = det A' = det D = –det E = –det B.

Now let A be any n-by-n matrix over R. We define the ijth *minor* of A to be the determinant $M[i;j]$ of the $(n-1)$-by-$(n-1)$ matrix obtained by deleting the ith row and jth column of A.

THEOREM 12. Let A be a square matrix and let C and M be the matrices of cofactors and minors of A, respectively. Then

$$C[i;j] = (-1)^{i+j} M[i;j].$$

Proof. The preceding example contains all the essential ideas of the general case. Let A' be the matrix obtained from A by setting $A[i;j]$ equal to 1 and all other entries in $A[i;]$ and $A[;j]$ equal to 0. Let B be obtained from A by deleting the ith row and jth column. Then $C[i;j] = \det A'$ and $M[i,j] = \det B$. Performing $n - j$ column interchanges and then $n - i$ row interchanges transforms A' into the matrix E, which has the form

$$\begin{bmatrix} & & 0 \\ & & \cdot \\ & B & \cdot \\ & & \cdot \\ & & 0 \\ 0 & \ldots & 0 \quad 1 \end{bmatrix}$$

We have $\det A' = (-1)^{n-j}(-1)^{n-i}\det E = (-1)^{2n-i-j}\det E = (-1)^{i+j} \det E$, since $i + j \equiv 2n-i-j \pmod 2$. Now, by an argument analogous to the previous one, $\det E = \det B$. Thus $C[i;j] = \det A' = (-1)^{i+j}\det E = (-1)^{i+j}\det B = (-1)^{i+j}M[i;j]$. □

We can now write the determinant of the n-by-n matrix A in terms of the determinants of n submatrices of A of size $(n-1)$-by-$(n-1)$. For any i and j we have

$$\det A = \sum_{t=1}^{n} (-1)^{i+t} A[i;t]M[i;t],$$

$$\det A = \sum_{t=1}^{n} (-1)^{t+j} A[t;j]M[t;j].$$

These formulas are referred to as the expansion of det A by the minors of the ith row and the jth column, respectively. Again using our example

$$A = \begin{bmatrix} 3 & -1 & 4 & 2 \\ 1 & 2 & 5 & -3 \\ 7 & 2 & -2 & 1 \\ 8 & -6 & 3 & 2 \end{bmatrix}$$

and expanding det A by the minors of the third row, we get

$$\det A = (-1)^4 7 \begin{vmatrix} -1 & 4 & 2 \\ 2 & 5 & -3 \\ -6 & 3 & 2 \end{vmatrix} + (-1)^5 2 \begin{vmatrix} 3 & 4 & 2 \\ 1 & 5 & -3 \\ 8 & 3 & 2 \end{vmatrix}$$

$$+ (-1)^6(-2) \begin{vmatrix} 3 & -1 & 2 \\ 1 & 2 & -3 \\ 8 & -6 & 2 \end{vmatrix} + (-1)^7 1 \begin{vmatrix} 3 & -1 & 4 \\ 1 & 2 & 5 \\ 8 & -6 & 3 \end{vmatrix}$$

$$= 7(109)-2(-121)-2(-60)-1(-17)$$

$$= 1142.$$

The formula $A^{-1} = (\det A)^{-1} A^*$ can be used to compute inverses in $M_n(R)$. For example, if

$$A = \begin{bmatrix} a & b \\ c & d \end{bmatrix},$$

then the matrices M, C, and A^* are given by

$$M = \begin{bmatrix} d & c \\ b & a \end{bmatrix}, \quad C = \begin{bmatrix} d & -c \\ -b & a \end{bmatrix},$$

$$A^* = \begin{bmatrix} d & -b \\ -c & a \end{bmatrix}.$$

Thus, if $\det A = ad - bc$ is a unit in R, we have

$$A^{-1} = (ad - bc)^{-1} \begin{bmatrix} d & -b \\ -c & a \end{bmatrix}.$$

If n is large, however, this method requires the evaluation of a great many determinants and is quite time consuming. If R happens to be the integers, a field, or the polynomial ring $F[X]$ over a field, then there is a much more efficient algorithm for computing inverses of matrices. This algorithm will be discussed in Section 6.2.

The monadic primitive APL operation ⌹ computes approximate inverses in $M_n(\mathbf{R})$.

```
      ⎕←A←2 2⍴3 2 7 6
3 2
7 6
      ⎕←B←⌹A
 1.5    ¯0.5
¯1.75    0.75
      A+.×B
 1.000000000E0       4.440892099E¯16
¯1.994801444E¯15   1.000000000E0
```

Note that although the value printed for B is the exact inverse of A, the value stored in the computer is not exact, so $A+.\times B$ is not quite the identity matrix.

The procedures *ZMATINV* and *ZNMATINV* compute inverses in $M_n(\mathbf{Z})$ and $M_n(\mathbf{Z}_m)$, respectively. With *ZNMATINV* the value $\underline{N}$ of m may not exceed 10^7.

```
      ⎕←C←2 2⍴3 2 7 5
3 2
7 5
      ⎕←D←ZMATINV C
 5 ¯2
¯7  3
      C+.×D
1 0
0 1
      ⎕←E←2 2⍴2 1 4 3
2 1
4 3
      N←5
      ⎕←F←ZNMATINV E
4 2
3 1
      E ZNMATPROD F
1 0
0 1
```

EXERCISES

1 What are the results of applying each of the integer row operations $O_1(3,-1)$, $O_2(1,2)$, and $O_3(1,3,-2)$ to the matrix

$$U = \begin{bmatrix} 7 & -2 & 1 \\ -3 & 2 & 6 \\ 4 & -1 & -5 \end{bmatrix} ?$$

2 Considering the matrix

$$P = \begin{bmatrix} 3 & 1 & 6 \\ 2 & 0 & 4 \\ 1 & 5 & 3 \end{bmatrix}$$

to be an element of $M_3(\mathbf{Z}_7)$, apply each of the row operations $O_1(2,4)$, $O_2(2,1)$, and $O_3(2,3,3)$ to P.

3 Complete the proof of Theorem 1.

4 Complete the proof of Theorem 2.

5 Construct the elementary matrices in $M_3(\mathbf{Z})$ corresponding to the row operations $O_1(3,-1)$, $O_2(1,3)$, and $O_3(1,3,-2)$. Compute the products of these elementary matrices with the matrix U in Exercise 1. Compare the results with your answers to Exercise 1.

6 Compute the inverses of the elementary matrices obtained in Exercise 5.

7 Show that the map taking an element σ in Σ_n to the permutation matrix $B(\sigma)$ is an injective homomorphism of Σ_n into the group $GL_n(R)$.

8 Suppose *SIGMA* is a vector that is a permutation of ⍳*N*. Write an APL expression for the permutation matrix corresponding to *SIGMA*.

9 Let A be in $M_n(R)$ and let σ be in Σ_n. Define B to be the n-by-n matrix such that $B[i;j] = A[i\sigma;j]$. Show $\det B = (\operatorname{sgn} \sigma) \det A$.

10 Suppose C is the matrix of cofactors of the square matrix A. Show that C^t is the matrix of cofactors of A^t.

11 Let *A* be an *N*-by-*N* matrix and let *I* and *J* be elements of ⍳*N*. Write an APL expression for the matrix *B* obtained by deleting the *I*th row and *J*th column from *A*.

12 Compute the determinant of the matrix U in Exercise 1, first using the expansion by minors of the third row and then using the expansion by minors of the second column.

13 For each of the following pairs of a ring R and a matrix A in $M_n(R)$, show that A is a unit in $M_n(R)$ and use the formula $A^{-1} = (\det A)^{-1} A^*$ to compute A^{-1}.

(a) $R = \mathbf{Z}$, $$A = \begin{bmatrix} 6 & 7 \\ 7 & 8 \end{bmatrix}.$$

(b) $R = \mathbf{Q}$, $$A = \begin{bmatrix} 3 & 9 \\ 2 & 7 \end{bmatrix}.$$

(c) $R = \mathbf{Z}$, $$A = \begin{bmatrix} 1 & 3 & 1 \\ 1 & 2 & 0 \\ -1 & -5 & -2 \end{bmatrix}.$$

(d) $R = \mathbf{Z}_9$, $$A = \begin{bmatrix} 4 & 3 \\ 5 & 2 \end{bmatrix}.$$

(e) $R = \mathbf{Z}[X]$, $$A = \begin{bmatrix} 1 + 2X & X^2 \\ -4 & 1 - 2X \end{bmatrix}$$

14 Define column operations. Show that a column operation can be applied to a matrix A by applying a row operation to A^t and taking the transpose of the result.

15 Show that every elementary matrix can be obtained from the identity matrix by a column operation. (Thus there is no need to distinguish between "row elementary matrices" and "column elementary matrices".)

16 State and prove the analogues of Theorems 1, 2, 3, 6, and 7 and Corollary 8 for column operations.

17 Show that column operations commute with row operations. That is, suppose O is a row operation and O' is a column operation. Prove that applying O and then O' gives the same result as applying O' and then O.

18 Let A and B be square matrices over R. Show that the determinant of the block matrix

$$\begin{bmatrix} A & C \\ 0 & B \end{bmatrix}$$

is $(\det A)(\det B)$. Here C is any matrix over R with the appropriate shape and 0 denotes a matrix of zeros.

19 Let A be an n-by-n matrix with entries in R. Let P denote the set of all subsets of $S = \{1, \ldots, n\}$. For Y in P, let $d(Y)$ be the determinant of the submatrix of A consisting of the entries in the first $r = |Y|$ rows and the columns whose indices are in Y. If $Y = \{y_1, \ldots, y_r\}$ and $y_1 < y_2 < \ldots < y_r$, show that

$$d(Y) = \sum_{i=1}^{r} (-1)^{i+r} A[r; y_i] d(Y - \{y_i\}).$$

Explain how to compute $\det A = d(S)$ by starting with $d(\phi) = 1$. Show that the total number of arithmetic operations involved is less than some constant times $n2^n$. (This is the method used in *FRDET*.)

20 Use monadic ⌹, *ZMATINV*, or *ZNMATINV*, as appropriate, to compute the inverses of the matrices in Exercise 13a–d.

8. FIELDS OF FRACTIONS

Let R be a ring. In Section 2 we noted that the map $h:\mathbf{Z} \rightarrow R$ taking m in $\mathbf{Z}$ to $m1$ in R is a ring homomorphism. We defined the characteristic of R to be the nonnegative generator of the kernel of h. Thus, if R has characteristic 0, then h is an injection and the image of $\mathbf{Z}$ under h is a subring of R isomorphic to $\mathbf{Z}$. If R happens to be a field, we can say even more.

THEOREM 1. If F is a field of characteristic 0, then F contains a subfield isomorphic to $\mathbf{Q}$.

Proof. Let $h:\mathbf{Z}\to F$ map m to $m1$. We would like to extend h to a map of $\mathbf{Q}$ into F. Each element of $\mathbf{Q}$ can be written as a/b, with a and b in $\mathbf{Z}$ and $b \neq 0$. By assumption, h is injective, and so $h(b) \neq 0$. It is therefore reasonable to try to map a/b to $h(a)h(b)^{-1} = h(a)/h(b)$ in F. But is this well defined? Suppose $a/b = c/d$, with a, b, c, d all in $\mathbf{Z}$ and b and d not 0. Then $ad = bc$ and hence $h(a)h(d) = h(b)h(c)$. The element $h(b)h(d)$ of F is not 0, so we may divide by it to obtain $h(a)/h(b) = h(c)/h(d)$. Thus $a/b \mapsto h(a)/h(b)$ does define a map h' of $\mathbf{Q}$ into F. For all a/b and c/d in $\mathbf{Q}$, we have

$$h'\left(\frac{a}{b} + \frac{c}{d}\right) = h'\left(\frac{ad+bc}{bd}\right) = \frac{h(ad+bc)}{h(bd)}$$

$$= \frac{h(a)h(d) + h(b)h(c)}{h(b)h(d)} = \frac{h(a)}{h(b)} + \frac{h(c)}{h(d)}$$

$$= h'(a/b) + h'(c/d).$$

Also,

$$h'\left(\frac{a}{b} \times \frac{c}{d}\right) = h'\left(\frac{ac}{bd}\right) = \frac{h(ac)}{h(bd)} = \frac{h(a)h(c)}{h(b)h(d)}$$

$$= \frac{h(a)}{h(b)} \times \frac{h(c)}{h(d)} = h'(a/b)h'(c/d).$$

Therefore h' is a ring homomorphism. If $h'(a/b) = 0$, then $h(a)/h(b) = 0$ and so $h(a) = 0$. This implies that $a = 0$. Thus the kernel of h' is trivial and the image of $\mathbf{Q}$ under h' is a subfield of F isomorphic to $\mathbf{Q}$. □

Theorem 1 states that any field that contains a copy of $\mathbf{Z}$ also contains a copy of $\mathbf{Q}$. The purpose of this section is to obtain a similar result for any integral domain. An *embedding* of one ring R in another ring S is an injective homomorphism of R into S. Suppose R is an integral domain. It is reasonable to ask the following questions.

1. Can R be embedded in a field F?
2. If so, can we choose F in such a way that whenever R can be embedded in a field K, then F can be embedded in K?

We will be able to answer both these questions affirmatively.

THEOREM 2. Every integral domain can be embedded in a field.

Proof. Let R be an integral domain. In constructing a field F and an embedding of R into F, we will be guided by the example of the familiar embedding of $\mathbf{Z}$ into $\mathbf{Q}$. Elements of $\mathbf{Q}$ can be written as quotients of integers, but not in a unique manner. This suggests that elements of F should be described as quotients of elements of R. Since division of one element of R by another is not defined, we must introduce "formal quotients". This is done as follows. Let M be the set of all ordered pairs (a,b), with a and b in R and $b \neq 0$. We define a relation $\sim$ on M by saying that $(a,b) \sim (c,d)$ if and only if $ad = bc$.

LEMMA 3. The relation $\sim$ is an equivalence relation on M.

Proof. The statement $(a,b) \sim (a,b)$ is equivalent to $ab = ba$, which holds because R is commutative. Thus $\sim$ is reflexive. If $(a,b) \sim (c,d)$, then $ad = bc$. Therefore $cb = da$ and $(c,d) \sim (a,b)$. Hence $\sim$ is symmetric. Now suppose $(a,b) \sim (c,d)$ and $(c,d) \sim (e,f)$. Then $ad = bc$ and $cf = de$. Multiplying the first equality by f and the second by b, we obtain

$$daf = bcf = dbe.$$

Since $d \neq 0$ and R is an integral domain, the equality $daf = dbe$ implies $af = be$. Thus $(a,b) \sim (e,f)$ and $\sim$ is transitive. □

Let F be the set of equivalence classes of $\sim$ on M. For (a,b) in M, we will denote the equivalence class containing (a,b) by $[a,b]$. We think of $[a,b]$ as the formal quotient of a and b. We must now define binary operations $+$ and $\times$ on F in such a way that $(F,+,\times)$ is a field. In any field we have

$$\frac{u}{v} + \frac{x}{y} = \frac{uy + vx}{vy},$$

$$\left(\frac{u}{v}\right)\left(\frac{x}{y}\right) = \frac{ux}{vy}.$$

This suggests the following definitions.

$$[a,b] + [c,d] = [ad + bc, bd].$$

$$[a,b] \times [c,d] = [ac, bd].$$

However, we must show that these operations are well defined. Suppose $[a,b] = [a',b']$ and $[c,d] = [c',d']$. Then $ab' = ba'$ and $cd' = dc'$. Therefore $ab'cd' = a'bc'd$ and, hence, $[ac,bd] = [a'c',b'd']$. This means that the product of two elements of F does not depend on the particular representation of the factors, so $\times$ is well defined as a binary operation on F. We also have

$$(ad + bc)b'd' = ab'dd' + bb'cd' = ba'dd' + bb'dc'$$
$$= bd(a'd' + b'c').$$

Thus $[ad + bc, bd] = [a'd' + b'c', b'd']$ and + is well defined too.

We must now verify that $(F, +, \times)$ satisfies all of the field axioms. This is not difficult, and we leave most of the work as an exercise. The additive identity element is $0 = [0,1]$ and $[a,b] = 0$ if and only if $a = 0$. The additive inverse of $[a,b]$ is $[-a,b]$. The multiplicative identity is $1 = [1,1]$ and, if $[a,b] \neq 0$, then the multiplicative inverse of $[a,b]$ is $[b,a]$.

All that remains is to define a map h of R into F and check that h is an embedding. For a in R, set $h(a) = [a,1]$ in F. Then

$$h(a+b) = [a+b,1] = [a,1] + [b,1] = h(a) + h(b),$$

$$h(ab) = [ab,1] = [a,1] \times [b,1] = h(a) \times h(b).$$

Therefore h is a ring homomorphism. If $h(a) = [a,1] = 0$, then $a = 0$. Thus h is an embedding. This completes the proof of Theorem 2. □

The field F constructed in the proof of Theorem 2 is called the *field of fractions* of R. Sometimes F is referred to as the *field of quotients* of R. It is important to be aware of the difference between the terms "field of quotients" and "quotient field". The latter refers to a quotient ring R/I of R that happens to be a field. To avoid any confusion, we will always use the term "field of fractions".

The field of fractions of $\mathbf{Z}$ is $\mathbf{Q}$. After the integers, the next most commonly encountered integral domains are the polynomial rings $K[X]$, where K is a field. The quotient field of $K[X]$ is called the *field of rational functions* in one variable over K. It is denoted $K(X)$. Elements of $K(X)$ are written as quotients of polynomials in $K[X]$. For example,

$$\frac{2 - 3X + 4X^2 - X^4}{7X + 6X^2 + 3X^3}$$

is an element of $\mathbf{Q}(X)$.

We have answered the first of our two questions concerning embeddings of integral domains in fields. It is now time to answer the second.

THEOREM 4. Let R be an integral domain with field of fractions F. If R can be embedded in a field K, then F can also be embedded in K.

Proof. In Theorem 1 we have already proved a special case of this result, and the proof of Theorem 1 may be used here with only minor changes. Let $h: R \longrightarrow K$ be an embedding. The embedding $h': F \longrightarrow K$ is given by $h'([a,b]) = h(a)/h(b)$. As in the proof of Theorem 1, we must show that h' is well defined. The details of this verification and the proof that h' is an embedding are left as exercises. □

EXERCISES

1 Complete the proof of Theorem 2 by showing that the set F of formal quotients is a field.

2 Let a be a nonzero element of $\mathbf{Z}_{30}$. Show that there is a ring homomorphism of $\mathbf{Z}_{30}$ into a field such that a is not mapped to 0. Describe the set of positive integers n such that a similar statement can be made about $\mathbf{Z}_n$.

3 Fill in the details in the proof of Theorem 4.

4 In this section we defined *the* field of fractions of an integral domain R. It is possible to generalize our definition slightly. We say that *a* field of fractions for R is a pair (F,h) consisting of a field F and an embedding $h:R \longrightarrow F$ that satisfy the following conditions. Whenever $f:R \longrightarrow K$ is an embedding of R into a field K, then there exists an embedding $g:F \longrightarrow K$ such that the diagram

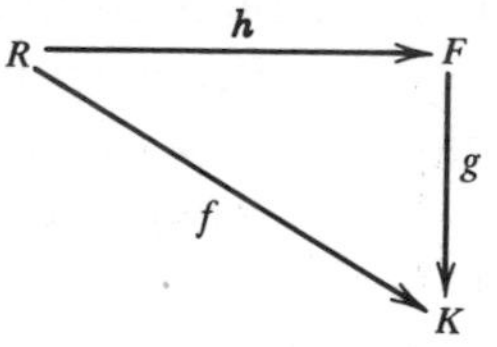

is commutative. Show that any two fields of fractions (F_1,h_1) and (F_2,h_2) are isomorphic in the sense that there is an isomorphism $\theta : F_1 \longrightarrow F_2$ such that the diagram is commutative.

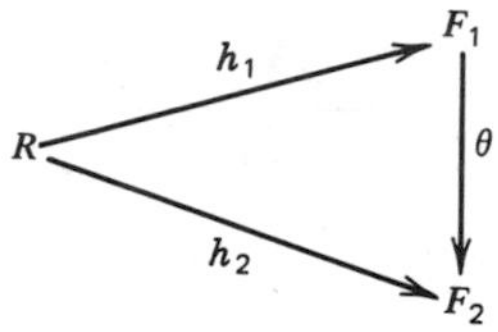

9. EUCLIDEAN DOMAINS

In Chapter 2 we studied the ring $\mathbf{Z}$ extensively. Looking back, we can see that one of the properties of the integers that was very important for establishing the existence of greatest common divisors and for proving the existence and uniqueness of factorizations into primes was the division property. The division property states that, given two integers a and b with $b \neq 0$, there exist unique integers q and r such that $a = qb + r$ and $0 \leq r < |b|$. We called q the integral quotient of a and b and r the remainder when

a is divided by b. The procedures *ZQUOT* and *ZREM* in *CLASSLIB* compute integer quotients and remainders, respectively.

In this section we will introduce a class of rings for which many of the results in Chapter 2 hold. The definition of this class of rings depends on a generalization of the division property. In **Z** we can divide by a nonzero integer and get a small remainder, where smallness is measured by the absolute value function. In our more general situation, smallness in a ring R is measured by a function from $R - \{0\}$ to the set of nonnegative integers.

Let R be an integral domain. A *Euclidean norm* on R is a function N from R- $\{0\}$ to **Z** such that

1. $0 \leq N(a) \leq N(ab)$ for all a,b in $R - \{0\}$.
2. If a and b are in R with $b \neq 0$, then there exist q and r in R such that $a = qb + r$ and either $r = 0$ or $N(r) < N(b)$.

A *Euclidean domain* is an integral domain possessing a Euclidean norm.

THEOREM 1. The ring **Z** is a Euclidean domain.

Proof. For any integer a, set $N(a) = |a|$. Then N is a Euclidean norm on **Z**. □

Note that condition (2) in the definition of a Euclidean norm does not require that q and r be unique. Even in **Z** we generally have two choices for r. For example, if $a = 7$ and $b = 5$, then

$$a = b + 2 = 2b + (-3)$$

and both $|2|$ and $|-3|$ are less than $|b|$. In **Z** we get uniqueness by adding the extra condition that r be nonnegative.

In the next example of a Euclidean domain q and r are always unique. Recall that the degree of a nonzero polynomial f is denoted $\deg(f)$.

THEOREM 2. If F is a field, then the function deg is a Euclidean norm on $F[X]$.

Proof. For any $f \neq 0$ in $F[X]$, the degree $\deg(f)$ of f is a nonnegative integer. If $g \neq 0$, then $\deg(fg) = \deg(f) + \deg(g) \geq \deg(f)$. The algorithm taught in high school for the division of polynomials in $\mathbf{R}[X]$ generalizes immediately to polynomials over any field and produces the required quotient and remainder. □

As an example, let us consider $f = 2X^2 + X + 3$ and $g = 4X^4 + 3X^3 + X^2 + 2X + 4$ in $\mathbf{Z}_5[X]$. Dividing,

$$
\begin{array}{r}
2X^2 + 3X + 1 \\
2X^2 + X + 3 \;\big)\; 4X^4 + 3X^3 + X^2 + 2X + 4 \\
4X^4 + 2X^3 + X^2 \qquad\qquad \\
\hline
X^3 + 0X^2 + 2X \quad\; \\
X^3 + 3X^2 + 4X \quad\; \\
\hline
2X^2 + 3X + 4 \\
2X^2 + X + 3 \\
\hline
2X + 1
\end{array}
$$

We find that $g = qf + r$, where $q = 2X^2 + 3X + 1$ and $r = 2X + 1$.

The procedures *RXQUOT* and *RXREM* compute quotients and remainders in $\mathbf{R}[X]$.

```
      A←¯3.2 4 7.4 2
      B←¯1.5 2.1 1
      A RXQUOT B
3.2 2
      B RXREM A
1.6 0.28
```

Here we see that the quotient and remainder when $-3.2 + 4X + 7.4X^2 + 2X^3$ is divided by $-1.5 + 2.1X + X^2$ are $3.2 + 2X$ and $1.6 + 0.28X$, respectively. It is important to remember that the results of *RXQUOT* and *RXREM* are generally only approximations of the true quotient and remainder. Both procedures are defined for arrays of polynomials. In fact, all the procedures with the suffixes *QUOT* and *REM* are defined for arrays over the appropriate Euclidean domain.

If n is a prime, then $\mathbf{Z}_n[X]$ is a Euclidean domain by Theorem 2. Even if n is composite, we can still perform division in $\mathbf{Z}_n[X]$ provided the leading coefficient of the divisor is a unit in $\mathbf{Z}_n$. (Why?) The procedures *ZNXQUOT* and *ZNXREM* compute quotients and remainders in $\mathbf{Z}_n[X]$, where n is given by $\underline{N}$, which must not exceed 10^7.

```
      N←5
      F←3 1 2
      G←4 2 1 3 4
      G ZNXQUOT F
1 3 2
      F ZNXREM G
1 2
      1 2 ZNXSUM 1 3 2 ZNXPROD F
4 2 1 3 4
```

In a Euclidean domain the units are easily determined.

THEOREM 3. Let R be a Euclidean domain with norm N. If $0 \neq b$ in R, then b is a unit in R if and only if $N(b) = N(1)$. If b is not a unit, then $N(ab) > N(a)$ for all $a \neq 0$.

Proof. Let a and b be nonzero elements of R. Then $N(ab) \geq N(a)$.

Suppose $N(ab) = N(a)$. We can find q and r in R such that $a = q(ab) + r$ and either $r = 0$ or $N(r) < N(ab) = N(a)$. But $r = a(1 - qb)$ and so, if $r \neq 0$, then $N(r) \geq N(a)$. Thus $r = 0$ and $a = aqb$. Since R is an integral domain, we may cancel the factor a and obtain $1 = qb$. Therefore b is a unit. Taking $a = 1$, we see that $N(b) = N(1)$ implies that b is a unit. But if b is any unit, then $bc = 1$ for some c in R and $N(1) \geq N(b)$. Since $N(b) = N(1b) \geq N(1)$, this means that $N(b) = N(1)$. Thus the units in R are the elements b with $N(b) = N(1)$. □

There is one more ring that we have already encountered that is a Euclidean domain.

THEOREM 4. The ring $\mathbf{Z}[i]$ of Gaussian integers is a Euclidean domain.

Proof. Since $\mathbf{Z}[i]$ is a subring of $\mathbf{C}$, $\mathbf{Z}[i]$ is certainly an integral domain. For any complex number $z = x + yi$, with x and y real, we defined the norm $N(z)$ to be $x^2 + y^2 = z\bar{z}$. This norm function satisfies the condition $N(zw) = N(z)N(w)$. (See Exercise 3.6.) If z and w are in $\mathbf{Z}[i]$ and $z \neq 0$, then $N(z) \geq 1$ and $N(zw) \geq N(w)$. All that remains is to prove that we can write $w = qz + r$ with $N(r) < N(z)$. Since $\mathbf{C}$ is a field, we can write $w/z = a + bi$, where a and b are in $\mathbf{R}$ (in fact, in $\mathbf{Q}$). Choose integers u and v such that $|a - u| \leq 1/2$ and $|b - v| \leq 1/2$. Set $q = u + vi$ and $r = w - qz$. Then

$$r = z(a + bi) - qz = z[(a - u) + (b - v)i].$$

Therefore

$$N(r) = N(z)N[(a - u) + (b - v)i] \leq N(z)\left(\frac{1}{4} + \frac{1}{4}\right) < N(z). \quad □$$

The library *CLASSLIB* contains procedures

GAUSSQUOT and *GAUSSREM*

for computing quotients and remainders in $\mathbf{Z}[i]$. For example, if $z = -2 + 3i$ and $w = 10 - 7i$, then the calculations

```
      Z←¯2 3
      W←10 ¯7
      □←Q←W GAUSSQUOT Z
¯3 ¯1
      □←R←Z GAUSSREM W
1 0
      R CSUM Q CPROD Z
10 ¯7
      CNORM Z
13
      CNORM R
1
```

show that $w = qz + r$, where $q = -3 - i$ and $r = 1$, and $N(r) < N(z)$.

In Chapter 2, one of the first facts we proved about $\mathbf{Z}$ was that every ideal, in fact, every subgroup of $(\mathbf{Z}, +)$, is generated by one element. In a commutative ring R, an ideal I is called *principal* if $I = Rx$ for some x in R. A *principal ideal domain* is an integral domain in which every ideal is principal. The phrase "principal ideal domain" is often abbreviated "PID".

THEOREM 5. Every Euclidean domain is a PID.

Proof. Let R be a Euclidean domain with norm N and let I be an ideal of R. If $I = \{0\}$, then $I = R0$, so I is principal. If $I \neq \{0\}$, then $\{N(x) \mid x \in I - \{0\}\}$ is a nonempty set of nonnegative integers and so has a smallest element. Choose x in $I - \{0\}$ such that $N(x) \leq N(y)$ for all y in $I - \{0\}$. Clearly, $I \geq Rx$. To prove $I = Rx$, we must prove $I \leq Rx$. Suppose y is in I. We can find q and r in R such that $y = qx + r$ and either $r = 0$ or $N(r) < N(x)$. Now $r = y - qx$ and so r is in I. Hence, if $r \neq 0$, then $N(r) \geq N(x)$. Thus $r = 0$ and $y = qx$ is in Rx. Therefore $I \subseteq Rx$. □

The converse of Theorem 5 is false. There do exist PID's that are not Euclidean domains. An example may be found in Motzkin.

The next result exhibits an important property of PID's and hence of Euclidean domains.

THEOREM 6. Let R be a PID and let $I_1 \subseteq I_2 \subseteq \ldots$ be an infinite sequence of ideals in R with each ideal contained in the next. Then there is an integer n such that $I_m = I_n$ for all $m \geq n$.

Proof. Let

$$I = \bigcup_{m=1}^{\infty} I_m .$$

We claim I is an ideal of R. Let a and b be elements of I and let r be in R. Then $a \in I_p$ and $b \in I_q$ for some p and q. We may assume $p \geq q$ and so, in fact, both a and b are in I_p. Therefore ra, $a + b$, and $a - b$ are in I_p and hence in I. Thus I is an ideal. But R is a PID, so $I = Rx$ for some x in I. There exists an n such that $x \in I_n$. Suppose $m \geq n$. Then

$$I_n \subseteq I_m \subseteq I = Rx \subseteq I_n .$$

Therefore $I_m = I_n$ for all $m \geq n$. □

A ring whose ideals satisfy the condition of Theorem 6, that is, a ring with no infinite strictly increasing sequence of ideals, is said to satisfy the *ascending chain condition* on ideals.

The class of Euclidean domains contains most rings in which we will want to perform computations. The existence of a Euclidean norm makes calculation much easier than in an arbitrary ring. Most of Chapter 6 is devoted to a study of algorithms for performing certain types of computations related to Euclidean domains. However, we should point out that the rings $\mathbf{Z}[X]$ and $F[X,Y]$, where F is a field, are important rings that are not Euclidean domains; in fact, they are not PID's.

We close this section with some facts about polynomials with coefficients in a field.

Let F be a field and let f be a polynomial in $F[X]$. An element c of F is called a *root* of f if $f(c) = 0$. Every element of F is a root of the zero polynomial. However, as the next result shows, nonzero polynomials in $F[X]$ cannot have many roots.

THEOREM 7. Let F be a field and let f be an element of $F[X]$ with $\deg(f) = n > 0$. Then f has at most n roots in F.

Proof. We proceed by induction on n. Suppose $n = 1$. Then $f = aX + b$, where $a \neq 0$. If $0 = f(c) = ac + b$, then $ac = -b$ and $c = -b/a$. Thus f has exactly one root in F. Suppose now that $n > 1$. Clearly, we may assume f has at least one root c in F. Dividing f by $X - c$, we can write $f = (X - c)q + r$, where q is in $F[X]$ and r is in F. The degree of q is $n - 1$. Evaluating f at c, we find that $0 = f(c) = (c - c)q(c) + r = r$. Thus $r = 0$ and $f = (X - c)q$. Suppose d is a root of f in F with $d \neq c$. Then $0 = f(d) = (d - c)q(d)$. Since $d - c \neq 0$ and F has no zero divisors, this means that $g(d) = 0$, so d is a root of q. By induction, q has at most $n - 1$ roots in F. Thus f has at most n roots in F. □

The following corollary of Theorem 7 will be important for the polynomial factoring algorithm described in Section 12.

COROLLARY 8. Let F be a field and let f and g be elements of $F[X]$ such that $\deg(f)$ and $\deg(g)$ are each at most n. If there exist distinct elements $a_0, \ldots, a_n$ of F such that $f(a_i) = g(a_i)$, $0 \leq i \leq n$, then $f = g$.

Proof. Let $h = f - g$. Then, for $0 \leq i \leq n$, we have $h(a_i) = f(a_i) - g(a_i) = 0$, so each a_i is a root of h. By Theorem 7, h must be a constant polynomial, and so $h = 0$. This means that $f = g$. □

EXERCISES

1 Let N be a Euclidean norm on the integral domain R. Suppose c is a nonnegative integer and $N'(x) = N(x) + c$ for all x in R. Show that N' is a Euclidean norm on R.

2 Let F be a field. For $0 \neq x$ in F, set $N(x) = 0$. Show that N is a Euclidean norm on F.

3 Let F be a field. Give an explicit description of the division algorithm in $F[X]$. That is, if $f = a_0 + a_1X + \ldots + a_nX^n$ and $g = b_0 + \ldots + b_mX^m$ are in $F[X]$ with $g \neq 0$, tell how to compute the coefficients of the polynomials q and r such that $f = qg + r$ and either $r = 0$ or $\deg(r) < \deg(g)$.

4 A *monic* polynomial is one whose leading coefficient is 1. Let

f and g be in $\mathbf{Z}[X]$, with g monic. Show that there exist q and r in $\mathbf{Z}[X]$ with $f = qg + r$ and either $r = 0$ or $\deg(r) < \deg(g)$.

5 Let z and w be Gaussian integers with $z \neq 0$. Determine the maximum number of pairs (q, r) such that $w = qz + r$ and $N(r) < N(z)$. State an additional condition that makes r unique.

6 Let $f = 1 + X + 3X^2 + 3X^3 + 4X^4 + 5X^5$ and $g = 5 + 2X + 3X^2$ in $\mathbf{Z}_7[X]$. Compute by hand the quotient and remainder when f is divided by g. Check your answer using *ZNXQUOT* and *ZNXREM*.

7 Let $z = 3 - 2i$ and $w = -11 + 7i$. Find all pairs (q, r) of Gaussian integers such that $w = qz + r$ and $N(r) < N(z)$.

8 Let I be the set of all polynomials $a_0 + a_1X + \ldots + a_n X^n$ in $\mathbf{Z}[X]$ with a_0 even. Show that I is an ideal of $\mathbf{Z}[X]$ and that I is not principal.

9 Let F be a field. Prove that $F[X,Y]$ is not a PID.

*10 Show that $\mathbf{Z}[X]$ satisfies the ascending chain condition for ideals.

11 Let R be a ring in which every ideal I is finitely generated in that there exist $x_1, \ldots, x_n$ in I such that I is the smallest ideal in R containing all of the x_i. Show that R satisfies the ascending chain condition for ideals.

12 Define the descending chain condition for ideals in a ring. Does $\mathbf{Z}$ satisfy this condition?

13 Show that the requirement in Theorem 8 that F be a field is necessary by determining the number of roots of $X^2 - 1$ in the ring $\mathbf{Z}_8$.

10. FACTORIZATION

Let a and b be elements of an integral domain R. We say *a divides b* in R if there is an element c such that $b = ac$. We also say that a is a *divisor* of b, that a is a *factor* of b, and that b is *divisible by a*. In $\mathbf{Z}$ two integers a and b can be divisors of each other if and only if $a = \pm b$. The following theorem generalizes this fact.

THEOREM 1. Let a and b be elements of the integral domain R. Then the following are equivalent.

(a) a divides b and b divides a in R.
(b) $Ra = Rb$.
(c) There is a unit u in R such that $b = ua$.

Proof. If a divides b, then $b = ac$ for some c in R. If r is any element of R, then $rb = (rc)a$, so $Ra \supseteq Rb$. If, in addition, b divides a, then $a =$

bd for some d in R and $Rb \supseteq Ra$. Thus $Ra = Rb$ and so (a) implies (b). Also, we have $a = bd = acd$. If $a \neq 0$, we may cancel the factor a in the equation $a = acd$ to obtain $1 = cd$, and so c and d are units. If $a = 0$, then $b = ac = 0$. Here $b = ua$ with $u = 1$. Hence (a) also implies (c). It is easy to see that (b) and (c) each imply (a). If $Ra = Rb$, then b is in Ra, and so $b = ca$ for some c. Similarly, $a = bd$ for some d. Thus (b) implies (a). Finally, if $b = ua$, with u a unit in R, then obviously a divides b. If v is the inverse of u, then $bv = uva = a$ and so (c) also implies (a). □

If a and b satisfy any one, and hence all, of the conditions in Theorem 1, we say a and b are *associates*. It is easy to show (see Exercise 1) that the relation "is an associate of" is an equivalence relation. In some integral domains there is a natural choice for a representative in each class of associate elements. For example, in **Z** each class contains a unique nonnegative element, while in the polynomial ring $F[X]$ over a field F every nonzero polynomial is an associate of a unique *monic* polynomial, one whose leading coefficient is 1. In $\mathbf{Z}[i]$ an element $z \neq 0$ has four associates, $z, iz, -z$, and $-iz$. It is less obvious here which element to choose as the representative for the class. See Exercise 3.

In an arbitrary integral domain R there is no notion of the relative size of two elements, but we can still define the concept of a greatest common divisor. If a and b are in R, then an element d of R is called a *greatest common divisor* (gcd) of a and b, provided the following conditions hold.

1. d is a divisor of both a and b.
2. If c is an element of R that divides both a and b, then c divides d.

THEOREM 2. Any two greatest common divisors of a and b are associates.

Proof. Suppose d_1 and d_2 are two greatest common divisors of a and b. Then d_2 divides both a and b and so d_1 divides d_2. By symmetry, d_2 divides d_1, so d_1 and d_2 are associates. □

In general, two elements of R need not have a greatest common divisor, but this cannot happen in a PID.

THEOREM 3. Let R be a PID. If a and b are in R, then a and b have a greatest common divisor d and $d = ra + sb$ for some r and s in R.

Proof. The ideal of R generated by a and b is $I = Ra + Rb$. Since R is a PID, we know that $I = Rd$ for some element d in R. Clearly, d has the form $ra + sb$. Also, since a and b are both in I, it follows that d divides both a and b are both in I, it follows that d divides both a and b. Finally, if c divides both a and b, then c divides $ra + sb = d$. Therefore d is a greatest common divisor of a and b. □

In Section 2.2 we learned how to compute greatest common divisors of integers by repeated computation of remainders. For example,

```
      57 ZREM 91
34
      34 ZREM 57
23
      23 ZREM 34
11
      11 ZREM 23
1
      1 ZREM 11
0
```

shows that gcd(57,91) = 1. This Euclidean algorithm works in any Euclidean domain. To compute the greatest common divisor of $1 + 2X + 3X^2 + 3X^3$ and $2 + X + 2X^2 + 2X^4$ in $\mathbf{Z}_5[X]$, we proceed as follows.

```
      N←5
      A←1 2 3 3
      B←2 1 2 0 2
      ⎕←C←A ZNXREM B
1 0 1
      ⎕←D←C ZNXREM A
3 4
      ⎕←E←D ZNXREM C
0
```

We normally replace the result $3 + 4X$ by its monic associate.

```
      ⎕←D←D ZNXPROD ZNINV ¯1↑D
2 1
```

Thus we obtain $\gcd(1 + 2X + 3X^2 + 3X^3, 2 + X + 2X^2 + 2X^4) = 2 + X$ in $\mathbf{Z}_5[X]$. Similarly, *RXREM* and *GAUSSREM* may be used to compute greatest common divisors in $\mathbf{R}[X]$ and $\mathbf{Z}[i]$, respectively. As usual, *RXREM* must be used with caution, since round-off errors can be disastrous.

The fact that Theorem 2.4.2 is called the Fundamental Theorem of Arithmetic indicates the importance of being able to factor positive integers into a product of primes in essentially one way. A prime p in $\mathbf{Z}$ is a nonzero nonunit that satisfies each of the following two conditions.

1. If $p = ab$ with a and b in $\mathbf{Z}$, then either a or b is a unit in $\mathbf{Z}$.
2. If p divides a product ab of integers, then either p divides a or p divides b.

In $\mathbf{Z}$ these conditions are equivalent. However, if we replace $\mathbf{Z}$ by an arbitrary integral domain, then conditions (1) and (2) are no longer equivalent.

Let R be an integral domain. An element $p \neq 0$ of R is said to be *irreducible* in R if p is not a unit and, whenever $p = ab$ with a and b in R, then either a or b is a unit. A *prime* in R is a nonzero nonunit p such that whenever p divides a product ab of elements in R, then either p divides a or p divides b. An easy induction argument shows that if a prime p divides a product $a_1a_2 \ldots a_n$ of elements of R, then p divides one of the factors a_i.

THEOREM 4. Every prime element in R is irreducible.

Proof. Suppose p is a prime in R and $p = ab$. Then certainly p divides ab and, therefore, p divides a or b. We may assume p divides a. In this case, $a = pc$ and $p = pcb$ for some c in R. Since $p \neq 0$, we may cancel the factor p to obtain $1 = cb$. Thus b is a unit. □

There exist integral domains that contain irreducible elements that are not primes. For example, let $R = \mathbf{Z}[\sqrt{-5}]$. It is not hard to show (see Exercise 22) that 3, $2 + \sqrt{-5}$, and $2 - \sqrt{-5}$ are irreducible in R. Now

$$(3)(3) = 9 = (2 + \sqrt{-5})(2 - \sqrt{-5}).$$

Thus 3 divides the product of $(2 + \sqrt{-5})$ and $(2 - \sqrt{-5})$. However, it is easy to see that 3 does not divide either factor, so 3 is not a prime in R.

In a PID the distinction between primes and irreducible elements vanishes.

THEOREM 5. If R is a PID, then every irreducible element in R is a prime.

Proof. Suppose R is a PID and p is irreducible in R. Assume that p divides ab. By Theorem 3, a and p have a greatest common divisor d, and d can be written as $ra + sp$ with r and s in R. If d is a unit, we may assume (see Exercise 2) that $d = 1$. Multiplying the equality $1 = ra + sp$ by b, we obtain $b = rab + spb$. Since p divides both ab and p, we see that p divides b. If, on the other hand, d is not a unit, then $p = cd$ for some c in R, and since p is irreducible, c must be a unit. Now d divides a, and so $a = qd = qc^{-1}p$ for some q in R. Therefore p divides a. We have shown that p divides a or p divides b, so p is prime. □

A *unique factorization domain* (UFD) is an integral domain R such that each nonzero element a of R that is not a unit can be factored as

$$a = p_1p_2 \ldots p_r,$$

where each p_i is irreducible. Moreover, we require this factorization to be unique up to order and associates. This means that if

$$a = q_1q_2 \ldots q_s$$

with each q_j irreducible, then $r = s$ and there is a permutation σ of $\{1, \ldots, r\}$ such that q_i is an associate of $p_{i\sigma}$. By Theorem 2.4.2, we know that $\mathbf{Z}$ is a UFD. The ring $\mathbf{Z}[\sqrt{-5}]$ is not a UFD, since

$$(3)(3) = (2 + \sqrt{-5})(2 - \sqrt{-5})$$

and 3 is not an associate of either of the elements $2 \pm \sqrt{-5}$, and 3 and $2 \pm \sqrt{-5}$ are all irreducible.

Theorem 5 remains true when we replace "PID" by "UFD".

THEOREM 6. Let R be a UFD. Every irreducible element in R is prime.

Proof. Let p be an irreducible element in R and suppose p divides uv. Then $uv = pw$ for some w in R. We must show that p divides u or p divides v. If u or v is 0, then clearly we are done. Thus we may assume neither u nor v is 0. If u is a unit, then $v = pwu^{-1}$ and p divides v. Hence we may also assume that neither u nor v is a unit. Since R is a UFD, we may write $u = u_1 \ldots u_r$, $v = v_1 \ldots v_s$, and $w = w_1 \ldots w_t$, where each u_i, v_j, and w_k is irreducible. Moreover, we have

$$pw_1 \ldots w_t = u_1 \ldots u_r v_1 \ldots v_s.$$

By the uniqueness of factorizations in R, we know that p is an associate of some u_i or v_j. If p is an associate of u_i, then p divides u_i and thus p divides u. If p is an associate of v_j, then p divides v. Therefore p divides either u or v. □

Now we come to the main result of this section.

THEOREM 7. Every PID is a UFD.

Proof. Let R be a PID. We will prove the existence and uniqueness of factorizations in R by a sequence of lemmas.

LEMMA 8. There do not exist infinite sequences $a_1, a_2, \ldots$ and $b_1, b_2, \ldots$ of elements of R such that each a_i is a nonzero nonunit, each b_j is nonzero, and

$$a_1 b_1 = a_1 a_2 b_2 = a_1 a_2 a_3 b_3 = \ldots.$$

Proof. Suppose such a pair of sequences existed. Cancelling the factor $a_1 \ldots a_i$ in the equality $a_1 \ldots a_i b_i = a_1 \ldots a_{i+1} b_{i+1}$, we see that $b_i = a_{i+1} b_{i+1}$, and so b_{i+1} divides b_i for all $i \geq 1$. Set $I_i = Rb_i$. Then $I_1 \subseteq I_2 \subseteq I_3 \subseteq \ldots$. By Theorem 9.7 there exists an integer n such that $I_m = I_n$ for all $m \geq n$. In particular, $I_n = I_{n+1}$, and so b_n and b_{n+1} must be associates. This means that a_{n+1} is a unit, which contradicts our assumption that each a_i is a nonunit. □

LEMMA 9. Let a be a nonzero nonunit in R. Then a is divisible by an irreducible element.

Proof. If a is irreducible, we are done, since a is divisible by itself. If a is not irreducible, we can write $a = a_1 b_1$, where a_1 and b_1 are nonzero nonunits. If b_1 is irreducible, then again we are done. If b_1 is not irreducible, we can write $b_1 = a_2 b_2$, where a_2 and b_2 are nonzero nonunits. Continuing in this way, we see that the assumption that a is not divisible by an irreducible element leads to the construction of a pair of sequences that Lemma 8 says cannot be constructed in R. □

LEMMA 10. Every nonzero nonunit in R can be factored into a product of irreducible elements.

Proof. Let a be a nonzero nonunit. By Lemma 9, we have $a = a_1 b_1$, where a_1 is irreducible. If b_1 is a unit, then a is itself irreducible (see Exercise 9), and we are done. If b_1 is not a unit, then $b_1 = a_2 b_2$, with a_2 irreducible. If b_2 is a unit, then b_1 is irreducible, and we are done. Otherwise, we can factor b_2 as $a_3 b_3$ with a_3 irreducible. If we are not to produce a pair of sequences of the type ruled out by Lemma 8, we must eventually arrive at a factorization $a = a_1 a_2 \ldots a_n b_n$, in which b_n and each a_i are irreducible. □

We complete the proof of Theorem 7 with the following.

LEMMA 11. The factorization in Lemma 10 is unique up to order and associates.

Proof. Suppose $p_1 \ldots p_r = q_1 \ldots q_s$, where each p_i and q_j is irreducible in R. If $r = 1$, then s must be 1, since p_1 is irreducible. Thus we may assume r and s are each at least 2 and proceed by induction on r. By Theorem 5, p_1 is a prime. Since p_1 divides the product $q_1 \ldots q_s$, we know that p_1 divides some q_j. Renumbering the q's, we may assume p_1 divides q_1. But q_1 is irreducible, so $q_1 = up_1$, where u is a unit in R. Replacing q_2 by its associate $u^{-1}q_2$, we have $p_1 \ldots p_r = p_1 q_2 \ldots q_s$. Thus $p_2 \ldots p_r = q_2 \ldots q_s$. By induction, $r = s$, and we may rearrange the q's so that p_i and q_i are associates, $2 \le i \le r$. Thus the factorization $p_1 \ldots p_r$ is unique up to order and associates. □

Theorem 7 together with Theorem 9.5 tells us that every Euclidean domain is a UFD. Thus the rings $F[X]$, where F is a field, and $\mathbf{Z}[i]$ are UFD's. Let us investigate factorization in these rings in more detail.

If F is a field, then any polynomial f in $F[X]$ of positive degree can be factored as $f = df_1 f_2 \ldots f_r$, where d is in F and each f_i is a monic irreducible polynomial. The f_i are unique up to order. There are infinitely many monic irreducible polynomials in $F[X]$. If a is in F, then $X + a$ is irreducible. Hence, if F is an infinite field, then there are infinitely many monic irreducible polynomials of degree 1. If F is finite, it can be shown (see Exercise 16) that irreducible polynomials can be found in $F[X]$ with arbitrarily large degree.

If p is a prime and m is a positive integer such that p^m is not too large, say $p^m < 1000$, then the monic irreducible polynomials in $\mathbf{Z}_p[X]$ with degree at most m can be listed with a simple variation of the sieve technique used in Section 2.4 to list primes in $\mathbf{Z}$. For example, if

```
      ⎕IO←0
      N←2
      ⎕←A←⌽⍉2 2 2 2⊤2↓⍳16
0 1 0 0
1 1 0 0
0 0 1 0
1 0 1 0
0 1 1 0
1 1 1 0
0 0 0 1
1 0 0 1
0 1 0 1
1 1 0 1
0 0 1 1
1 0 1 1
0 1 1 1
1 1 1 1
```

then the rows of A list the 14 nonconstant polynomials in $\mathbf{Z}_2[X]$ with degree not exceeding 3. All of these polynomials are monic. The first row of A represents the polynomial X, which is irreducible. We can sieve out the multiples of this polynomial by

```
      Y←∨/0≠A[0;] ZNXREM A
      ⎕←A←Y⌿A
1 1 0 0
1 0 1 0
1 1 1 0
1 0 0 1
1 1 0 1
1 0 1 1
1 1 1 1
```

Here we compute the remainder when each row of A is divided by the first row of A in $Z_2[X]$. The vector Y is the characteristic vector of the set of rows in A that are not divisible by the first row. We use Y to select the rows of A we wish to keep.

The new first row of A corresponds to $1 + X$ and is also irreducible. Sieving out its multiples,

```
      Y←∨/0≠A[0;] ZNXREM A
      ⎕←A←Y⌿A
1 1 1 0
1 1 0 1
1 0 1 1
```

we obtain a matrix whose rows list polynomials of degree 2 and 3. Any polynomial of degree 2 or 3 that is not irreducible must have a factor of degree 1. Since we have eliminated all polynomials with a linear factor, these three polynomials must be irreducible. Thus the irreducible polynomials in $\mathbf{Z}_2[X]$ with degree at most 3 are X, $1 + X$, $1 + X + X^2$, $1 + X + X^3$, and $1 + X^2 + X^3$.

If M is a positive integer, then $A \leftarrow ZNXMONIC\ M$ is a matrix whose rows list the monic polynomials of degree at most M with coefficients in $\mathbf{Z}_n[X]$, where n is $\underline{N}$.

```
      N̲
2
      ZNXMONIC 3
 0 1 0 0
 1 1 0 0
 0 0 1 0
 1 0 1 0
 0 1 1 0
 1 1 1 0
 0 0 0 1
 1 0 0 1
 0 1 0 1
 1 1 0 1
 0 0 1 1
 1 0 1 1
 0 1 1 1
 1 1 1 1
```

If $\underline{N}$ is prime, then $ZNXIRRED$ uses the sieve method to produce a matrix listing the monic irreducible polynomials up to a given degree.

```
      ⎕←B←ZNXIRRED 3
 0 1 0 0
 1 1 0 0
 1 1 1 0
 1 1 0 1
 1 0 1 1
```

If f is a polynomial of degree m in $\mathbf{Z}_p[X]$, p being a prime, then one method of factoring f is simply to divide f by all monic irreducible polynomials in $\mathbf{Z}_p[X]$ whose degrees do not exceed $m/2$. For example, to factor $f = 1 + X + X^4 + X^6$ in $\mathbf{Z}_2[X]$, we divide f by the irreducible polynomials listed in the matrix B.

```
        B ZNXREM 1 1 0 0 1 0 1
1 0
0 0
0 0
0 0
0 1
```

The second, third, and fourth rows of *B* give a remainder of 0, and $f = (1+X)(1+X+X^2)(1+X+X^3)$.

If p^m is large, we cannot expect to have available the list of monic irreducible polynomials of degree up to $m/2$ in order to factor a given polynomial in $\mathbf{Z}_p[X]$ of degree m. There is a better algorithm for factoring polynomials in $\mathbf{Z}_p[X]$. This algorithm, which is based on the theory of finite fields, is discussed in Section 7.4 and is used in the procedure *ZNXFACTOR*. If the vector *F* lists the coefficients of a polynomial f in $\mathbf{Z}_p[X]$ of positive degree, then *G←ZNXFACTOR F* is a matrix whose rows list the monic irreducible factors of f. The global variable *N̲*, which gives the value of p, must be a prime less than 10^7.

```
        N̲←2
        ZNXFACTOR 1 1 0 0 1 0 1
1 1 0 0
1 1 1 0
1 1 0 1
        N̲←1019
        ZNXFACTOR 752 346 713 1
311   1   0
661 402   1
```

In the second example we obtained the factorization

$$752 + 346X + 713X^2 + X^3 = (311 + X)(661 + 402X + X^2)$$

in $\mathbf{Z}_{1019}[X]$.

The algorithm used in *ZNXFACTOR* is different in one respect from any of the other algorithms we have encountered so far in that it is probabilistic. This means that at certain points in the computation random choices are made, in this case of elements of ⍳ *N̲*. It is possible that these choices will not lead to the factorization. However, the probability that this will happen for any particular polynomial is less than 10^{-10}. Although *ZNXFACTOR* may report failure to find a factorization, it will never give an incorrect factorization.

In $\mathbf{C}[X]$ every irreducible polynomial has degree 1. This fact, which is known as the Fundamental Theorem of Algebra, is really a theorem in analysis and is usually not proved in introductory algebra texts. Another way to state this result is as follows. Every element in $\mathbf{C}[X]$ of positive degree has a root in $\mathbf{C}$. In $\mathbf{R}[X]$ irreducible polynomials have degree 1 or 2. If f is in $\mathbf{R}[X]$ and $\deg(f) > 0$, then f has a root c in $\mathbf{C}$ by the Fundamental Theorem of Algebra. If c is in $\mathbf{R}$, then $X - c$ divides f in $\mathbf{R}[X]$. If c is not in $\mathbf{R}$, then $\bar{c} \neq c$ and $f(\bar{c}) = \overline{f(c)} = \bar{0} = 0$, since f has real coefficients. Therefore f is divisible in $\mathbf{C}[X]$ by $X - c$ and $X - \bar{c}$ and hence by $g = (X - c)(X - \bar{c})$. But $g = X^2 - (c + \bar{c})X + c\bar{c}$ has real coefficients, so g divides f in $\mathbf{R}[X]$.

The procedure *RXFACTOR* computes approximate factorizations in $\mathbf{R}[X]$. If *F* is a vector listing the coefficients of a polynomial f in $\mathbf{R}[X]$ of positive degree, then *G←RXFACTOR F* is a matrix whose rows list approximations to the monic irreducible factors of f in $\mathbf{R}[X]$.

```
      RXFACTOR ¯2 ¯1 ¯3 0 ¯1 1
¯2  1  0
 1  1  1
 1  0  1
```

Here we have obtained the exact factorization

$$-2 - X - 3X^2 - X^4 + X^5 = (-2 + X)(1 + X + X^2)(1 + X^2)$$

in $\mathbf{R}[X]$.

Techniques for approximate factorization in $\mathbf{R}[X]$ are properly part of the branch of mathematics called numerical analysis and will not be discussed in detail in this book. More information can be found in books on numerical analysis.

Let us now turn to the problems of factorization and the determination of primes in $\mathbf{Z}[i]$. We will show that it is possible to reduce these problems to the corresponding ones in $\mathbf{Z}$. As usual, the norm of the complex number z will be written $N(z)$.

THEOREM 12. If z is in $\mathbf{Z}[i]$ and $N(z) = p$ is a prime in $\mathbf{Z}$, then z is a prime in $\mathbf{Z}[i]$.

Proof. Suppose $N(z) = p$ is a prime in $\mathbf{Z}$ and $z = uv$ with u and v in $\mathbf{Z}[i]$. Then $p = N(z) = N(u)N(v)$ and so either $N(u) = 1$ or $N(v) = 1$. But elements in $\mathbf{Z}[i]$ with norm 1 are units, and so z is a prime. □

By Theorem 12, the numbers $1 + i$, $1 - 2i$, and $-2 + 3i$ are all primes in $\mathbf{Z}[i]$.

THEOREM 13. Let z be a prime in $\mathbf{Z}[i]$. There is a prime p in $\mathbf{Z}$ such that z divides p in $\mathbf{Z}[i]$.

Proof. The norm $N(z) = z\bar{z}$ is an integer greater than 1 and can be factored as a product $p_1 \ldots p_r$ with each p_i a prime in $\mathbf{Z}$. Since z divides $N(z)$, it follows that z divides some p_i. □

Theorem 13 reduces the problem of finding the primes in $\mathbf{Z}[i]$ to the factorization in $\mathbf{Z}[i]$ of the primes in $\mathbf{Z}$.

THEOREM 14. Let p be a prime in $\mathbf{Z}$. Then either p is a prime in $\mathbf{Z}[i]$ or $p = uv$, with u and v conjugate primes in $\mathbf{Z}[i]$.

Proof. Suppose $p = uv$ in $\mathbf{Z}[i]$. Then $p^2 = N(p) = N(u)N(v)$. If neither u nor v is a unit, then $N(u) = N(v) = p$. In this case, both u and v are primes in $\mathbf{Z}[i]$ by Theorem 8 and, since $p = N(u) = u\bar{u}$, we see that $v = \bar{u}$. □

If a prime p in $\mathbf{Z}$ is not prime in $\mathbf{Z}[i]$, then by Theorem 14 there exist integers x and y such that $p = (x + yi)(x - yi) = x^2 + y^2$. Thus, whether or not p factors in $\mathbf{Z}[i]$ depends on whether or not p can be written as a sum of two integer squares. Since the factorization of 2 as $(1 + i)(1 - i)$ is easily checked, we may assume p is odd.

THEOREM 15. If p is a prime in $\mathbf{Z}$ and $p \equiv 3 \pmod 4$, then p is a prime in $\mathbf{Z}[i]$.

Proof. Suppose $p = x^2 + y^2$ with x and y in $\mathbf{Z}$. Modulo 4 the only squares are 0 and 1. Thus $x^2 + y^2$ is congruent modulo 4 to 0, 1, or 2. If p is a prime and $p \equiv 3 \pmod 4$, then p is not the sum of two integer squares and p is prime in $\mathbf{Z}[i]$ by Theorem 14. □

This leaves the primes in $\mathbf{Z}$ congruent to 1 modulo 4. A theorem of the French mathematician Pierre de Fermat (1601?–1665) states that every prime of this type has a nontrivial factorization in $\mathbf{Z}[i]$. Before proving Fermat's result, we will present an example showing one way to find the factorization using APL.

The integer 337 is prime and congruent to 1 modulo 4. If we wish to express 337 as $x^2 + y^2$, with x and y in $\mathbf{Z}$, then we may assume $0 < x < y$, and so $x < \sqrt{337/2} = 12.9\ldots$. Thus we may proceed as follows.

```
      ⎕IO←1
      U←(337-(⍳12)*2)*0.5
      ⎕←X←(U=⌊U)⍳1
9
      ⎕←Y←⌊U[X]
16
```

Here we find that $337 = (9 + 16i)(9 - 16i)$. This method is not efficient for large primes. A better method is described in the paper by Brillhart cited in the bibliography.

THEOREM 16 ***(Fermat).*** Let p be a prime in $\mathbf{Z}$ with $p \equiv 1 \pmod 4$. Then there exist integers x and y with $p = x^2 + y^2$.

Proof. By Theorem 14 it is enough to show that p is not a prime in $\mathbf{Z}[i]$. Let $u = (1)(2)(3) \ldots (p-1)/2$. Since $p \equiv 1 \pmod 4$, it follows that $(p-1)/2$ is even, and so

$$u = (-1)(-2)(-3) \ldots [-(p-1)/2].$$

But $-k \equiv p - k \pmod p$, and thus

$$u \equiv (p-1)(p-2)(p-3) \ldots [(p+1)/2] \pmod p.$$

Therefore

$$u^2 \equiv (1)(2)(3) \ldots \left(\frac{p-1}{2}\right)\left(\frac{p+1}{2}\right) \ldots (p-1) = (p-1)! \pmod p.$$

At this point we invoke Wilson's Theorem (Corollary 3.11.9), which states that $(p-1)! \equiv -1 \pmod p$.

By Wilson's Theorem, we have $u^2 \equiv -1 \pmod p$ or $u^2 + 1 = cp$ for some integer c. In $\mathbf{Z}[i]$ we can factor $u^2 + 1$ as $(u+i)(u-i)$. Since p divides neither $u + i$ nor $u - i$ in $\mathbf{Z}[i]$ but does divide their product, it follows that p cannot be a prime in $\mathbf{Z}[i]$. □

The procedure *GAUSSFACTOR* factors a nonzero nonunit in $\mathbf{Z}[i]$ into primes.

```
      GAUSSFACTOR  19 9
 1  1
 2  3
 1 ¯4
```

The product of the factors is in general only an associate of the original number.

EXERCISES

In the following exercises R is an integral domain.

1 Show "is an associate of" is an equivalence relation on R.

2 Suppose d is a greatest common divisor of the elements a and b in R. Show that any associate of d is also a greatest common divisor of a and b.

3 Show that any nonzero Gaussian integer is an associate of a unique Gaussian integer $a + bi$ with $a > 0$ and $b \geq 0$.

4 Compute the monic gcd of $4 + 6X + 3X^3 + 6X^4 + 2X^5$ and $5 +$

$4X^2 + 2X^3 + 3X^4 + 6X^5 + 3X^6$ in $\mathbf{Z}_7[X]$ using the Euclidean algorithm, as in the example in the text.

5 Compute a gcd of $33 - 4i$ and $-6 + 35i$ in $\mathbf{Z}[i]$ using the Euclidean algorithm. The procedure *GAUSSREM* may be used to compute remainders.

6 Give a formal description of the Euclidean algorithm in a general Euclidean domain.

7 Let F be a subfield of the field K. Then $F[X]$ is a subring of $K[X]$. Suppose f and g are in $F[X]$. Is the monic greatest common divisor of f and g in $F[X]$ necessarily the same as the monic greatest common divisor of f and g in $K[X]$?

*8 Let f and g be in $\mathbf{Z}[X]$ and suppose for all primes p the polynomials $\bar{f}$ and $\bar{g}$ in $\mathbf{Z}_p[X]$ represented by f and g, respectively, have a nonconstant common factor. Show that f and g have a nonconstant common factor in $\mathbf{Q}[X]$.

9 Let a be irreducible in R. Show that any associate of a is also irirreducible.

10 Do Exercise 9 with "irreducible" replaced by "prime".

11 For what Gaussian integers z is $\bar{z}$ an associate of z?

12 Suppose z is prime in $\mathbf{Z}[i]$. Show that $\bar{z}$ is also prime.

13 Show that in any UFD greatest common divisors always exist.

14 Give a definition for a least common multiple of two elements in R. Show that if R is a UFD, then least common multiples exist and the intersection of two principal ideals is principal.

15 Let f be a nonconstant polynomial in $\mathbf{Z}[X]$. Suppose p is a prime that does not divide the leading coefficient of f. Show that if f is irreducible modulo p, that is, in $\mathbf{Z}_p[X]$, then f is irreducible in $\mathbf{Z}[X]$. Give an example showing that the converse is false.

16 Prove that for any field F the ring $F[X]$ has infinitely many monic irreducible elements.

17 Set `N←5` and `A←ZNXMONIC 4`. Use the sieve technique of the text to determine the irreducible polynomials in $\mathbf{Z}_2[X]$ of degree at most 4. Check your result with *ZNXIRRED*.

18 Determine the monic irreducible polynomials in $\mathbf{Z}_3[X]$ of degree at most 3 using the approach of Exercise 17.

19 Factor 5, 29, and 653 into primes in $\mathbf{Z}[i]$ without using *GAUSSFACTOR*.

20 Factor $z = 23 + 11i$ into primes in $\mathbf{Z}[i]$ without using

`GAUSSFACTOR.`

[*Hint:* First factor $N(z)$ in $\mathbf{Z}$.)

21 Make a list of one representative from each class of associate primes in $\mathbf{Z}[i]$ having norm at most 50.

22 Let m be an integer that is not a square. For an element $z = a + b\sqrt{m}$ in $\mathbf{Z}[\sqrt{m}]$, with a and b in $\mathbf{Z}$, set $M(z) = a^2 - b^2 m$. Show for all z and w in $\mathbf{Z}[\sqrt{m}]$ we have $M(zw) = M(z)M(w)$. Prove that z is a unit in $\mathbf{Z}[\sqrt{m}]$ if and only if $M(z) = \pm 1$. Show that when $m < 0$, the ring $\mathbf{Z}[\sqrt{m}]$ has only finitely many units. Show also that if $m < 0$, then a nonzero element of $\mathbf{Z}[\sqrt{m}]$ has only finitely many divisors and describe a process for determining all divisors of an element in $\mathbf{Z}[\sqrt{m}]$.

23 By trial and error, find nonzero integers a and b satisfying $a^2 - 5b^2 = 1$. Thus $a + b\sqrt{5}$ is a unit in $\mathbf{Z}[\sqrt{5}]$. Show that the group of units in $\mathbf{Z}[\sqrt{5}]$ is infinite.

*24 For any positive integer m that is not a square, prove that the set of positive units in $\mathbf{Z}[\sqrt{m}]$ is a cyclic group.

25 Show that the ideal of $\mathbf{Z}[\sqrt{-5}]$ generated by 3 and $2 + \sqrt{-5}$ is not principal.

11. POLYNOMIAL RINGS OVER UFD's

In Section 9 we showed that every Euclidean domain is a PID and in Section 10 we proved that every PID is a UFD. This allowed us to conclude that the Euclidean domains $\mathbf{Z}[i]$ and $F[X]$, where F is a field, are all UFD's. In this section we will discuss the following important theorem, which implies that several other familiar rings are UFD's.

THEOREM 1. Let R be a UFD. Then $R[X]$ is also a UFD.

Taking R in Theorem 1 to be $\mathbf{Z}$, we see that $\mathbf{Z}[X]$ is a UFD. Also, since $F[X_1, \ldots, X_n]$ is the polynomial ring in one variable over $F[X_1, \ldots, X_{n-1}]$, a simple induction argument shows that $F[X_1, \ldots, X_n]$ is a UFD too. The rings $\mathbf{Z}[X]$ and $F[X_1, \ldots, X_n]$, $n > 1$, are not Euclidean domains. In fact, they are not even PID's.

The proof of Theorem 1 involves three basic ideas. By Theorem 8.2 we may consider R to be a subring of its field of fractions F. Thus $R[X]$ is a subring of $F[X]$. By Theorems 9.2, 9.5, and 10.7, we know that $F[X]$ is a UFD. Finally, using our knowledge about factorization in R and in $F[X]$, we can demonstrate unique factorization in $R[X]$. All of these ideas are present in the proof of Theorem 1 for the case $R = \mathbf{Z}$. Because of this, we will restrict ourselves to proving that $\mathbf{Z}[X]$ is a UFD and leave the generalization to the reader.

Let $f = a_0 + a_1X + \ldots + a_nX^n$ be a nonzero element of $\mathbf{Z}[X]$. The *content* of f is the integer $c(f) = \gcd(a_0, \ldots, a_n)$. We say f is *primitive* if $c(f) = 1$. We can write $f = c(f)g$, where g is a primitive polynomial in $\mathbf{Z}[X]$. In a moment we will show that irreducibility in $\mathbf{Z}[X]$ and in $\mathbf{Q}[X]$ are equivalent for primitive polynomials. This is not true for nonprimitive polynomials. For example, if $f = 2 + 2X^2$, then $2(1 + X^2)$ is a factorization of f as a product of two nonunits in $\mathbf{Z}[X]$. However, 2 is a unit in $\mathbf{Q}[X]$ and f is irreducible in $\mathbf{Q}[X]$.

LEMMA 2 ***(Gauss' Lemma).*** If f and g are primitive polynomials in $\mathbf{Z}[X]$, then fg is primitive.

Proof. Suppose f and g are primitive in $\mathbf{Z}[X]$ and fg is not primitive. Let p be a prime dividing $c(fg)$. For any polynomial h in $\mathbf{Z}[X]$, let $\bar{h}$ be the polynomial in $\mathbf{Z}_p[X]$ obtained by mapping the coefficients of h into their congruence class modulo p. The map $h \longmapsto \bar{h}$ is easily seen to be a homomorphism. (See Exercise 4.10.) Since p divides $c(fg)$, all the coefficients of fg are divisible by p. Thus $0 = \overline{fg}$. But $\mathbf{Z}_p[X]$ is an integral domain, and so either $\bar{f} = 0$ or $\bar{g} = 0$, that is, either p divides all the coefficients of f or p divides all the coefficients of g. This contradicts our assumption that $c(f) = c(g) = 1$. □

The following lemma forms the foundation for the proof that $\mathbf{Z}[X]$ is a UFD.

LEMMA 3. Let f be a primitive polynomial in $\mathbf{Z}[X]$. Then f is irreducible in $\mathbf{Z}[X]$ if and only if f is irreducible in $\mathbf{Q}[X]$.

Proof. The only primitive polynomials of degree 0 are ± 1. Since each of these is a unit in both $\mathbf{Z}[X]$ and $\mathbf{Q}[X]$, we may assume that f has positive degree and so is a nonunit in $\mathbf{Z}[X]$ and in $\mathbf{Q}[X]$. We will prove the lemma by showing that f has a nontrivial factorization in $\mathbf{Z}[X]$ if and only if f has a nontrivial factorization in $\mathbf{Q}[X]$.

Suppose f is not irreducible in $\mathbf{Z}[X]$. Then $f = uv$, with u and v nonunits in $\mathbf{Z}[X]$. If u is a unit in $\mathbf{Q}[X]$, then u is in $\mathbf{Q} \cap \mathbf{Z}[X] = \mathbf{Z}$, and so u divides $c(f) = 1$. But this means that $u = \pm 1$, contradicting our assumption that u is not a unit in $\mathbf{Z}[X]$. Therefore u and also v are nonunits in $\mathbf{Q}[X]$ and f is not irreducible in $\mathbf{Q}[X]$. Thus the irreducibility of f in $\mathbf{Q}[X]$ implies the irreducibility of f in $\mathbf{Z}[X]$.

Now suppose f is not irreducible in $\mathbf{Q}[X]$. Then there exist elements u and v in $\mathbf{Q}[X]$ such that $f = uv$ and both u and v have positive degree. Since u has the form

$$\frac{a_0}{b_0} + \frac{a_1}{b_1}X + \ldots + \frac{a_n}{b_n}X^n,$$

with each a_i and b_i in $\mathbf{Z}$, we can write

$$u = \frac{p(a'_0 + a'_1X + \ldots + a'_nX^n)}{q},$$

where p, q, and the a'_i are integers, p and q are both positive, and

$$\gcd(a'_0, \ldots, a'_n) = 1.$$

Thus $u = (p/q)u'$, where u' is a primitive polynomial in $\mathbf{Z}[X]$. Similarly, we can write $v = (r/s)v'$, where r and s are positive integers and v' is primitive in $\mathbf{Z}[X]$. Hence

$$f = uv = (p/q)u'(r/s)v' = (a/b)u'v',$$

where $a = pr$ and $b = qs$. Thus $bf = au'v'$. Now f is assumed primitive and $u'v'$ is primitive by Lemma 2. Therefore $c(bf) = b$ and $c(au'v') = a$, so $a = b$. It follows that $f = u'v'$ is a factorization of f in $\mathbf{Z}[X]$, with u' and v' nonunits, and so f is not irreducible in $\mathbf{Z}[X]$. Thus the irreducibility of f in $\mathbf{Z}[X]$ implies the irreducibility of f in $\mathbf{Q}[X]$. □

LEMMA 4. The irreducible elements in $\mathbf{Z}[X]$ are the primes in $\mathbf{Z}$ together with the irreducible primitive polynomials of positive degree.

Proof. Let f be a nonzero nonunit in $\mathbf{Z}[X]$. Then $f = c(f)g$, where g is primitive. If f is irreducible, then either $c(f)$ or g is a unit in $\mathbf{Z}[X]$. If $c(f) = 1$, then f is a primitive polynomial of positive degree. If g is a unit in $\mathbf{Z}[X]$, then f is an integer and any nontrivial factorization of f in $\mathbf{Z}$ is a nontrivial factorization of f in $\mathbf{Z}[X]$, and so f is a prime (or the negative of a prime).

All that remains is to show that each prime integer p is irreducible in $\mathbf{Z}[X]$. If $p = uv$, with u and v in $\mathbf{Z}[X]$, then u and v must both have degree 0, so u and v are in $\mathbf{Z}$. Since p is a prime, this means either u or v is a unit in $\mathbf{Z}$ and hence a unit in $\mathbf{Z}[X]$. □

LEMMA 5. Let f be a primitive polynomial of positive degree in $\mathbf{Z}[X]$. Then f can be factored into a product of irreducible elements in $\mathbf{Z}[X]$, and this factorization is unique up to order and associates.

Proof. First, we show that f can be factored into a product of primitive irreducibles. If f is irreducible in $\mathbf{Z}[X]$, we are done. If not, then $f = uv$, with u and v nonunits in $\mathbf{Z}[X]$. By Lemma 2, $c(f) = c(u)c(v) = 1$, and so u and v are both primitive. Since u and v are nonunits, they both have positive degree less than the degree of f. By induction on the degree of f, we may factor u and v into products of primitive irreducible polynomials, and this gives the required factorization of f.

Now suppose that $f = u_1 \ldots u_r = v_1 \ldots v_s$, with each u_i and v_j irreducible in $\mathbf{Z}[X]$. By Lemma 4, the u_i and v_j are primitive and so, by Lemma 3, they are all irreducible in $\mathbf{Q}[X]$. Since $\mathbf{Q}[X]$ is a UFD, we have $r = s$ and, with a suitable renumbering of the factors, u_i and v_i are associates in $\mathbf{Q}[X]$. This means that $v_i = (a_i/b_i)u_i$ for some a_i and b_i in $\mathbf{Z}$. Thus $b_i v_i = a_i u_i$. The content of $b_i v_i$ is $|b_i|$, while the content of $a_i u_i$ is $|a_i|$. Thus $|a_i| = |b_i|$ and $a_i/b_i = \pm 1$. Therefore u_i and v_i are associates in $\mathbf{Z}[X]$ and the factorization of f is unique up to order and associates. □

THEOREM 6. $\mathbf{Z}[X]$ is a UFD.

Proof. Let f be a nonzero nonunit in $\mathbf{Z}[X]$. Then $f = c(f)g$, where g is primitive. We can factor $c(f)$ into a product of prime integers and, by Lemma 5, we can factor g into a product of primitive irreducible polynomials. Thus we can factor f into a product of irreducible elements in $\mathbf{Z}[X]$. Suppose

$$f = p_1 \ldots p_m u_1 \ldots u_r = q_1 \ldots q_n v_1 \ldots v_s,$$

where the p_i and q_j are primes in $\mathbf{Z}$ and the u_i and v_j are primitive irreducible elements of $\mathbf{Z}[X]$. Then $c(f) = p_i \ldots p_m = q_1 \ldots q_n$, and so $m = n$ and we may assume $p_i = q_i$, $1 \le i \le m$. Thus $u_1 \ldots u_r = v_1 \ldots v_s$ and, by Lemma 5, $r = s$ and, after renumbering, $u_i = \pm v_i$. Thus the factorization of f is unique. □

As we have already stated, our proof of Theorem 6 can easily be generalized to prove Theorem 1. If R is a UFD, then by Exercise 10.14 greatest common divisors exist in R. This means that we can define the content of a polynomial f in $R[X]$ and we can talk about primitive polynomials in $R[X]$. The proofs of Lemmas 2 to 5 and Theorem 6 are changed only to the extent of replacing $\mathbf{Q}$ by the field of fractions F of R and allowing for the possibility that R has units other than ± 1.

Although we have shown that $\mathbf{Z}[X]$ is a UFD, we do not yet have any algorithms for deciding whether or not a given polynomial in $\mathbf{Z}[X]$ is irreducible or for obtaining the factorization when it is not irreducible. In Section 12 we will show that if R is a UFD with a finite set of units and there is an algorithm for factoring in R, then there is an algorithm for factoring in $R[X]$.

We close this section with some remarks about linear and quadratic factors and with an irreducibility test in $\mathbf{Z}[X]$.

THEOREM 7. Let R be an integral domain and let $f = a_0 + a_1 X + \ldots + a_n X^n$ be in $R[X]$. Suppose $n \ge 1$ and u, v, and w are in R. Then $X - w$ divides f in $R[X]$ if and only if $f(w) = 0$. If $u + vX$ divides f in $R[X]$, then u divides a_0 and v divides a_n in R.

Proof. Since $X - w$ is monic, we can divide f by $X - w$ to produce a quotient q in $R[X]$ and a remainder r in R such that $f = (X - w)q + r$. Now $f(w) = (w - w)q(w) + r = r$, and so $f(w) = 0$ if and only if $X - w$ divides f. If g is any nonzero element of $R[X]$, then the leading coefficient of $(u + vX)g$ is the product of v and the leading coefficient of g. Similarly, the constant term of $(u + vX)g$ is divisible by u. Thus, if $u + vX$ divides f, then u divides a_0 and v divides a_n. □

Suppose f is in $\mathbf{Z}[X]$ and f has positive degree. If the constant term of f is 0, then X is a factor of f. If the constant term of f is not 0, then Theorem 7 shows that any factor of f in $\mathbf{Z}[X]$ with degree 1 is one of a finite set of polynomials $u + vX$, where u divides the constant term of f and v divides the leading coefficient. Of course, we may assume that $v > 0$. For example, if $f = -6 + 7X + 17X^3 + 6X^4$ and $u + vX$ divides f, then u and v belong to the set $\{\pm 1, \pm 2, \pm 3, \pm 6\}$. Since f is primitive, u and v must be relatively prime. Assuming $v > 0$, we have 18 possibilities for $u + vX$. By trial and error, we find that

$$f = (-1 + 2X)(3 + X)(2 + X + 3X^2).$$

If $h = aX^2 + bX + c$ is in $\mathbf{Z}[X]$ and $a \neq 0$, then in $\mathbf{C}[X]$ we can factor h as $a(X - \alpha)(X - \beta)$, where

$$\alpha = \frac{-b + \sqrt{b^2 - 4ac}}{2a}, \quad \beta = \frac{-b - \sqrt{b^2 - 4ac}}{2a}.$$

Thus h is irreducible in $\mathbf{Q}[X]$ if and only if α and β are not in $\mathbf{Q}$, which is equivalent to $b^2 - 4ac$ not being a square in $\mathbf{Z}$. Since $1^2 - 4(3)(2) = -23$, the factor $2 + X + 3X^2$ of f is irreducible in $\mathbf{Q}[X]$ and, since it is primitive, it is irreducible in $\mathbf{Z}[X]$.

There are relatively few ways known for recognizing irreducible polynomials in $\mathbf{Q}[X]$ of degree greater than 3 without a great deal of calculation. One test that is sometimes useful is the following.

THEOREM 8 ***(Eisenstein's Criterion).*** Let $f = a_0 + a_1X + \ldots + a_nX^n$ be in $\mathbf{Z}[X]$. Assume that there is a prime p in $\mathbf{Z}$ such that p divides $a_0, \ldots, a_{n-1}$, but p does not divide a_n and p^2 does not divide a_0. Then f is irreducible in $\mathbf{Q}[X]$.

Proof. Since p does not divide a_n, p does not divide $c(f)$, and so we may divide f by $c(f)$ without changing the hypothesis. Thus we may assume f is primitive. If f is reducible in $\mathbf{Q}[X]$, we can factor f as gh in $\mathbf{Z}[X]$, with g and h of positive degree. Let $\bar{\ }: \mathbf{Z}[X] \to \mathbf{Z}_p[X]$ be defined by reduction of coefficients modulo p. Then $\bar{f} = \overline{gh}$ and $\bar{f}$ has the form uX^n for some nonzero element u of $\mathbf{Z}_p$. Every factor of uX^n with positive degree has a constant term of 0. Thus $\bar{g}$ and $\bar{h}$ each have 0 as their constant term. Therefore the constant terms of g and h are each divisible by p, so the constant term a_0 of f must be divisible by p^2, which violates our assumption. □

The polynomial $f = 6 + 4X + 2X^2 + X^3$ satisfies Eisenstein's Criterion with $p = 2$ and hence is irreducible in $\mathbf{Q}[X]$. Unfortunately, there are integer polynomials, such as $1 + X^2$, that do not satisfy Eisenstein's Criterion for any prime p and are nevertheless irreducible in $\mathbf{Q}[X]$.

As a final remark, we note that the Eisenstein of Eisenstein's Criterion is the German mathematician Ferdinand Eisenstein (1823-1852).

EXERCISES

1 Compute the content of the following polynomials in $\mathbf{Z}[X]$.
   (a) $63 - 105X + 78X^2$.
   (b) $-154 + 1001X + 182X^2 - 286X^3$.

2 The polynomial $f = (4 + 2i) - (3 + i)X + (-1 + 3i)X^2$ is in $R[X]$, where $R = \mathbf{Z}[i]$. Compute the content of f.

3 Considering the polynomial

$$g = (1 + X - X^2 - X^3) + (3 + 4X + 3X^2 + 2X^3)Y + (2 + 3X + X^2)Y^2$$

to be an element of $R[Y]$, where $R = \mathbf{Z}_5[X]$, compute the content of g.

4 Suppose f and g are in $\mathbf{Z}[X]$ and g is primitive. Prove that if g divides f in $\mathbf{Q}[X]$, then g divides f in $\mathbf{Z}[X]$.

5 Find all factors in $\mathbf{Z}[X]$ of degree 1 in the following polynomials.
   (a) $9 - 9X + 14X^2 - 8X^3$.
   (b) $10 + 9X + 6X^2 - 4X^3 - 3X^4$.
   (c) $2 + 3X - 3X^2 + 8X^3 - 6X^4$.

6 The polynomial $(X + X^2) + (4 + 2X + 4X^2)Y + 4X^2Y^2 + X^2Y^3$ in $\mathbf{Z}_5[X, Y]$ contains a factor which has degree 1 in Y. Find this factor.

12. INTERPOLATION

Let R be a commutative ring and let $a_0, \ldots, a_m$ and $b_0, \ldots, b_m$ be elements of R with the a_i distinct. The process of finding a polynomial f in $R[X]$ such that $f(a_i) = b_i$, $i = 0, \ldots, m$, is called *interpolation*. It may not be possible to find any f. For example, there is no polynomial in $\mathbf{Z}[X]$ whose value at 0 is 0 and whose value at 2 is 1. When f does exist, it is never unique. If $g = (X - a_0)(X - a_1) \ldots (X - a_m)$, then adding any multiple of g to f yields another polynomial with the same values at $a_0, \ldots, a_m$. The following theorem shows that if we assume that R is a field and that f has degree at most m, then the interpolation problem always has a unique solution.

THEOREM 1 ***(Lagrange Interpolation).*** Let $a_0, \ldots, a_m$ be distinct elements of a field F and let $b_0, \ldots, b_m$ be elements of F (not necessarily distinct). Then

$$f = \sum_{i=0}^{m} b_i \prod_{j \neq i} \frac{X - a_j}{a_i - a_j}$$

is the unique polynomial in $F[X]$ having degree at most m and taking on the value b_i at a_i, $0 \leq i \leq m$.

Proof. The uniqueness of f follows from Corollary 9.8. To see that $f(a_j) = b_j$, we define

$$g_i = \prod_{j \neq i} \frac{X - a_j}{a_i - a_j}$$

and note that $g_i(a_i) = 1$ and $g_i(a_j) = 0$ for all $j \neq i$. Then

$$f = \sum_{i=0}^{m} b_i g_i,$$

so

$$f(a_j) = \sum_{i=0}^{m} b_i g_i(a_j) = b_j g_j(a_j) = b_j.$$

Since each g_i has degree m, the degree of f is at most m. □

As an example of Lagrange interpolation, let us find the polynomial f in $\mathbf{R}[X]$ of degree at most 2 such that $f(1) = -8$, $f(3) = 2$, and $f(4) = 13$. By Theorem 1,

$$f = \frac{-8(X-3)(X-4)}{(1-3)(1-4)} + \frac{2(X-1)(X-4)}{(3-1)(3-4)} + \frac{13(X-1)(X-3)}{(4-1)(4-3)} =$$

$$= -\frac{4}{3}(X^2 - 7X + 12) - (X^2 - 5X + 4) + \frac{13}{3}(X^2 - 4X + 3)$$

$$= 2X^2 - 3X - 7.$$

There are other methods for performing interpolation. For example, we could try to find the polynomial f in the previous example by writing it in the form

$$f = c_0 + c_1(X-1) + c_2(X-1)(X-3).$$

Then $c_0 = f(1) = -8$, and

$$2 = f(3) = c_0 + 2c_1 = -8 + 2c_1.$$

Thus $c_1 = 5$. Finally,

$$13 = f(4) = c_0 + 3c_1 + 3c_2 = -8 + 15 + 3c_2,$$

giving $c_2 = 2$. Thus we obtain again

$$f = -8 + 5(X-1) + 2(X-1)(X-3) = 2X^2 - 3X - 7.$$

Another view of interpolation is discussed in Section 6.4.

The procedures *RXINTERP* and *ZXINTERP* perform interpolation in **R**[X] and **Z**[X], respectively. If *A* and *B* are vectors of length *N*+1, then *F*←*A RXINTERP B* is the vector of coefficients for the polynomial in **R**[X] of degree at most *N* whose value at *A*[*I*] is *B*[*I*]. Of course, the entries in *A* must be distinct.

```
      1 3 4 RXINTERP ¯8 2 13
¯7 ¯3 2
      ⎕←F←0 ¯1 3 5 RXINTERP 3.2 6.1 ¯2 10
3.2 ¯3.235833333 ¯0.1266666667 0.2091666667
      F RXEVAL 0 ¯1 3 5
3.2 6.1 ¯2 10
```

Here we have repeated the previous example and also found the cubic polynomial whose values at 0, −1, 3, and 5 are 3.2, 6.1, −2, and 10, respectively. As usual, the results of *RXINTERP* are subject to round-off error.

The arguments of *ZXINTERP* must have integer entries and, if the interpolated polynomial does not have integer entries, then a domain error is indicated.

```
      1 3 4 ZXINTERP ¯8 2 13
¯7 ¯3 2
```

The second arguments of *RXINTERP* and *ZXINTERP* may be matrices listing several vectors of values, so that more than one interpolation can be performed at one time. If *A* is a vector and *B* is a matrix with ρ*A* columns, then *F*←*A RXINTERP B* is the matrix whose *I*th row is the vector of coefficients for the polynomial in **R**[X] whose value at *A*[*J*] is *B*[*I*;*J*].

```
      ⎕←B←2 3⍴¯8 2 13 4 14 22
 ¯8   2  13
  4  14  22
      1 3 4 RXINTERP B
¯7 ¯3  2
 2  1  1
      ¯7 ¯3 2 RXEVAL 1 3 4
¯8 2 13
      2 1 1 RXEVAL 1 3 4
4 14 22
```

Interpolation can be used to factor polynomials in $\mathbf{Z}[X]$. Let f be a primitive polynomial in $\mathbf{Z}[X]$ with degree $n > 1$ and suppose f is not irreducible. Then f can be factored as gh in $\mathbf{Z}[X]$, where g and h have positive degree and one of them, say g, has degree not exceeding $n/2$. For any integer a we have $f(a) = g(a)h(a)$, and so $g(a)$ is a divisor of $f(a)$. If $f(a) = 0$, then $X - a$ is a factor of f and we may remove this factor before proceeding. If $f(a) \neq 0$, then $f(a)$ has only finitely many divisors, so there are only finitely many possibilities for $g(a)$. We can find a factor g of f as follows. Let m be the largest integer not exceeding $n/2$ and let $a_0, \ldots, a_m$ be distinct integers. If $f(a_i) = 0$ for some i, we take $g = X - a_i$. Otherwise, we consider all possible sequences $b_0, \ldots, b_m$, where b_i divides $f(a_i)$. There are only finitely many such sequences. For each sequence we interpolate a polynomial g of degree at most m satisfying $g(a_i) = b_i$, $0 \le i \le m$. If g has integral coefficients, we check whether or not g divides f.

Let us use the method of interpolation to factor $f = 4 + 11X - 13X^2 + 9X^3 - 6X^4$ in $\mathbf{Z}[X]$. Here $n = 4$ and $m = 2$. If we take $a_0 = -1$, $a_1 = 0$, and $a_2 = 1$, then $f(a_0) = -35$, $f(a_1) = 4$, and $f(a_2) = 5$. Thus b_0 is in $\{\pm 1, \pm 5, \pm 7, \pm 35\}$, b_1 is in $\{\pm 1, \pm 2, \pm 4\}$, and b_2 is in $\{\pm 1, \pm 5\}$. Since we may replace g by $-g$, we may assume that b_1, the constant term of g, is positive. Thus there are $8(3)(4) = 96$ possible choices for the sequence b_0, b_1, b_2.

One way of proceeding is to consider each of the 96 sequences one at a time. For example, suppose we have $b_0 = -5$, $b_1 = 2$, and $b_2 = -1$. We find the polynomial g in $\mathbf{R}[X]$ of degree at most 2 such that $g(-1) = -5$, $g(0) = 2$, and $g(1) = -1$ by interpolating.

```
      ⎕←G←¯1 0 1 RXINTERP ¯5 2 ¯1
2 2 5
```

Here we see that $g = 2 + 2X + 5X^2$ is, in fact, in $\mathbf{Z}[X]$. We now check whether g divides f.

```
      F←4 11 ¯13 9 ¯6
      G RXREM F
¯1.104 8.536
```

It does not.

Although it would be possible to continue in this manner with the remaining 95 cases, there is a better way. We can construct a 96-by-3 matrix listing the 96 choices as follows.

```
      B0←1 ¯1 5 ¯5 7 ¯7 35 ¯35
      B1←1 2 4
      B2←1 ¯1 5 ¯5
      T←8 3 4⊤⍳96
      B←⍉3 96⍴B0[T[0;]],B1[T[1;]],B2[T[2;]]
```

Here `B[I;]` is `B0[J],B1[K],B2[L]`, where `I = L+(4×K)+12×J` and `0 ≤ J ≤ 7`, `0 ≤ K ≤ 2`, and `0 ≤ L ≤ 3`. Next we interpolate in **R**[*X*] to find the polynomials whose values at −1, 0, and 1 are given by the rows of *B*

```
      G←¯1 0 1 RXINTERP B
```

and select the rows of *G* that have integer entries.

```
      G←(∧/G=⌊G)⌿G
```

(In this particular example we do not get rid of any rows of *G*.) Next, we keep only those rows of *G* that define divisors of *f* in **R**[*X*].

```
      G←(∧/0=G RXREM F)⌿G
```

Finally, we remove the rows of *G* that do not divide *f* in **Z**[*X*].

```
      G←(∧/Q=⌊Q←F RXQUOT G)⌿G
      G
 1  0  0
 1  3 ¯3
 4 ¯1  2
```

It is now easy to see that the factorization of *f* is $(1 + 3X - 3X^2)(4 - X + 2X^2)$. There was some risk in using *RXINTERP*, *RXREM*, and *RXQUOT* in the preceding computation due to possible round-off errors. However, because of the small numbers involved, the risk was small.

Interpolation may be used to factor polynomials over rings other than **Z**.

THEOREM 2. Let R be a UFD with a finite group of units that is explicitly known. Assume also that there is an algorithm for performing prime factorization in R. Then there is an algorithm for carrying out prime factorization in $R[X]$.

Proof. Let f be a polynomial in $R[X]$ of degree $n > 0$. If R is finite, then there are only finitely many polynomials in $R[X]$ of degree less than n, and we may try each one as a possible factor of f. Suppose R is infinite. Then we can always choose $m + 1$ distinct elements $a_0, \ldots, a_m$ of R, where m is the largest integer not exceeding $n/2$. If $f(a_i) = 0$, we have found the factor $X - a_i$ of f. Thus we may assume $f(a_i) \neq 0$, $0 \leq i \leq m$. Since there is a factorization algorithm in R, we can write $f(a_i)$ as $p_1 \ldots p_r$, where the p_j are primes in R. Any divisor of $f(a_i)$ is an associate of a product of a subset of the p_j. Since we know the units of R and there are only finitely many, we can list the divisors of $f(a_i)$. Interpolation now gives us a finite set of polynomials that must contain a factor of f provided f is not irreducible. □

An induction argument based on Theorem 2 shows that there is a factorization algorithm for $\mathbf{Z}[X_1, \ldots, X_n]$.

EXERCISES

1. Use Lagrange interpolation to find the polynomial f in $\mathbf{Q}[X]$ with degree at most 2 such that $f(-1) = -9$, $f(0) = 2$, and $f(2) = 18$. Check your answer using *RXINTERP*.
2. Let $a_0, \ldots, a_m$ be distinct elements of the field F and let $b_0, \ldots, b_m$ be in F. Suppose that f is the polynomial in $F[X]$ of degree at most m such that $f(a_i) = b_i$, $0 \leq i \leq m$. Show that $f - b_0 = (X - a_0)g$ for some g in $F[X]$ with degree at most $m - 1$. What are the values of g at $a_1, \ldots, a_m$? Describe a recursive procedure for computing f. Use this procedure to perform the interpolation in Exercise 1.
3. Find an element f of $\mathbf{Z}[X]$ such that $f(1) \equiv 8 \pmod{11}$, $f(3) \equiv 1 \pmod{11}$, and $f(7) \equiv 4 \pmod{11}$.
4. Let $F = \mathbf{Z}_3(X)$, the field of rational functions in one variable with coefficients in $\mathbf{Z}_3$. Find a polynomial g in $F[Y]$ such that $g(0) = X$, $g(1) = X^2 - 1$, and $g(2) = X^3 - X^2$.
5. Use the method of interpolation to factor $10 - 13X + X^2 + 11X^3 - 6X^4$ in $\mathbf{Z}[X]$.

*6 Factor $(3 + i) + (4 - i)X + (3 + 2i)X^2 + (2 + i)X^3 + iX^4$ in $\mathbf{Z}[i][X]$.

7 Let `A←((?3 3 3⍴201)-100)÷10` and consider A to represent an element of $M_3(\mathbf{R}[X])$. Construct the vector D of length 7 such that `D[I]` is `RDET A RXEVAL I` for I in `⍳7`. Compare

`(⍳7) RXINTERP D` and `RXDET A`.

Explain your observations.

5
MODULES

Let A be an abelian group written additively. In Section 3.2 we defined, for each integer n and each element a of A, the element na of A. That is, we defined a map $f: \mathbf{Z} \times A \longrightarrow A$. This map has many useful properties. (See, for example, Theorem 3.2.4 and Exercise 3.2.5.) If R is any ring, then a left module for R is an abelian group M together with a map from $R \times M$ to M satisfying conditions similar to those satisfied by the map f. This chapter is devoted to a general discussion of modules, with very few restrictions placed on the ring R. However, many of the most important examples of modules arise when R is $\mathbf{Z}$, a field, or the polynomial ring over a field. All of these rings are Euclidean domains. For rings of this special type it is possible to obtain much more detailed information about modules. This is done in Chapter 6.

1. DEFINITIONS

Throughout this section R will be a ring. A *left R-module* consists of an abelian group M together with a map from $R \times M$ to M, the image of the pair (r, u) being written ru. In addition, the following axioms must be satisfied for all r, s in R and all u, v in M.

1. $r(u + v) = ru + rv$.
2. $(r + s)u = ru + su$.
3. $r(su) = (rs)u$.
4. $1u = u$.

We often say that M is a left module *over R*.

Let us consider some examples of left modules.

Example 1. By the additive version of Theorem 3.2.4, every abelian group is a left $\mathbf{Z}$-module. Thus $\mathbf{Z}_{10}$, $\mathbf{Z}_2 \oplus \mathbf{Z}$, and $\mathbf{Z}$ itself are all left $\mathbf{Z}$-modules.

Example 2. Let R be a subring of a ring S. If r is in R and u is in S, then ru is certainly defined as an element of S. It is easy to check that S is a left

R-module. We always consider R to be a subring of $R[X]$. Therefore $R[X]$ is a left R-module.

Example 3. Let I be a left ideal of the ring R. If r is in R and u is in I, then ru is in I. It follows immediately from the ring axioms that I is a left R-module.

Example 4. Let R be any ring and let $M = \{0\}$ be an abelian group with one element. If we define $r0$ to be 0 for all r in R, then M becomes a left R-module. Such a module is said to be *trivial.*

Example 5. Let $R = M_2(\mathbf{Z})$, the ring of 2-by-2 integer matrices, and let $M = \mathbf{Z}^2 = \mathbf{Z} \oplus \mathbf{Z}$, the abelian group whose elements are integer vectors of length 2 with addition performed componentwise. For A in R and u in M, the matrix product $Au = A+.\times u$ is in M. We leave it as an exercise to check that this product makes M a left R-module. The following dialogue verifies special cases of the axioms $A(u + v) = Au + Av$ and $A(Bu) = (AB)u$.

```
      ⎕←A←2 2⍴2 ¯1 1 3
 2 ¯1
 1  3
      ⎕←B←2 2⍴4 1 ¯3 ¯2
 4  1
¯3 ¯2
      U←3 2
      V←¯1 4
      A+.×U
4 9
      ∧/(A+.×U+V)=(A+.×U)+A+.×V
1
      ∧/(A+.×B+.×U)=(A+.×B)+.×U
1
```

It is not hard to show that, for any ring S, the same construction makes the direct sum $S^n = S \oplus \ldots \oplus S$ with n summands into a left $M_n(S)$-module.

Example 6. The arrays *PLUS* and *TIMES* in *EXAMPLES* are the addition and multiplication tables, respectively, for a ring R with 8 elements. The matrix *MPLUS* is the addition table for an abelian group M of order 4. The matrix *RMOD* has shape 8 4 and has entries in ⍳4 in origin 0.

```
      ⍴RMOD
8 4
      ⎕IO←0
```

```
      ∧/,RMOD∊⍳4
1
```

Thus we may consider *RMOD* to define a map from $R \times M$ to M. The calculation

```
      RMOD[3;MPLUS[1;2]]=MPLUS[RMOD[3;1];RMOD[3;2]]
1
```

verifies the module axiom $r(u + v) = ru + rv$ for the case $r = 3, u = 1, v = 2$. Similarly,

```
      RMOD[4;RMOD[2;3]]=RMOD[TIMES[4;2];3]
1
```

checks the axiom $r(su) = (rs)u$ for $r = 4$, $s = 2$, and $u = 3$. We leave it to the reader to formulate each of the module axioms as a single APL statement and to verify that *PLUS*, *TIMES*, *MPLUS*, and *RMOD* satisfy all of these axioms.

The following theorem lists some elementary results that hold for any left module.

THEOREM 1. Let M be a left R-module. For all r in R and all u in M, we have

(a) $r0 = 0$.
(b) $0u = 0$.
(c) $(-r)u = -(ru)$.
(d) If r is a unit and $ru = 0$, then $u = 0$.

Proof. The proofs of (a), (b), and (c) require only slight modifications of the proof of Lemma 4.1.2a, b. Note that in (a) both zeros refer to the additive identity of M, while in (b) the first zero refers to the additive identity of R. Note also that in (c) the first minus sign refers to negation in R while the second minus sign refers to negation in M. To prove (d), assume r is a unit in R and $ru = 0$. Then

$$u = 1u = (r^{-1}r)u = r^{-1}(ru) = r^{-1}0 = 0. \quad \square$$

So far we have talked only about left modules. It is clear that the term "right R-module" should mean an abelian group M and a map $(u,r) \longmapsto ur$ of $M \times R$ to M such that the analogues of axioms 1 to 4 hold. But do we really need to distinguish between right and left modules? Suppose M is a left R-module. Can't we make M into a right R-module by defining ur to be ru for all u in M and all r in R? The answer is that we can if R is commutative, but we run into serious problems otherwise. For example, one of the right module axioms is $(ur)s = u(rs)$. If we have defined ur to be ru, then

$$(ur)s = s(ur) = s(ru) = (sr)u = u(sr).$$

Thus $(ur)s = u(sr)$, not $u(rs)$. Of course, if R is commutative, everything is all right. The following theorem gives the whole story. Recall that the opposite ring R^{op} of a ring R is obtained by defining a new multiplication $\circ$ on R, with $r \circ s = sr$.

THEOREM 2. Let M be a left R-module. If we define ur to be ru for all u in M and all r in R, then M becomes a right R^{op}-module.

Proof. The axiom that gave us trouble before was the third. Now we have $(ur)s = u(sr) = u(r \circ s)$. The other axioms are easily checked. □

If R is commutative, then R and R^{op} are the same ring. Thus every left R-module is a right R-module and every right R-module is a left R-module. In this case we are justified in referring simply to R-modules and using either right or left notation, whichever is most convenient. However, if R is not commutative, we must be careful to make the distinction between right and left R-modules. For example, let S be any ring, $R = M_n(S)$, and $M = S^n$. If A is in R and u is in M, then defining Au to be $A{+}.{\times}u$ makes M into a left R-module and defining uA to be $u{+}.{\times}A$ makes M into a right R-module. These two module structures on M are similar in many ways, but they *are* different and cannot be identified or considered the same.

A *vector space* is a module over a field. Historically, vector spaces were among the first modules to be considered. As we continue our study of modules, we will come across several examples of a concept with two names, one used when referring to modules over arbitrary rings and the other used only in connection with vector spaces. This is a case of the older vector space names coexisting with more modern terminology. Since it seems likely that the vector space names will continue to be widely used, we will perpetuate this dual terminology here.

If we are to continue the pattern of Chapter 3, where we defined groups, subgroups, and quotient groups, and of Chapter 4, where we defined rings, subrings, and quotient rings, then we should next define submodules and quotient modules. Let M be a left R-module. An *R-submodule* of M is a subgroup N of the abelian group $(M,+)$ such that ru is in N for all r in R and all u in N. These conditions ensure that N is a left R-module in its own right. If the ring R is clear from context, we simply call N a submodule of M. If R is a field, so that M is a vector space over R, then a submodule of M is usually called a *subspace*.

Here are some examples of submodules.

Example 7. If A is an abelian group, then any subgroup of A is a $\mathbf{Z}$-submodule of A.

Example 8. Let $M = \mathbf{Z}^2$ be made into a left $M_2(\mathbf{Z})$-module, as in Example

5. The set N of all vectors (x,y) in M with x and y both even is an $M_2(\mathbf{Z})$-submodule of M. However, the set L of (x,y) in M with x even is not an $M_2(\mathbf{Z})$-submodule, as the following calculation shows.

```
      ⎕←A←2 2⍴¯1 1 2 ¯1
¯1  1
 2 ¯1
      U←2 3
      A+.×U
1 1
```

Here U is in L, but $A+.\times U$ is not in L. It is true that L is a $\mathbf{Z}$-submodule of M.

Example 9. If we consider a ring R to be a left module over itself, then the submodules of R are the left ideals of R.

Given two submodules of a left R-module, we can construct two additional submodules.

THEOREM 3. Let U and V be submodules of a left R-module M. Then $U + V$ and $U \cap V$ are submodules of M.

Proof. By definition, $U + V = \{u + v \mid u \in U, v \in V\}$. Since U and V are subgroups of $(M,+)$, it follows that both $U + V$ and $U \cap V$ are subgroups of $(M,+)$. If $r \in R$, $u \in U$, and $v \in V$, then $r(u + v) = ru + rv$ is in $U + V$, and so $U + V$ is a submodule. If w is in $U \cap V$, then w is in U, and so rw is in U. Similarly, rw is in V, and so rw is in $U \cap V$. Hence $U \cap V$ is a submodule. □

Let X be a subset of a left R-module M. An element u of M is said to be an *R-linear combination* of the elements in X if u can be written as $r_1x_1 + \ldots + r_nx_n$, where each r_i is in R and each x_i is in X. Note that although X may be infinite, only finitely many elements of X are used to obtain u. It is easy to see that the set N of all R-linear combinations of the elements of X is a submodule of M. We say that N is the submodule *generated by* X and denote it by $<X>$. If $X = \{x\}$, then $<X> = Rx = \{rx \mid r \in R\}$. If $X = \{x_1, \ldots, x_n\}$ is finite, then

$$<X> = <x_1, \ldots, x_n> = Rx_1 + \ldots + Rx_n.$$

A submodule is called *cyclic* if it is generated by a single element and *finitely generated* if it has a finite generating set. In the context of vector spaces, one calls $<X>$ the subspace *spanned* by X, and one speaks of a subspace spanned by a finite set as being *finite dimensional.*

Let N be a submodule of a left R-module M. Since N is a subgroup of $(M,+)$, we can form the quotient group M/N as an abelian group.

LEMMA 4. If for all u in M and all r in R we define $r(N+u)$ to be $N+ru$, then M/N becomes a left R-module.

Proof. We must first show that this module multiplication is well defined. Suppose $N+u = N+v$. Then $u-v$ is in N. Since N is a submodule, it follows that $r(u-v) = ru - rv$ is in N and, hence, that $N+ru = N+rv$. The verification of the module axioms now becomes quite straightforward. For example, for all u and v in M and all r in R, we have

$$\begin{aligned} r[(N+u)+(N+v)] &= r(N+u+v) = N+r(u+v) \\ &= N+ru+rv = (N+ru)+(N+rv) \\ &= r(N+u)+r(N+v). \end{aligned}$$

Checking the remaining axioms is left as an exercise. □

The module M/N is called the *quotient module* of M modulo N. Since quotient modules can be formed modulo any submodule, it is best to think of submodules as the analogues of normal subgroups or (two-sided) ideals instead of as the analogues of subgroups or subrings. With vector spaces, the term "quotient space" is used instead of "quotient module".

After defining submodules and quotient modules, we take the obvious next step and define homomorphisms from one module to another. This is done only for modules over the same ring. Let M and N be left modules over the ring R. An *R-homomorphism* from M to N is a map $f:M \longrightarrow N$ such that for all r in R and all u, v in M we have

1. $(u+v)f = uf + vf$.
2. $(ru)f = r(uf)$.

Thus, in particular, f is a homomorphism of abelian groups. The map taking each element u of M to 0 in N is an R-homomorphism, as is the identity map on M. If L is a submodule of M, then $u \longmapsto L+u$ is an R-homomorphism of M onto M/L. If R is commutative and a in R, then the evaluation map at a, under which a polynomial g in $R[X]$ is mapped to its value $g(a)$, is an R-homomorphism of $R[X]$ onto R. As with group or ring homomorphisms, the composition of two R-homomorphisms is again an R-homomorphism. If V and W are vector spaces over the same field F, then an F-homomorphism from V to W is usually called a *linear transformation.*

If $f:M \longrightarrow N$ is an R-homomorphism, then the *kernel* of f is, as usual, $\{0\}f^{-1}$, the set of all elements u in M such that $uf = 0$.

THEOREM 5. Let $f:M \longrightarrow N$ be an R-homomorphism and let K and L be submodules of M and N, respectively. Then Kf is a submodule of

N and Lf^{-1} is a submodule of M. In particular, the kernel of f is a submodule of M. If f is bijective, then f^{-1} is an R-homomorphism from N to M.

Proof. By Theorem 3.4.1, Kf and Lf^{-1} are subgroups of $(N,+)$ and $(M,+)$, respectively. If u is in Kf, then $u = vf$ for some v in K. Then, for any r in R, we have $ru = r(vf) = (rv)f$ and, since rv is in K, it follows that ru is in Kf. Thus Kf is a submodule of N. Similarly, if x is in Lf^{-1}, then $y = xf$ is in L. Therefore $(rx)f = r(xf) = ry$ is in L. Hence rx is in Lf^{-1} and Lf^{-1} is a submodule of M. The kernel of f is the inverse image of the trivial submodule $\{0\}$ of N. Thus the kernel of f is a submodule of M. Finally, suppose f is bijective. Then, by Theorem 3.4.3, f^{-1} is a homomorphism of $(N, +)$ onto $(M, +)$. If u is in N, then $u = vf$, where $v = uf^{-1}$. For r in R we have $(rv)f = r(vf) = ru$. Therefore $(ru)f^{-1} = rv = r(uf^{-1})$. Thus f^{-1} is an R-homomorphism. □

A bijective R-homomorphism is called an *R-isomorphism*, and two R-modules are *R-isomorphic* if there is an R-isomorphism from one to the other. The three basic theorems on group isomorphisms carry over to R-isomorphisms. The First Isomorphism Theorem becomes the following theorem.

THEOREM 6. Let $f{:}M \longrightarrow N$ be a surjective R-homomorphism with kernel K. Then $K + u \longmapsto uf$ defines an R-isomorphism g of M/K onto N.

Proof. By Theorem 3.5.7, we know that g is well defined and g is an isomorphism of abelian groups. If r is in R and u is in M, then

$$\begin{aligned}[r(K+u)]\,g &= (K+ru)g = (ru)f \\ &= r(uf) = r[(K+u)g].\end{aligned}$$

Therefore g is an R-isomorphism. □

The other two isomorphism theorems are covered by the following theorem.

THEOREM 7. Let U, V, and W be submodules of the left R-module M with $V \subseteq W$. Then

(a) $(U+V)/U$ is R-isomorphic to $V/(U \cap V)$.
(b) W/V is a submodule of M/V, and $(M/V)/(W/V)$ is R-isomorphic to M/W.

Proof. See Exercise 19. □

Let M and N be left R-modules. In Section 3.6 we defined $M \oplus N$ to be the abelian group with $M \times N$ as its set of elements and $(u_1,v_1) + (u_2,v_2) = (u_1 + u_2, v_1 + v_2)$. We can make $M \oplus N$ into a left R-module by defining $r(u,v)$ to be (ru,rv) for all r in R and all (u,v) in $M \times N$. We call this module the *external direct sum* of M and N.

As with groups, there is a notion of an internal direct sum of modules. If U and V are submodules of a module M, then the map $f:(u,v) \longmapsto u + v$ is an R-homomorphism of $U \oplus V$ into M. We say M is the *internal direct sum* of U and V and write $M = U \oplus V$ if f is an R-isomorphism. This is equivalent to saying that $M = U + V$ and $U \cap V = \{0\}$.

EXERCISES

1 Verify that the axioms for a left module hold in Examples 2 to 5.

2 Formulate each of the four left module axioms as a single APL proposition using the four matrices in Example 6. Enter these propositions at a terminal and check that they do hold for this example.

3 Complete the proof of Theorem 1.

4 Let R be the ring and M the R-module of Example 6. Construct the multiplication table for R^{op} and the matrix describing the structure of M as a right R^{op}-module. Formulate the right module axioms as APL propositions and verify them for M and R^{op}.

5 Let R be a ring and M an abelian group, and suppose that $(r,u) \longmapsto ru$ is a map of $R \times M$ to M satisfying the first three axioms for a left module but not necessarily the fourth. Set $K = \{u \in M \mid 1u = u\}$ and $L = \{u \in M \mid 1u = 0\}$. Show that as an abelian group $M = K \oplus L$. Show also that K is a left R-module and $ru = 0$ for all r in R and all u in L.

6 Let $f:R \longrightarrow S$ be a ring homomorphism and let M be a left S-module. Show how M can be made into a left R-module.

7 Let V be a vector listing the elements of a subset of ⍳4 in origin 0. Write an APL proposition corresponding to the statement that V is a submodule of the module in Example 6.

8. Show that the module in Example 6 has no proper nontrivial submodules.

9 Prove that the subset N in Example 8 is a submodule.

10 Prove the statement made in Example 9.

11 Let X be a subset of a left R-module M. Show that the set N of all R-linear combinations of the elements in X is a submodule of M. Is N adequately defined when $X = \emptyset$? What is a reasonable definition of $\langle \emptyset \rangle$?

12 What is the submodule of the module in Example 5 generated by $(3,0)$?

13 Complete the proof of Theorem 4.

14. Let N be a submodule of a left R-module M. Show that there is a 1-1 correspondence between the submodules of M/N and the submodules of M that contain N. (Compare Theorem 3.5.9.)

15 Complete the proof of Theorem 2.

16 Let M be a left module over the commutative ring R. Show that for any r in R the map $u \mapsto ru$ is an R-homomorphism of M into itself. Thus $\{ru \mid u \in M\}$ is a submodule of M.

17 Prove that $\mathbf{Q}$ is not finitely generated as a $\mathbf{Z}$-module.

18 Let H be a vector of length 4 with components in ⍳4 in origin 0. Write an APL proposition corresponding to the assertion that H is a module homomorphism of the module in Example 6 into itself.

19 Prove Theorem 7 by making the appropriate changes in the solutions to Exercises 15 and 16 of Section 3.5.

20 An R-module M satisfies the *ascending chain condition* (ACC) on submodules if there are no infinite sequences $U_1 \subset U_2 \subset \ldots$ of submodules, with each U_i a proper subset of U_{i+1}. Show that M satisfies the ACC for submodules if and only if every submodule of M is finitely generated. (See Theorem 4.9.6 and Exercise 4.9.11.)

21 Suppose an R-module M satisfies the ACC on submodules and let U be a submodule of M. Show that M/U satisfies the ACC on submodules.

2. FREE MODULES

Throughout this section we will assume that R is a nontrivial ring, that is, $1 \neq 0$. Let $x_1, \ldots, x_n$ be a sequence of elements of a left R-module M. We say that $x_1, \ldots, x_n$ are *linearly independent* over R if the only way 0 can be written as an R-linear combination $r_1x_1 + \ldots + r_nx_n$ of the x_i is the obvious way: with $r_1 = r_2 = \ldots = r_n = 0$. For example, the real numbers 1 and $\sqrt{2}$ are linearly independent over $\mathbf{Q}$. To see this, suppose r and s are rational numbers and $r1 + s\sqrt{2} = 0$. If $s \neq 0$, we can divide by s to obtain $\sqrt{2} = -r/s$. However, $\sqrt{2}$ is irrational and $-r/s$ is a rational number. Thus s must be 0, and so $r = 0$ too. The equation $(\sqrt{2})1 + (-1)\sqrt{2} = 0$ shows that 1 and $\sqrt{2}$ are not linearly independent over $\mathbf{R}$. We say that 1 and $\sqrt{2}$ are *linearly dependent* over $\mathbf{R}$. If i is a complex number with $i^2 = -1$, then a similar argument shows that 1 and i are linearly independent over $\mathbf{R}$ but not over $\mathbf{C}$.

THEOREM 1. Let M be a left R-module generated by $x_1, \ldots, x_n$. If $x_1, \ldots, x_n$ are linearly independent over R, then each element of M can be written as an R-linear combination of the x_i in exactly one way.

Proof. Suppose $r_1, \ldots, r_n$ and $s_1, \ldots, s_n$ are in R and $r_1x_1 + \ldots r_nx_n = s_1x_1 + \ldots + s_nx_n$. Then

$$(r_1 - s_1)x_1 + \ldots + (r_n - s_n)x_n = 0.$$

Since the x_i are linearly independent over R, this means that for $1 \leq i \leq n$ we have $r_i - s_i = 0$ and, hence, $r_i = s_i$. □

We can extend the notion of linear independence to arbitrary subsets of a module M as follows. A subset X of M is said to be *linearly independent* over R if every finite sequence $x_1, \ldots, x_n$ of distinct elements of X is linearly independent over R. By this definition, the empty set $\emptyset$ is always linearly independent. (See Exercise 7.)

A *basis* for M is a linearly independent generating set. We say that M is a *free R-module* if M has a basis. For example, in R^n let $x_i = (0, \ldots, 0, 1, 0, \ldots 0)$, where each component is 0 except for the ith, which is 1. Since $1 \neq 0$, the x_i are distinct. If $(r_1, \ldots, r_n)$ is in R^n, then $(r_1, \ldots, r_n) = r_1x_1 + \ldots + r_nx_n$, so the x_i generate R^n. Moreover, if $r_1x_1 + \ldots + r_nx_n = 0$, then each r_i is 0, so the x_i are linearly independent. Thus $x_1, \ldots, x_n$ is a basis for R^n. We call this the *standard basis* for R^n. Not every left R-module is free. In the **Z**-module $\mathbf{Z}_2$ we have $2u = 0$ for all u in $\mathbf{Z}_2$. Thus no nonempty subset of $\mathbf{Z}_2$ is linearly independent, and $\mathbf{Z}_2$ does not have a basis.

As a further example, let us consider the **Z**-submodule M of $\mathbf{Z}^3$ generated by

$$u_1 = (3, -1, 2),$$
$$u_2 = (0, 2, 5),$$
$$u_3 = (0, 0, 4).$$

We will show that u_1, u_2, u_3 are linearly independent over **Z**, so M is a free **Z**-module with basis u_1, u_2, u_3. To check linear independence, we assume r_1, r_2, r_3 are integers such that $r_1u_1 + r_2u_2 + r_3u_3 = 0$. The first component of $r_1u_1 + r_2u_2 + r_3u_3$ is $3r_1$, so r_1 must be 0. The second component of $r_1u_1 + r_2u_2 + r_3u_3$ is thus $2r_2$ and, therefore, $r_2 = 0$. Finally, the third component of $r_1u_1 + r_2u_2 + r_3u_3$ is $4r_3$, and $r_3 = 0$. Thus u_1, u_2, u_3 are linearly independent and M is a free **Z**-module.

THEOREM 2. Let M be a free left R-module with basis X and let N be a left R-module. If $f: X \longrightarrow N$ is any map, then there is a unique R-homomorphism $g: M \longrightarrow N$ such that $xg = xf$ for all x in X.

Proof. We will prove the theorem only in the case in which X is finite, leaving the infinite case as an exercise. Suppose $|X| = n$ and $X = \{x_1, \ldots, x_n\}$. Set $x_i f = y_i$. If u is in M, then u can be written uniquely as $r_1x_1 + \ldots + r_nx_n$, where each r_i is in R. If g exists, then

$$\begin{aligned} ug &= (r_1x_1 + \ldots + r_nx_n)g = r_1(x_1g) + \ldots + r_n(x_ng) \\ &= r_1(x_1f) + \ldots + r_n(x_nf) = r_1y_1 + \ldots + r_ny_n. \end{aligned}$$

Thus g is unique, and all that remains is to show that the map g taking $u = r_1x_1 + \ldots + r_nx_n$ to $r_1y_1 + \ldots + r_ny_n$ is an R-homomorphism. Suppose $v = s_1x_1 + \ldots + s_nx_n$, with each s_i in R. Then

$$\begin{aligned} (u+v)g &= \left[\sum_i (r_i + s_i)x_i\right]g = \sum_i (r_i + s_i)y_i \\ &= \sum_i r_iy_i + \sum_i s_iy_i = ug + vg. \end{aligned}$$

Similarly, $(ru)g = r(ug)$ for all r in R. □

COROLLARY 3. Let N be a left R-module generated by n elements. Then N is isomorphic to a quotient module of R^n.

Proof. Let N be generated by $y_1, \ldots, y_n$ and let $x_1, \ldots, x_n$ be the standard basis for R^n. By Theorem 2, the map g taking $(r_1, \ldots, r_n) = r_1x_1 + \ldots + r_nx_n$ to $r_1y_1 + \ldots + r_ny_n$ is an R-homomorphism of R^n into N. Since the y_i generate N, the map g is actually surjective so, by Theorem 1.6, N is isomorphic to R^n/K, where K is the kernel of g. □

We have seen a special case of Corollary 3 in Exercise 3.6.5.

COROLLARY 4. If N is a left R-module possessing a basis with n elements, then N is isomorphic to R^n.

Proof. Let $y_1, \ldots, y_n$ be a basis for N and let $g: R^n \longrightarrow N$ map $(r_1, \ldots, r_n)$ to $r_1y_1 + \ldots + r_ny_n$. Then, as in the proof of Corollary 3, g maps R^n onto N. Suppose $(r_1, \ldots, r_n)$ is in the kernel of g. Then $r_1y_1 + \ldots + r_ny_n = 0$ and, by the linear independence of the y_i, this implies that $r_1 = r_2 = \ldots = r_n = 0$. Thus the kernel of g is trivial and g is injective. Therefore g is an R-isomorphism. □

Corollary 4 shows that every free left R-module N with a finite basis is isomorphic to R^n for some integer n. If R is finite, then $|N| = |R|^n$ and, as $|R| > 1$, the integer n is uniquely determined. It is natural to ask whether n is always unique. Put another way, can R^m be isomorphic to R^n as a left R-module when $m \neq n$? We will provide an answer to this question later in this section.

For each positive integer n the module R^n is a free module having a basis with n elements. Free modules with infinite bases exist also, and every R-module is isomorphic to a quotient module of a free R-module. (See Exercises 8 and 9.) For vector spaces, things are especially nice.

THEOREM 5. Let V be a vector space over the field F. Then V is a free F-module.

Proof. We will assume that V is finite dimensional. The theorem is true without this assumption, but the proof in the infinite dimensional case is beyond the scope of this text.

Since V is finite dimensional, there is a finite subset X of V that spans V. Among all such sets X choose one such that $n = |X|$ is as small as possible. If $n = 0$, then $X = \emptyset$ and X is a basis for V. (See Exercises 7 and 1.11.) Thus we may assume $n \geq 1$. Let $X = \{x_1, \ldots, x_n\}$ and suppose X is not a basis for V. Then the elements of X must be linearly dependent over F, and so there exist elements $r_1, \ldots, r_n$ of F such that $r_1 x_1 + \ldots + r_n x_n = 0$ and not all of the r_i are 0. Let r_i be the first nonzero coefficient. Since F is a field, r_i has a multiplicative inverse in F. Multiplying by r_i^{-1}, we obtain

$$x_i + r_i^{-1} r_{i+1} x_{i+1} + \ldots + r_i^{-1} r_n x_n = 0$$

or

$$x_i = -r_i^{-1} r_{i+1} x_{i+1} - \ldots - r_i^{-1} r_n x_n .$$

Therefore x_i is in $<X - \{x_i\}>$, and V is spanned by $X - \{x_i\}$, contradicting our choice of X. Thus X is a basis for V and V is a free F-module. □

Let A be an m-by-n matrix with entries in R. Each row of A is an element of R^n, so we may consider the submodule of R^n generated by the rows of A. We will denote this submodule by $S_R(A)$ or, simply, $S(A)$ when the ring R is clear from context. If we use APL indexing for A, then a typical element of $S_R(A)$ is $r_1 A[1;] + \ldots + r_m A[m;]$ or $u+.\times A$, where $u = (r_1, \ldots, r_m)$ is in R^m. Since the rows of the n-by-n identity matrix I are the standard basis for R^n, we have $S_R(I) = R^n$. In the following discussion we will usually suppress the symbol $+.\times$ for matrix multiplication.

THEOREM 6. Let A and B be l-by-m and m-by-n matrices over R, respectively. Then $S_R(AB) \subseteq S_R(B)$. If $l = m$ and A is a unit in $M_m(R)$, then $S(AB) = S(B)$. If C is any matrix over R with n columns such that $S_R(C) \subseteq S(B)$, then there exists a matrix D such that $C = DB$.

Proof. The ith row of AB is $A[i;]B$, which is in $S_R(B)$. Thus $S_R(AB) \subseteq S_R(B)$. If A is a unit in $M_m(R)$, then

$$S_R(B) = S_R(A^{-1}(AB)) \subseteq S_R(AB),$$

and so $S_R(AB) = S_R(B)$. Finally, suppose C is a matrix with n columns such that $S_R(C) \subseteq S_R(B)$. Then each row of C is in $S_R(B)$ and, for each row index i of C, there is a vector $D[i;]$ such that $C[i;] = D[i;]B$. However, this means that $C = DB$. □

COROLLARY 7. If R is commutative and n is a positive integer, then a matrix A in $M_n(R)$ is a unit if and only if $S_R(A) = R^n$.

Proof. If A is a unit and I is the n-by-n identity matrix, then

$$R^n = S_R(I) = S_R(A^{-1}A) \subseteq S_R(A),$$

and so $S_R(A) = R^n$. On the other hand, suppose A is in $M_n(R)$ and $S_R(A) = R^n$. Then $S_R(I) \subseteq S_R(A)$ and, by Theorem 6, there is a matrix C such that $CA = I$. Now C is in $M_n(R)$ and, if we take determinants, we obtain (det C) (det A) = det I = 1. Thus det A is a unit in R and A is a unit in $M_n(R)$ by Theorem 4.7.11. □

Let us illustrate the usefulness of Corollary 7 with an example in which $R = \mathbf{Z}$. Let A be the following matrix.

```
      ⎕←A←3 3⍴1 3 1 2 5 ¯1 1 2 ¯1
 1  3  1
 2  5 ¯1
 1  2 ¯1
```

The determinant of A is -1.

```
      ZDET A
¯1
```

Since -1 is a unit in $\mathbf{Z}$, the rows of A generate $\mathbf{Z}^3$.

We can derive other important facts from Corollary 7.

COROLLARY 8. If R is commutative, then every generating set of R^n has at least n elements.

Proof. Suppose $u_1, \ldots, u_m$ generate R^n and $m < n$. Let A be the n-by-n matrix such that $A[i;] = u_i$ for $1 \leq i \leq m$ and $A[i;]$ is the zero vector for $m < i \leq n$. Then $S_R(A) = R^n$ and so, by Corollary 7, we know that A is a unit in $M_n(R)$. However, $A[n;]$ is the zero vector, so det $A = 0$, which is not a unit in R. This contradicts Theorem 4.7.11. □

COROLLARY 9. If R is commutative, then every basis of R^n has n elements. If R^m is isomorphic to R^n, then $m = n$.

Proof. First, we will show that R^n has no infinite bases. Suppose Y is an infinite basis for R^n. Let $x_1, \ldots, x_n$ be the standard basis. Then Y generates R^n and each x_i is an R-linear combination of a finite sequence of elements of Y. Thus there is a finite subset Y_0 of Y such that x_i is in $< Y_0 >$

for each i. But this means that $< Y_0 > = R^n$. If y is in $Y - Y_0$, then y can be expressed as a linear combination of the elements in Y_0. Therefore $Y_0 \cup \{y\}$ is a linearly dependent set and Y cannot be a basis for R^n.

Now suppose $y_1, \ldots, y_m$ is a basis for R^n. By Corollary 4, $R^n \cong R^m$. By symmetry, we may assume $m \leq n$. If $m < n$, then R^n is generated by fewer than n elements, contradicting Corollary 8. □

The assumption of commutativity in Corollary 9 is essential. Exercise 3.11 demonstrates the existence of a noncommutative ring R such that R^n is isomorphic to R as a left R-module for all $n \geq 1$.

If M is a finitely generated free module over a commutative ring R, then M is isomorphic to R^n for a unique value of n. We call n the *rank* of M. In the case of a finite dimensional vector space V over a field F, we usually speak of the *dimension* of V instead of the rank and denote it $\dim_F(V)$.

Let M be a free left module over a commutative ring R with basis $x_1, \ldots, x_n$. If y is in M, then there exist unique elements $r_1, \ldots, r_n$ of R such that $y = r_1x_1 + \ldots + r_nx_n$. The vector $u = (r_1, \ldots, r_n)$ in R^n is called the *coordinate vector* of y with respect to $x_1, \ldots, x_n$. It is clear that u is the image of y under the R-isomorphism of M onto R^n taking $x_1, \ldots, x_n$ to the standard basis of R^n. Now suppose $y_1, \ldots, y_n$ is another basis for M. Let P be the n-by-n matrix over R such that $P[i;]$ is the coordinate vector of y_i with respect to $x_1, \ldots, x_n$. That is,

$$y_i = \sum_j P_{ij}x_j, \qquad 1 \leq i \leq n.$$

We call P the *transition matrix* from the basis $x_1, \ldots, x_n$ to the basis $y_1, \ldots, y_n$.

LEMMA 10. The matrix P is a unit in $M_n(R)$. Moreover, P^{-1} is the transition matrix from $y_1, \ldots, y_n$ to $x_1, \ldots, x_n$. Let z be in M and let u and v be the coordinate vectors for z with respect to $x_1, \ldots, x_n$ and $y_1, \ldots, y_n$, respectively. Then $u = vP$ and $v = uP^{-1}$.

Proof. Let Q be the transition matrix from $y_1, \ldots, y_n$ to $x_1, \ldots, x_n$. Then we have

$$y_i = \sum_j P_{ij}x_j, \qquad 1 \leq i \leq n,$$

$$x_i = \sum_j Q_{ij}y_j, \qquad 1 \leq i \leq n.$$

Therefore

$$x_i = \sum_j Q_{ij} \sum_k P_{jk}x_k$$

$$= \sum_k \left(\sum_j Q_{ij}P_{jk} \right) x_k .$$

Since x_i can be written as a linear combination of $x_1, \ldots, x_n$ in only one way, it follows that

$$\sum_j Q_{ij}P_{jk} = 0, \qquad i \neq k,$$

$$\sum_j Q_{ij}P_{ji} = 1.$$

Thus QP is I, the n-by-n identity matrix. Similarly, $PQ = I$, so Q and P are inverses of each other in $M_n(R)$. If $z = v_1y_1 + \ldots v_ny_n$, then

$$z = \sum_i v_iy_i = \sum_i v_i \sum_j P_{ij}x_j$$

$$= \sum_j \left(\sum_i v_iP_{ij} \right) x_j .$$

Therefore $z = u_1x_1 + \ldots + u_nx_n$, where

$$u_j = \sum_i v_iP_{ij} .$$

Thus, if $u = (u_1, \ldots, u_n)$ and $v = (v_1, \ldots, v_n)$, then $u = vP$. Similarly, $v = uQ = uP^{-1}$. □

The next lemma is a partial converse of Lemma 10.

LEMMA 11. Let P be a unit in $M_n(R)$, let $x_1, \ldots, x_n$ be a basis of M, and set

$$y_i = \sum_j P_{ij}x_j, \qquad 1 \leq i \leq n.$$

Then $y_1, \ldots, y_n$ is a basis for M.

Proof. Let $Q = P^{-1}$. Then, for $1 \leq i \leq n$, we have

$$\sum_j Q_{ij}y_j = \sum_j Q_{ij} \sum_k P_{jk}x_k$$

$$= \sum_k \left(\sum_j Q_{ij}P_{jk}\right) x_k$$

$$= \sum_k (QP)_{ik}x_k = x_i,$$

since $QP = I$. Thus $< y_1, \ldots, y_n >$ contains each x_i, and so $y_1, \ldots, y_n$ generate M.

We must now show that $y_1, \ldots, y_n$ are linearly independent over R. Suppose $r_1, \ldots, r_n$ are in R and $r_1y_1 + \ldots + r_ny_n = 0$. Then

$$0 = \sum_i r_iy_i = \sum_i r_i \sum_j P_{ij}x_j$$

$$= \sum_j \left(\sum_i r_iP_{ij}\right) x_j.$$

Since $x_1, \ldots, x_n$ are linearly independent over R, this means that

$$\sum_i r_iP_{ij} = 0, \qquad 1 \le j \le n.$$

Thus, if $u = (r_1, \ldots, r_n)$, then $uP = 0$. But then $u = u(PQ) = (uP)Q = 0Q = 0$, since $PQ = I$. Therefore $r_1 = \ldots = r_n = 0$ and $y_1, \ldots, y_n$ are linearly independent. □

The following theorem combines Lemmas 10 and 11 with facts previously obtained to form the main result of this section.

THEOREM 12. Suppose R is a commutative ring and P is in $M_n(R)$. Then the following are equivalent.

(a) P is a unit in $M_n(R)$.
(b) $\det P$ is a unit in R.
(c) The rows of P generate R^n.
(d) The rows of P are a basis for R^n.

Proof. We know that (a) and (b) are equivalent by Theorem 4.7.11. Let $x_1, \ldots, x_n$ be the standard basis of R^n. Then, using APL indexing for P, we have

$$P[i;] = \sum_j P[i;j]x_j.$$

Lemmas 10 and 11 now show that (a) and (d) are equivalent. Finally, Corollary 7 states that (a) and (c) are equivalent. □

Note that if P satisfies any one and hence all of the conditions of

Theorem 12, then P is the transition matrix from the standard basis of R^n to the basis consisting of the rows of P.

Theorem 12 has two important corollaries.

COROLLARY 13. If M is a free module of rank n over a commutative ring R, then elements $x_1, \ldots, x_n$ of M form a basis for M if and only if $x_1, \ldots, x_n$ generate M.

Proof. By the equivalence of conditions (c) and (d) of Theorem 12, the corollary holds in R^n and hence in any module isomorphic to R^n. (See Exercise 10.) □

COROLLARY 14. Suppose R is a commutative ring and M and N are free R-modules of rank n. Then any surjective R-homomorphism from M to N is injective.

Proof. Let $x_1, \ldots, x_n$ be a basis of M and suppose $f:M \to N$ is a surjective R-homomorphism. Set $y_i = x_i f$, $1 \le i \le n$. Then $x_1, \ldots, x_n$ generate M, and N, which is the image of M under f, is generated by $y_1, \ldots, y_n$. By Corollary 13, this means that $y_1, \ldots, y_n$ form a basis of N. Now suppose $u = r_1x_1 + \ldots + r_nx_n$ is in the kernel of f. Then $0 = uf = r_1y_1 + \ldots + r_ny_n$. Since $y_1, \ldots, y_n$ are linearly independent over R, this means that $r_1 = \ldots = r_n = 0$, and so $u = 0$. Thus the kernel of f is trivial and f is injective. □

We need to point out that, under the assumptions of Corollary 14, injective homomorphisms from M to N need not be surjective. For example, the map $f: \mathbf{Z} \to \mathbf{Z}$ with $xf = 2x$ is an injective $\mathbf{Z}$-homomorphism that is not surjective.

The final results of this section concern submodules of finitely generated modules.

THEOREM 15. Let R be a ring in which every left ideal is finitely generated. Then, for all integers $n \ge 1$, every submodule of R^n is finitely generated.

Proof. Let U be a submodule of R^n. For $1 \le i \le n$, let U_i be the subset of U consisting of the elements whose first $i-1$ components are 0. Each U_i is a submodule of U. Now let S_i be the set of ith components of the elements in U_i. If a and b are in S_i, then there exist elements $u = (0, \ldots, 0, a, \ldots)$ and $v = (0, \ldots, 0, b, \ldots)$ in U_i. If r is in R, then $ru = (0, \ldots, 0, ra, \ldots)$ and $u + v = (0, \ldots, 0, a + b, \ldots)$ are in U_i, and so ra and $a + b$ are in S_i. Therefore S_i is a left ideal of R. By assumption, S_i is finitely generated. Let a_{ij}, $1 \le j \le n_i$, be a set of generators of S_i. For each pair (i,j), with $1 \le j \le n_i$, select an element $u_{ij} = (0, \ldots, 0, a_{ij}, \ldots)$ in U_i whose ith component is a_{ij}. We will now show that U is generated by the u_{ij}.

Suppose w is in $U = U_1$. Let a be the first component of w. Then a is in S_1, and so there exist $r_1, \ldots, r_{n_1}$ such that $a = r_1 a_{11} + \ldots + r_{n_1} a_{1n_1}$. The first component of $r_1 u_{11} + \ldots + r_{n_1} u_{1n_1}$ is a, and so the first component of

$$w_2 = w - r_1 u_{11} - \ldots - r_{n_1} u_{1n_1}$$

is 0. That is, w_2 is in U_2. Now we consider the second component b of w_2. Since b is in S_2, there are elements $s_1, \ldots, s_{n_2}$ of R such that $s_1 a_{21} + \ldots + s_{n_2} a_{2n_2} = b$. Therefore

$$w_3 = w_2 - s_1 a_{21} - \ldots - s_{n_2} a_{2n_2}$$

is in U_3. Proceeding in the same manner, we are able to continue subtracting elements in the submodule V generated by the u_{ij} from w until we get the zero vector. Thus w is in V and $V = U$. Therefore U is finitely generated. □

COROLLARY 16. If R is a PID, then every submodule of R^n can be generated by n elements.

Proof. If R is a PID, then in the proof of Theorem 15 we may take each n_i to be 1, so the submodule U of R^n is generated by $u_{11}, \ldots, u_{nn}$. □

It follows from Corollary 16 that every **Z**-submodule of $\mathbf{Z}^3$ can be generated by three elements and that every **R**-subspace of $\mathbf{R}^4$ can be generated by four elements.

EXERCISES

In the following exercises R is a nontrivial ring and M is a left R-module. Procedures such as *ZDET* and *ZMATINV* may be used in solving the computational exercises.

1 Let X be a linearly independent subset of M. Prove that any subset of X is linearly independent.

2 Prove that ϕ is a linearly independent subset of M.

3 Show that $\{0\}$ is a linearly dependent subset of M.

4 Let Y be a linearly dependent subset of M. Show that any subset of M containing Y is linearly dependent.

5 Let A be the matrix

$$\begin{bmatrix} 2 & 1 & -3 & 6 & 0 \\ 0 & 3 & 5 & -2 & 1 \\ 0 & 0 & 0 & -1 & 3 \end{bmatrix}.$$

Show that the rows of A are linearly independent over $\mathbf{Z}$.

6 Let A be the matrix in Exercise 5. Show that $u = (6, 0, -14, 18, 5)$ is in $S_{\mathbf{Z}}(A)$ and find the coordinate vector of u with respect to the basis of $S_{\mathbf{Z}}(A)$ consisting of the rows of A.

7 Let X be any set. Show that there is a free left R-module having a basis that can be put in 1–1 correspondence with the elements of X. (*Hint*. Consider certain functions from X to R.)

8 Let A be the matrix of Exercise 5 and let

$$B = \begin{bmatrix} 4 & -1 & -11 & 13 & 2 \\ -2 & 2 & 8 & -11 & 10 \\ 6 & 3 & -9 & 17 & 3 \end{bmatrix}.$$

Find a 3-by-3 integer matrix C such that $B = CA$.

9 Show that $\mathbf{Q}$ is not a free $\mathbf{Z}$-module.

10 Suppose M is free with basis $x_1, \ldots, x_n$ and assume $f: M \longrightarrow N$ is an R-isomorphism of left modules. Show that N is free with basis $y_1 f, \ldots, y_n f$.

11 Explain the difference between the following two assertions.

(a) The elements $x_1, \ldots, x_n$ form a basis of M.

(b) The module M is the internal direct sum of the submodules $Rx_1, \ldots, Rx_n$.

12 In each of the following show that the rows of P are a basis for R^n.

(a) $R = \mathbf{Z}$, $n = 2$, $P = \begin{bmatrix} 3 & 4 \\ 5 & 7 \end{bmatrix}$.

(b) $R = \mathbf{Z}$, $n = 3$, $P = \begin{bmatrix} 1 & 0 & -1 \\ 1 & 7 & 3 \\ 0 & 2 & 1 \end{bmatrix}$.

(c) $R = \mathbf{Z}_{13}$, $n = 3$, $P = \begin{bmatrix} 3 & 1 & 7 \\ 2 & 5 & 1 \\ 4 & 0 & 3 \end{bmatrix}$.

(d) $R = \mathbf{Z}_2[X]$, $n = 3$, $P = \begin{bmatrix} 1 & 1 & X \\ X & 1+X & 1+X^2 \\ 1 & 1+X & 1 \end{bmatrix}$.

13 Let P_n be the set of polynomials in $\mathbf{R}[X]$ of the form $a_0 + a_1 X + \ldots + a_n X^n$. Show that $1, X, \ldots, X^n$ is a basis of P_n as an $\mathbf{R}$-module.

Define $X^{(i)}$ to be $X(X-1)\ldots(X-i+1)$, the ith *factorial power* of X. Show that $1, X^{(1)}, \ldots, X^{(n)}$ is a basis for P_n. Take $n = 3$ and compute the transition matrices from $1, X, X^2, X^3$ to $1, X^{(1)}, X^{(2)}, X^{(3)}$ and from $1, X^{(1)}, X^{(2)}, X^{(3)}$ to $1, X, X^2, X^3$.

14 Suppose P is the transition matrix from the basis $x_1, \ldots, x_n$ of M to the basis $y_1, \ldots, y_n$ and Q is the transition matrix from $y_1, \ldots, y_n$ to the basis $z_1, \ldots, z_n$. What is the transition matrix from $x_1, \ldots, x_n$ to $z_1, \ldots, z_n$?

15 Let P be the matrix in Exercise 12b. Find the transition matrix from the rows of P to the rows of the transpose P^t.

16 Let U be the set of vectors (a, b, c) in $\mathbf{Z}^3$ such that $2a - 3b + 5c = 0$. Show that U is a $\mathbf{Z}$-submodule of $\mathbf{Z}^3$ and find a two-element generating set for U.

3. ENDOMORPHISM RINGS

Let R be a ring. In this section we will study the collection of all R-homomorphisms from one R-module to another. Of particular interest will be the structure of the set of all R-homomorphisms of a given R-module into itself.

Let M and N be left R-modules. The set of all R-homomorphisms of M into N is denoted $\mathrm{Hom}_R(M, N)$. This set is always nonempty. (Why?) We can define a binary operation + on $\mathrm{Hom}_R(M, N)$ by $x(f+g) = xf + xg$ for all f, g in $\mathrm{Hom}_R(M, N)$ and all x in M. Clearly, $f+g$ is a map from M to N, but we must check that $f+g$ is an R-homomorphism. If x, y are in M and r is in R, then

$$
\begin{aligned}
(x+y)(f+g) &= (x+y)f + (x+y)g \\
&= xf + yf + xg + yg \\
&= xf + xg + yf + yg = x(f+g) + y(f+g),
\end{aligned}
$$

and

$$
\begin{aligned}
(rx)(f+g) &= (rx)f + (rx)g = r(xf) + r(xg) \\
&= r(xf + xg) = r[x(f+g)].
\end{aligned}
$$

Therefore $f+g$ is in fact an element of $\mathrm{Hom}_R(M, N)$. We can also try to define a multiplication of elements in $\mathrm{Hom}_R(M, N)$ by elements of R using the formula $x(rf) = r(xf)$. However, if R is not commutative, then the object rf is usually not in $\mathrm{Hom}_R(M, N)$. If x, y are in M, then

$$(x + y)(rf) = r[(x + y)f] = r(xf + yf)$$
$$= r(xf) + r(yf) = x(rf) + y(rf).$$

Thus rf is a homomorphism of abelian groups. However, if s is in R, then

$$(sx)(rf) = r[(sx)f] = r[s(xf)] = (rs)(xf).$$

But if rf is to be an R-homomorphism, then $(sx)(rf)$ should be

$$s[x(rf)] = s[r(xf)] = (sr)(xf).$$

Since $(rs)(xf)$ and $(sr)(xf)$ will not generally be equal unless R is commutative, it follows that rf will normally be in $\mathrm{Hom}_R(M, N)$ only when R is commutative.

THEOREM 1. Suppose that R is a ring and that M and N are left R-modules. The set $\mathrm{Hom}_R(M, N)$ is an abelian group under addition. If R is commutative, then $\mathrm{Hom}_R(M, N)$ is an R-module.

Proof. The axioms for an abelian group are easily checked. The additive identity element of $\mathrm{Hom}_R(M, N)$ is the map $x \longmapsto 0$, and the additive inverse of an R-homomorphism f is the map $-f$ defined by $x(-f) = -(xf)$. Suppose that R is commutative and that r, s are in R. Then, for any f, g in $\mathrm{Hom}_R(M, N)$ and any x in M, we have

$$x[r(f + g)] = r[x(f + g)] = r(xf + xg)$$
$$= r(xf) + r(xg) = x(rf) + x(rg)$$
$$= x(rf + rg).$$

Therefore $r(f + g) = rf + rg$. Similarly, $(r + s)f = rf + sf$. Also,

$$x[r(sf)] = r[x(sf)] = r[s(xf)]$$
$$= (rs)(xf) = x[(rs)f].$$

Hence $r(sf) = (rs)f$, and $\mathrm{Hom}_R(M, N)$ is an R-module. □

If M is a left R-module, then an R-homomorphism of M into itself is called an *R-endomorphism* of M. The set of all R-endomorphisms of M is $\mathrm{Hom}_R(M, M)$, which is usually denoted $\mathrm{End}_R(M)$. Besides the operations of addition defined previously, there is another binary operation on $\mathrm{End}_R(M)$—composition.

THEOREM 2. Let M be a left module over the ring R. Under the operations of addition and composition, $\mathrm{End}_R(M)$ is a ring.

Proof. By Theorem 1, $\mathrm{End}_R(M)$ is an abelian group under addition. Composition of functions on a set is always associative, and so $\mathrm{End}_R(M)$ is a semigroup under multiplication. The identity function on M is an R-

endomorphism and, hence, $\text{End}_R(M)$ has a multiplicative identity. All that remains is to check the distributive laws. If f, g, and h are in $\text{End}_R(M)$ and x is in M, then

$$\begin{aligned} x[f(g+h)] &= (xf)(g+h) = (xf)g + (xf)h \\ &= x(fg) + x(fh) = x(fg+fh). \end{aligned}$$

Therefore $f(g+h) = fg + fh$. Similarly, $(f+g)h = fh + gh$ and thus $\text{End}_R(M)$ is a ring. □

If R is commutative, then $\text{End}_R(M)$ is an R-module by Theorem 1 and a ring by Theorem 2. An *R-algebra* is a ring S that is also an R-module such that for all r in R and all a, b in S we have $r(ab) = (ra)b = a(rb)$.

THEOREM 3. If M is a left R-module over a commutative ring R, then $\text{End}_R(M)$ is an R-algebra.

Proof. Let f, g be in $\text{End}_R(M)$ and let r be in R. Then, for any x in M,

$$\begin{aligned} x[(rf)g] &= [x(rf)]g = [r(xf)]g \\ &= r[(xf)g] = r[x(fg)] \\ &= x[r(fg)]. \end{aligned}$$

Thus $(rf)g = r(fg)$. Similarly, $f(rg) = r(fg)$, and so $\text{End}_R(M)$ is an R-algebra. □

When R is a commutative ring, there are many examples of R-algebras besides the endomorphism rings of R-modules. These include $R[X]$, $M_n(R)$, and R itself. The next section will discuss algebras in greater detail.

The main goal of this section is to provide a description of $\text{Hom}_R(M, N)$ when M and N are finitely generated free modules. Any such module is isomorphic to R^m for some integer m, and so we begin with $\text{Hom}_R(R^m, R^n)$.

LEMMA 4. Let A be an m-by-n matrix over R. The map taking u in R^m to uA is in $\text{Hom}_R(R^m, R^n)$.

Proof. Suppose u and v are in R^m and r is in R. Then $uA = u+.\times A$ is in R^n, and so $u \longmapsto uA$ is a map from R^m to R^n. Moreover, $(u+v)A = uA + vA$ and $(ru)A = r(uA)$. Therefore $u \longmapsto uA$ is an R-homomorphism. □

LEMMA 5. If $f: R^m \longrightarrow R^n$ is an R-homomorphism, then there is an m-by-n matrix A over R such that $uf = uA$ for all u in R^m.

Proof. Let $x_1, \ldots, x_m$ be the standard basis of R^m and let A be the m-by-n matrix whose ith row is the vector $x_i f$ in R^n. If $u = (r_1, \ldots, r_m)$ is in R^m, then

$$uf = \left(\sum_i r_i x_i\right) f = \sum_i r_i(x_i f) = \sum_i r_i A[i;] = uA. \quad \square$$

The map $f:\mathbf{Z}^3 \longrightarrow \mathbf{Z}^3$ given by $(x, y, z)f = (2x - y, x + z, 3x + y - z)$ is easily seen to be a $\mathbf{Z}$-homomorphism. Since

$$(1, 0, 0)f = (2, 1, 3),$$
$$(0, 1, 0)f = (-1, 0, 1),$$
$$(0, 0, 1)f = (0, 1, -1),$$

$(x, y, z)f$ is $(x, y, x) +.\times A$, where A is the matrix

$$\begin{bmatrix} 2 & 1 & 3 \\ -1 & 0 & 1 \\ 0 & 1 & -1 \end{bmatrix}.$$

Lemmas 4 and 5 may be combined to give the following important result.

THEOREM 6. If R is a ring, then $\mathrm{Hom}_R(R^m, R^n)$ is isomorphic as an abelian group to the set M of m-by-n matrices over R. If R is commutative, then $\mathrm{Hom}_R(R^m, R^n)$ and M are isomorphic as R-modules.

Proof. Let $x_1, \ldots, x_m$ be the standard basis of R^m. For f in $\mathrm{Hom}_R(R^m, R^n)$, let $f\sigma$ be the m-by-n matrix A whose ith row is $x_i f$. By Lemma 5, $uf = uA$ for all u in R^m. Suppose g is also in $\mathrm{Hom}_R(R^m, R^n)$ and $B = g\sigma$. Then

$$x_i(f + g) = x_i f + x_i g = A[i;] + B[i;] = (A + B)[i;].$$

Therefore $(f + g)\sigma = A + B = f\sigma + g\sigma$, and σ is a homomorphism of abelian groups. If $f\sigma$ is the zero matrix, then f maps each of the x_i to zero in R^n. Thus f maps every linear combination of the x_i to zero, so f is the zero homomorphism. Therefore the kernel of σ is trivial and σ is injective. If A is any m-by-n matrix over R, then by Lemma 4 the map $f:u \longmapsto uA$ is in $\mathrm{Hom}_R(R^m, R^n)$. Since $x_i f = x_i A = A[i;]$, we have $f\sigma = A$ and σ is surjective. Thus σ is an isomorphism of abelian groups. If R is commutative, then, for any r in R and any f in $\mathrm{Hom}_R(R^m, R^n)$, the map rf is in

$$\mathrm{Hom}_R(R^m, R^n)$$

and

$$x_i(rf) = r(x_i f) = rA[i;] = (rA)[i;],$$

so $(rf)\sigma = rA = r(f\sigma)$. Therefore σ is an R-homomorphism. □

COROLLARY 7. The rings $\mathrm{End}_R(R^n)$ and $M_n(R)$ are isomorphic. If R is commutative, then $\mathrm{End}_R(R^n)$ and $M_n(R)$ are isomorphic as R-algebras.

Proof. Let σ be the map from $\mathrm{End}_R(R^n) = \mathrm{Hom}_R(R^n, R^n)$ to $M_n(R)$ defined in the proof of Theorem 6. Then σ is an isomorphism of abelian groups and, if R is commutative, an isomorphism of R-modules. All that remains is to show that $(fg)\sigma = (f\sigma) +.\times(g\sigma)$ for all f and g in $\mathrm{End}_R(R^n)$ and that σ maps 1 in $\mathrm{End}_R(R^n)$ to 1 in $M_n(R)$. Let $A = f\sigma$ and $B = g\sigma$. Then

$$\begin{aligned} x_i(fg) &= (x_if)g = (x_iA)B \\ &= x_i(AB) = (AB)[i;]. \end{aligned}$$

Therefore $(fg)\sigma = AB$. Finally, if f is the identity endomorphism of R^n, then $x_if = x_i = I[i;]$, where I is the n-by-n identity matrix. Thus $f\sigma = I$ and σ is a ring isomorphism. □

Suppose M and N are finitely generated free modules over R. Thus $M \cong R^m$ and $N \cong R^n$ for some integers m and n. It is easy to show (see Exercise 3) that $\mathrm{Hom}_R(M, N) \cong \mathrm{Hom}_R(R^m, R^n)$ and so, by Theorem 6, the R-homomorphisms of M into N may be put in 1–1 correspondence with m-by-n matrices over R. It is important to make this correspondence explicit. Let $x_1, \ldots, x_m$ be a basis of M and $y_1, \ldots, y_n$ a basis of N. Suppose f is in $\mathrm{Hom}_R(M, N)$. Then the equations

$$x_if = \sum_{j=1}^{n} A[i;j]\,y_j, \quad 1 \le i \le m,$$

uniquely determine an m-by-n matrix A over R. The ith row of A is the coordinate vector of x_if with respect to $y_1, \ldots, y_n$. We call A the *matrix of* f with respect to the bases $x_1, \ldots, x_m$ and $y_1, \ldots, y_n$.

LEMMA 8. If u in R^m is the coordinate vector of an element x in M with respect to $x_1, \ldots, x_m$, then uA is the coordinate vector of xf with respect to $y_1, \ldots, y_n$.

Proof. Suppose $u = (r_1, \ldots, r_m)$. Then $x = r_1x_1 + \ldots + r_mx_m$ and

$$\begin{aligned} xf &= \Big(\sum_i r_ix_i\Big) f = \sum_i r_i(x_if) \\ &= \sum_i r_i \sum_j A[i;j]\,y_j \\ &= \sum_j \Big(\sum_i r_iA[i;j]\Big)\, y_j \end{aligned}$$

and

$$\sum_i r_iA[i;j]$$

is the jth component of uA. □

If $M = N$ so that f is in $\mathrm{End}_R(M)$, then it is standard practice to take

$y_1, \ldots, y_n$ to be $x_1, \ldots, x_m$ and to speak of A as the matrix of f with respect to $x_1, \ldots, x_m$.

The matrix of an element of $\text{Hom}_R(M, N)$ depends on the choice of bases, and we need to know now the matrix changes when we change the bases in M and N. In Section 2 we made the assumption that R is commutative when discussing change of basis in free modules, and we will make the same assumption here.

THEOREM 9. Let R be a commutative ring and let M and N be R-modules with bases $x_1, \ldots, x_m$ and $y_1, \ldots, y_n$, respectively. Suppose P is the transition matrix from $x_1, \ldots, x_m$ to another basis $x'_1, \ldots, x'_m$ of M and suppose Q is the transition matrix from $y_1, \ldots, y_n$ to another basis $y'_1, \ldots, y'_n$ of N. If f is in $\text{Hom}_R(M, N)$ and A is the matrix of f with respect to $x_1, \ldots, x_m$ and $y_1, \ldots, y_n$, then PAQ^{-1} is the matrix of f with respect to $x'_1, \ldots, x'_m$ and $y'_1, \ldots, y'_n$.

Proof. Let x be in M and let u be the coordinate vector of x with respect to $x'_1, \ldots, x'_m$. By Lemma 2.10, the coordinate vector of x with respect to $x_1, \ldots, x_m$ is UP. By the definition of A, the coordinate vector of xf with respect to $y_1, \ldots, y_n$ is $(uP)A$. Again by Lemma 2.10, the coordinate vector of xf with respect to $y'_1, \ldots, y'_n$ is $((uP)A)Q^{-1} = u(PAQ^{-1})$. Thus PAQ^{-1} is the matrix of f with respect to $x'_1, \ldots, x'_m$ and $y'_1, \ldots, y'_n$. □

COROLLARY 10. If A is the matrix of an element f in $\text{End}_R(M)$ with respect to $x_1, \ldots, x_m$, then PAP^{-1} is the matrix of f with respect to $x'_1, \ldots, x'_m$.

Proof. In Theorem 9 we have $Q = P$. □

Let us consider an example of the use of Theorem 9. Let A be the matrix

```
      ⎕←A←3 4⍴1 ¯2 0 5 3 ¯1 2 1 ¯4 1 1 3
 1 ¯2  0  5
 3 ¯1  2  1
¯4  1  1  3
```

and let f be the homomorphism from $\mathbf{Z}^3$ to $\mathbf{Z}^4$ taking U to `U+.×A`. For example, if

```
      U←3 ¯2 1
```

then the image of U under f is

```
      U+.×A
¯7 ¯3 ¯3 16
```

Then A is the matrix of f with respect to the standard bases of $\mathbf{Z}^3$ and $\mathbf{Z}^4$. Now let

```
      ⎕←P←3 3⍴6 5 2 10 5 3 3 2 1
 6 5 2
10 5 3
 3 2 1
```

The determinant of P is -1.

```
      ZDET P
¯1
```

Thus, by Theorem 2.12, the rows of P are a basis for $\mathbf{Z}^3$ and P is the transition matrix from the standard basis of $\mathbf{Z}^3$ to this new basis. Similarly, if

```
      ⎕←Q←4 4⍴6 4 0 ¯1 0 ¯1 1 1 3 2 0 ¯1 ¯1 ¯1 ¯2 1 1
 6  4 0 ¯1
 0 ¯1 1  1
 3  2 0 ¯1
¯1 ¯2 1  1
```

then the calculation

```
      ZDET Q
1
```

shows that the rows of Q are a basis of $\mathbf{Z}^4$. By Theorem 9, the matrix of f with respect to the rows of P and the rows of Q is

```
      ⎕←B←P+.×A+.×ZMATINV Q
 45  ¯23  ¯74  35
 75  ¯40 ¯124  53
 22  ¯11  ¯37  16
```

As a check, let

```
      U←3 ¯4 6
      ⎕←V←U+.×P
¯4 7 0
```

Then U is the coordinate vector for V with respect to the rows of P. Now the image of V under f is

```
      ⎕←W←V+.×A
17 1 14 ¯13
```

and, according to Theorem 9, the coordinate vector of W with respect to the rows of Q should be

```
      ⎕←X←U+.×B
¯25 25 52 ¯11
```

This means that W should equal X+.×Q,

```
      X+.×Q
17 1 14 ¯13
```

which, indeed, it does.

EXERCISES

1 Fill in the details in the proof of Theorem 1.

2 What are the orders of the following abelian groups?

(a) $\mathrm{Hom}_{\mathbf{Z}}(\mathbf{Z}_4, \mathbf{Z}_8)$.
(b) $\mathrm{Hom}_{\mathbf{Z}}(\mathbf{Z}_8, \mathbf{Z}_4)$.
(c) $\mathrm{Hom}_{\mathbf{Z}}(\mathbf{Z}_4 \times \mathbf{Z}_2, \mathbf{Z}_4 \times \mathbf{Z}_4)$.
(d) $\mathrm{Hom}_{\mathbf{Z}}(\mathbf{Z}_6, \mathbf{Z}_8)$.

3 Suppose M_1, M_2, N_1, N_2 are R-modules with $M_1 \cong M_2$ and $N_1 \cong N_2$. Show that $\mathrm{Hom}_R(M_1, N_1)$ and $\mathrm{Hom}_R(M_2, N_2)$ are isomorphic as abelian groups and, if R is commutative, as R-modules. Prove also that $\mathrm{End}_R(M_1)$ and $\mathrm{End}_R(M_2)$ are isomorphic rings.

4 Let A be an abelian group. Show that the group of units in the ring $\mathrm{End}_{\mathbf{Z}}(A)$ is $\mathrm{Aut}(A)$.

5 Let R be a commutative ring. Prove that R, $R[X]$, and $M_n(R)$ are all R-algebras.

6 Let M be an R-module that is nontrivial and *simple* in the sense that the only submodules of M are $\{0\}$ and M. Show that $\mathrm{End}_R(M)$ is a division ring.

7 Let A be an m-by-n matrix over R and let f be the element of $\mathrm{Hom}_R(R^m, R^n)$ taking u to uA. Show that the image of f is $S_R(A)$.

8 Let P be the matrix

$$\begin{bmatrix} 2 & 4 & 3 \\ 1 & 5 & 1 \\ 3 & 0 & 2 \end{bmatrix}$$

in $M_3(\mathbf{Z}_{11})$. Show that the rows of P are a basis for $M = (\mathbf{Z}_{11})^3$. Let $f: M \longrightarrow M$ be the map taking (x, y, z) to $(y+z, 0, x-z)$. Show that

f is in $\mathrm{End}_{\mathbf{Z}}(M)$. What is the matrix of f with respect to the standard basis of M? What is the matrix of f with respect to the rows of P?

9 Let $f:\mathbf{Q}^3 \longrightarrow \mathbf{Q}^4$ take (x, y, z) to

$$\left(y - \frac{3}{2}z, x + 4y, \frac{1}{2}y + \frac{1}{3}z, x - z\right).$$

Show that f is in $\mathrm{Hom}_{\mathbf{Q}}(\mathbf{Q}^3, \mathbf{Q}^4)$ and determine the matrix of f with respect to the standard bases of $\mathbf{Q}^3$ and $\mathbf{Q}^4$.

10 Let R be a commutative ring, $\mathbf{N}$ the set of positive integers, and M the set of functions $f:\mathbf{N} \longrightarrow R$ such that $\{n \in \mathbf{N} \mid f(n) \neq 0\}$ is finite. Show that M is an R-module. For each i in $\mathbf{N}$, let f_i be the element in M taking i to 1 and j to 0 for all $j \neq i$. Show that $\{f_i \mid i \in \mathbf{N}\}$ is a basis for M. Prove that $N = M \oplus M$ is isomorphic to M as an R-module.

11 Let R, M, and N be as in Exercise 10 and let $\theta : N \longrightarrow M$ be a fixed R-isomorphism. Define maps α, β, γ, δ from N to N as follows. For $u = (x, y)$ in N,

$$\begin{aligned} u\alpha &= (0, u\theta), \\ u\beta &= (u\theta, 0), \\ u\gamma &= y\theta^{-1}, \\ u\delta &= x\theta^{-1}. \end{aligned}$$

Prove that $\alpha, \beta, \gamma, \delta$ are in $S = \mathrm{End}_R(N)$ and that in $M_2(S)$

$$\begin{bmatrix} \alpha & 0 \\ \beta & 0 \end{bmatrix} \begin{bmatrix} \gamma & \delta \\ 0 & 0 \end{bmatrix} = \begin{bmatrix} 1 & 0 \\ 0 & 1 \end{bmatrix}.$$

Show also that (γ, δ) is a basis for $S \oplus S$. Conclude that $S \oplus S$ is isomorphic to S as a left S-module and hence that S^n is isomorphic to S for all $n \geq 1$.

4. ALGEBRAS

Throughout this section R will be a commutative ring. In Section 3 we defined an R-algebra to be a ring S that is also an R-module and in which $r(ab) = (ra)b = a(rb)$ for all r in R and all a and b in S. Among the examples of R-algebras we have seen are R, $R[X]$, $M_n(R)$, and $\mathrm{End}_R(M)$, where M is any R-module. We will construct additional examples in this section.

Let S be an R-algebra and suppose I is an ideal of S. Then I is also an R-submodule of S. For, if r is in R and a is in I, then $ra = (r1)a$ is in I because I is an ideal. It follows that S/I is a ring because I is an ideal of S and

S/I is an R-module because I is a submodule of S. Moreover, for all r in R and all a and b in S, we have

$$\begin{aligned} r[(I+a)(I+b)] = r[I+ab] &= I + r(ab) \\ &= [r(I+a)]\ (I+b) \\ &= (I+a)\ [r(I+b)] \end{aligned}$$

and so S/I is an R-algebra, the *quotient algebra* of S modulo I. The natural map from S to S/I is an *algebra homomorphism*, that is, it is compatible with both the ring and the module structures.

As an example of a quotient algebra, we may take $S = R[X]$ and $I = \langle f \rangle$, the set of all multiples in S of some polynomial f. Then I is an ideal and S/I is an R-algebra.

THEOREM 1. Suppose f in $R[X]$ has positive degree n and a leading coefficient that is a unit in R. Then, as an R-module, $R[X]/\langle f \rangle$ is free of rank n.

Proof. Let U be the subset of $R[X]$ consisting of the polynomials of degree at most $n - 1$, including the zero polynomial. Then U is an R-submodule of $R[X]$ that is free with basis $1, X, \ldots, X^{n-1}$. We will show that under the natural map from $R[X]$ to $R[X]/\langle f \rangle$ the submodule U is mapped isomorphically onto $R[X]/\langle f \rangle$. Suppose g is in $R[X]$. Since the leading coefficient of f is a unit in R, we can perform a division of g by f and obtain polynomials g and r such that $g = qf + r$ and r is in U. Then $\langle f \rangle + g = \langle f \rangle + r$, and U is mapped onto $R[X]/\langle f \rangle$. Now suppose some nonzero element h of U is mapped to 0 in $R[X]/\langle f \rangle$. Then h is in $\langle f \rangle$ and is divisible by f. Therefore $h = fu$ for some nonzero u in $R[X]$. But, again, because the leading coefficient of f is a unit in R,

$$\begin{aligned} n - 1 \geq \deg(h) = \deg(fu) &= \deg(f) + \deg(u) \\ &= n + \deg(u) \geq n, \end{aligned}$$

which is a contradiction. Thus $R[X]/\langle f \rangle$ is isomorphic as an R-module to U and so is free of rank n. □

Suppose f is the polynomial $3 - X + 2X^2 - X^3$ in $\mathbf{Z}[X]$. Then, by Theorem 1, the algebra $\mathbf{Z}[X]/\langle f \rangle$ is a free $\mathbf{Z}$-module with basis $\langle f \rangle + 1$, $\langle f \rangle + X$, $\langle f \rangle + X^2$.

Algebras of the type discussed in Theorem 1 are of interest when we are trying to factor polynomials. The reason is given by the following theorem and its corollary.

THEOREM 2. Let S be a commutative ring and let a be a nonzero nonunit in S. Then a is prime if and only if $S/\langle a \rangle$ is an integral domain.

Proof. Recall that a is prime if, whenever a divides a product then it divides one of the factors. For s in S let $\bar{s}$ denote the image of s in $S/\langle a \rangle$.

First, suppose that $S/\langle a \rangle$ is an integral domain and that a divides bc, where b and c are in S. Then, in $S/\langle a \rangle$, we have $0 = \overline{bc} = \bar{b}\bar{c}$. Since $S/\langle a \rangle$ is an integral domain, either $\bar{b} = 0$ or $\bar{c} = 0$. That is, either b is in $\langle a \rangle$ or c is in $\langle a \rangle$. Therefore a is prime. Now suppose a is prime. Since a is not a unit, $\langle a \rangle \neq S$ and $1 \neq 0$ in $S/\langle a \rangle$. If $\overline{bc} = 0$ in $S/\langle a \rangle$, then a divides bc, so a divides either b or c. Therefore $\bar{b} = 0$ or $\bar{c} = 0$ and $S/\langle a \rangle$ is an integral domain. □

COROLLARY 3. Let R be a UFD and let f be a polynomial of positive degree in $R[X]$. Then f is irreducible in $R[X]$ if and only if $R[X]/\langle f \rangle$ is an integral domain.

Proof. Since f has positive degree, f is a nonzero nonunit in $R[X]$. By Theorem 4.11.1, $R[X]$ is a UFD and by Theorems 4.10.4 and 4.10.6 the concepts of primality and irreducibility are the same in UFD's. Thus the corollary follows immediately from Theorem 2. □

Corollary 3 can be rephrased as follows. Suppose we can find two polynomials g and h in $R[X]$ such that f divides gh but f does not divide either g or h. Then either $\gcd(f, g)$ or $\gcd(f, h)$ is a nontrivial factor of f.

If R is a field, we can strengthen Corollary 3.

COROLLARY 4. Let F be a field and let f in $F[X]$ have positive degree. Then f is irreducible if and only if $F[X]/\langle f \rangle$ is a field.

Proof. If $S = F[X]/\langle f \rangle$ is a field, then S is certainly an integral domain and f is irreducible by Corollary 3. Now suppose f is irreducible and let g be a polynomial in $F[X]$ representing a nonzero element of S. We may assume that $\deg(g) < \deg(f)$. Let h be the monic gcd of f and g. Then $\deg(h) \leq \deg(g)$ and h divides f. Therefore, since f is irreducible, h must be 1. Thus, since $F[X]$ is a PID, there exist u and v in $F[X]$ such that $1 = uf + vg$. The element of S represented by v is an inverse for the element represented by g. Hence S is a field. □

The procedures *ZNXIRRED* or *ZNXFACTOR* may be used to find irreducible elements in $\mathbf{Z}_p[X]$, p a prime. Corollary 4 then shows how to construct finite fields that do not have prime cardinality. For example, the calculation

```
      N←3
      ZNXIRRED 2
0 1 0
1 1 0
2 1 0
1 0 1
2 1 1
2 2 1
```

determines the monic irreducible polynomials in $\mathbf{Z}_3[X]$ of degree at most 2. The first quadratic one is $f = 1 + X^2$. By Corollary 4, $S = \mathbf{Z}_3[X]/<f>$ is a field and, by Theorem 1, we know that S is a free $\mathbf{Z}_3$-module of rank 2. Therefore $|S| = |\mathbf{Z}_3|^2 = 9$.

CLASSLIB contains procedures for working in algebras $R[X]/<f>$, where f is monic and R is one of the rings $\mathbf{Z}$, $\mathbf{Z}_n$, or $\mathbf{R}$. We will illustrate these procedures with an example in which R is $\mathbf{Z}_5$ and f is $2 + 4X + X^2 + X^3$.

```
      N←5
      F←2 4 1 1
```

We can represent elements of $S = \mathbf{Z}_5[X]/<f>$ by polynomials, remembering that two polynomials g and h represent the same element of S if and only if f divides $g - h$ or, equivalently, if and only if g and h give the same remainder when divided by f. The computations

```
      G←2 1 3
      H←3 4 3 1 3
      F ZNXREM G ZNXDIFF H
0
      F ZNXREM G
2 1 3
      F ZNXREM H
2 1 3
```

verify in two different ways that $2 + X + 3X^2$ and $3 + 4X + 3X^2 + X^3 + 3X^4$ represent the same element of S. We can use *ZNXSUM*, *ZNXDIFF*, and *ZNXPROD* to perform the ring operations in S and reduce as necessary modulo f using *ZNXREM*. For example, the calculation

```
      A←4 2 3
      F ZNXREM 2 ZNXSUM (2×A) ZNXSUM A ZNXPROD A
0
```

shows that the element a of S represented by $4 + 2X + 3X^2$ satisfies $2 + 2a + a^2 = 0$. The element of S represented by a polynomial g is a unit in S if and only if $\gcd(f, g) = 1$. We can use *ZNXGCD* to try to compute inverses in S.

```
      F ZNXGCD A
1
      S
2 4 1
      F ZNXREM A ZNXPROD S
1
```

Here we see that a is a unit and a^{-1} is given by the global variable, S computed by *ZNXGCD*, which is $2 + 4X + X^2$.

The process of reducing modulo f using *ZNXREM* is somewhat time consuming. If many calculations in S are to be performed, there is a more efficient way to reduce modulo f. Since f is cubic, every element of S can be represented by a polynomial of degree at most 2. The product of two polynomials of degree 2 has degree 4 and, if we always reduce modulo f after every multiplication, we need never work with polynomials of degree greater than 4. We start by making a matrix R whose rows give the remainders modulo f of $1, X, X^2, X^3$, and X^4. One way to do this is

```
      ⎕←P←(⍳5)∘.=⍳5                    ⎕←R←F ZNXREM P
 1 0 0 0 0                             1 0 0
 0 1 0 0 0                             0 1 0
 0 0 1 0 0                             0 0 1
 0 0 0 1 0                             3 1 4
 0 0 0 0 1                             2 2 2
```

Now, if G is a vector of length 5 representing a polynomial in $\mathbf{Z}_5[X]$ of degree at most 4, then `5|G+.×R` is the same as `F ZNXREM G`, and the former takes less time to compute.

```
      G←1 2 3 2 1                      F ZNXREM G
      5|G+.×R                          4 1 3
 4 1 3
```

Note that the first three rows of R are the identity matrix and need not be computed. If we replace R by

```
      ⎕←R0←3 0↓R
 3 1 4
 2 2 2
```

then `F ZNXREM G` can be computed as follows.

```
      5|(3↑G)+(3↓G)+.×R0
4 1 3
```

In *CLASSLIB* the procedures with the prefix *ZNXF* may be used to compute in $\mathbf{Z}_n[X]/<f>$, where f is monic. As usual, the value of n must be assigned to N. In addition, the procedure *ZNXFINIT* must be used to initialize a matrix *ZNXRT* that is used like *R0* above to reduce modulo f. (The letters *RT* stand for remainder table.)

```
      ZNXFINIT F
      ZNXRT
3 1 4
2 2 2
```

Provided *A* and *B* represent arrays of polynomials in $\mathbf{Z}_n[X]$ of degree less that the degree of *f*, then *A ZNXFPROD B* produces the same result as *F ZNXREM A ZNXPROD B*.

```
      A←1 2 3
      B←2 1 4
      F ZNXREM A ZNXPROD B
4
      A ZNXFPROD B
4
```

The procedures *ZNXFSUM*, *ZNXFDIFF*, *ZNXFPOWER*, and *ZNXFINV* may also be used for computations in $\mathbf{Z}_n[X]/<f>$. Procedures with the prefixes *ZXF* and *RXF* perform computations in $\mathbf{Z}[X]/<f>$ and $R[X]/<f>$, respectively, where *f* is a monic polynomial.

In Section 4.4 we defined, for any commutative ring *R*, evaluation maps of $R[X]$ into *R*. These are the maps obtained by fixing an element *a* of *R* and mapping a polynomial *f* to its value *f*(*a*). We can extend this concept to maps from $R[X]$ into any *R*-algebra. Suppose *S* is an *R*-algebra and *a* is an element of *S*. If $f = c_0 + c_1X + \ldots + c_nX^n$ is in $R[X]$, then $c_0 + c_1a + \ldots c_na^n$ is a well-defined element of *S*, which we will call the value of *f* at *a* and denote *f*(*a*). It is easy to show that the map $f \longmapsto f(a)$ is an *R*-algebra homomorphism. As an example, let us take $R = \mathbf{Z}$, $S = M_2(\mathbf{Z})$,

$$A = \begin{bmatrix} 2 & -1 \\ 1 & 3 \end{bmatrix}$$

and $f = 4 - 2X + X^2$. Then

$$f(A) = 4I - 2A + A^2 = 4\begin{bmatrix} 1 & 0 \\ 0 & 1 \end{bmatrix} - 2\begin{bmatrix} 2 & -1 \\ 1 & 3 \end{bmatrix} + \begin{bmatrix} 3 & -5 \\ 5 & 8 \end{bmatrix}$$

$$= \begin{bmatrix} 3 & -3 \\ 3 & 6 \end{bmatrix}.$$

We can define a map from *R* to any *R*-algebra *S* by mapping *r* in *R* to *r*1, where 1 is the multiplicative identity of *S*. This map is also an algebra homomorphism. If it is an embedding, that is, an injective map, then we often identify *r* and *r*1 and consider *R* to be a subalgebra of *S*.

The remainder of this section will be devoted to another method for constructing R-algebras that are finitely generated free R-modules.

THEOREM 5. Let S be an R-algebra with an R-basis $x_1, \ldots, x_n$. Then the multiplication in S is determined by the module structure and the n^2 products $x_i x_j$, $1 \le i, j \le n$.

Proof. If a and b are in S, we can find $r_1, \ldots, r_n$ and $s_1, \ldots, s_n$ in R such that $a = r_1 x_1 + \ldots r_n x_n$ and $b = s_1 x_1 + \ldots s_n x_n$. Then the axioms for an R-algebra imply that

$$ab = \left(\sum_i r_i x_i\right) \left(\sum_j s_j x_j\right)$$
$$= \sum_{i,j} (r_i x_i)(s_j x_j) = \sum_{i,j} r_i s_j (x_i x_j).$$

Thus, if we know the structure of S as an R-module and are given the products $x_i x_j$, we can compute ab. □

The most natural way to describe the products $x_i x_j$ of Theorem 5 is to express them as linear combinations of the basis elements, that is, by giving elements e_{ijk} of R such that

$$x_i x_j = \sum_k e_{ijk} x_k .$$

If this is done, then the formula for multiplication in S is

$$\left(\sum_i r_i x_i\right) \left(\sum_j s_j x_j\right) = \sum_{i,j} r_i s_j (x_i x_j)$$
$$= \sum_{i,j} r_i s_j \sum_k e_{ijk} x_k \qquad (*)$$
$$= \sum_k \left(\sum_{i,j} r_i s_j e_{ijk}\right) x_k .$$

We cannot choose the e_{ijk} arbitrarily and get an R-algebra.

THEOREM 6. Let S be a free R-module with a basis $x_1, \ldots, x_n$ and for $1 \le i, j, k \le n$, let e_{ijk} be in R. Then formula (*) defines a multiplication on S that satisfies both of the distributive laws and the property $r(uv) = (ru)v = u(rv)$ for all r in R and all u, v in S. Moreover, the following are equivalent.

(a) The multiplication is associative.
(b) $x_i(x_j x_k) = (x_i x_j)x_k$ for all i, j, k.
(c) For all i, j, k, m

$$\sum_{\ell} e_{jk\ell}\, e_{i\ell m} = \sum_{\ell} e_{ij\ell}\, e_{\ell km}\,.$$

Proof. Let $u = r_1x_1 + \ldots + r_nx_n$, $v = s_1x_1 + \ldots + s_nx_n$, and $w = t_1x_1 + \ldots + t_nx_n$, with all r_i, s_i, and t_i in R. Then

$$\begin{aligned} u(v+w) &= \left(\sum_i r_ix_i\right)\left(\sum_j (s_j+t_j)x_j\right) \\ &= \sum_{i,j} r_i(s_j+t_j)\,(x_ix_j) \\ &= \sum_{i,j} r_is_j(x_ix_j) + \sum_{i,j} r_it_j(x_ix_j) \\ &= uv + uw. \end{aligned}$$

Similarly, we can verify by direct calculation that $(u+v)w = uw + vw$ and $r(uv) = (ru)v = u(rv)$ for all r in R. Clearly, condition (a) implies condition (b). To see that (b) implies (a), we compute

$$\begin{aligned} u(vw) &= \left(\sum_i r_ix_i\right)\left[\left(\sum_j s_jx_j\right)\left(\sum_k t_kx_k\right)\right] \\ &= \left(\sum_i r_ix_i\right)\left(\sum_{j,k} s_jt_k(x_jx_k)\right) \\ &= \sum_{i,j,k} r_is_jt_k\,[x_i(x_jx_k)] \end{aligned}$$

and

$$\begin{aligned} (uv)w &= \left[\left(\sum_i r_ix_i\right)\left(\sum_j s_jx_j\right)\right]\left(\sum_k t_kx_k\right) \\ &= \left(\sum_{i,j} r_is_j(x_ix_j)\right)\left(\sum_k t_kx_k\right) \\ &= \sum_{i,j,k} r_is_jt_k\,[(x_ix_j)x_k]. \end{aligned}$$

Thus, if $x_i(x_jx_k) = (x_ix_j)x_k$ for all i, j, k, then we have $u(vw) = (uv)w$ and the associative law holds. Now

$$
\begin{aligned}
x_i(x_jx_k) &= x_i \sum_{\ell} e_{jk\ell}x_{\ell} \\
&= \sum_{\ell} e_{jk\ell}x_ix_{\ell} \\
&= \sum_{\ell} e_{jk\ell} \sum_{m} e_{i\ell m}\, x_m \\
&= \sum_{m} \left(\sum_{\ell} e_{jk\ell}\, e_{i\ell m} \right) x_m
\end{aligned}
$$

and

$$
\begin{aligned}
(x_ix_j)x_k &= \left(\sum_{\ell} e_{ij\ell}\, x_{\ell} \right) x_k \\
&= \sum_{\ell} e_{ij\ell}\, x_{\ell}\, x_k \\
&= \sum_{\ell} e_{ij\ell} \sum_{m} e_{\ell km}\, x_m \\
&= \sum_{m} \left(\sum_{\ell} e_{ij\ell}\, e_{\ell km} \right) x_m .
\end{aligned}
$$

Since $x_1, \ldots, x_n$ are a basis for S, we see that $x_i(x_jx_k) = (x_ix_j)x_k$ if and only if

$$\sum_{\ell} e_{jk\ell}\, e_{i\ell m} = \sum_{\ell} e_{ij\ell}\, e_{\ell km}$$

for all m. Thus (b) and (c) are equivalent. □

The elements e_{ijk} are called the *structure constants* for the multiplication defined by (*).

For any commutative ring R there is an important R-algebra called the *quaternion algebra* over R. This is the algebra Q with an R-basis $1, i, j, k$ whose elements multiply according to the following table.

×	1	i	j	k
1	1	i	j	k
i	i	-1	k	$-j$
j	j	$-k$	-1	i
k	k	j	$-i$	-1

This table can be remembered easily if one notes that 1 is the multiplicative identity, $i^2 = j^2 = k^2 = -1$ and, if (a, b, c) is any cyclic permutation of (i, j, k), then $ab = c$ and $ba = -c$. It is standard practice to identify an element r in R with $r1$ in S, thereby making R a subring of S.

To verify, using Theorem 6, that the quaternion algebra Q is really an R-algebra we must show that multiplication is associative for triples of basis elements. There are $4^3 = 64$ such triples, and checking each one by hand is somewhat tedious. The exercises describe alternative approaches to verifying associativity.

One reason for the importance of quaternion algebras is given by the next theorem.

THEOREM 7. The quaternion algebra over $\mathbf{R}$ is a noncommutative division ring.

Proof. Let Q be the algebra of real quaternions. Then direct computation shows that for all real numbers a, b, c, d

$$(a + bi + cj + dk)(a - bi - ci - dk) = a^2 + b^2 + c^2 + d^2.$$

If a, b, c, d are not all 0, then $a^2 + b^2 + c^2 + d^2 \neq 0$, and it is easily checked that

$$\frac{a - bi - cj - dk}{a^2 + b^2 + c^2 + d^2}$$

is a two-sided inverse for $a + bi + cj + dk$. Since $ij \neq ji$, the algebra Q is certainly noncommutative. □

The ring of real quaternions is our first example of a noncommutative division ring.

Let us turn now to some examples that show how to compute in algebras defined by arrays of structure constants. Let S be a $\mathbf{Z}$-algebra with basis $x_1, \ldots, x_n$ and structure constants e_{ijk} relative to this basis. There is an isomorphism of S as a $\mathbf{Z}$-module onto $\mathbf{Z}^n$ in which x_i maps to the ith standard basis vector in $\mathbf{Z}^n$. This isomorphism allows us to use the structure constants to define an algebra structure on $\mathbf{Z}^n$ that is isomorphic to that of S. Thus, without loss of generality, we may assume that $S = \mathbf{Z}^n$ and $x_1, \ldots, x_n$ is the standard basis of $\mathbf{Z}^n$. Formula (*) now gives the product

$$(u_1, \ldots, u_n)(v_1, \ldots, v_n) = (w_1, \ldots, w_n),$$

where

$$w_k = \sum_{i,j} u_i v_j e_{ijk}.$$

Suppose the e_{ijk} are given by an APL array E of rank 3 and U and V are integer vectors of length n. Then the product of U and V in the algebra defined by E is `V+.×U+∘.×E`.

Suppose E is the following rank 3 array.

```
      ⎕←E←2 2 2⍴19 ¯26 10 ¯13 10 ¯13 5 ¯6
19 ¯26
10 ¯13

10 ¯13
 5  ¯6
```

It is not hard to show (see Exercise 7a) that E satisfies condition (c) of Theorem 6, and so E defines an associative multiplication on $\mathbf{Z}^2$. The computation

```
      U←¯2 1                          V+.×U+.×E
      V←3 4                     ¯144 197
```

shows that the product of $(-2, 1)$ and $(3, 4)$ is $(-144, 197)$.

It is not obvious that the multiplication defined by E has an identity element. However, from

```
      I←¯1 2                          Y+.×I+.×E
      X←1 0                     0 1
      Y←0 1                           I+.×Y+.×E
      X+.×I+.×E                 0 1
1 0
      I+.×X+.×E
1 0
```

we can see that $(-1, 2)$ acts as a two-sided identity on the standard basis of $\mathbf{Z}^2$, and it follows easily that $(-1, 2)$ is the identity element for this multiplication.

The procedures in *CLASSLIB* with prefix *ZA* can be used to compute in a **Z**-algebra defined by an array of structure constants. The procedures assume that the structure constants are given by the array *ZSC*, which can be initialized with *ZAINIT*.

```
      ZAINIT E
      ZSC
19 ¯26
10 ¯13

10 ¯13
 5  ¯6
```

Now we can use *ZAPROD* to compute products.

```
      U ZAPROD V
¯144 197
```

Addition is just ordinary addition of vectors but may be performed by using *ZASUM* if desired.

The procedures with the prefixes *ZNA* and *RA* may be used to perform computations in algebras over $\mathbf{Z}_n$ and $\mathbf{R}$, respectively.

EXERCISES

1 Let $S = \mathbf{Z}_7[X]/<f>$, where $f = 6 + 6X + 4X^2 + 5X^3 + X^4$, and let u be the element of S represented by $2 + 3X^2 + 6X^3$. For which of the following polynomials g does $g(u) = 0$?

(a) $5 + 6X + 2X^2 + X^3$.
(b) $1 + 5X + X^2$.
(c) $1 - X^4$.

2 Let S and u be as in Exercise 1. Show that the set I of polynomials g in $\mathbf{Z}_7[X]$ such that $g(u) = 0$ is an ideal of $\mathbf{Z}_7[X]$ and therefore consists of all multiples of a unique monic polynomial h. Find h.

3 Make comparisons of the execution times for the statements

```
F ZNXREM G ZNXPROD H
```

and

```
G ZNXFPROD H
```

Here we assume *ZNXFINIT F* has already been executed and that ρG and ρH are at most the degree of F.

4 Let S be as in Exercise 1 and let v in S be represented by $1 + X^2$. Show that v is a unit in S and find v^{-1}.

5 Let G be a finite group and let S be an R-module. Suppose for each g in G we have an element x_g of S such that the map $g \mapsto x_g$ is $1 - 1$ and $\{x_g \mid g \in G\}$ is a basis for S. For g, h in G define $x_g x_h = x_{gh}$. Show that condition (b) of Theorem 6 is satisfied and that the multiplication on S defined by formula (*) makes S into an R-algebra. Show that S is isomorphic to the group ring of G over R defined in Exercise 4.11.15.

6 The matrices

$$X_1 = \begin{bmatrix} 1 & 0 \\ 0 & 0 \end{bmatrix}, \; X_2 = \begin{bmatrix} 0 & 1 \\ 0 & 0 \end{bmatrix}, \; X_3 = \begin{bmatrix} 0 & 0 \\ 1 & 0 \end{bmatrix}, \; X_4 = \begin{bmatrix} 0 & 0 \\ 0 & 1 \end{bmatrix}$$

form an R-basis of $M_2(R)$. Determine the structure constants for $M_2(R)$ with respect to this basis.

7 Let E be an n-by-n-by-n array of real numbers. Formulate the following as APL propositions:

(a) Condition (c) of Theorem 6.

(b) The commutativity of multiplication defined on $\mathbf{R}^n$ by E.

(c) The vector I is a two-sided identity for the multiplication.

8 Let $a = 2 - 3i + j - 2k$ and $b = -1 + 2i + j - 3k$ in the algebra of real quaternions. Compute a^2, $ab + ba$, and a^{-1}.

9 Construct the array E of structure constants for the quaternion algebras relative to the basis $1, i, j, k$. Show that E satisfies the APL proposition of Exercise 7a.

10 Let $X = \{\pm 1, \pm i, \pm j, \pm k\}$ in the algebra Q of real quaternions. Show that X is closed under multiplication in Q and construct a binary operation table T for multiplication in X. Show that T is associative and so in particular multiplication in Q is associative for triples of basis elements. Thus Q satisfies condition (b) of Theorem 6.

11 What are the units in the integer quaternion algebra?

12 Find the order of the group of units in the quaternion algebra over $\mathbf{Z}_3$.

13 Let p be a prime. Show that $a^p = a$ for all a in $\mathbf{Z}_p$. Let S be a commutative $\mathbf{Z}_p$-algebra. Prove that the map $x \mapsto x^p$ is an algebra homomorphism of S into itself.

14 Is $\mathbf{Z}[X]/\langle 2X^2 \rangle$ finitely generated as a $\mathbf{Z}$-module?

15 Let S be the field $\mathbf{Z}_3[X]/\langle f \rangle$, where $f = 1 + X^2$. What is the multiplicative order of the element in S represented by X? Find an element of multiplicative order 8 in S.

16 Suppose S is an R-algebra that is a finitely generated free R-module. Show that the map $r \mapsto r1$ of R into S is an algebra embedding of R into S.

6
MODULES OVER EUCLIDEAN DOMAINS

Among the objects traditionally studied in an introductory algebra course are finitely generated abelian groups, finite dimensional vector spaces, and pairs (V, f) consisting of a finite dimensional vector space V and a linear transformation f of V into itself. These three types of algebraic objects correspond to modules over three different rings, the integers, a field, and the polynomial ring over a field. Since each of these rings is a Euclidean domain, it is possible to obtain most of the important results about the objects listed as corollaries of theorems about modules over Euclidean domains. This approach gives unity to the discussion and demonstrates the power of abstraction and the axiomatic approach in algebra. Throughout this chapter R will be a Euclidean domain with Euclidean norm N.

1. ROW EQUIVALENCE

In this section we begin a discussion that will provide further information about the submodules of R^n and the units in $M_n(R)$. Initially we will use only the fact that R is a PID. Later the discussion will depend heavily on the fact that R is a Euclidean domain. We will draw on many results proved in Chapters 4 and 5. Of particular importance will be the row operations over R defined in Section 4.7.

Let n be a positive integer and let U be a submodule of R^n. By Theorem 4.9.6, we know that R is a PID and so, by Corollary 5.2.16, it follows that U can be generated by n elements. Thus there is an n-by-n matrix A over R such that $U = S_R(A)$. This fact is so important that we will review briefly the steps by which it was obtained. For $1 \leq i \leq n$ let U_i be the set of elements in U whose first $i - 1$ components are 0. Each U_i is a submodule of U, and $U = U_1 \supseteq \ldots \supseteq U_n$. Let S_i be the set of ith components of the elements in U_i. Then S_i is an ideal of R and so has the form Ra_i for some a_i in S_i. The element a_i is not unique but, by Theorem 4.10.1, it is determined up to an associate. For each i let u_i be an element of U_i having a_i as its ith component. Then U is generated by $u_1, \ldots, u_n$. If we take A to be the matrix whose ith row is u_i, then A has the form

$$\begin{bmatrix} a_1 & & & & \\ & a_2 & & ? & \\ & & \cdot & & \\ & & & \cdot & \\ & 0 & & \cdot & \\ & & & & a_n \end{bmatrix},$$

where all the entries below the main diagonal are 0, and $U = S_R(A)$. We will call a_i the ith *standard invariant* of U. It must be remembered, however, that the a_i are determined only up to associates. The word "standard" calls attention to the fact that these invariants are relative to the standard basis of R^n.

If $a_i = 0$, we can choose u_i to be 0. Clearly, we may delete rows of A that are zero without changing $S_R(A)$. If this is done, we obtain a matrix A_1 with $U = S_R(A_1)$, and A_1 has the form

$$\begin{bmatrix} a_{j_1} & & & & \\ & a_{j_2} & & ? & \\ & & \cdot & & \\ & & & \cdot\ \cdot & \\ & 0 & & \vdots & \\ & & & & a_{j_r} \end{bmatrix},$$

where $a_{j_1}, \ldots, a_{j_r}$ are the nonzero standard invariants of U and all entries below the "staircase" are zero.

The preceding discussion leads naturally to the following definition. An m-by-n matrix B over R is in *row-echelon form* if there exist integers r and $j_1, \ldots, j_r$ such that

1. $0 \le r \le m$ and $1 \le j_1 < j_2 < \ldots < j_r \le n$.
2. $B[i;]$ is 0 for $i > r$.
3. If $1 \le i \le r$, then $B[i;j_i] \neq 0$ and $B[i;j] = 0$ for $1 \le j < j_i$.

Condition (3) states that $B[i;j_i]$ is the first nonzero entry in $B[i;]$ and condition (1) states that these first nonzero entries occur in later columns as i increases. Thus B has basically the same form as A_1, except that rows consisting of zeros are allowed, providing they come after all nonzero rows. We will call the entries $B[i;j_i]$, $1 \le i \le r$, the *corner entries* of B.

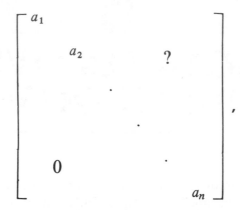

The matrix

$$B = \begin{bmatrix} 0 & 2 & 1 & -3 & 2 & 4 \\ 0 & 0 & -3 & 4 & 1 & 0 \\ 0 & 0 & 0 & 0 & 1 & -2 \\ 0 & 0 & 0 & 0 & 0 & 0 \end{bmatrix}$$

is in row-echelon form and its corner entries are 2, –3, and 1.

THEOREM 1. Let U be a submodule of R^n and let $a_{j_1}, \ldots, a_{j_r}$ be the nonzero standard invariants of U with $j_1 < j_2 < \ldots < j_r$. There exists an r-by-n matrix B such that

(a) B is in row-echelon form.
(b) The ith corner entry of B is a_{j_i} and occurs in column j_i.
(c) $U = S_R(B)$.

Proof. The matrix B is simply the preceding matrix A_1. □

The following important result is essentially the converse of Theorem 1.

THEOREM 2. Let B be an m-by-n matrix over R that is in row-echelon form with corner entries $B[i;j_i]$, $1 \le i \le r$. Set $U = S_R(B)$. Then

(a) The nonzero rows of B are a basis for U.
(b) If $a_1, \ldots, a_n$ are the standard invariants of U, then $a_j = 0$ if j is not in $\{j_1, \ldots, j_r\}$ and a_{j_i} is an associate of $B[i;j_i]$.

Proof. Clearly, the nonzero rows of B generate U so, to prove (a), we need only show that the first r rows of B are linearly independent over R. Suppose there exists elements $c_1, \ldots, c_r$ of R such that $c_1B[1;] + \ldots + c_rB[r;] = 0$ and not all of the c_i are 0. Let c_i be the first nonzero coefficient. The j_ith component of $c_iB[i;] + \ldots + c_rB[r;]$ is

$$c_iB[i;j_i] + c_{i+1}\, B[i+1,j_i] + \ldots + c_rB[r;j_i],$$

which must be 0. But if $k > i$, then $B[k;j_i] = 0$, so we have

$$c_iB[i;j_i] = 0.$$

Since $B[i;j_i] \neq 0$, this means $c_i = 0$, contradicting our choice of i. Thus the nonzero rows of B are a basis for U.

Let U_j be the submodule of U consisting of the elements of U whose first $j-1$ components are 0. A slight modification of the argument in the preceding paragraph shows that U_j is generated by the rows $B[i;]$ for which $j_i \geq j$. For suppose $c_1B[1;] + \ldots + c_rB[r;]$ is in U_j and $c_i \neq 0$ for some i with $j_i < j$. Choose i as small as possible. Then, as before,

$$c_iB[i;j_i] + \ldots + c_rB[r;j_r] = 0$$

and $B[k;j_i] = 0$ for $k > i$. Thus $c_i = 0$, which we assumed not to be the case. Now let S_j be the set of jth components of the elements of U_j and let i be the smallest index such that $j_i \geq j$. Then U_j is generated by $B[i;], \ldots, B[r;]$, so S_j is generated by $B[i;j], \ldots, B[r;j]$. But if $k > i$, then $j < j_k$ and $B[k;j] = 0$. Thus S_j is generated by $B[i;j]$. Therefore a_j and $B[i;j]$ are associates. If $j \neq j_i$, then $a_j = B[i;j] = 0$. □

COROLLARY 3. Every submodule of R^n is a free R-module.

Proof. By Theorems 1 and 2a, every submodule of R^n has a basis. □

COROLLARY 4. Suppose A and B are matrices over R such that

(a) Both A and B have n columns.
(b) Both A and B are in row-echelon form.
(c) $S_R(A) = S_R(B)$.

Then A and B have the same number r of nonzero rows and the ith corner entries of A and of B occur in the same column and are associates, $1 \leq i \leq r$.

Proof. Let U be $S_R(A)$ and let $a_1, \ldots, a_n$ be the standard invariants for U. By Theorem 2, the jth column of A contains a corner entry if and only if $a_j \neq 0$. If $a_j \neq 0$, then the corner entry in the jth column is an associate of a_j. Since $U = S_R(B)$ also, the corner entries A and B are associates and occur in the same columns. □

If B is in row-echelon form, it is easy to decide whether a given element x of R^n is in $S_R(B)$. Consider the following example in $\mathbf{Z}^4$. Let

$$B = \begin{bmatrix} 1 & -1 & 3 & 0 \\ 0 & 2 & 4 & -2 \\ 0 & 0 & 0 & 3 \end{bmatrix}$$

and $x = (4, -10, 0, 9)$. Is x in $U = S_{\mathbf{Z}}(B)$? That is, can we find an integer vector $c = (c_1, c_2, c_3)$ such that x is $c+.\times B$? Computing the first, second, and fourth components of $c+.\times B$, we have

$$\begin{aligned} c_1 &= 4, \\ -c_1 + 2c_2 &= -10, \\ -2c_2 + 3c_3 &= 9. \end{aligned}$$

Substituting $c_1 = 4$ in the second equation, we get $-4 + 2c_2 = -10$ or $c_2 = -3$. Substituting in the third equation, we obtain $6 + 3c_3 = 9$. Thus $c_3 = 1$ and $c = (4, -3, 1)$. We must still check that $c+.\times B$ is really x, since it is possible that the third components of $c+.\times B$ and x may not be the same. However, x is, in fact, $(4, -3, 1)+.\times B$.

If we use the same procedure to try to write $y = (-3, 4, 1, -2)$ as a **Z**-linear combination $(d_1, d_2, d_3)+.\times B$ of the rows of B, then we are led to the equations

$$\begin{aligned} d_1 &= -3, \\ -d_1 + 2d_2 &= 4, \\ -2d_2 + 3d_3 &= -2. \end{aligned}$$

Substituting $d_1 = -3$ in the second equation gives $3 + 2d_2 = 4$ or $2d_2 = 1$. This equation has no solutions in $\mathbf{Z}$ and so y is not in U.

The fact that B is in row-echelon form was very important in making the preceding computations so easy. If

$$A = \begin{bmatrix} 5 & -7 & 11 & 5 \\ 6 & -4 & 22 & -2 \\ 10 & -10 & 30 & 3 \end{bmatrix},$$

then the preceding method may be used to show that each row of A is in $S_{\mathbf{Z}}(B)$, and so $S_{\mathbf{Z}}(A) \subseteq S_{\mathbf{Z}}(B)$. It is actually the case that $S_{\mathbf{Z}}(A) = S_{\mathbf{Z}}(B)$, but it is not obvious how to write the rows of B as **Z**-linear combinations of the rows of A. By Theorem 1, there exists a matrix C in row-echelon form such that $S_{\mathbf{Z}}(C) = S_{\mathbf{Z}}(A)$ but, so far, we do not have a procedure for calculating such a C.

In Section 4.7 we introduced row operations and studied their effect on the determinant of a square matrix. We showed that a row operation O can be applied to an m-by-n matrix A by forming $E+.\times A$, where E is the m-by-m elementary matrix corresponding to O. We also proved that elementary matrices are units in $M_m(R)$. It turns out that row operations play a key role in our solution to the problem of finding, for any matrix A over R, a matrix C in row-echelon form such that $S_R(C) = S_R(A)$.

Let A and B be m-by-n matrices over R. We say that A and B are *row equivalent* over R if there exists a sequence of matrices

$$A = A_0, A_1, \ldots, A_t = B$$

such that for $1 \leq i \leq t$ the matrix A_i is obtained from A_{i-1} by an elementary row operation over R. Thus A is row equivalent to B if and only if B can be obtained from A by a sequence of row operations.

THEOREM 5. Row equivalence is an equivalence relation on the set of m-by-n matrices over R.

Proof. If A is an m-by-n matrix over R, then A can be obtained from A by the empty sequence of row operations, so row equivalence is reflexive. If $A = A_0, A_1, \ldots, A_t = B$ is a sequence of matrices with A_i obtained from A_{i-1} by a row operation, then by Theorem 4.7.2 there exists a row operation that converts A_i back to A_{i-1}. Thus, if A is row equivalent to B, then B is row equivalent to A, so row equivalence is symmetric. Finally, if A is row equivalent to B and B is row equivalent to C, it is obvious how to transform A into C using row operations. First, transform A into B; then transform B into C. Therefore row equivalence is transitive. □

THEOREM 6. Two m-by-n matrices A and B over R are row equivalent if and only if there is an m-by-m matrix E such that E is a product of elementary matrices and $B = E+.\times A$.

Proof. Suppose A and B are row equivalent so that there are matrices $A = A_0, A_1, \ldots, A_t = B$ such that for $1 \leq i \leq t$ there is a row operation O_i transforming A_{i-1} to A_i. Let E_i be the elementary matrix obtained by applying O_i to the m-by-m identity matrix. Then, suppressing the symbol $+.\times$, we have $A_i = E_i A_{i-1}$ and $B = E_t E_{t-1} \ldots E_1 A = EA$, where $E = E_t E_{t-1} \ldots E_1$ is a product of elementary matrices.

Conversely, suppose $B = EA$, where $E = E_t E_{t-1} \ldots E_1$ and each E_i is an elementary matrix. Set $A_0 = A$ and $A_i = E_i A_{i-1}$, $1 \leq i \leq t$. Then $A_t = B$ and each A_i is obtained from A_{i-1} by a row operation. Thus B is row equivalent to A. □

COROLLARY 7. If two matrices A and B are row equivalent, then $S_R(A) = S_R(B)$. Moreover, the rows of A are a basis for $U = S_R(A)$ if and only if the rows of B are a basis for U.

Proof. Suppose A and B are row equivalent m-by-n matrices. By Theorem 6, there is a matrix $E = E_t E_{t-1} \ldots E_1$ such that each E_i is an elementary matrix and $B = EA$. Each E_i is a unit in $M_m(R)$, and so E is a unit in $M_m(R)$ and $A = E^{-1}B$. By Theorem 5.2.6, $S_R(A) = S_R(B)$. If the rows of A are a basis for $U = S_R(A)$, then

$$B[i;] = \sum_j E[i;j]A[j;]$$

and so, by Lemma 5.2.11, the rows of B are a basis for U. By symmetry, it follows that if the rows of B are a basis for U, then so are the rows of A. □

So far we have really used only the fact that R is a PID. The next theorem uses our hypothesis that R has a Euclidean norm.

THEOREM 8. Every matrix with entries in R is row equivalent over R to a matrix in row-echelon form.

Proof. Let A be an m-by-n matrix over R. We will prove the theorem by describing a procedure for finding a sequence of row operations that transforms A into a matrix in row-echelon form. The procedure is recursive in the sense that it reduces the problem to finding a row-echelon form matrix row equivalent to a certain $(m-1)$-by-n matrix. Thus the proof of the validity of the procedure is by induction on m.

If $A = 0$ or if $m = 1$, then A is in row-echelon form to begin with and we are done. Thus we may assume that $m \geq 2$ and $A \neq 0$. Let j be the number of the first column of A containing a nonzero entry. Our first task is to show that using row operations we can change A so that it has the form

$$B = \left[\quad {}^{|*}\!\!\!\rceil \qquad ? \quad\right],$$

where the entry $*$ is nonzero and in the jth column. Let $a = A[i;j]$ be an entry in $A[;j]$ that is nonzero and for which the norm $N(a)$ is as small as possible. Assume for the moment that a divides every entry in $A[;j]$. Then, for each k with $1 \leq k \leq m$ and $k \neq i$, we can write $A[k;j] = e_k a$ with e_k in R. If, for each such k, we subtract $e_k A[i;]$ from $A[k;]$, then we obtain a matrix in which the only nonzero entry in the jth column is a in the ijth position. Interchanging rows 1 and i, we have a matrix of the required form.

But suppose a does not divide every entry in $A[;j]$. Then, for some $k \neq i$, we can write $A[k;j]$ as $qa + r$, where $r \neq 0$ and $N(r) < N(a)$. If we subtract $qa[i;]$ from $A[k;]$, then the new entry in the kjth position is r, which has norm less than $N(a)$. At this point, we replace i by k and a by r and repeat the process. Either a divides every entry in the jth column or else a row operation can be found that produces a matrix with a nonzero entry in the jth column that has norm less than $N(a)$. Since the values of $N(a)$ are nonnegative integers, they cannot continue decreasing indefinitely. Eventually we will reach the situation in which a divides all the entries in the jth column and so obtain a matrix B of the required form that is row equivalent to A.

Now let C be the $(m - 1)$-by-n matrix consisting of all the rows of B except the first (`1 0↓B` in APL notation). By induction on m, we can find a sequence of row operations that transforms C into a matrix D in row-echelon form. Since row operations do not change columns that contain only zeros, D has the form

$$D = \left[\,0 \mid D'\,\right],$$

where D' has $n - j$ columns and is in row-echelon form. We can consider the row operations that transform C to D to be acting on the last $m - 1$ rows of B. The result of these operations is to transform B into

$$\begin{bmatrix} & * & ? \; \dots \; ? \\ 0 & & D' \end{bmatrix}$$

which is in row-echelon form. Thus we have found a matrix in row-echelon form that is row equivalent to A. □

The procedure presented in the proof of Theorem 8 is called *row reduction.* We will illustrate it with a few examples. First, suppose

```
      ⎕←A←3 4⍴5 ¯7 11 5 6 ¯4 22 ¯2 10 ¯10 30 3
 5  ¯7  11  5
 6  ¯4  22 ¯2
10 ¯10  30  3
```

so that A is one of the matrices that appeared in our earlier example. The entry in the first column with the smallest absolute value is `A[1;1]`, but it does not divide every entry of `A[;1]`. Performing a row operation,

```
      ⎕IO←1
      A[2;]←A[2;]-A[1;]
      A
 5  ¯7  11  5
 1   3  11 ¯7
10 ¯10  30  3
```

we obtain a nonzero entry in the first column that has absolute value less that 5. The entry `A[2;1]` now divides all entries in `A[;1]`. Three more row operations

```
      A[1;]←A[1;]-5×A[2;]
      A[3;]←A[3;]-10×A[2;]
      A[1 2;]←A[2 1;]
      A
1   3  11 ¯7
0 ¯22 ¯44 40
0 ¯40 ¯80 73
```

yield a matrix with only one nonzero entry in the first column. Now we need only work with the submatrix `A[2 3;]`. Three additional row operations are needed to fix up the second column.

```
      A[3;]←A[3;]-2×A[2;]
      A
 1    3   11   ¯7
 0  ¯22  ¯44   40
 0    4    8   ¯7
      A[2;]←A[2;]+6×A[3;]
      A
 1    3   11   ¯7
 0    2    4   ¯2
 0    4    8   ¯7
      A[3;]←A[3;]-2×A[2;]
      A
 1    3   11   ¯7
 0    2    4   ¯2
 0    0    0   ¯3
```

Now A is in row-echelon form.

When R is a field, row reduction works much faster. For example, consider the matrix

```
      ⎕←U←3 3⍴2 4 1 6 5 2 4 3 5
 2 4 1
 6 5 2
 4 3 5
```

to represent an element of $M_3(\mathbf{Z}_{11})$. The multiplicative inverse of 2 in $\mathbf{Z}_{11}$ is 6. Multiplying `U[1;]` by 6 and reducing modulo 11,

```
      U[1;]←11|6×U[1;]
      U
 1 2 6
 6 5 2
 4 3 5
```

we make `U[1;1]` equal to 1. It is now easy to complete the reduction of the first column.

```
      U[2 3;]←11|U[2 3;]-6 4∘.×U[1;]
      U
 1  2  6
 0  4 10
 0  6  3
```

Note the way in which several row operations of type 3 can be performed with one APL statement. We can finish the reduction of the entire matrix in a similar manner.

```
      U[2;]←11|3×U[2;]
      U
1 2 6
0 1 8
0 6 3
      U[3;]←11|U[3;]-6×U[2;1]
      U
 1  2  6
 0  1  8
 0  0 10
      U[3;]←11|10×U[3;]
      U
1 2 6
0 1 8
0 0 1
```

Over a field we may always make the corner entries of a matrix in row-echelon form equal to 1, and this is normally done.

As our last example, let us consider

$$\begin{bmatrix} 1+X^2 & 1+X \\ X-X^2 & 1+X+X^2 \end{bmatrix}$$

in $M_2(\mathbf{Z}_3[X])$. Adding the first row to the second, we get

$$\begin{bmatrix} 1+X^2 & 1+X \\ 1+X & -1-X+X^2 \end{bmatrix}$$

Subtracting $-1+X$ times the second row from the first gives us

$$\begin{bmatrix} -1 & X-X^2-X^3 \\ 1+X & -1-X-X^2 \end{bmatrix}.$$

Finally, adding $1+X$ times the first tow to the second, we obtain

$$\begin{bmatrix} -1 & X-X^2-X^3 \\ 0 & -1-X^2+X^3-X^4 \end{bmatrix},$$

which is in row-echelon form.

EXERCISES

1 Let U and V be submodules of R^n with $U \subseteq V$. Suppose that $(a_1, \ldots, a_n)$ and $(b_1, \ldots, b_n)$ are the standard invariants for U and V, respectively. Prove that b_i divides a_i, $1 \leq i \leq n$. Show also that if a_i and b_i are associates for all i, then $U = V$.

2 Let $U = S_{\mathbf{Z}}(A)$, where

$$A = \begin{bmatrix} 6 & 1 \\ 0 & 8 \end{bmatrix}.$$

Show that there is no submodule V of $\mathbf{Z}^2$ such that V contains U and V has standard invariants $(3, 4)$.

3 Compute the number of 2-by-3 matrices over $\mathbf{Z}_n$ that are in row-echelon form. (The ring $\mathbf{Z}_n$ is not an integral domain unless n is a prime, but the definition of row-echelon form makes sense for matrices over any ring.)

4 For each of the following matrices A, determine the rank of $U = S_{\mathbf{Z}}(A)$, a basis for U, and the standard invariants of U.

(a) $$A = \begin{bmatrix} 2 & 9 \\ -5 & 4 \end{bmatrix}.$$

(b) $$A = \begin{bmatrix} 6 & -8 & 15 \\ 10 & -14 & 26 \\ -15 & 9 & 21 \end{bmatrix}$$

5 Let $R = \mathbf{Z}_7$ and let

$$A = \begin{bmatrix} 2 & 6 & 5 & 4 \\ 3 & 0 & 3 & 2 \\ 5 & 4 & 0 & 2 \end{bmatrix},$$

considered as a matrix over R. Find the dimension of $U = S_R(A)$ and a basis for U.

6 Let $R = \mathbf{Z}[i]$ and

$$A = \begin{bmatrix} 3 + 5i & -4 + 3i \\ 3 + i & -1 + 3i \end{bmatrix}.$$

Determine the standard invariants for $S_R(A)$.

7 Let $R = \mathbf{Z}_2[X]$ and

$$A = \begin{bmatrix} X + X^2 & X + X^2 & 1 \\ X^3 & 1 & 1 + X \\ X^2 & 1 & 1 \end{bmatrix}.$$

Determine the standard invariants for $S_R(A)$.

8 Let A be as in Exercise 4a. Is $(3, 17)$ in $S_{\mathbf{Z}}(A)$? Is $(-2, 8)$ in $S_{\mathbf{Q}}(A)$?

9 Let A be as in Exercise 4b. Are $(3, 0, 1)$ and $(2, -1, 7)$ in $S_{\mathbf{Z}}(A)$?

10 Let R and A be as in Exercise 5. Are $(1, 1, 5, 5)$ and $(3, 2, 4, 6)$ in $S_R(A)$?

11 Let R and A be as in Exercise 6. Is $(2i, 1)$ in $S_R(A)$?

12 Let R and A be as in Exercise 7. Is $(X, X, 1 + X + X^2)$ in $S_R(A)$?

2. ROW EQUIVALENCE, CONTINUED

In this section we will continue the investigations begun in Section 1. Among the important topics we will consider are the problem of determining a unique representative for each row-equivalence class of matrices over R, a converse of Corollary 1.7, a proof that the group $GL_n(R)$ of units in $M_n(R)$ is generated by elementary matrices, and an algorithm for computing inverses of matrices.

Let A be a matrix over R. By Theorem 1.8, there is a matrix B in row-echelon form that is row equivalent to A. Unfortunately, B is generally not unique. For example, the matrices

$$\begin{bmatrix} 2 & 1 \\ 0 & 3 \end{bmatrix}, \quad \begin{bmatrix} -2 & -1 \\ 0 & 3 \end{bmatrix}, \quad \begin{bmatrix} 2 & 4 \\ 0 & -3 \end{bmatrix}$$

are all in row-echelon form and are row equivalent over $\mathbf{Z}$ to one another. It is very useful to be able to add extra conditions on B in such a way that B becomes unique. The manner in which this is done depends on the ring R. The case in which R is a field is the easiest to handle.

Let F be a field. A matrix B with entries in F is said to be *row reduced* over F if the following conditions are satisfied.

1. B is in row-echelon form.
2. Each corner entry of B is 1.
3. If $B[i;j]$ is a corner entry, then $B[k;j] = 0$ for all $k \neq i$.

Condition (3) states that not only are entries below a corner entry equal to 0 but all of the entries above a corner entry are also 0. For example, the matrix

$$A = \begin{bmatrix} 1 & 0 & 1/2 \\ 0 & 1 & 2 \end{bmatrix}$$

is row reduced over $\mathbf{Q}$ but neither

$$B = \begin{bmatrix} 2 & 0 & 1 \\ 0 & 1 & 2 \end{bmatrix}$$

nor

$$C = \begin{bmatrix} 1 & 1 & 5/2 \\ 0 & 1 & 2 \end{bmatrix}$$

is row reduced over **Q**, since $B[1;1]$ is not 1 and $C[1;2]$ is not 0.

THEOREM 1. Let A be a matrix over a field F. Then A is row equivalent to exactly one matrix that is row reduced over F.

Proof. By Theorem 1.8, there is a matrix B in row-echelon form that is row equivalent to A. As we have already noted, we can multiply the ith nonzero row of B by the reciprocal of its first nonzero entry $B[i;j_i]$ and so make all of the corner entries 1. To obtain a matrix that is row reduced over F, we have only to change B so all of the entries above each corner 1 are zeros. There are no entries above the first corner entry $B[1;j_1]$. Subtracting $B[1;j_2]$ times $B[2;]$ from $B[1;]$ makes $B[1;j_2]$ zero. Now subtracting $B[1;j_3]$ times $B[3;]$ from $B[1;]$ and $B[2;j_3]$ times $B[3;]$ from $B[2;]$ makes $B[1;j_3]$ and $B[2;j_3]$ zero without affecting the earlier columns. Continuing in this manner, we get a matrix that is row reduced over F and row equivalent to A.

To prove uniqueness, suppose B and C are each row reduced over F and also row equivalent to each other. The corner entries of B and C occur in the same columns $j_1 < j_2 < \ldots < j_r$ and $S_F(B) = S_F(C)$. Suppose $B \neq C$. Then, for some i, the rows $B[i;]$ and $C[i;]$ are different. If $i > r$, then $B[i;] = C[i;] = 0$. Thus $1 \leq i \leq r$. The entries $B[i;j_i]$ and $C[i;j_i]$ are both 1, while $B[i,j_k] = C[i;j_k] = 0$ if $k \neq i$. Thus, if $x = B[i;] - C[i;]$, then x is in $S_F(B)$ and the j_kth component of x is 0, $1 \leq k \leq r$. Since x is in $S_F(B)$, we can find $e_1, \ldots, e_r$ in F such that

$$x = \sum_{k=1}^{r} e_k B[k;].$$

However, the only nonzero entry in $B[;j_k]$ is $B[k;j_k]$, so the j_kth component of the sum on the right is e_k. Therefore $e_k = 0$ for all k and $x = 0$. Thus $B = C$. □

The proof of Theorem 1 actually tells us how to find the row-reduced matrix row equivalent to a given matrix. For example, suppose $F = \mathbf{Z}_5$ and

```
      ⎕←A←B←3 3⍴2 3 3 3 1 4 4 2 4
2 3 3
3 1 4
4 2 4
```

To transform B into a matrix row equivalent to A and row reduced over $\mathbf{Z}_5$, we proceed as follows.

```
      B[1;]←5|3×B[1;]
      B
1 4 4
3 1 4
4 2 4
      B[2 3;]←5|B[2 3;]-3 4∘.×B[1;]
      B
1 4 4
0 4 2
0 1 3
      B[2;]←5|4×B[2;]
      B
1 4 4
0 1 3
0 1 3
      B[1 3;]←5|B[1 3;]-4 1∘.×B[2;]
      B
1 0 2
0 1 3
0 0 0
```

Over the integers the situation is a little more complicated. Let A be an integer matrix. It is not always possible to find a matrix B that is row equivalent to A over **Z** and is also row reduced over **Q**. We say that B is *row reduced* over **Z** provided the following hold.

1. B is in row-echelon form.
2. All corner entries of B are positive.
3. If $B[i;j]$ is a corner entry and $1 \leq k < i$, then $0 \leq B[k;j] < B[i;j]$.

For example,

$$\begin{bmatrix} 2 & 1 & 5 & 0 \\ 0 & 3 & -2 & 1 \\ 0 & 0 & 0 & 2 \end{bmatrix}$$

is row reduced over **Z**. With this definition the analogue of Theorem 1 holds.

THEOREM 2. Let A be an integer matrix. Then A is row equivalent over **Z** to exactly one matrix that is row reduced over **Z**.

Proof. We can find a matrix B in row-echelon form row equivalent to A over **Z**. By multiplying rows of B by -1 if necessary, we may assume all the corner entries $B[i;j_i]$ are positive. Write $B[1;j_2]$ in the form $aB[2;j_2] + b$, where a and b are integers and $0 \leq b < B[2;j_1]$. Subtracting a times $B[2;]$ from $B[1;]$, we make $B[1;j_2]$ equal to b and so $0 \leq B[1;j_2] < B[2;j_2]$.

Proceeding with the entries above each succeeding corner entry, we obtain a matrix B row equivalent to A and row reduced over $\mathbf{Z}$. The uniqueness of B we leave to the reader. (See Exercise 1.) □

We will also need a definition of "row reduced" for matrices of polynomials. Let F be a field. A matrix B with entries in $F[X]$ is *row reduced* over $F[X]$ if the following conditions hold.

1. B is in row-echelon form.
2. The corner entries of B are monic polynomials.
3. If $B[i;j]$ is a corner entry and $1 \leq k < i$, then either $B[k;j]$ is 0 or the degree of $B[k;j]$ is less than the degree of $B[i;j]$.

The matrix

$$\begin{bmatrix} 1+X & 1-X & 2X+X^2 \\ 0 & 2+X^2 & 0 \\ 0 & 0 & 3+X^3 \end{bmatrix}$$

is row reduced over $\mathbf{Z}_5[X]$.

THEOREM 3. Let F be a field and let A be a matrix with entries in $F[X]$. Then A is row equivalent over $F[X]$ to a unique matrix that is row reduced over $F[X]$.

Proof. See Exercise 2. □

Let A be a matrix over R. We know how to find a matrix B that is row equivalent to A and in row-echelon form. If R happens to be $\mathbf{Z}$, a field F or the ring $F[X]$, then we can choose B to be row reduced over R. However, we often need more. By Theorem 1.6, there is a matrix E that is a product of elementary matrices such that $B = E+.\times A$. We sometimes need to know one such matrix E. The procedures described in the proofs of Theorems 1.8, 1, and 2 allow us to solve this problem. We know that $E = E_t E_{t-1} \ldots E_1$ is the product of the elementary matrices corresponding to the row operations that transform A into B. If we apply these same row operations to the m-by-m identity matrix I, where m is the number of rows of A, then the result is $E+.\times I$ or E. Thus, to compute both B and E, we first set $B = A$ and $E = I$. Then, as we apply row operations to B in order to row reduce it, we apply the same operations to E. In this way the equation $B = E+.\times A$ remains correct throughout the computation. Here is an example with $R = \mathbf{Z}$.

```
      ⎕←A←B←2 2⍴5 4 3 3              ⎕←E←(⍳2)∘.=⍳2
5 4                                 1 0
3 3                                 0 1
      B[1;]←B[1;]-B[2;]                   E[1;]←E[1;]-E[2;]
      B                                   E
2 1                                  1 ¯1
3 3                                  0  1
      B[2;]←B[2;]-B[1;]                   E[2;]←E[2;]-E[1;]
      B                                   E
2 1                                  1 ¯1
1 2                                 ¯1  2
      B[1;]←B[1;]-2×B[2;]                 E[1;]←E[1;]-2×E[2;]
      B                                   E
 0 ¯3                                3 ¯5
 1  2                               ¯1  2
      ⎕←B←B[2 1;]                         ⎕←E←E[2 1;]
 1  2                               ¯1  2
 0 ¯3                                3 ¯5
      B[2;]←-B[2;]                        E[2;]←-E[2;]
      B                                   E
1 2                                 ¯1 2
0 3                                 ¯3 5
                                          E+.×A
                                    1 2
                                    0 3
```

The matrix B is now reduced over **Z** and B is $E+{\circ}{\times}A$.

The workspace *CLASSLIB* contains procedures for row reducing matrices over **R**, **Z**, $\mathbf{Z}_n$, $\mathbf{R}[X]$, and $\mathbf{Z}_n[X]$. In the case of $\mathbf{Z}_n[X]$, the integer n must be a prime. For example, if A is an integer matrix, then

```
ZROWREDUCE A
```

is the unique matrix B row equivalent to A over **Z** and row reduced over **Z**. In addition, *ZROWREDUCE* computes a global variable $\underline{R}$ that is a product of elementary integer matrices such that B is $\underline{R}+{\circ}{\times}A$. Using the preceding example, we have

```
      A                                   R̲
5 4                                 ¯1 2
3 3                                 ¯3 5
      ⎕←B←ZROWREDUCE A                    R̲+.×A
1 2                                 1 2
0 3                                 0 3
```

The other procedures for row reduction work in a similar manner.

Row operations can be used to compute inverses in $M_n(R)$. However, before learning how, we must prove two theorems.

THEOREM 4. Let A and B be m-by-n matrices over R. Then A and B are row equivalent over R if and only if $S_R(A) = S_R(B)$.

Proof. If A and B are row equivalent, then by Corollary 1.7 we know that $S_R(A) = S_R(B)$. Thus we have only to show the converse.

Set $U = S_R(A)$ and suppose $S_R(B) = U$ also. We must show that A and B are row equivalent. By Theorems 1.5 and 1.8, we may assume that both A and B are in row-echelon form. From Corollary 1.4 it follows that A and B have the same number r of nonzero rows and the ith corner entries of A and B occur in the same column j_i and are associates, $1 \le i \le r$. Multiplying the rows of A by units in R if necessary, we may assume $A[i;j_i] = B[i;j_i]$ for each i. If $r = 0$, then $A = B = 0$. Thus we may assume $r \ge 1$. The vector $x = A[1;] - B[1;]$ is in U and has zeros for its first j_1 components. By the proof of Theorem 2, we know that x is in the submodule of R^n spanned by $B[2;], \ldots, B[r;]$. Thus we can find $e_2, \ldots, e_r$ in R such that

$$x = \sum_{i=2}^{r} e_i B[i;].$$

Now $A[1;] = B[1;] + x$ and, if we add $e_i B[i;]$ to $B[1;]$ for $2 \le i \le r$, then the new B is row equivalent to the old and $B[1;]$ is now equal to $A[1;]$. Let A' and B' be the matrices obtained by deleting the first rows of A and B, respectively. Then $S_R(A') = S_R(B') = U'$, where U' is the submodule of U consisting of the elements whose first j_1 components are 0. By induction on m, the matrices A' and B' are row equivalent and, hence, so are A and B. □

Now we can provide additional information about the units in $M_n(R)$.

THEOREM 5. Let A be a matrix in $M_n(R)$. Then the following are equivalent.

(a) A is a unit in $M_n(R)$.
(b) $\det A$ is a unit in R.
(c) $S_R(A) = R^n$.
(d) The rows of A are a basis for R^n.
(e) A is row equivalent to the n-by-n identity matrix.
(f) A is a product of elementary matrices.

Proof. By Theorem 5.2.12, conditions (a), (b), (c), and (d) are equivalent. Let I be the n-by-n identity matrix. Since $S_R(I) = R^n$, the equivalence of (d) and(e) follows from Theorem 4. Thus all that remains is to show that (f) is equivalent to the others. Now elementary matrices are units, so (f) implies (a). Suppose (e) holds. Then there is a matrix $E = E_t E_{t-1} \ldots E_1$ such that each E_i is an elementary matrix and $EA = I$.

Multiplying on the left by E^{-1}, we get $A = E^{-1} = E_1^{-1} \ldots E_t^{-1}$. By Theorem 4.7.4, each E_i^{-1} is an elementary matrix, so (f) holds. □

Let A be a matrix in $M_n(R)$. Theorem 5 gives us a new method for deciding whether or not A is a unit and, if it is, for computing A^{-1}. First, row reduce A to a matrix B in row-echelon form using the procedure of Theorem 1.8. If B has fewer than n nonzero rows or if one of the corner entries of B is not a unit in R, then B is not row equivalent to I and A is not a unit. However, if B has n nonzero rows and all corner entries are units in R, we may row reduce B to I using the procedure in the proof of Theorem 1. Applying to I the row operations used to obtain I from A, we get A^{-1}.

For example, suppose $R = \mathbf{Z}$ and

```
      ⎕←A←3 3⍴1 1 3 3 2 11 ¯1 1 ¯6
  1   1   3
  3   2  11
 ¯1   1  ¯6
```

Row reducing *A* using *ZROWREDUCE*,

```
      ZROWREDUCE A                        R+.×A
1 0 0                               1 0 0
0 1 0                               0 1 0
0 0 1                               0 0 1
      R                                   A+.×R
 23  ¯9  ¯5                         1 0 0
 ¯7   3   2                         0 1 0
 ¯5   2   1                         0 0 1
```

we see that *A* is row equivalent to the identity, so the inverse of *A* is the matrix *R* computed by *ZROWREDUCE*. This is the method used by

ZMATINV.

```
      ZMATINV A
 23  ¯9  ¯5
 ¯7   3   2
 ¯5   2   1
```

We close this section with a two-part problem that illustrates the kind of computations we can now carry out. Let

```
      ⎕←A←3 4ρ¯1 3 ¯5 3 ¯3 11 ¯16 8 2 ¯4 9 ¯4
¯1   3  ¯5   3
¯3  11 ¯16   8
 2  ¯4   9  ¯4
      ⎕←B←3 4ρ¯3 ¯3 ¯9 ¯15 11 13 32 58 ¯7 ¯9 ¯20 ¯39
¯3  ¯3  ¯9 ¯15
11  13  32  58
¯7  ¯9 ¯20 ¯39
      C←5 1 17 10
```

First, we want to decide whether or not C is in $S_{\mathbf{Z}}(A)$ and, if so, to find an integer vector X such that C is $X+.\times A$. Second, we would like to know whether or not A and B are row equivalent over $\mathbf{Z}$ and, if so, to find an element E of $GL_3(\mathbf{Z})$ such that B is $E+.\times A$. Row reducing A,

```
      ⎕←D←ZROWREDUCE A                       R̲
 1  1  3  1                            3  0  2
 0  2 ¯1  2                            2  0  1
 0  0  0  3                            5 ¯1  1
```

we can now use the method described in Section 1 to see if C is in $S_{\mathbf{Z}}(D) = S_{\mathbf{Z}}(A)$ and, if so, to write C as $Y+.\times D$. The value of $Y[1]$ must be

$$C[1]\div D[1;1],$$

or 5. Subtracting,

```
      ⎕←C1←C-5×D[1;]
0 ¯4 2 5
```

we see that $Y[2]$ must be $C1[2]\div D[1;2]$, or -2. Subtracting again,

```
      ⎕←C2←C1-¯2×D[2;]
0 0 0 9
```

we find that $C2$ is $3\times D[3;]$ and so $Y[3]$ is 3 and C is in $S_{\mathbf{Z}}(D)$.

```
      Y←5 ¯2 3
      Y+.×D
5 1 17 10
```

However, we were asked to write C in the form $X+.\times A$. We know that C is $Y+.\times D$ and D is $\underline{R}+.\times A$. Therefore C is $Y+.\times\underline{R}+.\times A$, which is $(Y+.\times\underline{R})+.\times A$. Thus we may take X to be $Y+.\times\underline{R}$.

```
      ⎕←X←Y+.×R̲                             X+.×A
26 ¯3 11                              5 1 17 10
```

In order to decide whether A and B are row equivalent, we must row reduce B. Before doing so, we save the matrix $\underline{R}$ obtained before.

```
      P←R̲
      ZROWREDUCE B
1  1  3  1
0  2 ¯1  2
0  0  0  3
```

```
      R̲
 5  4  4
 5  2  1
¯4 ¯3 ¯3
```

Here we see that B is also row equivalent to D, so A and B are row equivalent. Also, D is `R̲+.×B`, so B is `Q+.×D`, where Q is the inverse of $\underline{R}$. Thus B is `Q+.×P+.×A` or `(Q+.×P)+.×A`.

```
      ⎕←Q←ZMATINV R̲
¯3   0  ¯4
11   1  15
¯7  ¯1 ¯10
      ⎕←E←Q+.×P
¯29    4  ¯10
110  ¯15   38
¯73   10  ¯25
```

```
      E+.×A
¯3  ¯3  ¯9 ¯15
11  13  32  58
¯7  ¯9 ¯20 ¯39
```

We have found E such that B is `E+.×A`. Since both Q and P are units in $M_3(\mathbf{Z})$, it follows that E is also a unit.

EXERCISES

1 Complete the proof of Theorem 2.

2 Prove Theorem 3.

3 For each of the following matrices A find a unit E in $M_3(\mathbf{Z})$ such that `E+.×A` is row reduced over $\mathbf{Z}$. Do the computations by hand and check your work using *ZROWREDUCE*.

(a)
$$A = \begin{bmatrix} 5 & -2 & 1 \\ 6 & -3 & 0 \\ -2 & 1 & 3 \end{bmatrix}.$$

(b)
$$A = \begin{bmatrix} 3 & 1 & 2 & 1 \\ -2 & 0 & -1 & 3 \\ 5 & 4 & -3 & -2 \end{bmatrix}.$$

(c)
$$A = \begin{bmatrix} 15 \\ 21 \\ 35 \end{bmatrix}.$$

4 For each of the following matrices A over $\mathbf{Z}_7$ find an element E of $GL_3(\mathbf{Z}_7)$ such that $E+.\times A$ is row reduced over $\mathbf{Z}_7$. Do the computations by hand and check your work using *ZNROWREDUCE*.

(a)
$$A = \begin{bmatrix} 1 & 6 & 0 \\ 2 & 5 & 1 \\ 3 & 2 & 5 \end{bmatrix}$$

(b)
$$A = \begin{bmatrix} 4 & 1 & 2 & 5 \\ 6 & 3 & 0 & 1 \\ 2 & 5 & 3 & 4 \end{bmatrix}$$

5 Let A be the following matrix over $R = \mathbf{Z}_3[X]$.

$$\begin{bmatrix} 1+X & -1+X \\ 1+X^2 & X^2 \end{bmatrix}$$

Find a unit E in $M_2(R)$ such that $E+.\times A$ is row reduced over R. Do the computations by hand and check your work using *ZNXROWREDUCE*.

6 Let A be a square matrix over a Euclidean domain. Describe a procedure for computing det A using row reduction. Use this procedure to compute the determinant of the matrix

$$\begin{bmatrix} 9 & 1 & 7 \\ 2 & 5 & 4 \\ 3 & 6 & 8 \end{bmatrix}$$

over $\mathbf{Z}_{11}$.

7 Compute the inverse of

$$A = \begin{bmatrix} 1 & 1 & -2 \\ 2 & 3 & -5 \\ 1 & 0 & -2 \end{bmatrix}$$

in $M_3(\mathbf{Z})$ by row reducing A to the identity matrix I and applying the same row operations to I. Check your work using *ZMATINV*.

8 Determine the rank of

$$A = \begin{bmatrix} 2 & 4 & -6 \\ 18 & -9 & 6 \end{bmatrix}$$

modulo p for several small primes p. What is the set of primes modulo which A has rank 1?

9 Let `C←¯16 35 ¯23 52` and

```
      ⎕←A←3 4⍴2 9 ¯3 8 4 ¯1 6 ¯7 ¯3 2 5 0
 2  9 ¯3  8
 4 ¯1  6 ¯7
¯3  2  5  0
```

Find an integer vector X such that C is `X+.×A`.

10 Find a unit E in $M_3(\mathbf{Z})$ such that $B = E+.\times A$, where

$$A = \begin{bmatrix} 3 & -1 & 5 & 4 \\ -2 & 6 & 1 & 0 \\ 4 & 2 & 3 & 2 \end{bmatrix},$$

$$B = \begin{bmatrix} -3 & 3 & 3 & 2 \\ -7 & -29 & -18 & -10 \\ 5 & 21 & 14 & 8 \end{bmatrix}.$$

*11 Find an appropriate definition of "row reduced over $\mathbf{Z}[i]$" such that every matrix over $\mathbf{Z}[i]$ is row equivalent to exactly one row-reduced matrix.

12 Let A be a matrix with rational entries and set $m = \dim_{\mathbf{Q}}(S_{\mathbf{Q}}(A))$ and $n = \dim_{\mathbf{R}}(S_{\mathbf{R}}(A))$. Can it happen that $m \neq n$?

13 Let A and B be row-equivalent matrices over R. Show that the gcd of the entries of A is the same (up to associates) as the gcd of the entries of B.

*14 Generalize Exercise 13 as follows. Let A be an m-by-n matrix over R and let k be an integer between 1 and the smaller of m and n. Set $D_k(A)$ equal to the gcd of the determinants of the k-by-k submatrices of A. Show that if B is row equivalent to A, then $D_k(B) = D_k(A)$.

15 Show that $(x_1, \ldots, x_n)$ in R^n is part of a basis for R^n if and only if $\gcd(x_1, \ldots, x_n) = 1$. *Hint.* Row reduce the n-by-1 matrix

$$\begin{bmatrix} x_1 \\ \cdot \\ \cdot \\ \cdot \\ x_n \end{bmatrix}.$$

16 Find a unit in $M_3(\mathbf{Z})$ with $(6, 10, 15)$ as its first row.

3. VECTOR SPACES

At this point we have already spent a considerable amount of time studying finitely generated free R-modules, their submodules, and their endomorphisms. In Sections 5.2 and 5.3 we began with very few assumptions about the ring R and later proved some theorems that required R to be a PID. In the first two sections of this chapter we obtained further results assuming that R is a Euclidean domain. Now we will specialize even further and assume that R is a field. Before doing so, however, it will be convenient to list some important facts that we have already established.

THEOREM 1. Let R be a nontrivial commutative ring and let n be a positive integer. Then

- **(a)** Every basis of R^n has n elements.
- **(b)** Elements $u_1, \ldots, u_n$ of R^n form a basis for R^n if and only if they generate R^n.
- **(c)** If f in $\text{End}_R(R^n)$ is surjective, then f is injective.
- **(d)** If R is a PID, then every submodule of R^n is a free module of rank at most n.

Proof. This theorem is a restatement of Corollaries 9, 13, 14, and 16 of Section 5.2. □

If R is a Euclidean domain, then the row-reduction procedures of the previous two sections provide a means for answering inclusion and membership questions about submodules of R^n from a knowledge of generating sets for those submodules. That is, given $x, y_1, \ldots, y_r$ and $z_1, \ldots, z_s$ in R^n, we can decide whether x is in $M = \langle y_1, \ldots, y_r \rangle$ and whether M is contained in $\langle z_1, \ldots, z_s \rangle$.

Throughout this section F will be a field and, unless otherwise indicated, all vector spaces will be over F. By Theorem 5.2.5, we know that vector spaces are free modules, so the conclusions of Theorem 1 apply to any vector space of dimension n over F.

THEOREM 2. Let $v_1, \ldots, v_m$ be linearly independent elements of the vector space V and suppose v_{m+1} is in $V - \langle v_1, \ldots, v_m \rangle$. Then $v_1, \ldots, v_{m+1}$ are linearly independent.

Proof. Suppose $c_1, \ldots, c_{m+1}$ are elements of F such that $c_1 v_1 + \ldots + c_{m+1} v_{m+1} = 0$. If $c_{m+1} \neq 0$,

$$v_{m+1} = -\frac{1}{c_{m+1}}(c_1 v_1 + \ldots + c_m v_m),$$

contradicting our assumption that v_{m+1} is not in $\langle v_1, \ldots, v_m \rangle$. Thus

$c_{m+1} = 0$ and $c_1v_1 + \ldots + c_m v_m = 0$. By the linear independence of $v_1, \ldots, v_m$, this implies that $c_1 = c_2 = \ldots = c_{m+1} = 0$. □

COROLLARY 3. Let V be a vector space of dimension n and let $v_1, \ldots, v_m$ be linearly independent elements of V. Then there exists elements $v_{m+1}, \ldots, v_n$ of V such that $v_1, \ldots, v_n$ is a basis for V.

Proof. Let $W_1 = < v_1, \ldots, v_m >$. If $W_1 = V$, then $v_1, \ldots, v_m$ is a basis for V and $m = n$. If $W_1 \neq V$, choose v_{m+1} in $V - W_1$ and set $W_2 = < v_1, \ldots, v_{m+1} >$. If $W_2 \neq V$, choose v_{m+2} in $V - W_2$. Continuing in this way, we must reach a basis for V. Otherwise, after $n - m + 1$ steps, we will have produced $n + 1$ linearly independent elements of V, which do not exist by Theorem 1d. □

COROLLARY 4. If V is a finite dimensional vector space and W is a subspace of V with $\dim_F W = \dim_F V$, then $W = V$.

Proof. Let $v_1, \ldots, v_n$ be a basis for W. Then $n = \dim_F V$. If $W \neq V$, we can choose v_{n+1} in $V - W$ and obtain $n + 1$ linearly independent elements of V, again contradicting Theorem 1d. □

COROLLARY 5. If V is a finite dimensional vector space, then any spanning set for V contains a subset that is a basis for V.

Proof. Let $n = \dim_F V$ and let X be a spanning set for V. Since no linearly independent subset of V contains more than n elements, we can find a linearly independent subset Y of X such that Y is not contained properly in any other linearly independent subset of X. Set $W = < Y >$. If $W \neq V$, then X is not contained in W. Thus there exists an element x in $X - W$. By Theorem 2, $Y \cup \{x\}$ is a linearly independent subset of X properly containing Y. Thus $W = V$ and Y is a basis for V. □

COROLLARY 6. Let V be a vector space of dimension n and let $v_1, \ldots, v_n$ be elements of V. Then the following are equivalent.

(a) $v_1, \ldots, v_n$ span V.
(b) $v_1, \ldots, v_n$ are linearly independent.
(c) $v_1, \ldots, v_n$ form a basis for V.

Proof. By definition, (c) implies (a) and (b). By Theorem 1a, (a) implies (c). Finally, if (b) holds, then Corollary 4 shows that $< v_1, \ldots, v_n > = V$, and so (b) implies (c). □

Theorem 2, from which Corollaries 3 to 6 follow, is not true in free modules over more general commutative rings. For example, in $\mathbf{Z}^2$ the elements $v_1 = (2, 0)$ and $v_2 = (0, 2)$ are linearly independent and $v_3 = (1, 0)$ is not in $< v_1, v_2 >$. However, $v_1 - 2v_3 = 0$, so v_1, v_2, v_3 are not linearly independent over $\mathbf{Z}$.

Suppose we have elements $v_1, \ldots, v_m$ in F^n that are linearly independent. Corollary 3 states that we can extend this sequence to a basis of F^n. How is this done in practice? For example, if

```
      ⎕←A←3 5⍴3 4 2 4 1 2 1 4 0 2 1 3 1 3 2
3 4 2 4 1
2 1 4 0 2
1 3 1 3 2
```

then modulo 5 the rows of A are linearly independent, as the following computation shows.

```
      N←5
      ⎕←B←ZNROWREDUCE A

1 3 0 0 1
0 0 1 0 0
0 0 0 1 2
```

How can we find two more elements in $(\mathbf{Z}_5)^5$ that, together with the rows of A, form a basis for $(\mathbf{Z}_5)^5$? If we let

```
      ⎕←C←(⍳5)∘.=⍳5
1 0 0 0 0
0 1 0 0 0
0 0 1 0 0
0 0 0 1 0
0 0 0 0 1
```

```
      C[1 3 4;]←B
      C
1 3 0 0 1
0 1 0 0 0
0 0 1 0 0
0 0 0 1 2
0 0 0 0 1
```

then C is in row-echelon form and the rows of C are a basis for $(\mathbf{Z}_5)^5$. Since A is row equivalent to B, it follows that C is row equivalent to D, where

```
      D←C
      D[1 3 4;]←A
      D
3 4 2 4 1
0 1 0 0 0
2 1 4 0 2
1 3 1 3 2
0 0 0 0 1
```

Thus the rows of D form a basis for $(\mathbf{Z}_5)^5$ that includes the rows of A. This method may be used over any field F and with any value of n. That is, if A is an m-by-n matrix over F whose rows are linearly independent, we row reduce A to a matrix B in row-echelon form and let J be the set of integers j such that $B[;j]$ contains a corner entry. Then the rows of A together with the standard basis vectors u_j of F^n with j in $\{1, \ldots, n\} - J$ form a basis for F^n.

Corollary 5 suggests problems of the following type. Suppose A is an m-by-n matrix with entries in F. How can we select a subset of the rows of A that is a basis for $S_F(A)$? For example, let

```
      N̲←11
      ⎕←A←4 4⍴2 1 3 8 7 4 1 6 9 0 0 3 4 5 2 8
2 1 3 8
7 4 1 6
9 0 0 3
4 5 2 8
```

where A is considered to be an element of $M_4(\mathbf{Z}_{11})$, and set $U = S_{\mathbf{Z}_{11}}(A)$. The nonzero rows of

```
      ZNROWREDUCE A
 1  0  0  4
 0  1  0 10
 0  0  1  4
 0  0  0  0
```

form a basis for U. However, Corollary 5 says that we can also find a basis that consists of rows of A. One method for finding such a basis is to row reduce the first m rows for $2 \leq m \leq 4$. We have already done the case $m = 4$. With $m = 2, 3$, we get

```
      ZNROWREDUCE 2 4↑A
1 0 0 4
0 1 3 0
      ZNROWREDUCE 3 4↑A
1 0 0 4
0 1 3 0
0 0 0 0
```

Thus the first two rows of A are linearly independent and generate a subspace that contains `A[3;]` but not `A[4;]`. Therefore the rows of `A[1 2 4;]` form a basis for U. In the next section we will present an alternative solution to this problem that requires only one row reduction.

Theorem 2 and its corollaries deal with subspaces of a finite dimensional vector space V. Now we turn our attention to quotient spaces and homomorphic images of V.

THEOREM 7. Let V be a finite dimensional vector space and let W be a subspace of V. Then

$$\dim_F W + \dim_F (V/W) = \dim_F V.$$

Proof. Let $n = \dim_F V$ and $m = \dim_F W$ and let $v_1, \ldots, v_m$ be a basis for W. By Corollary 3, we can find $v_{m+1}, \ldots, v_n$ in V such that $v_1, \ldots, v_n$ is a basis for V. For any v in V let $\bar{v}$ be the image of v under the natural map from V to V/W. Thus $\bar{v} = W + v$. Since V is spanned by $v_1, \ldots, v_n$, it follows that V/W is spanned by $\bar{v}_1, \ldots, \bar{v}_n$. But $\bar{v}_1 = \ldots = \bar{v}_m = 0$, so V/W is spanned by $\bar{v}_{m+1}, \ldots, \bar{v}_n$. Suppose $a_{m+1}, \ldots, a_n$ are elements of F such that $a_{m+1}\bar{v}_{m+1} + \ldots + a_n\bar{v}_n = 0$. Let $w = a_{m+1}v_{m+1} + \ldots + a_n v_n$. Then $\bar{w} = 0$, so w is in W. Thus we can find $a_1, \ldots, a_m$ in F such that $w = a_1 v_1 + \ldots + a_m v_m$. But then

$$a_1 v_1 + \ldots + a_m v_m - a_{m+1} v_{m+1} - \ldots - a_n v_n = 0.$$

Since the v_i are linearly independent, we must have $a_1 = \ldots = a_m = a_{m+1} = \ldots = a_n = 0$. Therefore $\bar{v}_{m+1}, \ldots, \bar{v}_n$ are linearly independent, so

$$\bar{v}_{m+1}, \ldots, \bar{v}_n$$

is a basis for V/W. Hence $\dim_F(V/W) = n - m$ and

$$\dim_F W + \dim_F(V/W) = m + (n - m) = n = \dim_F V. \quad \square$$

COROLLARY 8. Let V and W be vector spaces over F with V finite dimensional. If $f: V \longrightarrow W$ is a linear transformation with kernel K and image U, then

$$\dim_F K + \dim_F U = \dim_F V.$$

Proof. By the First Isomorphism Theorem, U is isomorphic to V/K. In particular, $\dim_F U = \dim_F(V/K)$. The corollary now follows immediately from Theorem 7. $\square$

COROLLARY 9. If f is a linear transformation of the finite dimensional vector space V into itself, then f is surjective if and only if f is injective.

Proof. Let K be the kernel of f and U the image of f. By Corollary 9, we know that $\dim_F K = 0$ if and only if $\dim_F U = \dim_F V$. But this says that $K = \{0\}$ if and only if $U = V$. $\square$

Corollary 9 should be considered analogous to the fact that surjectivity and injectivity are equivalent for maps of a given finite set into itself.

COROLLARY 10. If U and V are finite dimensional subspaces of a vector space, then

$$\dim_F(U + V) + \dim_F(U \cap V) = \dim_F U + \dim_F V.$$

Proof. By the Second Isomorphism Theorem, $(U + V)/V$ is isomorphic to $U/(U \cap V)$. Therefore

$$\dim_F [(U + V)/V] = \dim_F [U/(U \cap V)].$$

But, by Theorem 7,

$$\dim_F [(U + V)/V] = \dim_V (U + V) - \dim_F V,$$

$$\dim_F [U/(U \cap V)] = \dim_F U - \dim_F (U \cap V).$$

The result now follows immediately. □

In Section 5.3 we studied homomorphisms between free modules. Suppose U and V are vector spaces of dimension m and n, respectively, over F. By Theorem 5.3.4, $\mathrm{Hom}_F(U, V)$ is isomorphic as a vector space to the set of m-by-n matrices over F. In particular,

$$\dim_F(\mathrm{Hom}_F(U, V)) = mn.$$

If $u_1, \ldots, u_m$ is a basis for U and $v_1, \ldots, v_n$ is a basis for V, then for $1 \leq i \leq m$ and $1 \leq j \leq n$ we can define f_{ij} in $\mathrm{Hom}_F(U, V)$ by

$$u_i f_{ij} = v_j,$$

$$u_k f_{ij} = 0, \quad k \neq i.$$

The elements f_{ij} form a basis for $\mathrm{Hom}_F(U, V)$.

Let V be any vector space over F. The vector space $\mathrm{Hom}_F(V, F)$ is called the *dual space* and will be denoted V^*. Elements of V^* are sometimes called *linear functionals* on V. Suppose V is finite dimensional. Then the preceding remark implies that $\dim_F V^* = \dim_F V$ as $\dim_F F = 1$. Moreover, if $v_1, \ldots, v_n$ is a basis for V, then there is a basis $f_1, \ldots, f_n$ of V^* such that $v_i f_i = 1$ and $v_i f_j = 0, i \neq j$. We call $f_1, \ldots, f_n$ the *dual basis* of $v_1, \ldots, v_n$. Even though V and V^* are isomorphic, there is no isomorphism of V onto V^* that is natural in the sense of not requiring a choice of basis to be made in V. We can also form $V^{**} = (V^*)^*$. When V is finite dimensional, V^{**} is also isomorphic to V, and this time the isomorphism is natural.

THEOREM 11. Let V be a finite dimensional vector space. For v in V and f in V^*, define $T_v(f)$ to be vf. Then T_v is in V^{**} and $v \longmapsto T_v$ is an isomorphism of V onto V^{**}.

Proof. By definition, $T_v(f)$ is in F, so T_v maps V^* to F. If g is also in V^* and a, b are in F, then

$$T_v(af + bg) = v(af + bg) = a(vf) + b(vg) = aT_v(f) + bT_v(g).$$

Therefore T_v is in $\mathrm{Hom}_F(V^*, F) = V^{**}$. If w is another element of V, then for all f in V^* we have

$$T_{v+w}(f) = (v+w)f = vf + wf$$
$$= T_v(f) + T_w(f) = (T_v + T_w)(f).$$

Therefore $T_{v+w} = T_v + T_w$. Similarly, $T_{av} = aT_v$ for every every a in F. Therefore $v \longmapsto T_v$ is a linear transformation of V into V^{**}. Suppose v is in V and $T_v = 0$. Then $vf = 0$ for all v in V^*. If $v \neq 0$, we can find a basis $v_1 = v$, $v_2, \ldots, v_n$ of V. Let $f_1, \ldots, f_n$ be the corresponding dual basis of V^*. Then $vf_1 = 1$, contradicting $T_v = 0$. Thus $v = 0$ and $v \longmapsto T_v$ is injective. Since $\dim_F V = \dim_F V^{**}$, it follows (See Exercise 7) that this injection is an isomorphism. □

The map T_v in Theorem 11 is called *evaluation* at v. Theorem 11 says that each element of V^{**} is evaluation at a unique v in V.

EXERCISES

1 Give an example of a generating set for $\mathbf{Z}^2$ that does not contain a basis.

2 Show that $\mathbf{Z}^2$ contains proper $\mathbf{Z}$-submodules isomorphic to $\mathbf{Z}^2$.

3 Let A be the matrix

$$\begin{bmatrix} 3 & 1 & 5 & 2 & 6 \\ 2 & 4 & 1 & 3 & 2 \\ 5 & 0 & 6 & 6 & 4 \\ 0 & 2 & 0 & 1 & 3 \\ 3 & 6 & 1 & 5 & 4 \end{bmatrix},$$

considered over $\mathbf{Z}_7$. Find a set of rows of A that forms a basis for $S_{\mathbf{Z}_7}(A)$.

4 Show that the rows of the matrix

$$A = \begin{bmatrix} 3 & 2 & 6 & 5 & 1 \\ 5 & 7 & 10 & 2 & 6 \\ 7 & 1 & 3 & 1 & 3 \end{bmatrix}$$

are linearly independent over $\mathbf{Z}_{11}$ and find two more vectors that, together with the rows of A, form a basis for $(\mathbf{Z}_{11})^5$.

5 Let U and V be subspaces of a vector space of dimension 10. If $\dim_F U = 6$ and $\dim_F V = 7$, what are the possible values for $\dim_F(U \cap V)$?

6 Let $U = S_{\mathbf{Z}_5}(A)$ and $V = S_{\mathbf{Z}_5}(B)$, where

$$A = \begin{bmatrix} 2 & 4 & 3 & 4 \\ 1 & 0 & 3 & 3 \end{bmatrix},$$

$$B = \begin{bmatrix} 3 & 4 & 1 & 3 \\ 2 & 1 & 4 & 4 \end{bmatrix}.$$

Compute $\dim_{\mathbf{Z}_5} U$, $\dim_{\mathbf{Z}_5} V$, $\dim_{\mathbf{Z}_5} (U + V)$, and $\dim_{\mathbf{Z}_5} (U \cap V)$.

7 Let U and V be vector spaces of the same finite dimension. Show that f in $\mathrm{Hom}_F(U, V)$ is injective if and only if it is surjective.

8 Give an example of an element of $\mathrm{End}_{\mathbf{Z}}(\mathbf{Z}^2)$ that is injective but not surjective.

9 How many bases does $(\mathbf{Z}_p)^n$ have, p a prime?

10 Let $u = (1, 0)$ and $v = (0, 2)$ in $\mathbf{Z}^2$. Show that there is an element of $\mathrm{End}_{\mathbf{Z}}(\mathbf{Z}^2)$ that maps u to v but no element of $\mathrm{End}_{\mathbf{Z}}(\mathbf{Z}^2)$ that maps v to u.

11 Let $v_1, \ldots, v_n$ and $w_1, \ldots, w_n$ be bases of the vector space V and let $f_1, \ldots, f_n$ and $g_1, \ldots, g_n$ be the corresponding dual bases of V^*. Suppose P is the transition matrix from the v's to the w's. What is the transition matrix from the f's to the g's?

12 Let U and V be vector spaces and let T be in $\mathrm{Hom}_F(U, V)$. For f in V^* set $fT^* = T \circ f$. Show that T^* is in $\mathrm{Hom}_F(V^*, U^*)$. Suppose $u_1, \ldots, u_m$ is a basis for U and $v_1, \ldots, v_n$ is a basis for V. Let $g_1, \ldots, g_m$ and $f_1, \ldots, f_n$ be the bases of U^* and V^* dual to the u's and the v's, respectively. Suppose A is the matrix for T with respect to the u's and the v's. What is the matrix for T^* with respect to the f's and the g's? (We call T^* the *contragredient* of T.)

13 Let W be a subspace of the vector space V over the field F. The *annihilator* $\mathrm{Ann}(W)$ of W is the set of all f in V^* such that $wf = 0$ for all w in W. Show that $\mathrm{Ann}(W)$ is a subspace of V^*.

14 Let V and W be as in Exercise 13 and assume that V is finite dimensional. Prove that the function that maps f in V^* to the restriction $f|_W$ is a linear transformation of V^* onto W^* with kernel $\mathrm{Ann}(W)$. Conclude that $\dim_F W + \dim_F \mathrm{Ann}(W) = \dim_F V$.

15 Let U and W be subspaces of the finite dimensional vector space V. Show that

$$\mathrm{Ann}(U + W) = \mathrm{Ann}(U) \cap \mathrm{Ann}(W),$$
$$\mathrm{Ann}(U \cap V) = \mathrm{Ann}(U) + \mathrm{Ann}(W).$$

4. SOLVING LINEAR SYSTEMS USING ROW OPERATIONS

Let R be a commutative ring. By a *system of linear equations* over R we mean a system of simultaneous equations of the form

$$\begin{aligned} a_{11}x_1 + a_{12}x_2 + \ldots + a_{1n}x_n &= b_1, \\ a_{12}x_1 + a_{22}x_2 + \ldots + a_{2n}x_n &= b_2, \\ &\vdots \\ a_{m1}x_1 + a_{m2}x_2 + \ldots + a_{mn}x_n &= b_m, \end{aligned} \qquad (*)$$

where the a_{ij} and the b_i are in R. A *solution* to the system (*) is a vector $X = (x_1, \ldots, x_n)$ in R^n whose components satisfy the equations. If we let A be the matrix of a_{ij}'s and B the vector $(b_1, \ldots, b_n)$, we can write (*) as $A+.\times X = B$ or, simply, $AX = B$. We call A the *matrix of coefficients* for the system. If $B = 0$, we say that the system is *homogeneous.* The APL proposition ∧/B=A+.×X corresponds to the assertion that X is a solution of the linear system over **R** with A as the matrix of coefficients and B as the vector of constant terms.

Linear systems need not have any solutions. For example, the system

$$x + y = 0,$$

$$x + y = 1,$$

has no solutions, provided R is nontrivial, and the system

$$2x + 6y = 5$$

has no solutions in **Z**, even though it has infinitely many solutions in **Q**. A linear system (*) that has no solutions in R^n is said to be *inconsistent.* A homogeneous system $AX = 0$ is always consistent, since $X = 0$ is a solution.

Linear systems arise in many different contexts, and their solutions can be interpreted in many different ways. Before discussing techniques for solving linear systems, we will examine a few examples that illustrate the importance of such systems.

Example 1. Suppose f is in $\mathrm{Hom}_R(R^n, R^m)$. By Theorem 5.3.4, there is a unique n-by-m matrix A such that the image of X in R^n under f is $X+.\times A$. If we want to know the kernel of f, we are looking for vectors X in R^n such that $X+.\times A = 0$. Since the transpose of a vector in itself, we can write this condition as $A^t+.\times X = 0$. Thus, computing the kernel of f is equivalent to solving the homogeneous linear system whose matrix of coefficients is

A^t. (The fact that the matrix is A^t and not A is the price we pay for writing our homomorphisms on the right.)

Example 2. Let f and A be as in Example 1 and let B be an element of R^m. A natural question to ask is whether or not B is in the image of R^n under f. In other words, is there a vector X in R^n such that $X+.\times A = B$ or, equivalently, $A^tX = B$? Thus B is in the image of R^n if and only if the linear system $A^tX = B$ is consistent.

Example 3. The homogeneous system $AX = 0$ can be written

$$\sum_{j=1}^{n} x_j A[;j] = 0$$

This system has a nontrivial solution if and only if the columns of A are linearly dependent over R. Thus, in solving the system $AX = 0$, we are answering the following question. Are the columns of A linearly independent and, if not, what are all the ways the zero vector can be written as a linear combination of the columns of A?

Example 4. Let A and B be the integer matrices

$$A = \begin{bmatrix} 2 & -1 & 3 \\ 1 & 5 & -2 \end{bmatrix}, \quad B = \begin{bmatrix} 3 & 2 & -7 \\ 4 & -1 & 2 \end{bmatrix}$$

and let $U = S_Z(A)$ and $V = S_Z(B)$. Suppose we want to describe $U \cap V$. A typical element of U is

$$(x, y)+.\times A = (2x + y, -x + 5y, 3x - 2y)$$

and a typical element of V is

$$(z, w)+.\times B = (3z + 4w, 2x - w, -7z + 2w).$$

If C is an element of $U \cap V$, then there must be integers x, y, z, w such that $(x, y)+.\times A = C = (z, w)+.\times B$. We can write the vector equality

$$(x, y)+.\times A = (z, w)+.\times B$$

in the form

$$\begin{aligned} 2x + y - 3z - 4w &= 0, \\ -x + 5y - 2z + w &= 0, \\ 3x - 2y + 7z - 2w &= 0. \end{aligned}$$

Each vector C in $U \cap V$ gives rise to one or more solutions to this system. Conversely, if (x, y, z, w) is a solution to this system, then $C = (x, y) + . \times A$ is in $U \cap V$. If A and B are any matrices over R with n rows, then finding $S_R(A) \cap S_R(B)$ is equivalent to solving the homogeneous system whose matrix of coefficients is $A^t, (-B^t)$, where we use the APL notation for catenation to describe the matrix with n rows whose columns consist of the rows of A followed by the negatives of the rows of B.

***Example** 5.* Interpolation of polynomials may be viewed as solving linear systems. Suppose we are given elements $a_0, \ldots, a_m$ and $b_0, \ldots, b_m$ of R and we want to find a polynomial $f = c_0 + c_1X + \ldots + c_nX^n$ such that $f(a_i) = b_i$, $0 \leq i \leq m$. Then the vector $(c_0, \ldots, c_n)$ is a solution of the system

$$\begin{gathered} c_0 + a_0c_1 + a_0^2c_2 + \ldots + a_0^nc_n = b_0, \\ c_0 + a_1c_1 + a_1^2c_2 + \ldots + a_1^nc_n = b_1, \\ \vdots \\ c_0 + a_mc_1 + a_m^2c_2 + \ldots + a_m^nc_n = b_m. \end{gathered}$$

If $m = n$, then the matrix of this system,

$$\begin{bmatrix} 1 & a_0 & a_0^2 & \ldots & a_0^n \\ 1 & a_0 & a_1^2 & \ldots & a_1^n \\ & & \vdots & & \\ 1 & a_n & a_n^2 & \ldots & a_n^n \end{bmatrix},$$

is called a *Vandermonde matrix*. [Alexandre T. Vandermonde (1735-1796) was a French mathematician.]

The general form of the set of all solutions to a linear system is not hard to describe.

THEOREM 1. Let A be an m-by-n matrix over a commutative ring R and let B be a vector of length m over R. The set U of all solutions of the system $AX = 0$ is a submodule of R^n. If the system $AX = B$ is consistent, then the set of all solutions of $AX = B$ is a coset of U in R^n.

Proof. Let $T: R^n \longrightarrow R^m$ be the R-homomorphism taking X to

$X+, \times A^t$. Then X is in U if and only if $XT = 0$. Therefore U is the kernel of T and so is a submodule of R^n. The system $AX = B$ is consistent if and only if B is in the image of T. If $AX = B$ is consistent, then the set of solutions is $\{B\}T^{-1}$, the inverse image of $\{B\}$ under T. If $\{B\}T^{-1}$ is nonempty, then, by Theorem 3.5.3, $\{B\}T^{-1}$ is the coset of $U + X_0$, where X_0 is any element of $\{B\}T^{-1}$. □

If in Theorem 1 the ring R is a PID, then U is a free module and can be described by giving a basis $u_1, \ldots, u_r$. Once we have found one solution X_0 of $AX = B$, then any solution X of this system has the form $X = X_0 + c_1u_1 + \ldots + c_ru_r$, where the c_i are unique elements of R.

We can now add to our list of properties that characterize invertible matrices over a field.

THEOREM 2. Let F be a field and let A be in $M_n(F)$. The following are equivalent.

(a) $\det A \neq 0$.
(b) The system $AX = B$ is consistent for every B in F^n.
(c) The only solution of the system $AX = 0$ is $X = 0$.

Proof. Let $T: F^n \longrightarrow F^n$ be the linear transformation taking X to $AX = XA^t$. Condition (b) asserts that T is surjective and condition (c) states that T is injective. These two conditions are equivalent by Corollary 6.3.9. Now suppose $\det A \neq 0$. Then $X = A^{-1}B$ is a solution of $AX = B$, and (b) holds. If $\det A = 0$, then $\det A^t = 0$ and, by Theorem 5.2.12, the rows of A^t do not form a basis of F^n. But then Corollary 6.3.6 shows that the rows of A^t are linearly dependent. Thus there exists a nonzero vector X in F^n such that $XA^t = AX = 0$. □

COROLLARY 3. Let $a_0, \ldots, a_n$ be elements of a field F and let A be the Vandermonde matrix

$$\begin{bmatrix} 1 & a_0 & a_0^2 & \cdots & a_0^n \\ 1 & a_1 & a_1^2 & \cdots & a_1^n \\ & & \vdots & & \\ 1 & a_n & a_n^2 & \cdots & a_n^n \end{bmatrix}$$

If $a_0, \ldots, a_n$ are distinct, then $\det A \neq 0$.

Proof. Let $B = (b_0, b_1, \ldots, b_n)$ be in F^n. Then, by the discussion in Example 5, solving the system $AX = B$ is equivalent to finding a polynomial

f in $F[X]$ of degree at most n such that $f(a_i) = b_i$, $0 \le i \le n$. By the Lagrange Interpolation Theorem (Theorem 4.12.1), such a polynomial exists if the a_i are distinct. In this case the system $AX = B$ is consistent for all B and, by Theorem 2, $\det A \neq 0$. □

Given a linear system $AX = B$, it would be useful to have a method for obtaining another system $A'Y = B'$ with the following properties:

1. The solutions of $A'Y = B'$ are easy to describe.
2. The solutions of $AX = B$ can be constructed in a simple manner from the solutions of $A'Y = B'$.

In this section we will discuss one method for getting such a system $A'Y = B'$. This method uses row reduction and is quite satisfactory when R is a field. However, when R is not a field, a more general approach must be used. This second approach is described in Section 7.

The following theorem provides the foundation for our first method of solving linear systems.

THEOREM 4. Let A be an m-by-n matrix over R, let B be in R^m, and let P be in $GL_m(R)$. Set $A' = PA$ and $B' = PB$. Then the systems $AX = B$ and $A'X = B'$ have the same solutions.

Proof. Suppose X in R^n satisfies $AX = B$. Then $A'X = (PA)X = P(AX) = PB = B'$. Similarly, if X satisfies

$$A'X = B', \text{ then } AX = (P^{-1}A')X = P^{-1}(A'X) = P^{-1}B' = B. \quad \square$$

Let A and B be as in Theorem 4. In APL notation the matrix A, B is m-by-$(n + 1)$ and is formed from A by adding B as the $(n + 1)$th column. We call A, B the *augmented matrix* of the system $AX = B$.

COROLLARY 5. If the augmented matrices for two systems $AX = B$ and $A'X = B'$ are row equivalent, then the systems have the same solutions.

Proof. If A, B is row equivalent to A', B', then there is an invertible matrix P such that $A', B' = P(A, B)$. But $P(A, B) = (PA), PB$, so $A' = PA$ and $B' = PB$. By Theorem 4, the systems $AX = B$ and $A'X = B'$ have the same solutions. □

Let us illustrate Corollary 5 with an example. Suppose we wish to solve

$$\begin{aligned} 3x + 2y + 8z &= 9, \\ 4x + 6y + 3z &= 1, \\ 7x + 4y + 7z &= 10, \end{aligned}$$

considered as a system over $\mathbf{Z}_{11}$. If we row reduce the augmented matrix for this system,

```
        N̲←11
        ⎕←A←3 4⍴3 2 8 9 4 6 3 1 7 4 7 10
  3   2   8   9
  4   6   3   1
  7   4   7 10
        ZNROWREDUCE A
 1 0 2 3
 0 1 1 0
 0 0 0 0
```

we see that our system has the same solutions as the system

$$\begin{aligned} x \quad + 2z &= 3, \\ y + z &= 0, \\ 0 &= 0. \end{aligned}$$

The solutions are now obvious. We may choose z arbitrarily and then set $x = 3 - 2z$ and $y = -z$.

If we now consider the system

$$\begin{aligned} 8x + 3y + 8z &= 8, \\ 7x + 5y + 10z &= 1, \\ 3x + 3y + 10z &= 7, \end{aligned}$$

over $\mathbf{Z}_{11}$, then the same procedure

```
        ⎕←A←3 4⍴8 3 8 8 7 5 10 1 3 3 10 7
  8   3   8   8
  7   5 10   1
  3   3 10   7
        ZNROWREDUCE A
 1 0 4 0
 0 1 3 0
 0 0 0 1
```

shows that the system has the same solutions as

$$\begin{aligned} x \quad + 4z &= 0, \\ y + 3z &= 0, \\ 0 &= 1. \end{aligned}$$

This system is clearly inconsistent and thus so is the original system.

We can now describe how to solve any linear system over a field.

THEOREM 6. Let $AX = B$ be a linear system over a field F with m equations in n variables. Assume further that A, B is row reduced over F and that the corner entries of A, B occur in columns $j_1 < j_2 < \ldots < j_r$. Set $K = \{1, \ldots, n\} - \{j_1, \ldots, j_r\}$. The system $AX = B$ is consistent if and only if $j_r \leq n$. If this is the case, then one solution is given by $C = (c_1, \ldots, c_n)$, where $c_{j_i} = b_i$, $1 \leq i \leq r$, and $c_k = 0$ for k in K. If k is in K, then let U_k be the vector $(u_{k1}, \ldots, u_{kn})$, where $u_{kk} = 1$, $u_{k\ell} = 0$ for ℓ in $K - \{k\}$ and $u_{kj_i} = -A_{ik}$, $1 \leq i \leq r$. The set $\{U_k \mid k \in K\}$ is a basis for the set of solutions of $AX = 0$.

Proof. If $j_r = n + 1$, then the rth equation in the system $AX = B$ is $0x_1 + \ldots + 0x_n = 1$, which certainly has no solutions in F^n. However, if $j_r \leq n$, then the last $m - r$ equations are of the form $0 = 0$ and so may be ignored. For $1 \leq i \leq r$ the variable x_{j_i} occurs with a nonzero coefficient only in the ith equation, which we may write in the form

$$x_{j_i} = b_i - \sum_{k \in K} A_{ik} x_k . \tag{**}$$

Thus we may choose arbitrary values for the variables x_k with k in K and use (**) to solve for the x_{j_i}. Taking $x_k = 0$ for all k in K gives the solution C. In solving the homogeneous system $AX = 0$ we may set $b_i = 0$ in (**). Now the vector U_k is the solution corresponding to the choice of $x_k = 1$ and $x_\ell = 0$ for ℓ in $K - \{k\}$. The form of (**) implies that the only solution of $AX = 0$ with $x_k = 0$ for all k in K is the trivial solution $X = 0$. Suppose for each k in K we have an element c_k in F such that

$$\sum_{k \in K} c_k U_k = 0 .$$

For ℓ in K the ℓth component of the sum on the left is

$$\sum_{k \in K} c_k u_{k\ell} = c_\ell .$$

Thus $c_\ell = 0$ for all ℓ in K, and the U_k are linearly independent. Finally, suppose $Y = (y_1, \ldots, y_n)$ is any solution of $AX = 0$. We will show that $Y = Z$, where

$$Z = \sum_{k \in K} y_k U_k .$$

If ℓ is in K, then the ℓth component of Z is y_ℓ. Thus $Y - Z$ is a solution of $AX = 0$ whose ℓth component is 0 for all ℓ in K. By our previous remark, this means that $Y - Z = 0$ or $Y = Z$. Therefore the U_k form a basis for the set of solutions of $AX = 0$. □

Let us illustrate Theorem 6 with an example over **Q**. The augmented matrix for the system

$$\begin{aligned} x - 2y \quad + 3w &= 2, \\ z - 5w &= 1, \end{aligned}$$

is

$$\begin{bmatrix} 1 & -2 & 0 & 3 & 2 \\ 0 & 0 & 1 & -5 & 1 \end{bmatrix},$$

which is row reduced over **Q**. In the notation of Theorem 6 we have $j_1 = 1$, $j_2 = 3$, and $K = \{2, 4\}$. Thus $C = (2, 0, 1, 0)$ is a solution of the system, and the vectors $U_2 = (2, 1, 0, 0,)$ and $U_4 = (-3, 0, 5, 1)$ form a basis for the vector space of solutions to the corresponding homogeneous system

$$\begin{aligned} x - 2y \quad + 3w &= 0, \\ z - 5w &= 0. \end{aligned}$$

Let us now use the method of Theorem 6 to solve the system

$$\begin{aligned} 2x_1 + 6x_2 + x_3 \qquad\qquad + 4x_5 \qquad\quad &= 1, \\ 3x_1 + 2x_2 + 2x_3 + x_4 + 6x_5 + x_6 &= 5, \\ 4x_1 + 5x_2 \qquad\quad + 3x_4 + 5x_5 \qquad\quad &= 6, \\ x_1 + 3x_2 + 2x_3 + 3x_4 + 3x_5 + 4x_6 &= 5, \end{aligned}$$

considered as a system over $\mathbf{Z}_7$. Entering the coefficients and the vector of constant terms and row reducing the augmented matrix, we get

```
      N←7
      ⎕←A←4 6⍴2 6 1 0 4 0 3 2 2 1 6 1 4 5 0 3 5 0 1 3 2 3 3 4
2 6 1 0 4 0
3 2 2 1 6 1
4 5 0 3 5 0
1 3 2 3 3 4
      B←1 5 6 5
      ⎕←A1←ZNROWREDUCE A,B
1 3 0 6 0 4 2
0 0 1 2 0 2 0
0 0 0 0 1 1 1
0 0 0 0 0 0 0
```

The last corner entry of *A*1 is not in the last column, so the system is consistent. A solution is given by *C*, constructed as follows.

```
      C←6ρ0                                 7|A+.×C
      C[1 3 5]←3↑A1[;7]     1 5 6 5
      C
2 0 0 0 1 0
```

We can now construct a matrix *U* whose rows form a basis for the solutions of the corresponding homogeneous system.

```
      U←3 6ρ0
      U[;2 4 6]←(ι3)∘.=3
      U
 0 1 0 0 0 0
 0 0 0 1 0 0
 0 0 0 0 0 1
      U[;1 3 5]←7|-⍉A1[1 2 3;2 4 6]
      U
 4 1 0 0 0 0
 1 0 5 1 0 0
 3 0 5 0 6 1
```

We leave it as an exercise to verify that this construction is the same as that described in Theorem 6.

We can now, as promised, provide a better algorithm for solving the following problem. Let A be a matrix over the field F. Choose a set of rows of A that is a basis for $S_F(A)$.

THEOREM 7. Let A be an m-by-n matrix over the field F and let B be the row-reduced matrix that is row equivalent to A^t. Then the row $A[i;]$ is a linear combination of $A[1;], \ldots, A[i-1;]$ if and only if $B[;i]$ does not contain a corner entry. Suppose the corner entries of B occur in columns $j_1 < \ldots < j_r$. Then $A[j_1;], \ldots, A[j_r;]$ form a basis for $S_F(A)$.

Proof. In Example 3 we noted that $x_1A[1;] + \ldots + x_mA[m;] = 0$ if and only if $X = (x_1, \ldots, x_m)$ satisfies $A^tX = 0$. Now $A[i;]$ is a linear combination of $A[1;], \ldots, A[i-1;]$ if and only if we can find such an X with $x_i = 1$ and $x_k = 0$ for $k > i$. The system $A^tX = 0$ has the same solutions as $BX = 0$. If $B[;i]$ does not contain a corner entry, we may set $x_i = 1$ and $x_j = 0$ for all other j such that $B[;j]$ does not contain a corner entry and obtain a solution X of $BX = 0$. If $j_\ell > i$, then $x_{j_\ell} = -B[\ell;i] = 0$, so $x_k = 0$ for all $k > i$. Therefore $A[i;]$ is a linear combination of $A[1;], \ldots, A[i-1;]$.

Now suppose $B[;i]$ does contain a corner entry $B[k;i] = 1$. If $X =$

$(x_1, \ldots, x_m)$ satisfies $BX = 0$ and $x_k = 0$ for all $k > i$, then

$$0 = \sum_k B[j;k]\, x_k = x_i,$$

since $x_k = 0$ for $k > i$ and $B[j;k] = 0$ for $k < i$. Thus $A[i;]$ is not a linear combination of $A[1;], \ldots, A[i-1;]$. Therefore $A[j_1;], \ldots, A[j_r;]$ span $S_F(A)$, and each of these rows is not in the subspace spanned by the preceding rows. Hence $A[j_1;], \ldots, A[j_r;]$ are linearly independent and form a basis for $S_F(A)$. □

In Section 3 we determined that the first, second, and fourth rows of the matrix

```
      ⎕←A←4 4⍴2 1 3 8 7 4 1 6 9 0 0 3 4 5 2 8
2 1 3 8
7 4 1 6
9 0 0 3
4 5 2 8
```

over $\mathbf{Z}_{11}$ form a basis for $S_{\mathbf{Z}_{11}}(A)$. Using Theorem 5, we can obtain this same result as follows.

```
      N←11
      ⎕←B←ZNROWREDUCE ⍉A
1 0 3 0
0 1 2 0
0 0 0 1
0 0 0 0
```

The corner entries of B occur in columns 1, 2, and 4.

EXERCISES

1 Show that the Vandermonde determinant

$$\begin{vmatrix} 1 & x & x^2 \\ 1 & y & y^2 \\ 1 & z & z^2 \end{vmatrix}$$

is equal to $(y - x)(z - x)(z - y)$.

2 Use the method of Theorem 6 to determine which of the following systems are consistent over $\mathbf{Z}_5$. For each consistent system find one solution and a basis for the solution space of the corresponding homogeneous system.

(a) $2x_1 + 3x_2 = 4,$
$4x_1 + x_2 = 2.$

(b) $2x_1 + 3x_2 = 4,$
$4x_1 + x_2 = 3.$

(c) $2x_1 + x_2 + 2x_3 = 4,$
$3x_1 + 2x_2 + 2x_3 = 2,$
$4x_1 + 3x_2 + 4x_3 = 1.$

(d) $3x_1 + 4x_2 + x_3 + 4x_4 = 4,$
$2x_1 + x_2 + 2x_3 + 4x_4 = 0,$
$4x_1 + 2x_2 + 3x_3 + 2x_4 = 0.$

3 For the following system over **Q**, use the method of Theorem 6 to find one solution and a basis for the solution space of the corresponding homogeneous equation.

$$3x_1 + x_2 + \frac{5}{2}x_3 = 0,$$

$$2x_1 + 2x_2 + 3x_3 = -4/3,$$

$$-6x_1 + 3x_2 = -5.$$

4 Let A be a matrix that is row reduced over **R**. Assume that the vector J lists the columns containing the corner entries of A. Using one or more APL statements, describe the construction of a matrix U whose rows form a basis for the solution space of the system ∧/0=A+.×X over **R**.

5 Let $F = \mathbf{Z}_{11}$ and let g be the element of $\text{Hom}_F(F^5, F^6)$ whose matrix with respect to the standard bases is

$$\begin{bmatrix} 3 & 10 & 1 & 9 & 4 & 5 \\ 5 & 2 & 4 & 8 & 2 & 8 \\ 8 & 1 & 5 & 6 & 1 & 9 \\ 2 & 3 & 7 & 9 & 3 & 10 \\ 1 & 7 & 6 & 8 & 10 & 9 \end{bmatrix}$$

Find a basis for the kernel of g.

6 The coefficient matrix for the integral system

$$2x_1 + 2x_2 + x_3 + x_4 = 0,$$
$$3x_2 + x_3 - x_4 = 0.$$

is row reduced over **Z**. Show that any solution of this system with $x_4 = 0$ is an integral multiple of $U = (-1, -2, 6, 0)$ and any solution with $x_3 = 0$ is an integral multiple of $V = (-5, 2, 0, 6)$. Show that the submodule of $\mathbf{Z}^4$ consisting of the solutions of this system contains $< U, V >$ properly. Conclude that the method of Theorem 6 cannot be applied conveniently to systems over $\mathbf{Z}$.

7 Let A be in $M_n(R)$, where R is commutative, and assume det A is a unit in R. Prove that any system $AX = B$ has a unique solution $X = (x_1, \ldots, x_n)$, where $x_i = (\det A^{(i)})/(\det A)$ and $A^{(i)}$ is obtained from A by replacing $A[;i]$ by B. [This result is known as *Cramer's Rule*. Gabriel Cramer (1704-1752) was a Swiss mathematician.]

8 Suppose we are given the structure constants e_{ijk}, $1 \le i, j, k \le n$, for an algebra A over a commutative ring R with respect to a basis $x_1, \ldots, x_n$ of A. Let $1 = u_1x_1 + \ldots + u_nx_n$ in A, where the u_i are in R. Show that we can find the u_i by solving a linear system over R. Suppose $v_1, \ldots, v_n$ are given elements of R. Prove that we can find the inverse of $x = v_1x_1 + \ldots + v_nx_n$ in A by solving a linear system over R, assuming x has an inverse.

9 Let A and B be matrices over a commutative ring R whose rows form bases for two submodules U and V, respectively, of R^n. Show that $U \cap V$ is isomorphic to the module of solutions to the system $CX = 0$, where $C = A^t, (-B^t)$ is formed by catenating the transposes of A and $-B$. Describe the isomorphism explicitly. Use this method to find a basis for $S_F(C_1) \cap S_F(C_2)$, where $F = \mathbf{Z}_7$ and

$$C_1 = \begin{bmatrix} 2 & 1 & 4 & 0 \\ 3 & 2 & 3 & 2 \\ 1 & 2 & 1 & 2 \end{bmatrix}, \quad C_2 = \begin{bmatrix} 4 & 2 & 1 & 3 \\ 2 & 3 & 2 & 1 \\ 1 & 3 & 1 & 2 \end{bmatrix}.$$

10 Let F be a field and let V be a subspace of F^n. Show that there is a homogeneous linear system $AX = 0$ over F whose solution space is V. If $F = \mathbf{Q}$, prove that A can be chosen to have integral entries.

11 Let $F = \mathbf{Z}_{13}$. Find a linear system over F whose solution space is $S_F(A)$, where

$$A = \begin{bmatrix} 2 & 1 & 9 & 7 & 4 \\ 8 & 3 & 10 & 5 & 12 \\ 7 & 6 & 2 & 11 & 7 \end{bmatrix}.$$

12 Let F be a field and let U_1 and U_2 be subspaces of F^n. Suppose U_1 and U_2 are the solution spaces of the systems $A_1X = 0$ and $A_2X = 0$, respectively. Describe a linear system whose solution space is $U_1 \cap U_2$. Use this method to find a basis for $S_F(C_1) \cap$

$S_F(C_2)$, where F, C_1, and C_2 are as in Exercise 9.

13 Let F be a field and let $AX = B$ and $A'X = B'$ be two linear systems over F, each with m equations in n unknowns. Suppose both systems are consistent and have the same solutions. Prove that the augmented matrices A, B and A', B' are row equivalent.

5. FINITELY GENERATED MODULES

In this section we will be interested in the following questions. Suppose we are given two finitely generated R-modules M_1 and M_2. How can we decide if M_1 and M_2 are isomorphic? For example, the seven groups

$$\mathbf{Z}_{24} \qquad \mathbf{Z}_2 \oplus \mathbf{Z}_3 \oplus \mathbf{Z}_4$$
$$\mathbf{Z}_2 \oplus \mathbf{Z}_{12} \qquad \mathbf{Z}_2 \oplus \mathbf{Z}_2 \oplus \mathbf{Z}_6$$
$$\mathbf{Z}_3 \oplus \mathbf{Z}_8 \qquad \mathbf{Z}_2 \oplus \mathbf{Z}_2 \oplus \mathbf{Z}_2 \oplus \mathbf{Z}_3$$
$$\mathbf{Z}_4 \oplus \mathbf{Z}_6$$

are all **Z**-modules of order 24. What are the isomorphisms, if any, among these groups? By Corollary 3.6.2 we know that $\mathbf{Z}_m \oplus \mathbf{Z}_n \cong \mathbf{Z}_{mn}$ whenever $\gcd(m, n) = 1$. Therefore

$$\mathbf{Z}_3 \oplus \mathbf{Z}_8 \cong \mathbf{Z}_{24},$$
$$\mathbf{Z}_2 \oplus \mathbf{Z}_3 \oplus \mathbf{Z}_4 \cong \mathbf{Z}_4 \oplus \mathbf{Z}_6 \cong \mathbf{Z}_2 \oplus \mathbf{Z}_{12},$$
$$\mathbf{Z}_2 \oplus \mathbf{Z}_2 \oplus \mathbf{Z}_2 \oplus \mathbf{Z}_3 \cong \mathbf{Z}_2 \oplus \mathbf{Z}_2 \oplus \mathbf{Z}_6.$$

It is a fact that there are no additional isomorphisms among these groups.

One could also ask if any abelian group of order 24 must be isomorphic to one of the listed groups. As a corollary of the main theorem of this section, we will show that any finite abelian group is isomorphic to a direct sum of cyclic groups, and in Section 6 we will prove that any two such direct sums are isomorphic if and only if they can be shown to be isomorphic by repeated use of Corollary 3.6.2.

Before we can discuss algorithms for determining the isomorphism of two given modules, we must clarify what it means to be "given" a module. Let M be an R-module generated by elements $u_1, \ldots, u_n$. The map $f: R^n \longrightarrow M$ taking $(a_1, \ldots, a_n)$ to $a_1u_1 + \ldots + a_nu_n$ is a surjective R-homomorphism. Therefore M is isomorphic to R^n/K, where K is the kernel of f. If R is a PID, then K is finitely generated and there is a matrix A over R with n columns such that $K = S_R(A)$. We will normally consider a module M to be given if we know a matrix A such that $M \cong R^n/S_R(A)$, where

n is the number of columns of A. Such a description of M is called a *finite presentation.*

Let us consider some examples with $R = \mathbf{Z}$. Since $\mathbf{Z}_n$ is $\mathbf{Z}/n\mathbf{Z}$, a finite presentation of $\mathbf{Z}_n$ is given by the 1-by-1 matrix whose entry is n. What is a finite presentation for $G = \mathbf{Z}_4 \oplus \mathbf{Z}_6$? The elements of G are pairs

$$([a]_4, [b]_6),$$

where $[m]_n$ denotes the congruence class $m + n\mathbf{Z}$. The map

$$(a, b) \longrightarrow ([a]_4, [b]_6)$$

is a homomorphism of $\mathbf{Z}^2$ onto G. The kernel consists of all (a, b) with $a \equiv 0 \pmod 4$ and $b \equiv 0 \pmod 6$. Thus $G \cong \mathbf{Z}^2/S_{\mathbf{Z}}(A)$, where

$$A = \begin{bmatrix} 4 & 0 \\ 0 & 6 \end{bmatrix}.$$

Similarly, $\mathbf{Z}_2 \oplus \mathbf{Z}_3 \oplus \mathbf{Z}_4$ is isomorphic to $\mathbf{Z}^3/S_{\mathbf{Z}}(D)$, where

$$D = \begin{bmatrix} 2 & 0 & 0 \\ 0 & 3 & 0 \\ 0 & 0 & 4 \end{bmatrix}.$$

If we allow R to be any ring, then there is no known algorithm for deciding whether $R^m/S_R(A)$ is isomorphic to $R^n/S_R(B)$. However, when R is a Euclidean domain, such an algorithm exists. This algorithm is one of the most important algorithms in algebra.

Let A and B be two m-by-n matrices with entries in R. We say that A and B are *equivalent* over R if B can be obtained from A by a sequence of row and column operations over R. For example, if

```
      ⎕←A←B←2 2⍴5 ¯1 2 4
5  ¯1
2   4
      B[;1]←B[;1]+5×B[;2]
      B
 0   ¯1
22    4
      B[2;]←B[2;]+4×B[1;]
      B
 0   ¯1
22    0
      ⎕←B←⌽B
¯1    0
 0   22
```

then the final matrix B is obtained from A by an integer column operation, an integer row operation, and another integer column operation. Thus A and B are equivalent over **Z**. Clearly, equivalence of matrices is an equivalence relation.

THEOREM 1. Let A and B be equivalent m-by-n matrices over R. Then $B = PAQ$, where P and Q are units in $M_m(R)$ and $M_n(R)$, respectively.

Proof. Row operations can be performed by multiplying on the left with elementary matrices, and column operations can be performed by multiplying on the right with elementary matrices. Therefore

$$B = P_r \ldots P_2 P_1 A Q_1 Q_2 \ldots Q_s,$$

where the P_i and Q_j are elementary matrices. If $P = P_r \ldots P_2 P_1$ and $Q = Q_1 Q_2 \ldots Q_s$, then $B = PAQ$ and P and Q are units in $M_m(R)$ and $M_n(R)$, respectively. □

The matrix P in Theorem 1 can be obtained by applying the row operations corresponding to the elementary matrices $P_1, P_2, \ldots, P_r$ to the identity matrix. Similarly, Q can be obtained by applying the column operations corresponding to the Q_j to the identity matrix. In our preceding example we can compute P and Q as follows.

```
      P←Q←(⍳2)∘.=⍳2
      P[2;]←P[2;]+4×P[1;]
      P
1 0
4 1
      Q[;1]←Q[;1]+5×Q[;2]
```

```
      ⎕←Q←⌽Q
0 1
1 5
      P+.×A+.×Q
¯1  0
 0 22
```

The next result provides a basis for showing that two quotient modules of R^n are isomorphic.

THEOREM 2. Let N be a submodule of an R-module M and let f be an R-automorphism of M. Then Nf is a submodule of M and $M/(Nf)$ is isomorphic to M/N.

Proof. By Theorem 5.1.5, we know that Nf is a submodule of M. Let $g{:}M \longrightarrow M/N$ be the natural map and set $h = f^{-1} \circ g$. Then, since both f^{-1} and g are surjective, h maps M onto M/N. If u is in Nf, then $u = vf$ for some v in N. Thus $uh = v(f \circ f^{-1} \circ g) = vg = 0$. Therefore Nf is contained in the kernel of h. Conversely, if u is in the kernel of h, then $(uf^{-1})g = 0$ and uf^{-1} is in the kernel of g, which is N. Hence $u = (uf^{-1})f$ is in Nf and the kernel of h is Nf. By the First Isomorphism Theorem, $M/(Nf)$ is isomorphic to M/N. □

COROLLARY 3. Let A be an m-by-n matrix over R and let P and

Q be units in $M_m(R)$ and $M_n(R)$, respectively. Then $R^n/S_R(A)$ and $R^n/S_R(PAQ)$ are isomorphic.

Proof. By Theorem 5.2.6, we have $S_R(PAQ) = S_R(AQ)$. Now the map $f{:}X \longrightarrow XQ$ of R^n into itself is an endomorphism of R^n and, by the isomorphism of Corollary 5.3.7, f is invertible and so f is an automorphism of R^n. The rows of AQ are the images under f of the rows of A and hence $S_R(AQ) = S_R(A)f$. The isomorphism of $R/S_R(AQ)$ with $R/S_R(A)$ now follows from Theorem 2. □

COROLLARY 4. Let A and B be equivalent m-by-n matrices over R. Then $R^n/S_R(A)$ and $R^n/S_R(B)$ are isomorphic.

Proof. By Theorem 1, there are invertible matrices P and Q such that $B = PAQ$. By Corollary 3, the modules $R^n/S_R(A)$ and $R^n/S_R(B)$ are isomorphic. □

There are certain n-column matrices D for which the structure of $R^n/S_R(D)$ is easy to describe. The next theorem generalizes some earlier remarks.

THEOREM 5. Let D be the m-by-n matrix

$$\begin{bmatrix} d_1 & & & & & & & \\ & d_2 & & & & & 0 & \\ & & \cdot & & & & & \\ & & & \cdot & & & & \\ & & & & \cdot & & & \\ & & & & & d_r & & \\ & & & & & & 0 & \\ & & & & & & & \cdot \\ & 0 & & & & & & \cdot \\ & & & & & & & \cdot \end{bmatrix}$$

where $d_1, \ldots, d_r$ are elements of R. Then $R^n/S_R(D)$ is isomorphic to

$$M = (R/Rd_1) \oplus \ldots \oplus (R/Rd_r) \oplus R^{n-r}.$$

Proof. If $1 \le i \le r$, then set $R_i = R/Rd_i$, and if $r < n$, then set $R_{r+1} = \ldots = R_n = R$. Thus we can write $M = R_1 \oplus \ldots \oplus R_n$. The map f taking $(a_1, \ldots, a_n)$ in R^n to

$$([a_1]_{d_1}, \ldots, [a_r]_{d_r}, a_{r+1}, \ldots, a_n)$$

is a homomorphism of R^n onto M. (Here we are writing $[a_i]_{d_i}$ for $a_i + Rd_i$, $1 \leq i \leq r$.) The kernel K of f is the set of vectors $(a_1, \ldots, a_r, 0, \ldots, 0)$ such that d_i divides a_i, $1 \leq i \leq r$. Thus $K = S_R(D)$ and $M \cong R^n/S_R(D)$. □

A matrix D with entries in R is said to be *reduced* if it has the form

$$\begin{bmatrix} d_1 & & & & & & \\ & d_2 & & & & 0 & \\ & & \ddots & & & & \\ & & & d_r & & & \\ & & & & 0 & & \\ & 0 & & & & \ddots & \\ & & & & & & \end{bmatrix}$$

where the d_i are nonzero elements of R and d_i divides d_{i+1}, $1 \leq i < r$. We will show that, whenever R is a Euclidean domain, any matrix over R is equivalent to a reduced matrix. Before proving this result, let us illustrate the basic idea with an example over the integers. Let

```
      ⎕←A←3 3⍴¯7 15 ¯7 4 8 ¯2 ¯5 11 ¯5
¯7  15  ¯7
 4   8  ¯2
¯5  11  ¯5
```

The following sequence of integer row and column operations converts A into a matrix that is reduced over **Z**.

```
      A[3;]←A[3;]-3×A[2;]
      A
 ¯7  15  ¯7
  4   8  ¯2
¯17 ¯13   1
      A[1 2;]←A[1 2;]+7 2∘.×A[3;]
      A
¯126  ¯76    0
 ¯30  ¯18    0
 ¯17  ¯13    1
      A[;1 2]←A[;1 2]+A[;3]∘.×17 13
      A
¯126  ¯76    0
 ¯30  ¯18    0
   0    0    1
      A[1 3;]←A[3 1;]
      A[;1 3]←A[;3 1]
      A
   1    0    0
   0  ¯18  ¯30
   0  ¯76 ¯126
      A[3;]←A[3;]-4×A[2;]
      A
 1   0   0
 0 ¯18 ¯30
 0  ¯4  ¯6
      A[2;]←A[2;]-4×A[3;]
      A
1  0  0
0 ¯2 ¯6
0 ¯4  6
      A[3;]←A[3;]-2×A[2;]
      A
1  0  0
0 ¯2 ¯6
0  0  6
      A[;3]←A[;3]-3×A[;2]
      A
1  0  0
0 ¯2  0
0  0  6
```

THEOREM 6. Let R be a Euclidean domain. Every matrix with entries in R is equivalent over R to a reduced matrix.

Proof. Let N be the Euclidean norm on R. Thus $N(a)$ is a nonnegative integer for all $a \neq 0$ in R. Let A be an m-by-n matrix over R. We will describe a procedure for converting A into a reduced matrix using row and column operations. As with the row-reduction procedure of Theorem 1.8, this procedure is recursive, and the proof of its correctness proceeds by induction on m.

If $A = 0$, then A is reduced and we are done. Therefore we may assume $A \neq 0$.

LEMMA 7. There is a matrix B equivalent to A over R such that B has the form

$$\begin{bmatrix} x & 0 & \cdots & 0 \\ 0 & & & \\ \vdots & & C & \\ 0 & & & \end{bmatrix},$$

where $x \neq 0$ and x divides every entry in C.

Proof. Since $A \neq 0$, there is a nonzero entry in A. Among all nonzero entries in A, let x be one such that $N(x)$ is minimal. We will prove the lemma by induction on $N(x)$. Let $x = A[i;j]$.

CASE 1. Suppose x divides every entry in A. [Note that this case must happen if $N(x) = N(1)$.] Interchanging $A[1;]$ and $A[i;]$ and then interchanging $A[;1]$ and $A[;j]$, we may assume $i = j = 1$. For $i > 1$ we can write $A[i;1]$ as $e_i x$ for some e_i in R. Subtracting e_i times $A[1;]$ from $A[i;]$, we convert A into a matrix of the form

$$\begin{bmatrix} x & ? & \cdots & ? \\ 0 & & & \\ \vdots & & ? & \\ 0 & & & \end{bmatrix}$$

and x still divides every entry in A. For $j > 1$ let $A[1;j] = f_j x$ with f_j in R. Subtracting f_j times $A[;1]$ from $A[;j]$, we now have A in the form

$$\begin{bmatrix} x & 0 & \cdots & 0 \\ 0 & & & \\ \vdots & & C & \\ 0 & & & \end{bmatrix}$$

and x divides every entry in C.

CASE 2. There is some entry $A[k;\ell]$ not divisible by x. Suppose $\ell = j$ so that x and $A[k;\ell]$ are in the same column. Write $A[k;\ell] = qx + r$, where $r \neq 0$ and $N(r) < N(x)$. Subtracting q times $A[i;]$ from $A[k;]$ makes $A[k;\ell]$ now equal to r. By induction on $N(x)$, we can convert A to the required form. If $i = k$, we proceed in a similar manner. Thus we may assume that $i \neq k$ and $j \cong \ell$ and that x divides $A[i;\ell]$. Write $A[i;\ell]$ as ex, with e in R, and subtract $e - 1$ times $A[;j]$ from $A[;\ell]$. Now $A[i;\ell]$ is $ex - (e-1)x = x$, so we can replace j by ℓ and continue as in the $\ell = j$ case. □

Having proved Lemma 7, we may assume A has the form

$$\begin{bmatrix} d_1 & 0 \ldots 0 \\ 0 & \\ \cdot & C \\ \cdot & \\ \cdot & \\ 0 & \end{bmatrix},$$

where d_1 divides every entry in C. By induction on m, the matrix C is equivalent to a reduced matrix

$$C' = \begin{bmatrix} d_2 & & & & & & \\ & \cdot & & & & 0 & \\ & & \cdot & & & & \\ & & & d_r & & & \\ & & & & 0 & & \\ & 0 & & & & \cdot & \\ & & & & & & \cdot \end{bmatrix}$$

It follows (why?) that A is equivalent to

$$D = \begin{bmatrix} d_1 & & & & & & \\ & d_2 & & & & 0 & \\ & & \cdot & & & & \\ & & & d_r & & & \\ & & & & 0 & & \\ & 0 & & & & \cdot & \\ & & & & & & \cdot \end{bmatrix}$$

This matrix D is reduced if d_1 divides d_2. The entries in C' are R-linear combinations of the entries in C. Since d_1 divides all of the entries in C, it follows that d_1 divides d_2. Thus A is equivalent to a reduced matrix. □

Theorem 6 has some very important corollaries.

COROLLARY 8. Let M be a finitely generated module over the Euclidean domain R. There exist nonnegative integers r and s and nonzero nonunits $d_1, \ldots, d_r$ of R such that d_i divides d_{i+1}, $1 \leq i < r$, and

$$M \cong (R/Rd_1) \oplus \ldots \oplus (R/Rd_r) \oplus R^s.$$

Proof. Let M be generated by n elements. Then $M \cong R^n/S_R(A)$, where A is some m-by-n matrix over R. By Theorem 6, there is a matrix D equivalent to A over R such that

$$D = \begin{bmatrix} d_1 & & & & & \\ & \ddots & & & 0 & \\ & & d_r & & & \\ & & & 0 & & \\ & 0 & & & \ddots & \end{bmatrix},$$

where the d_i are nonzero and d_i divides d_{i+1}, $1 \leq i < r$. By Corollary 4 and Theorem 5,

$$M \cong R^n/S_R(A) \cong R^n/S_R(D) \cong (R/Rd_1) \oplus \ldots \oplus (R/Rd_r) \oplus R^s,$$

where $s = n - r$. If some d_i is a unit, then $Rd_i = R$ and $R/Rd_i = \{0\}$. In this case the summand may be omitted. □

The isomorphism $M \cong (R/Rd_1) \oplus \ldots \oplus (R/Rd_r) \oplus R^s$ in Corollary 8 is called a *cyclic decomposition* of M. The quotient modules R/Rd_i are said to be *cyclic*, since they are generated by one element.

COROLLARY 9. Let A be a finitely generated abelian group. Then there exist nonnegative integers r and s and integers $d_1, \ldots, d_r$ greater than 1 such that

$$A \cong \mathbf{Z}_{d_1} \oplus \ldots \oplus \mathbf{Z}_{d_r} \oplus \mathbf{Z}^s.$$

Proof. The ring $\mathbf{Z}$ is a Euclidean domain. Applying Corollary 8 to A, we see that there are integers $r, s, d_1, \ldots, d_r$ such that $|d_i| \geq 2$ for all i, d_i divides d_{i+1}, $1 \leq i < r$, and

$$A \cong (\mathbf{Z}/\mathbf{Z}d_1) \oplus \ldots \oplus (\mathbf{Z}/\mathbf{Z}d_r) \oplus \mathbf{Z}^s.$$

Since $\mathbf{Z}d = \mathbf{Z}(-d)$, we may assume the d_i are positive. □

In Corollary 9 the integers $r, s, d_1, \ldots, d_r$ are actually uniquely determined by A. We will prove this fact in the next section.

COROLLARY 10. If F is a field and M is a finitely generated module over $R = F[X]$, then

$$M \cong (R/Rf_1) \oplus \ldots \oplus (R/Rf_r) \oplus R^s$$

where f_i are monic polynomials of positive degree and f_i divides f_{i+1}, $1 \leq i < r$.

Proof. This follows from Corollary 8. We may multiply the f_i by any unit of R and thus we may assume each f_i is monic. If some $f_i = 1$, then R/Rf_i has order 1 and may be omitted. □

The workspace *CLASSLIB* contains procedures for reducing matrices over $\mathbf{Z}$ and $\mathbf{Z}_n[X]$. Here n must be a prime. As an example, we can use *ZREDUCE* on the preceding matrix *A*.

```
      A                        ZREDUCE A
 ¯7  15  ¯7               1 0 0
  4   8  ¯2               0 2 0
 ¯5  11  ¯5               0 0 6
```

By Theorem 5, $\mathbf{Z}^3/S_{\mathbf{Z}}(A)$ is isomorphic to $\mathbf{Z}_2 \oplus \mathbf{Z}_6$. The procedure *ZREDUCE* computes two invertible integer matrices R̲ and S̲ such that *ZREDUCE A* is R̲+.×A+.×S̲.

```
      R̲                         R̲+.×A+.×S̲
 ¯1   3    0              1 0 0
  4   1   ¯6              0 2 0
  9   1  ¯13              0 0 6
      S̲
 0   0   1
 0   1  ¯3
 1  ¯9   8
```

The matrix

$$B = \begin{bmatrix} 1 - X & 2 \\ 1 & 1 - X \end{bmatrix}$$

over $R = \mathbf{Z}_7[X]$ can be reduced as follows.

```
      DAZV B←2 2 2ρ1 ¯1 2 0 1 0 1 ¯1
 1 ¯1    2  0
 1  0    1 ¯1
      N←7
      DAZV ZNXREDUCE B
1 0 0   0 0 0
0 0 0   6 5 1
```

Thus $R^2/S_R(B)$ is isomorphic to R/Rf, where $f = 6 + 5X + X^2$.

We close this section with another important corollary of Theorem 6.

COROLLARY 11. Let A be an n-by-n matrix over $\mathbf{Z}$. Then $S_{\mathbf{Z}}(A)$ has finite index in $\mathbf{Z}^n$ if and only if $\det A \neq 0$. If $\det A \neq 0$, then

$$|\mathbf{Z}^n : S_{\mathbf{Z}}(A)| = |\det A|.$$

Proof. We can find a reduced matrix

$$D = \begin{bmatrix} d_1 & & & & & & \\ & \cdot & & & & 0 & \\ & & \cdot & & & & \\ & & & \cdot & & & \\ & & & & d_r & & \\ & & & & & 0 & \\ & 0 & & & & & \cdot \\ & & & & & & & \cdot \\ & & & & & & & & \cdot \end{bmatrix}$$

row equivalent to A. The d_i may be assumed positive. There exist units P and Q in $M_n(\mathbf{Z})$ such that $D = PAQ$. Since $|\det P| = |\det Q| = 1$, we have

$$|\det A| = \det D = \begin{cases} 0, & r < n, \\ d_1 \ldots d_n, & r = n. \end{cases}$$

Now $\mathbf{Z}^n/S_{\mathbf{Z}}(A) \cong \mathbf{Z}^n/S_{\mathbf{Z}}(D)$. If $r < n$ then $\mathbf{Z}^n/S_{\mathbf{Z}}(D)$ is infinite while, if $r = n$, then $\mathbf{Z}/S_{\mathbf{Z}}(D) \cong \mathbf{Z}_{d_1} \oplus \ldots \oplus \mathbf{Z}_{d_n}$ has order $d_1 \ldots d_n$. □

EXERCISES

1 Find all isomorphisms derivable from Corollary 3.6.2 among the following groups.

$\mathbf{Z}_{36}$	$\mathbf{Z}_2 \oplus \mathbf{Z}_2 \oplus \mathbf{Z}_9$
$\mathbf{Z}_2 \oplus \mathbf{Z}_{18}$	$\mathbf{Z}_2 \oplus \mathbf{Z}_3 \oplus \mathbf{Z}_6$
$\mathbf{Z}_3 \oplus \mathbf{Z}_{12}$	$\mathbf{Z}_3 \oplus \mathbf{Z}_3 \oplus \mathbf{Z}_4$
$\mathbf{Z}_4 \oplus \mathbf{Z}_9$	$\mathbf{Z}_2 \oplus \mathbf{Z}_2 \oplus \mathbf{Z}_3 \oplus \mathbf{Z}_3$
$\mathbf{Z}_6 \oplus \mathbf{Z}_6$	

2 By considering elements of largest order, show that $\mathbf{Z}_{24}$, $\mathbf{Z}_2 \oplus \mathbf{Z}_{12}$, and $\mathbf{Z}_2 \oplus \mathbf{Z}_2 \oplus \mathbf{Z}_6$ are nonisomorphic.

3 Reduce by hand each of the following matrices over the integers to a reduced matrix.

(a) $\begin{bmatrix} 10 & 6 \\ 8 & 4 \end{bmatrix}$ (b) $\begin{bmatrix} 2 & -1 & 3 \\ 4 & 3 & -5 \end{bmatrix}$

(c) $\begin{bmatrix} 3 & -2 & 4 \\ 0 & -1 & 1 \\ 2 & 3 & 2 \end{bmatrix}$

4 For each matrix A in Exercise 3 find invertible integer matrices P and Q such that PAQ is reduced.

5 Execute `ZREDUCE ?4 4⍴10` ten times. What do the results indicate about the probability that $\mathbf{Z}^n/S_{\mathbf{Z}}(A)$ is cyclic, where A is a randomly chosen square integer matrix?

6 Reduce the matrix

$$B = \begin{bmatrix} 1+X & 3 & 2+X \\ X & 4+2X & 1 \\ 4 & 1+3X & 2X \end{bmatrix}$$

over $R = \mathbf{Z}_5[X]$ to a reduced matrix using *ZNXREDUCE*. What is the order of $R^3/S_{\mathbf{R}}(B)$?

7 Reduce the matrix

$$\begin{bmatrix} 3+i & 2+4i \\ 2i & -1+3i \end{bmatrix}$$

to a reduced matrix over $\mathbf{Z}[i]$.

8 Let m and n be integers and set $d = \gcd(m, n)$ and $\ell = \operatorname{lcm}(m, n)$. Show that

$$\begin{bmatrix} m & 0 \\ 0 & n \end{bmatrix} \quad \text{and} \quad \begin{bmatrix} d & 0 \\ 0 & \ell \end{bmatrix}$$

are equivalent over $\mathbf{Z}$.

9 Let $R = \mathbf{Z}[i]$ and $z = 2 + 3i$. Find a $\mathbf{Z}$-basis for Rz. Show that $|R/Rz|$ is finite and determine its order.

10 Generalize Exercise 9 to the case in which $z = a + bi$ is any nonzero element of $\mathbf{Z}[i]$.

11 Suppose G is an abelian group generated by elements a, b, c such that

$$\begin{aligned} 2a - 3b + 4c &= 0, \\ 4a + 5b - 6c &= 0, \\ -3a + 2b + 3c &= 0. \end{aligned}$$

Show that G is finite with order dividing 128.

12 Show that every square matrix over a Euclidean domain is equivalent to its transpose.

13 Let

$$A = \begin{bmatrix} 3 & 1 & -1 & 0 \\ -2 & 0 & 1 & 1 \\ 0 & 3 & 2 & 1 \end{bmatrix}$$

and

$$B = \begin{bmatrix} 4 & -7 & -7 & -2 \\ 1 & 4 & 2 & 2 \\ -7 & 2 & 5 & 3 \end{bmatrix}.$$

Set $U = S_{\mathbf{Z}}(A)$ and $V = S_{\mathbf{Z}}(B)$.

(a) Show that the rows of A are a basis for U.

(b) Show that $V \subseteq U$ and find an integer matrix C such that $B = C+.\times A$.

(c) Show that U/V is isomorphic to $\mathbf{Z}^3/S_{\mathbf{Z}}(C)$ and determine $|U/V|$.

14 Suppose a sequence of row and column operations reduces a square matriz Z to the identity matrix I. Is it true that the same sequence of row and column operations will transform I into A^{-1}?

15 Suppose the integer matrix A is equivalent to the reduced matrix

$$D = \begin{bmatrix} d_1 & & & & & \\ & \cdot & & & 0 & \\ & & \cdot & & & \\ & & & d_r & & \\ & & & & 0 & \\ & 0 & & & & \cdot \\ & & & & & & \cdot \end{bmatrix}.$$

Prove that $|d_1|$ is the gcd of the entries of A.

16 Show that a 2-by-2 integer matrix A is determined up to equivalence by $|\det A|$ and the gcd of the entries of A.

6. UNIQUENESS OF CYCLIC DECOMPOSITIONS

In this section we will complete our discussion of finitely generated modules over a Euclidean domain by sketching a proof of the following theorem, which shows that the cyclic decompositions given by Corollary 5.8 are essentially unique.

THEOREM 1. Let R be a Euclidean domain, let r, s, t, u be nonnegative integers and let $d_1, \ldots, d_r$ and $e_1, \ldots, e_t$ be nonzero nonunits of R such that d_i divides d_{i+1}, $1 \leq i < r$, and e_i divides e_{i+1}, $1 \leq i < t$. If

$$(R/Rd_1) \oplus \ldots \oplus (R/Rd_r) \oplus R^s$$

is isomorphic as an R-module to

$$(R/Re_1) \oplus \ldots \oplus (R/Re_t) \oplus R^u,$$

then $r = t$, $s = u$ and d_i is an associate of e_i for $1 \leq i \leq r$.

In order to prove Theorem 1 we must consider various submodules of a given R-module M. We will define these submodules and state some properties of them. The details will be left as exercises.

An element x of M is called a *torsion element* if $ax = 0$ for some nonzero a of R. The set of all torsion elements of M will be denoted $T(M)$.

LEMMA 2. The set $T(M)$ is a submodule of M. If M and N are isomorphic modules, then $T(M) \cong T(N)$ and $M/T(M) \cong N/T(N)$. For any R-modules U and V we have $T(U \oplus V) = T(U) \oplus T(V)$ and

$$(U \oplus V)/T(U \oplus V) \cong [U/T(U)] \oplus [V/T(V)].$$

Moreover, $T(R) = \{0\}$ and, if $d \neq 0$, then $T(R/Rd) = R/Rd$. □

For any a in R let $aM = \{ax \mid x \in M\}$.

LEMMA 3. The set aM is a submodule of M. If b is in R, then $a(bM) = (ab)M$. If M and N are isomorphic modules, then $aM \cong aN$ and $M/aM \cong N/aN$. For any R-modules U and V we have $a(U \oplus V) = (aU) \oplus (aV)$ and

$$(U \oplus V)/a(U \oplus V) \cong (U/aU) \oplus (V/aV).$$

Moreover, $a(R/Rd) = Rf/Rd$, where $f = \gcd(a, d)$ and

$$(R/Rd)/a(R/Rd) \cong R/Rf. \quad □$$

We have already seen a special case of the following lemma in Lemma 5.4.4.

LEMMA 4. Let p be a prime in R. Then R/Rp is a field. □

The next lemma is a special case of a more general result. (See Exercise 2.)

LEMMA 5. For all a in R we may give M/aM the structure of a module over R/Ra by defining

$$(b + Ra)(x + aM) = bx + aM. \quad \square$$

LEMMA 6. Suppose p and d are in R with p prime and d nonzero. Set $M = R/Rd$. If j is a positive integer, then the dimension of $(p^{j-1}M)/(p^jM)$ over the field $F = R/Rp$ is 1 or 0 according to whether p^j does or does not divide d.

Proof. Since $p^jM = p(p^{j-1}M)$, the quotient $V = (p^{j-1}M)/(p^jM)$ is a vector space over F by Lemmas 4 and 5. By Lemma 3,

$$p^{j-1}M = (Re)/Rd,$$
$$p^jM = (Rf)/Rd,$$

where $e = \gcd(p^{j-1}, d)$ and $f = \gcd(p^j, d)$. If p^j does not divide d, then e and f are associates and $Re = Rf$. Thus $p^{j-1}M = p^jM$ and V is trivial. If p^j does divide d, we may take $e = p^{j-1}$ and $f = p^j$. In this case V is isomorphic by the Third Isomorphism Theorem to $(Re)/(Rf)$, which is nontrivial and generated by $e + Rf$. Thus $\dim_F V = 1$. □

Now we can prove Theorem 1. Suppose

$$M = (R/Rd_1) \oplus \ldots \oplus (R/Rd_r) \oplus R^s$$

is isomorphic to

$$N = (R/Re_1) \oplus \ldots \oplus (R/Re_t) \oplus R^u.$$

Then, by Lemma 2,

$$T(M) \cong (R/Rd_1) \oplus \ldots \oplus (R/Rd_r)$$

is isomorphic to

$$T(M) \cong (R/Re_1) \oplus \ldots \oplus (R/Re_t).$$

Also

$$R^s \cong M/T(M) \cong N/T(N) \cong R^u.$$

Thus $s = u$ by Corollary 5.2.9. Now let p be a prime and set $F = R/R_p$. By Lemmas 3 and 6, the dimension over F of $T(M)/pT(M)$ is at most r. If we choose p dividing d_1, which we can do because d_1 is a nonzero nonunit, then the dimension of $T(M)/pT(M)$ is r. Similarly, t is the largest dimension of $T(N)/pT(N)$ for any prime p of R. But $T(M)/pT(M)$ and $T(N)/pT(N)$ are isomorphic, and so $r = t$. Now fix a prime p of R. For any $j \geq 1$ the dimension over F of

$$p^{j-1}T(M)/p^jT(M) \cong p^{j-1}T(N)/p^j(TN)$$

is the number of d_i that are divisible by p^j and also the number of e_i that are divisible by p^j. Since d_i divides d_{i+1} and e_i divides e_{i+1}, it follows that for each i the elements d_i and e_i are divisible by the same prime powers. Thus d_i and e_i are associates. □

Theorem 1 has many important consequences.

THEOREM 7 ***(Fundamental Theorem of Finitely Generated Abelian Groups).*** Let G be a finitely generated abelian group. There exist unique nonnegative integers r and s and unique integers $d_1, \ldots, d_r$ greater than 1 with d_i dividing d_{i+1} such that G is isomorphic to

$$\mathbf{Z}_{d_1} \oplus \ldots \oplus \mathbf{Z}_{d_r} \oplus \mathbf{Z}^s.$$

Proof. Existence of r, s and $d_1, \ldots, d_r$ is given by Corollary 5.9 and uniqueness by Theorem 1. □

THEOREM 8. Suppose the m-by-n matrix A over R is equivalent to the reduced matrix

$$D = \begin{bmatrix} d_1 & & & & & 0 \\ & \cdot & & & & \\ & & \cdot & & & \\ & & & d_r & & \\ & & & & 0 & \\ & 0 & & & & \cdot \\ & & & & & \cdot \end{bmatrix}$$

Then r is unique and $d_1, \ldots, d_r$ are determined up to associates.

Proof. We have

$$R^n/S_R(A) \cong R^n/S_R(D) \cong (R/Rd_1) \oplus \ldots \oplus R^{n-r}.$$

Suppose $d_1, \ldots, d_t$ are units and $t_{t+1}, \ldots, d_r$ are nonzero nonunits. Then, by Theorem 1, $n - r$ and $r - t$ are determined and $d_{t+1}, \ldots, d_r$ are determined up to associates. Therefore $t = n - (n - r) - (r - t)$ is determined and, since all units are associates of 1, we see that $d_1, \ldots, d_t$ are determined up to associates. □

The elements $d_1, \ldots, d_r$ of Theorem 8 are called the *elementary divisors* of A and r is called the *rank* of A. The rank of the free module $S_R(A)$ is called the *row rank* of A and the rank of the module $S_R(A^t)$ the module spanned by the columns of A, is called the *column rank* of A.

THEOREM 9. For any matrix A over R the ranks of A and of A^t are the same and up to associates A and A^t have the same elementary divisors. The rank, the row rank, and the column rank of A are all equal.

Proof. Assume that A is an m-by-n matrix and let A be equivalent to the reduced matrix

$$D = \begin{bmatrix} d_1 & & & & & \\ & \cdot & & & 0 & \\ & & \cdot & & & \\ & & & \cdot & & \\ & & & d_r & & \\ & & & & 0 & \\ & 0 & & & & \cdot \\ & & & & & & \cdot \end{bmatrix}$$

Thus $D = P_k \ldots P_1 A Q_1 \ldots Q_l$, where the P_i and Q_j are elementary matrices. Then $D^t = Q_l^t \ldots Q_1^t A^t P_1^t \ldots P_k^t$. Since the transpose of an elementary matrix is elementary, A^t is equivalent to D^t, which is reduced. Thus A and A^t have the same rank and up to associates the same elementary divisors. By the proof of Corollary 5.3, there is an automorphism of R^n taking $S_R(A)$ onto $S_R(D)$. Thus $S_R(A)$ and $S_R(D)$ have the same rank. Therefore the row rank of A is r. Similarly, the column rank of A, which is the row rank of A^t, is equal to the rank of D^t, which is r. □

There is one more question related to finitely generated modules over Euclidean domains that we should answer. Let M be a module generated by $x_1, \ldots, x_n$ and let $\tau: R^n \longrightarrow M$ be the map taking $(a_1, \ldots, a_n)$ to $a_1x_1 + \ldots + a_nx_n$. If we know a matrix A such that $S_R(A)$ is the kernel of τ, we can determine the structure of M by finding the reduced matrix

$$D = \begin{bmatrix} d_1 & & & & & \\ & \cdot & & & 0 & \\ & & \cdot & & & \\ & & & \cdot & & \\ & & & d_r & & \\ & & & & 0 & \\ & 0 & & & & \cdot \\ & & & & & & \cdot \\ & & & & & & & \cdot \end{bmatrix}$$

equivalent to A. We know that M is isomorphic to

$$N = (R/Rd_1) \oplus \ldots \oplus (R/Rd_r) \oplus R^{n-r}.$$

But how can we establish an explicit isomorphism between M and N?

The matrix D has the form PAQ, where P and Q are invertible matrices over R. Let $U = Q^{-1}$. For $1 \leq i \leq n$ define

$$y_i = \sum_{j=1}^{n} U_{ij}x_j.$$

THEOREM 10. Under the assumptions just described, M is the internal direct sum

$$(Ry_1) \oplus \ldots \oplus (Ry_n).$$

If $1 \leq i \leq r$, then Ry_i is isomorphic to R/Rd_i and, if $r < i \leq n$, then Ry_i is isomorphic to R.

Proof. To simplify the notation in the arguments that follow, let us extend the definition of matrix multiplication to cover the product of an array over R and an array with entries in M. The result will be an array with entries in M. Suppose $C = (c_1, \ldots, c_n)$ is a vector in R^n and E is an m-by-n matrix over R. If $V = (v_1, \ldots, v_n)$ is in M^n, then CV will denote

$$\sum_{j=1}^{n} c_i v_i,$$

and EV will be the element of M^m whose ith component is

$$\sum_{j=1}^{n} E_{ij}v_j.$$

The usual properties of matrix multiplication are satisfied. (See Exercise 4.)

The vectors $X = (x_1, \ldots, x_n)$ and $Y = (y_1, \ldots, y_n)$ are in M^n and $Y = UX$. Multiplying on the left by $Q = U^{-1}$, we obtain

$$QY = Q(UX) = (QU)X = X.$$

Thus the x_i can be expressed as linear combinations of the y_i, so $y_1, \ldots, y_n$ generate M. Therefore

$$M = (Ry_1) + \ldots + (Ry_n).$$

To show that M is the direct sum of the submodules Ry_i, we must show that whenever we have elements z_i in Ry_i, $1 \leq i \leq n$, such that $z_1 + \ldots + z_n = 0$, then $z_i = 0$ for each i.

If $C = (c_1, \ldots, c_n)$ is in R^n, then the image of C under τ is CX. If

C is in the kernel of τ, then C is a linear combination of the rows of A and there is a vector B over R such that $C = BA$. Since the rows of A are in the kernel of τ, we have $AX = (0, \ldots, 0)$, the zero element of M^n. Now

$$DY = (PAQ)(UX) = (PAQU)X = PAX = 0.$$

If $1 \le i \le r$, then the ith component of DY is $d_i y_i$, so $d_i y_i = 0$, $i \le i \le r$. Suppose z_i is in Ry_i for $1 \le i \le n$ and $z_1 + \ldots + z_n = 0$. Then $z_i = c_i y_i$ for some c_i in R. Let $C = (c_1, \ldots, c_n)$. We have

$$CY = c_1 y_1 + \ldots + c_n y_n = z_1 + \ldots + z_n = 0.$$

But $CY = C(UX) = (CU)X$, so there is a vector B over R such that $CU = BA$. Therefore

$$C = CUQ = BAQ = (BP^{-1})(PAQ) = (BP^{-1})D$$

and C is a linear combination of the rows of D. Hence $c_i = 0$ for $r < i \le n$ and d_i divides c_i for $1 \le i \le r$. Therefore $z_i = c_i y_i = 0$ for all i and

$$M = (Ry_1) \oplus \quad \oplus (Ry_n).$$

Suppose c is in R. If $1 \le i \le r$, then $cy_i = 0$ if and only if d_i divides c, so Ry_i is isomorphic to R/Rd_i. If $r < i \le n$, then $cy_i = 0$ only if $c = 0$, so Ry_i is isomorphic to R. □

Let us illustrate Theorem 10 with an example in which $R = \mathbf{Z}$. Suppose G is an abelian group of order 64 that we know is generated by elements a, b, and c. Assume also that the following equations hold in G.

$$7a + 5b + c = 0,$$
$$a + 3b - c = 0,$$
$$3a - 7b + 9c = 0.$$

Let $\tau : \mathbf{Z}^3 \longrightarrow G$ map (i, j, k) to $ia + jb + kc$. Then the rows of

```
      ⎕←A←3 3⍴7 5 1 1 3 ¯1 3 ¯7 9
7  5  1
1  3 ¯1
3 ¯7  9
```

are in the kernel of τ. Since A has determinant $64 = |G|$,

```
      ZDET A
64
```

$S_{\mathbf{Z}}(A)$ has index 64 in $\mathbf{Z}^3$, so the kernel of τ is $S_{\mathbf{Z}}(A)$. From

```
      ⎕←D←ZREDUCE A
1  0  0
0  4  0
0  0 16
```

we see that $G \cong \mathbf{Z}_4 \oplus \mathbf{Z}_{16}$. Now *ZREDUCE* computed two matrices $\underline{R}$ and $\underline{S}$ in $GL_3(\mathbf{Z})$ such that D is $\underline{R}+.\times A+.\times \underline{S}$.

```
       R                              S
 1     0     0              0     0     1
¯2     7     1              0     1     1
 5   ¯13    ¯2              1    ¯5   ¯12
```

Let

```
      ⎕←U←ZMATINV S
 7  5  1
¯1  1  0
 1  0  0
```

According to Theorem 10, if we set

$$x = 7a + 5b + c,$$

$$y = -a + b,$$

$$z = a,$$

then G is the internal direct sum $\langle x \rangle \oplus \langle y \rangle \oplus \langle z \rangle$ and x, y, and z have orders 1, 4, and 16, respectively. This means $x = 0$, and so $G = \langle y \rangle \oplus \langle z \rangle$.

EXERCISES

1 Prove Lemmas 2, 3, and 4.

2 Suppose M is an R-module and I is an ideal of R such that $au = 0$ for all a in I and all u in M. Prove that defining $(b + I)u = bu$ makes M into a module over $S = R/I$. Derive Lemma 5 as a corollary.

3 For any matrix A over R and any positive integer k let $I_k(A)$ be the ideal of R generated by the determinants of the k-by-k submatrices of A. Show that if A and B are equivalent, then $I_k(A) = I_k(B)$. Show also that if D is a reduced matrix with nonzero diagonal entries $d_1, \ldots, d_r$, then $I_k(D)$ is the ideal generated by $d_1 \ldots d_k$ if $k \leq r$ and $I_k(D) = \{0\}$ if $k > r$. Give an alternate proof of Theorem 8.

4 Let M be an R-module. If $C = (c_1, \ldots, c_n)$ is in R^n and $V = (v_1, \ldots, v_n)$ is in M^n, define CV to be the element

$$\sum_{i=1}^{n} c_i v_i.$$

Extend this definition to give the product AW of an array A over R and an array W with entries in M, provided the dimension of A along the last axis is the same as the dimension of W along the first axis. Let A and B be arrays over R and let U and W be arrays over M. Prove the following identities, assuming the indicated products are defined.

$$(A + B)W = AW + BW,$$

$$A(U + W) = AU + AW.$$

Show that $A(BW) = (AB)W$ provided B has rank at least 2.

5 Let G be an abelian group generated by elements a, b, c, d and suppose the kernel of the map $(i, j, k, \ell) \mapsto ia + jb + kc + \ell d$ of $\mathbf{Z}^4$ onto G is $S_{\mathbf{Z}}(A)$, where

$$A = \begin{bmatrix} 7 & -2 & 5 & -5 \\ 5 & -1 & 4 & 1 \\ -2 & -5 & 5 & 4 \end{bmatrix}$$

Show that $G \cong \mathbf{Z}_3 \oplus \mathbf{Z}_{12} \oplus \mathbf{Z}$. Find elements x, y, and z of G such that x has order 3, y has order 12, and G is the internal direct sum $\langle x \rangle \oplus \langle y \rangle \oplus \langle z \rangle$.

7. SOLVING LINEAR SYSTEMS USING ROW AND COLUMN OPERATIONS

The method of solving linear systems given in Section 4 is adequate for systems over fields but not for systems over other rings such as $\mathbf{Z}$. In this section we will describe additional techniques that are sufficient for solving systems over any Euclidean domain. These techniques involve column operations as well as row operations. In addition to Theorem 4.4, we will need the following result.

THEOREM 1. Let A be an m-by-n matrix over a commutative ring R, let B be in R^m, and let Q be in $GL_n(R)$. Set $A' = AQ$. There is a 1–1 correspondence between the solutions of $AX = B$ and the solutions of $A'Y = B$. This correspondence is given by $X = QY$ and $Y = Q^{-1}X$.

Proof. Suppose $AX = B$. Then $A'(Q^{-1}X) = (AQ)(Q^{-1}X) = AX = B$.

Thus $Q^{-1}X$ is a solution of $A'Y = B$. Conversely, if $A'Y = B$, then $A(QY) = (AQ)Y = A'Y = B$, and so QY is a solution of $AX = B$. □

The next theorem is the main result of this section.

THEOREM 2. Let $AX = B$ be a linear system with m equations in n variables over a Euclidean domain R. Suppose A is equivalent over R to the reduced matrix D with elementary divisors $d_1, \ldots, d_r$. Let $D = PAQ$, where P is in $GL_m(R)$ and Q is in $GL_n(R)$. Set $B' = PB = (b'_1, \ldots, b'_m)$. The system $AX = B$ is consistent if and only if $b'_i = 0$, $r < i \leq n$, and d_i divides b'_i, $1 \leq i \leq r$. If this is the case, then a solution is given by $C = QE$, where $E = (b'_1/d_1, \ldots, b'_r/d_r, 0, \ldots, 0)$. The last $n - r$ columns of Q are a basis for the module of solutions of $AX = 0$.

Proof. The system $DY = B'$ consists of the equations $d_i u_i = b'_i$, $1 \leq i \leq r$, and $0 = b'_i$, $r < i \leq n$. Clearly, this system is consistent if and only if d_i divides b'_i, $1 \leq i \leq r$, and $b'_i = 0$, $r < i \leq n$. If this condition is satisfied, then the vector $E = (b'_1/d_1, \ldots, b'_r/d_r, 0, \ldots, 0)$ in R^n satisfies $DE = B'$. Now $D = PAQ$ and $B' = PB$, so Theorem 4.4 tells us that $DY = B'$ and $(AQ)Y = B$ have the same solutions. By Theorem 1, there is a 1–1 correspondence between the solutions of $(AQ)Y = B$ and those of $AX = B$. Therefore $AX = B$ is consistent if and only if d_i divides b'_i, $1 \leq i \leq r$, and $b'_i = 0$, $r < i \leq n$. In this case QE is a solution of $AX = B$.

The solutions of $DY = 0$ are the vectors $Y = (y_1, \ldots, y_n)$ such that $y_1 = \ldots = y_r = 0$. Thus the last $n - r$ standard basis vectors for R^n span the set of solutions of $DY = 0$. The product of Q and the ith standard basis vector of R^n is $Q[;i]$, so the last $n - r$ columns of Q are solutions of $AX = 0$. If X in R^n satisfies $AX = 0$, then $Y = Q^{-1}X$ satisfies $DY = 0$. Thus the first r components of Y are 0, and $X = QY$ is a linear combination of the last $n - r$ columns of Q. Since Q is invertible, the columns of Q are linearly independent, and the last $n - r$ columns of Q are a basis for the set of solutions of $AX = 0$. □

Let us use Theorem 2 to solve the following system over **Z**.

$$\begin{aligned} -5x + 2y + 4z + w &= 8, \\ 27x + 10y + 2z + 7w &= 6, \\ -20x - 6y \qquad - 4w &= -10. \end{aligned}$$

Entering the matrix of coefficients and the vector of constant terms

```
      ⎕←A←3 4⍴¯5 2 4 1 27 10 2 7 ¯20 ¯6 0 ¯4
 ¯5   2   4   1
 27  10   2   7
¯20  ¯6   0  ¯4
      B←8 6 ¯10
```

and reducing the coefficient matrix,

```
      ⎕←D←ZREDUCE A
1 0 0 0
0 2 0 0
0 0 6 0
      R̲
1 0 0
4 0 1
1 1 2
```

```
      ⎕←B1←R̲+.×B
8 22 ¯6
      S̲
0  0  0  1
0  1 ¯8 ¯4
0  0  1  3
1 ¯2 12  1
```

we find that, in the notation of Theorem 2, $d_1 = 1, d_2 = 2, d_3 = 6$, and B' is $(8, 22, -6)$. Since d_i divides b_i', $1 \leq i \leq 3$, we get the solution

```
      ⎕←C←S̲+.×8 11 ¯1 0
0 19 ¯1 ¯26
      A+.×C
8 6 ¯10
```

Any solution of the corresponding homogeneous system is an integer multiple of the last column of S.

Theorem 2 has an important corollary.

COROLLARY 3. Let A be an m-by-n matrix over a Euclidean domain R. If $m < n$, then the system $AX = 0$ has a nontrivial solution.

Proof. Let A be equivalent over R to the reduced matrix D with elementary divisors $d_1, \ldots, d_r$. By Theorem 2, the solutions of the system $AX = 0$ form a submodule of R^n of rank $n - r$. Since $r \leq m < n$, it follows that $n - r$ is positive, so nontrivial solutions exist. □

There are procedures in *CLASSLIB* for solving linear systems over the rings **R**, **Z**, and $\mathbf{Z}_n$. For example, if `A` is an integer matrix and `B` is an integer vector of length `1↑⍴A`, then `C←A ZLSYS B` is a solution to the system `∧/B=A+.×X` over **Z**, provided this system is consistent. In addition, `ZLSYS` constructs a global variable `W̲`, which is an integer matrix whose rows are a basis for the solutions of the homogeneous system `∧/0=A+.×X`. Using `ZLSYS` to solve the linear system over **Z** discussed previously,

```
      A
 ¯5   2  4   1
 27  10  2   7
¯20  ¯6  0  ¯4
      B
8 6 ¯10
```

```
      A ZLSYS B
0 19 ¯1 ¯26
      W̲
1 ¯4 3 1
```

we get the solution obtained earlier.

The procedure *RLSYS* solves linear systems over **R** using the methods of Section 4. It is used the same way as *ZLSYS*.

```
      A RLSYS B                                W
26  ¯85  77  0                          1  ¯4   3   1
      A+.×26  ¯85  77  0
8  6  ¯10
```

To solve linear systems over $\mathbf{Z}_n$, n a prime, the procedure *ZNLSYS* may be used.

EXERCISES

1 Let R be an integral domain and let M be the submodule of R^n consisting of all solutions of some homogeneous linear system over R. Show that if u is in R^n and for some nonzero a in R the vector au is in M, then u is in M. Exhibit a submodule of $\mathbf{Z}^2$ that is not the set of solutions of any homogeneous system in two variables over $\mathbf{Z}$.

2 Let V be a subspace of $\mathbf{Q}^n$ of dimension m. Show that $\mathbf{Z}^n \cap V$ is a $\mathbf{Z}$-module of rank m.

3 Solve each of the following linear systems over $\mathbf{Z}$.

(a)
$$\begin{aligned} 2x_1 - x_2 &= 4 \\ 5x_1 - x_2 &= 9 \end{aligned}$$

(b)
$$\begin{aligned} -x_1 + 4x_2 &= 7 \\ 3x_1 - 13x_2 &= -23 \end{aligned}$$

(c)
$$\begin{aligned} 5x_1 - 2x_2 - 29x_3 &= 43 \\ 4x_1 - x_2 - 22x_3 &= 34 \\ 2x_1 + x_2 - 8x_3 &= 18 \end{aligned}$$

(d)
$$\begin{aligned} 4x_2 + x_3 - 2x_4 &= 9 \\ -2x_1 - 8x_3 + 11x_4 &= -16 \\ x_1 + x_2 + 5x_3 - 7x_4 &= 12 \end{aligned}$$

4 Let $R = \mathbf{Z}_5[X]$. Solve the system $AY = 0$ over R, where

$$A = \begin{bmatrix} 1 + X & 4 + X^2 & 4 + 2X^2 \\ X & 3 + X + X^2 & 1 + 2X^2 \\ X^2 & 1 + X + 2X^2 & 2 + X + X^3 \end{bmatrix}$$

5 Let $V = S_{\mathbf{Q}}(A)$, where

$$A = \begin{bmatrix} 3/2 & -1/3 & 2 & 7/8 \\ 1 & 1/5 & -1/2 & 2 \\ 4/3 & 9/2 & -1 & -2/5 \end{bmatrix}.$$

Determine a **Z**-basis for $V \cap \mathbf{Z}^4$. (*Hint.* Find an integer matrix C such that V is the set of solutions of the linear system $CX = 0$ over **Q**.)

*6 Show that Corollary 3 is true when R is any commutative ring.

7 Let f be the element of $\text{End}_{\mathbf{Z}}(\mathbf{Z}^4)$ whose matrix with respect to the standard basis is

$$\begin{bmatrix} 2 & -4 & 9 & 1 \\ -5 & -3 & -16 & -9 \\ 3 & -1 & 11 & 4 \\ 7 & -1 & 25 & 10 \end{bmatrix}.$$

Find a basis for the kernel of f and decide whether $(8, 0, 22, 12)$ is in the image of f.

8 Find a **Z**-basis for $S_{\mathbf{Z}}(A) \cap S_{\mathbf{Z}}(B)$, where

$$A = \begin{bmatrix} 3 & -1 & 2 & 4 \\ 5 & 2 & -7 & 1 \\ -2 & 4 & -1 & 3 \end{bmatrix}, \quad B = \begin{bmatrix} 2 & 5 & -3 & 8 \\ 7 & 1 & 0 & -6 \\ 3 & -4 & 5 & 3 \end{bmatrix}.$$

(*Hint.* See Exercise 4.9.)

7

FIELDS

So far in this book we have encountered relatively few fields. We have worked extensively with $\mathbf{Q}$, $\mathbf{R}$, $\mathbf{C}$, and $\mathbf{Z}_p$, p a prime. In Section 4.8 we defined fields of rational functions, and in Section 5.4 we constructed a field with nine elements. However, there are many other fields, and it is important to devote some time to studying the structure of fields. One valuable result of this study will be a complete description of all finite fields. Throughout this chapter, F will be a field.

1. EXTENSION FIELDS

Let F and K be fields with F contained in K. If we think of K being fixed and F varying, we usually say that F is a subfield of K. However, in this section we will be looking at a given field F and investigating the properties of fields that contain F. To emphasize this point of view, we will call K an *extension field*, or simply an *extension*, of F. (The terms "superfield" or "overfield" might also be used to describe the relation of K to F.) Since K contains F, it follows that K is an algebra over F and, in particular, K is a vector space over F. If K is finite dimensional over F, we say K is a *finite extension* of F. The *degree* of K over F is $\dim_F K$, which we denote by $[K:F]$.

THEOREM 1. If K is a finite extension of F and L is a finite extension of K, then L is a finite extension of F and $[L:F] = [L:K]\,[K:F]$.

Proof. Let $x_1, \ldots, x_m$ be a basis of K over F and let $y_1, \ldots, y_n$ be a basis of L over K. We will show that the mn products $z_{ij} = x_i y_j$ form a basis for L over F.

First, let us prove that the z_{ij} span L. If w is an element of L, then there exist elements $u_1, \ldots, u_n$ in K such that $w = u_1 y_1 + \ldots + u_n y_n$. Since each u_j is in K, there exist elements v_{ij} in F such that $u_j = v_{1j}x_1 + \ldots + v_{mj}x_m$. Then

$$w = \sum_j u_j y_j = \sum_j \left(\sum_i v_{ij} x_i \right) y_j = \sum_{i,j} v_{ij} z_{ij}.$$

Therefore the z_{ij} span L over F.

All that remains is to show that the z_{ij} are linearly independent. Suppose

$$0 = \sum_{i,j} v_{ij} z_{ij} = \sum_j \left(\sum_i v_{ij} x_i \right) y_j,$$

where the v_{ij} are in F. Set $u_j = v_{1j}x_1 + \ldots + v_{mj}x_m$. Then each u_j is in K and $u_1 y_1 + \ldots + u_n y_n = 0$. Since the y_j are linearly independent over K, this means that each u_j is 0. But the x_i are linearly independent over F, so $0 = v_{1j}x_1 + \ldots + v_{mj}x_m$ implies that each $v_{ij} = 0$. Thus the z_{ij} are linearly independent over F. □

COROLLARY 2. Let L be a finite extension of F and let K be a field with $F \subseteq K \subseteq L$. Then K is a finite extension of F and L is a finite extension of K. Thus $[L:F] = [L:K]\ [K:F]$.

Proof. We are given that L is a finite dimensional vector space over F. Since K is a subspace of L, it follows that K is finite dimensional. If $z_1, \ldots, z_n$ span L over F, then $z_1, \ldots, z_n$ certainly span L over K. Therefore L is a finite extension of K and Theorem 1 now applies. □

The field **C** is an extension of **R**. If $i^2 = -1$, then $1, i$ is a basis for **C** over **R**, so $[\mathbf{C}:\mathbf{R}] = 2$.

Let K be an extension of F and let a be an element of K. Since K is an algebra over F, we have the evaluation map defined in Section 5.4 taking $F[X]$ to K with a polynomial f being mapped to $f(a)$. This map is an algebra homomorphism whose kernel I_a is an ideal of $F[X]$. If f is in I_a, then $f(a) = 0$, and we say *a satisfies f* or a is a *root* of f. The image of $F[X]$ under the evaluation map at a is the smallest subring of K containing both F and a and is denoted $F[a]$. The intersection of all subfields of K that contain F and a is again a subfield of K containing F and a. This subfield is denoted $F(a)$. It is the smallest subfield containing F and a. We also call $F(a)$ the subfield obtained by *adjoining a* to F. Clearly, $F[a] \subseteq F(a)$.

The element a of K is *algebraic* over F if $I_a \neq 0$. Thus a is algebraic if and only if $f(a) = 0$ for some nonzero polynomial f in $F[X]$. If a is not a root of any nonzero polynomial in $F[X]$, we say a is *transcendental* over F. The real number $\sqrt{2}$ is algebraic over **Q**, since $\sqrt{2}$ is a root of the polynomial $X^2 - 2$ in $\mathbf{Q}[X]$. It can be shown that the real numbers π and e are transcendental over **Q**. Here $\pi = 3.14159\ldots$ is the ratio of the circumference of a circle to its diameter and $e = 2.71828\ldots$ is the base of the natural logarithms.

Suppose a is algebraic over F. Then I_a is a nonzero proper ideal of F[X] and therefore consists of all multiples of a unique monic polynomial f of positive degree. We call f the *minimal polynomial* of a over F.

THEOREM 3. Let a be an element of an extension field K of F and suppose a is algebraic over F. Then the minimal polynomial f of a over F is irreducible. Moreover, $F(a) = F[a]$ and $F(a)$ is isomorphic to $F[X]/I_a$. The degree $[F(a):F]$ is equal to the degree of f.

Proof. Suppose a is algebraic over F. If the minimal polynomial f of a factors as $f = gh$ in $F[X]$, then $0 = f(a) = g(a)h(a)$. Therefore either $g(a) = 0$ or $h(a) = 0$. Let us say $g(a) = 0$. Then g is in I_a; thus f divides g and h is a unit in $F[X]$. Therefore f is irreducible in $F[X]$. By Corollary 5.4.4, $F[X]/I_a$ is a field. But $F[X]/I_a$ is isomorphic to $F[a]$ and hence $F[a]$ is a field. Since F and a are both contained in $F[a]$, this means $F(a) \subseteq F[a]$. But then $F(a) = F[a]$. Let n be the degree of f. By Theorem 5.4.1, the dimension of $F(a)$ over F is n. That is, $[F(a):F] = n$. □

The following result summarizes some important properties of algebraic elements.

THEOREM 4. Let a be an element of an extension field K of F. Then the following are equivalent.

(a) a is algebraic over F.
(b) $F(a)$ is a finite extension of F.
(c) $F[a]$ is finite dimensional over F.
(d) $F(a) = F[a]$.
(e) Either $a = 0$ or a^{-1} is in $F[a]$.

Proof. By Theorem 3, condition (a) implies condition (b). If condition (b) holds, then $F(a)$ is finite dimensional and contains $F[a]$. Therefore condition (c) holds. If condition (c) holds and $n = \dim_F F[a]$, then the $n + 1$ elements $1, a, \ldots, a^n$ cannot be linearly independent over F. This implies that condition (a) is true. Thus conditions (a), (b), and (c) are equivalent.

We complete the proof by showing that conditions (a), (d), and (e) are equivalent. Theorem 3 shows that condition (a) implies condition (d). If condition (d) holds, then $F[a]$ is a field, and so condition (e) must be true. Finally, suppose condition (e) holds. If $a = 0$, then a is a root of the polynomial X in $F[X]$. If $a \neq 0$, then a^{-1} is in $F[a]$. Thus there exist $c_0, \ldots, c_n$ in F such that

$$a^{-1} = c_0 + c_1 a + \ldots + c_n a^n .$$

Multiplying by a, we get

$$1 = c_0 a + c_1 a^2 + \ldots + c_n a^{n+1} ,$$

and so a is a root of $c_n X^{n+1} + \ldots + c_0 X - 1$. Therefore, in either case, a is algebraic. □

Let us consider an example. The polynomial $f = 1 - X + X^3$ is irreducible in $\mathbf{Q}[X]$. Let a be a root of f in $\mathbf{C}$ and set $K = \mathbf{Q}(a)$. Then f is the minimal polynomial of a over $\mathbf{Q}$ and $[K:\mathbf{Q}] = 3$. Let $b = a^2$. Since b is in K, the elements $1, b, b^2, b^3$ cannot be linearly independent over $\mathbf{Q}$. Thus there exist rational numbers x, y, z, w, not all 0, such that $x + yb + zb^2 + wb^3 = 0$. Thus b is algebraic over $\mathbf{Q}$. To find the minimal polynomial of b, we need to know all rational solutions x, y, z, w of the equation $x + yb + zb^2 + wb^3 = 0$. Now $a^3 = a - 1$, so

$$b^2 = a^4 = a(a^3) = a(a-1) = a^2 - a,$$
$$b^3 = a^6 = (a^3)^2 = (a-1)^2 = a^2 - 2a + 1.$$

Therefore

$$\begin{aligned} 0 = x + yb + zb^2 + wb^3 &= x + ya^2 + z(a^2 - a) + w(a^2 - 2a + 1) \\ &= (x+w)1 - (z + 2w)a + (y + z + w)a^2 \end{aligned}$$

Since $1, a, a^2$ are linearly independent over $\mathbf{Q}$, we must have

$$\begin{aligned} x \qquad\quad + w &= 0, \\ z + 2w &= 0, \\ y + z + w &= 0. \end{aligned}$$

Using the techniques of Section 6.4 to solve this homogeneous system, we find that (x, y, z, w) must be a rational multiple of $(-1, 1, -2, 1)$. Thus b is a root of $g = -1 + X - 2X^2 + X^3$ and b is not a root of any nonzero polynomial in $\mathbf{Q}[X]$ of degree less than 3. Therefore g is the minimal polynomial of b over $\mathbf{Q}$.

The following theorem is an important consequence of Theorem 4.

THEOREM 5. Let L be an extension of F. The set K of all elements in L that are algebraic over F is a subfield of L containing F.

Proof. First, we note that if a is in F, than a is a root of the polynomial $X - a$ in $F[X]$, so a is in K. Now let a and b be any two elements of K. Then $F_1 = F(a)$ is a finite extension of F. Also, since b is a root of a nonzero polynomial in $F[X]$ and $F[X] \subseteq F_1[X]$, the field $F_2 = F_1(b)$ is a finite extension of F_1. Therefore F_2 is a finite extension of F. Suppose c is any one of the elements $a \pm b$, ab, or (if $a \neq 0$) a^{-1}. Then c is in F_2 and $F[c] \subseteq F_2$. Therefore $F[c]$ is finite dimensional over F and c is algebraic over F by Theorem 4. Thus K is a subfield of L. □

If $a_1, \ldots, a_n$ are elements of an extension field K of F, then $F(a_1, \ldots, a_n)$ is defined recursively by the formula

$$F(a_1, \ldots, a_n) = F(a_1, \ldots, a_{n-1})(a_n).$$

In the exercises an alternate definition is discussed. In particular, it can be shown that $F(a_1, \ldots, a_n)$ does not depend on the order of the a_i. If each a_i is algebraic over F, then a_i is algebraic over $F(a_1, \ldots, a_{i-1})$ and $F(a_1, \ldots, a_i)$ is a finite extension of $F(a_1, \ldots, a_{i-1})$. By induction, $F(a_1, \ldots, a_n)$ is a finite extension of F.

We say that an extension K of F is an *algebraic extension* if every element of K is algebraic over F.

THEOREM 6. Let K be an algebraic extension of F and let L be an algebraic extension of K. Then L is an algebraic extension of F.

Proof. Let u be an element of L. Then u is a root of a monic polynomial $a_0 + a_1X + \ldots + a_{n-1}X^{n-1} + X^n$ in $K[X]$. Let $M = F(a_0, \ldots, a_{n-1})$. Each a_i is in K and thus is algebraic over F. Therefore M is a finite extension of F. Now u is algebraic over M, so $M(u)$ is a finite extension of M and hence of F. Since $F(u) \subseteq M(u)$, $F(u)$ is a finite extension of F, and u is algebraic over F. □

The set of all complex numbers that are algebraic over **Q** is an algebraic extension of **Q** that is not a finite extension.

Let f be an element of $F[X]$ with positive degree n and suppose K is an extension of F. By Theorem 4.9.8, the number of distinct roots of f in K is at most n. If a is in K, then, by Theorem 4.11.7, a is a root of f if and only if $X - a$ divides f in $K[X]$. If a is a root of f, then the *multiplicity* of a as a root of f is the largest integer m such that $(X - a)^m$ divides f. A *multiple root* of f is a root of multiplicity greater than 1. If $a_1, \ldots, a_r$ are the distinct roots of f in K and a_i has multiplicity m_i, then

$$f = g \prod_{i=1}^{r} (X - a_i)^{m_i},$$

where g is a polynomial in $K[X]$ having no roots in K. We will call $m_1 + \ldots + m_r$ the number of roots of f in K counting multiplicities. This number is at most n.

Theorem 4.9.8 puts a significiant restriction on the structure of the multiplicative group of a field.

THEOREM 7. Let A be a finite subgroup of the multiplicative group of a field. Then A is cyclic.

Proof. By Corollary 6.5.9, A is isomorphic to a direct sum

$$B = \mathbf{Z}_{d_1} \oplus \ldots \oplus \mathbf{Z}_{d_r},$$

where each $d_i > 1$ and d_i divides d_{i+1}, $1 \leq i < r$. For any x in B we have

$dx = 0$, where $d = d_r$. This means that $a^d = 1$ for all a in A. (We are using additive notation in B and multiplicative notation in A.) Thus every element of A is a root of the polynomial $X^d - 1$. By Theorem 4.9.8, the number of roots of $X^d - 1$ in a given field is at most d. Theorefore $d_r = d \geq |A| = d_1 \ldots d_r$. This is impossible if $r > 1$, so $r = 1$ and A is cyclic. □

COROLLARY 8. The multiplicative group of a finite field is cyclic.

Proof. Clearly, the multiplicative group of a finite field is finite, so Theorem 7 applies. □

EXERCISES

1 Let K be a finite extension of F and suppose $[K:F]$ is a prime. Show that there are no subfields of K that contain F except F and K.

2 Let $K = \mathbf{Q}(a)$, where a is a root of $1 - X + X^3$ in $\mathbf{C}$. Find the minimal polynomials of $a + 1$ and $a^2 + a$.

3 The polynomial $f = 2 + X + X^2$ is irreducible in $R = \mathbf{Z}_3[X]$. Thus $F = R/Rf$ is a field with nine elements. Find a nonzero element of F whose multiplicative order is 8. (The procedures with prefix $ZNXF$ discussed in Section 5.4 may be of use.)

4 Given that π is transcendental over $\mathbf{Q}$, show that $\sqrt{\pi}$ and π^2 are transcendental over $\mathbf{Q}$.

5 Let K be the field of rational functions in one variable over $\mathbf{Q}$. Show that any element of K that is algebraic over $\mathbf{Q}$ belongs to $\mathbf{Q}$.

6 Find generators for the multiplicative groups of the fields $\mathbf{Z}_p$, $p = 2, 3, 5, 7, 11$.

2. SPLITTING FIELDS

Let f be a polynomial in $F[X]$ of positive degree n and let K be an extension of F. The number of roots of f in K counting multiplicities is at most n. In this section we will show that it is possible to choose K so that the number of roots of f in K counting multiplicities is exactly n.

First, we must show that we can construct extensions in which a given irreducible polynomial has a root.

THEOREM 1. Let f be a monic irreducible polynomial in $F[X]$. There exists an extension field K of F and an element a of K such that

(a) $K = F(a)$.
(b) a is algebraic over F with minimal polynomial f.

Proof. Let I be the ideal of all multiples of f in $F[X]$. Then, by Theorem 1 and Corollary 4 of Section 5.4, the quotient $K = F[X]/I$ is a field and an F-algebra of dimension $n = \deg(f)$. Now F is not strictly a subfield of K. However, we may identify an element c of F with $c1$ in K and consider F to be contained in K. Let a be the coset of I in $F[X]$ containing X. That is, $a = X + I$. Then

$$f(a) = f(X + I) = f(X) + I = I,$$

since f is in I. But I is the zero element of K, so $f(a) = 0$ in K. Therefore the minimal polynomial of a divides f. Since f is monic and irreducible, f must the the minimal polynomial of a over F. Finally, $[F(a):F] = n = [K:F]$, so $K = F(a)$. □

Now we can handle an arbitrary polynomial in $F[X]$.

THEOREM 2. Let f be a polynomial of positive degree n. There exists an extension K of F and an extension L of K such that

(a) f has a root in K.
(b) The number of roots of f in L counting multiplicities is n.
(c) $[K:F] \leq n$.
(d) $[L:F] \leq n!$.

Proof. Let g be a monic irreducible factor of f in $F[X]$. By Theorem 1, there is an extension K of F such that g has a root a in K and $[K:F] = \deg(g) \leq n$. The element a is a root of f and, in $K[X]$, we can write $f = (X - a)h$, where $\deg(h) = n - 1$. By induction, we can find an extension L of K such that the number of roots of h in L counting multiplicities is $n - 1$ and $[L:K] \leq (n - 1)!$. Then $[L:F] \leq n!$ and the number of roots of f in L counting multiplicities is n. □

Let f be a polynomial of positive degree n in $F[X]$. A *splitting field* for f over F is an extension L of F such that f has n roots in L counting multiplicities and $L = F(a_1, \ldots, a_n)$, where $a_1, \ldots, a_n$ are the roots of f in L with a root of multiplicity m repeated m times. In $L[X]$ we can write

$$f = c \prod_{i=1}^{n} (X - a_i),$$

where c is in F, but for no proper subfield K of L containing F can f be written as a product of linear polynomials in $K[X]$. Theorem 2 guarantees the existence of a splitting field L for f over F with $[L:F] \leq n!$ It is conceivable that f could have many splitting fields with very different structures. However, it is a fact that all splitting fields of f are isomorphic as F-algebras, but we will not prove this result here.

Let us consider an example. The polynomial $f = X^3 - 2$ is irreducible in $\mathbf{Q}[X]$. If $a = \sqrt[3]{2}$, the real cube root of 2, and $\omega = (-1 + \sqrt{3}i)/2$, then the roots of f in $\mathbf{C}$ are a, ωa, and $\omega^2 a$. The field $L = \mathbf{Q}(a, \omega a, \omega^2 a)$ is a splitting field for f over $\mathbf{Q}$. What is $[L:\mathbf{Q}]$? Since f is irreducible, $[\mathbf{Q}(a):\mathbf{Q}] = 3$ and 3 divides $[L:\mathbf{Q}]$ by Theorem 1.1. Now $L = \mathbf{Q}(a, \omega)$ and ω satisfies the polynomial $1 + X + X^2 = 0$. Thus $[L:\mathbf{Q}(a)] \leq 2$, so $[L:\mathbf{Q}] = 3$ or 6. Now $\mathbf{Q}(a)$ is contained in $\mathbf{R}$, but ω is not in $\mathbf{R}$. Therefore $L \neq \mathbf{Q}(a)$ and $[L:\mathbf{Q}] = 6$.

It is important to be able to tell whether a polynomial has multiple roots in a splitting field without actually constructing a splitting field. In order to do this, it turns out to be useful to define the derivative of a polynomial with coefficients an arbitrary field. In calculus the derivative is defined as a limit. This approach is not possible in our setting. However, in calculus it is shown that if $f = a_0 + a_1 X + \ldots + a_n X^n$ is in $\mathbf{R}[X]$, then the derivative f' of f is given by the formula

$$f' = a_1 + 2a_2 X + \ldots + na_n X^{n-1}. \qquad (*)$$

This formula makes sense even when the coefficients a_i are not real numbers but lie in any field. We will take (*) as the definition of f' for $f = a_0 + a_1 X + \ldots + a_n X^n$ in $F[X]$. Clearly, f' is again in $F[X]$.

LEMMA 3. If f and g are in $F[X]$ and c is in F, then

(a) $(f+g)' = f' + g'$.
(b) $(cf)' = cf'$.
(c) $(fg)' = f'g + fg'$.

Proof. These formulas are familiar from calculus. Exercise 2 outlines an approach to proving their correctness in this more general situation. □

The next result shows why we have introduced derivatives.

THEOREM 4. Let f be a polynomial in $F[X]$ of positive degree and let K be a splitting field of f over F. Then f has a multiple root in K if and only if f and f' have a common factor of positive degree in $F[X]$.

Proof. Suppose first that f has a multiple root a in K. Then $(X - a)^2$ divides f in $K[X]$ so we can write $f = (X - a)^2 g$, where g is in $K[X]$. Thus

$$f' = 2(X - a)g + (X - a)^2 g',$$

and $X - a$ is a common factor of f and f' in $K[X]$. But could it still happen that f and f' have no common factor of positive degree in $F[X]$? No. We can compute the greatest common divisor of f and f' using the Euclidean algorithm, and the result is the same whether we consider the polynomials to be in $F[X]$ or $K[X]$. Thus f and f' cannot be relatively prime in $F[X]$ and have the common factor $X - a$ in $K[X]$. Therefore f and f' have a nontrivial common factor in $F[X]$.

Now suppose that h is a common factor of f and f' in $F[X]$ of positive degree. The roots of h are roots of f, so h has a root a in K and a is also a root of f'. Let $f = (X - a)g$, where g is in $K[X]$. Then

$$f' = g + (X - a)g'.$$

Since $X - a$ divides f', it follows that $X - a$ divides g and thus $(X - a)^2$ divides f. Therefore a is a multiple root of f. □

Let us consider an example. Suppose $f = 25 - 30X + 19X^2 - 6X^3 + X^4$ in $\mathbf{R}[X]$. Then $f' = -30 + 38X - 18X^2 + 4X^3$. Computing $\gcd(f, f')$,

```
      F←25 ¯30 19 ¯6 1
      F1← 30 38  18 4
      F RXCGD F1
5 ¯3 1
```

we find that f and f' have a nontrivial common factor, and so f has multiple roots in any splitting field.

In $\mathbf{R}[X]$ the only polynomials whose derivatives are 0 are the constant polynomials. This result from calculus does not carry over to $F[X]$ in general. For example, if $f = 1 + X^2 + X^4$ in $\mathbf{Z}_2[X]$, then $f' = 2X + 4X^3 = 0$ as $2 = 0$ in $\mathbf{Z}_2$.

THEOREM 5. Let f be a polynomial in $F[X]$ and suppose $f' = 0$. Then either f is a constant polynomial or the characteristic p of F is a prime and f has the form $g(X^p)$ for some g in $F[X]$.

Proof. Let $f = a_0 + a_1X + \ldots + a_nX^n$. Then $0 = f' = a_1 + 2a_2X + \ldots + na_nX^{n-1}$. Therefore $ia_i = 0$, $1 \leq i \leq n$. If the characteristic p of F is 0, then for $1 \leq i \leq n$ we have $i \neq 0$ in F, so $a_i = 0$. Therefore $f = a_0$ is a constant polynomial. If p is a prime, then i is still not 0 in F unless p divides i. Therefore $a_i = 0$ if p does not divide i and

$$f = a_0 + a_pX^p + a_{2p}X^{2p} + \ldots = g(X^p),$$

where $g = a_0 + a_pX + a_{2p}X^2 + \ldots$. □

COROLLARY 6. Let f be an irreducible polynomial in $F[X]$ and suppose f has a multiple root in its splitting field. Then the characteristic p of F is a prime and $f = g(X^p)$ for some g in $F[X]$.

Proof. Since f has a multiple root, f and f' have a common factor of positive degree. Since f is irreducible, the only factors of f of positive degree are associates of f. Therefore f divides f'. But $\deg(f') < \deg(f)$, and so f' must be 0. By Theorem 3, p is a prime and $f = g(X^p)$. □

EXERCISES

1 Let f have positive degree n in $F[X]$. Suppose K is any splitting field of f over F. Show that $[K:F] \leq n!$.

2 Let $D:F[X] \longrightarrow F[X]$ be the derivative map. Show that D is a linear transformation, thereby establishing parts (a) and (b) of Lemma 3. Next prove part (c) for the special case $f = X^i$, $g = X^j$. Finally, prove (c) for all f and g.

3 Show that the chain rule $[f(g)]' \doteq f'(g)g'$ holds in $F[X]$ for any field F.

4 For each of the following polynomials f let L be the subfield of $\mathbf{C}$ generated by $\mathbf{Q}$ and the roots of f. Determine $[L:\mathbf{Q}]$.

(a) $-2 + X - 2X^2 + X^3$.
(b) $2 + 2X + X^3$.
*(c) $1 - 3X + X^3$.

5 Let F be a vector listing the coefficients of a polynomial f in $\mathbf{R}[X]$. Write an APL expression for the vector of coefficients of f'. Assume ⎕IO is 0.

6 For each of the following polynomials f in $\mathbf{Q}[X]$ determine whether f has a multiple root in a splitting field. (Is it "safe" to use $RXGCD$ to compute the greatest common divisor of f and f'?)

(a) $2 - X + 3X^2 - 2X^3 + X^4 + X^5$.
(b) $75 - 5X + 23X^2 + 5X^3 + X^4 + X^5$.
(c) $6 + 23X + 24X^2 + 15X^3 + 6X^4 + X^5$.

3. FINITE FIELDS

In this section we will give a complete description of all finite fields. If F is a finite field, then the characteristic of F is a prime p and F contains a subfield isomorphic to $\mathbf{Z}_p$. We may consider F to be an extension of $\mathbf{Z}_p$. If $[F:\mathbf{Z}_p] = n$, then $|F| = p^n$. Our main result will be that for each prime power p^n there is, up to isomorphism, exactly one field with p^n elements. We will construct finite fields as splitting fields of certain polynomials in $\mathbf{Z}_p[X]$.

THEOREM 1. Let p be a prime and let K be an extension of $\mathbf{Z}_p$ of degree n. Then K is a splitting field for $X^{p^n} - X$ over $\mathbf{Z}_p$ and K has the form $\mathbf{Z}_p(a)$ for some a in K. If f is an irreducible element of $\mathbf{Z}_p[X]$ of degree n, then K is a splitting field for f over $\mathbf{Z}_p$ and f divides $X^{p^n} - X$.

Proof. Let a be a nonzero element of K. The multiplicative group of

K has order $p^n - 1$, so $a^{p^n - 1} = 1$ or $a^{p^n - 1} - 1 = 0$. Multiplying by a, we get $a^{p^n} - a = 0$, so a is a root of $h = X^{p^n} - X$. Clearly, 0 is also a root of h. Therefore every element of K is a root of h. There are $p^n = \deg(h)$ roots of h in K, so there cannot be any more roots of h in an extension field of K. Since K is certainly generated as an extension of $\mathbf{Z}_p$ by the roots of h, it follows that K is a splitting field for h. If we choose a to be a generator of the multiplicative group of K, which is cyclic by Corollary 1.8, then $K = \mathbf{Z}_p(a)$.

Now suppose f in $\mathbf{Z}_p[X]$ is irreducible of degree n. Let L be a splitting field for f over K and let b be a root of f in L. Then $M = \mathbf{Z}_p(b)$ is an extension of $\mathbf{Z}_p$ of degree n and so, by the previous argument, every element of M is a root of h. But the roots of h in L are elements of K, so b is in K. Therefore $L = K$ and $K = \mathbf{Z}_p(b)$. Thus K is a splitting field for f over $\mathbf{Z}_p$. Since f is the minimal polynomial for b and b is a root of h, we see that f divides h. □

COROLLARY 2. For all a in $\mathbf{Z}_p$ we have $a^p - a = 0$. Moreover, in $\mathbf{Z}_p[X]$,

$$X^p - X = X(X-1)(X-2)\ldots(X-p+1).$$

Proof. By Theorem 1, with $n = 1$, every element of $\mathbf{Z}_p$ is a root of $h = X^p - X$. Therefore $g = X(X-1)\ldots(X-p+1)$ divides h. But g and h have the same degree and the same leading coefficient, so $g = h$. □

THEOREM 3. Let p be a prime and n a positive integer. There exist fields with p^n elements, and any two such fields are isomorphic.

Proof. Let $h = X^{p^n} - X$ and let L be a splitting field for h over $\mathbf{Z}_p$. Since $h' = -1$, we know by Theorem 2.2 that h has no multiple roots in L, so the set K of roots of h in L has p^n elements. Suppose a and b are in K. Then

$$(ab)^{p^n} = a^{p^n} b^{p^n} = ab$$

and ab is in K. By the Binomial Theorem,

$$(a+b)^p = a^p + \binom{p}{1} a^{p-1}b + \binom{p}{2} a^{p-2}b^2 + \ldots + b^p.$$

Now

$$\binom{p}{i} = \frac{p(p-1)\ldots(p-i+1)}{i!}$$

is divisible by p unless $i = 0$ or $i = p$. Since L has characteristic p, we have

$$(a+b)^p = a^p + b^p.$$

By induction, it follows that

$$(a+b)^{p^n} = a^{p^n} + b^{p^n} = a + b,$$

and so $a + b$ is in K too. By Corollary 2, every element of $\mathbf{Z}_p$ is in K, so K is a subfield of L. But L is generated as an extension of $\mathbf{Z}_p$ by the roots of h, so $L = K$ is a field with p^n elements.

Now let M be any other field with p^n elements. We may assume M is an extension of $\mathbf{Z}_p$. By Theorem 1, there is an element b such that $M = \mathbf{Z}_p(b)$. Let f be the minimal polynomial of b in $\mathbf{Z}_p[X]$. Again by Theorem 1, there is a root c of f in K. Since f is irreducible, we must have $K = \mathbf{Z}_p(c)$. But, by Theorem 1.3, this means that

$$K \cong \mathbf{Z}_p[X]/I \cong M,$$

where I is the ideal of all multiples of f in $\mathbf{Z}_p[X]$. Thus all fields with p^n elements are isomorphic. □

COROLLARY 4. If p is a prime and n is a positive integer, then there exist irreducible polynomials of degree n in $\mathbf{Z}_p[X]$. If f is such an irreducible polynomial, then f divides $X^{p^n} - X$ and f has no multiple roots in a splitting field. Every irreducible factor of $X^{p^n} - X$ has degree at most n.

Proof. By Theorem 3, there is a field K with p^n elements and, by Theorem 1, there is an element a of K such that $K = \mathbf{Z}_p(a)$. The minimal polynomial for a over $\mathbf{Z}_p$ is an irreducible polynomial of degree n in $\mathbf{Z}_p[X]$. If f in $\mathbf{Z}_p[X]$ is irreducible of degree n, then f divides $h = X^{p^n} - X$ by Theorem 1 and, since h has no multiple roots, neither does f. Let g be an irreducible factor of $X^{p^n} - X$. Then g has a root b in K. The minimal polynomial g_0 of b over $\mathbf{Z}_p$ has degree at most n. Since g_0 divides g and g is irreducible, the degree of g is at most n. □

Since any two fields with p^n elements are isomorphic, we often speak of *the* field of order p^n and denote it by GF(p^n). Here "GF" stands for "Galois field".

It is reasonable to ask when one finite field can be contained in another.

THEOREM 5. The field GF(p^n), p a prime, contains exactly one subfield with p^m elements for each divisor m of n and contains no other subfields.

Proof. Let $L = \text{GF}(p^n)$ and let K be a subfield of L. Then K contains $\mathbf{Z}_p$ and $|K| = p^m$, where $m = [K:\mathbf{Z}_p]$. Let $r = [L:K]$. Then $n = [L:\mathbf{Z}_p] = [L:K]\,[K:\mathbf{Z}_p] = rm$, and so m divides n. Every element of K is a root of $f = X^{p^m} - X$ and f has at most p^m roots in L. Therefore L has at most one subfield of order p^m. Now let M be the splitting field of f over L and let a be a root of f in M. Then

$$a^{p^m} = a,$$
$$a^{p^{2m}} = (a^{p^m})^{p^m} = a^{p^m} = a,$$

and so, by induction, $a^{p^n} = a^{p^{rm}} = a$. Thus a is a root of $h = X^{p^n} - X$. But

L is a splitting field for h, so all roots of f lie in L. The set K of roots of f in L is a subfield of L of order p^m. □

The workspace *CLASSLIB* provides several alternatives for performing calculations in finite fields. For the fields $\mathbf{Z}_p$ we can use the procedures with the prefix *ZN*, with *N̲* set to p. For GF(p^n) with $n > 1$ we can use *ZNXIRRED* to find an irreducible polynomial f of degree n in $\mathbf{Z}_p\,[X]$ and use the procedures with the prefix *ZNXF* after initializing with

```
ZNXFINIT F,
```

where *F* is the vector of coefficients of f. If p^n is small, we could construct addition and multiplication tables for GF(p^n) and use the procedures with the prefix *FR* after initializing with *FRINIT*. We could even determine structure constants for GF(p^n) as an algebra over $\mathbf{Z}_p$ and use the procedures with the prefix *ZNA*.

Let us illustrate these alternatives with GF(8). From

```
      N̲←2
      ZNXIRRED 3
 0 1 0 0
 1 1 0 0
 1 1 1 0
 1 1 0 1
 1 0 1 1
```

we see that $f = 1 + X + X^3$ is irreducible in $\mathbf{Z}_2[X]$. Thus we may take GF(8) to be $\mathbf{Z}_2[X]/\langle f \rangle$, where $\langle f \rangle$ is the ideal of $\mathbf{Z}_2[X]$ generated by f. The elements of GF(8) may be represented by polynomials of degree less than 3. After initializing,

```
      ZNXFINIT F←1 1 0 1
```

we can use the procedures with the prefix *ZNXF*.

```
      1 0 1 ZNXFPROD 1 1 1
0 1 1
      0 1 1 ZNXFPOWER 7
1
```

The elements of GF(8) are listed in the rows of the matrix *E* constructed as follows:

```
      ⎕IO←0
      ⎕←E←⌽⍉2 2 2⊤⍳8
0 0 0
1 0 0
0 1 0
1 1 0
0 0 1
1 0 1
0 1 1
1 1 1
```

This matrix was chosen because 0 and 1 are listed first. To make an addition table for GF(8), we first must construct the rank 3 array *A* such that *A*[*I*;*J*;] is the sum of *E*[*I*;] and *E*[*J*;]. Here is one way to construct *A*.

```
      DAZV S←8 8 3⍴,E
0 0 0  1 0 0  0 1 0  1 1 0  0 0 1  1 0 1  0 1 1  1 1 1
0 0 0  1 0 0  0 1 0  1 1 0  0 0 1  1 0 1  0 1 1  1 1 1
0 0 0  1 0 0  0 1 0  1 1 0  0 0 1  1 0 1  0 1 1  1 1 1
0 0 0  1 0 0  0 1 0  1 1 0  0 0 1  1 0 1  0 1 1  1 1 1
0 0 0  1 0 0  0 1 0  1 1 0  0 0 1  1 0 1  0 1 1  1 1 1
0 0 0  1 0 0  0 1 0  1 1 0  0 0 1  1 0 1  0 1 1  1 1 1
0 0 0  1 0 0  0 1 0  1 1 0  0 0 1  1 0 1  0 1 1  1 1 1
0 0 0  1 0 0  0 1 0  1 1 0  0 0 1  1 0 1  0 1 1  1 1 1
      DAZV R←TRAV 'S
0 0 0  0 0 0  0 0 0  0 0 0  0 0 0  0 0 0  0 0 0  0 0 0
1 0 0  1 0 0  1 0 0  1 0 0  1 0 0  1 0 0  1 0 0  1 0 0
0 1 0  0 1 0  0 1 0  0 1 0  0 1 0  0 1 0  0 1 0  0 1 0
1 1 0  1 1 0  1 1 0  1 1 0  1 1 0  1 1 0  1 1 0  1 1 0
0 0 1  0 0 1  0 0 1  0 0 1  0 0 1  0 0 1  0 0 1  0 0 1
1 0 1  1 0 1  1 0 1  1 0 1  1 0 1  1 0 1  1 0 1  1 0 1
0 1 1  0 1 1  0 1 1  0 1 1  0 1 1  0 1 1  0 1 1  0 1 1
1 1 1  1 1 1  1 1 1  1 1 1  1 1 1  1 1 1  1 1 1  1 1 1
      A←R ZNXFSUM S
```

Since *R*[*I*;*J*;] is *E*[*I*;] and *S*[*I*;*J*;] is *E*[*J*;], the array *A* has the desired property. Now, to get the addition table, we construct the matrix *PLUS* such that *A*[*I*;*J*;] is *E*[*PLUS*[*I*;*J*];].

```
      PLUS←2⊥⊖1 2 0 ⍉A
      A[3;5;]
0 1 1
      E[PLUS[3;5];]
0 1 1
```

To construct the multiplication table for GF(8), we need only replace *ZNXFSUM* by *ZNXFPROD* in the preceding calculation.

```
      TIMES←2⊥⊖1 2 0⍉R ZNXFPROD S
```

Now we can use the *FR* procedures to compute in GF(8).

```
        PLUS FRINIT TIMES
        3 FRPROD 5
4
        E[3;] ZNXFPROD E[5;]
0 0 1
        E[4;]
0 0 1
```

We close this section with a determination of the number $I(n, p)$ of monic irreducible polynomials of degree n in $\mathbf{Z}_p[X]$.

THEOREM 6. For all primes p and all positive integers n

$$p^n = \sum mI(m, p),$$

where the sum is over all positive divisors m of n.

Proof. Suppose m divides n. By Theorem 5, the field GF(p^n) contains a subfield K isomorphic to GF(p^m). If f in $\mathbf{Z}_p[X]$ is monic and irreducible of degree m, then f has m distinct roots in K by Theorem 1 and Corollary 4. Thus $mI(m, p)$ is the number of elements in GF(p^n) having minimal polynomial of degree m. If a is in GF(p^n), then the degree of the minimal polynomial of a divides n, and so the sum of the numbers $mI(m, p)$ as m ranges over the divisors of n is simply the order of GF(p^n). □

Taking $n = 1, 2, 3, 4$ in Theorem 6, we obtain

$$\begin{aligned}
p &= I(1, p),\\
p^2 &= I(1, p) + 2I(2, p),\\
p^3 &= I(1, p) + 3I(3, p),\\
p^4 &= I(1, p) + 2I(2, p) + 4I(4, p).
\end{aligned}$$

Solving these equations, we find

$$\begin{aligned}
I(1, p) &= p,\\
I(2, p) &= (p^2 - p)/2,\\
I(3, p) &= (p^3 - p)/3,\\
I(4, p) &= (p^4 - p^2)/4.
\end{aligned}$$

In general, it can be shown that

$$I(n, p) = \frac{1}{n} \sum \mu\left(\frac{n}{m}\right) p^m,$$

where again the sum is over the positive divisors of n and μ is the *Möbius function*, defined as follows:

$$\mu(k) = \begin{cases} 1, & \text{if } k = 1, \\ (-1)^r, & \text{if } k \text{ is a product of } r \text{ distinct primes,} \\ 0, & \text{if } k \text{ is divisible by the square of a prime.} \end{cases}$$

[The function μ is named for the German mathematician and astronomer August Ferdinand Möbius (1790-1868).]

EXERCISES

In the following exercises p is a prime.

1 Show that $X^{p^m} - X$ divides $X^{p^n} - X$ in $\mathbf{Z}_p[X]$ if and only if m divides n.

2 How many subfields do the following fields contain?

(a) $GF(2^6)$.
(b) $GF(3^7)$.
(c) $GF(5^{24})$.

3 Show that the map $\sigma: GF(p^n) \longrightarrow GF(p^n)$ with $\sigma(a) = a^p$ is an automorphism of $GF(p^n)$. What is the order of σ?

4 Let f in $\mathbf{Z}_p[X]$ be irreducible of degree n and let a be a root of f in $GF(p^n)$. Prove that the roots of f in $GF(p^n)$ are $a, a^p, a^{p^2}, \ldots, a^{p^{n-1}}$. Show that any automorphism of $GF(p^n)$ is a power of the map σ of Exercise 3.

5 Suppose f in $\mathbf{Z}_p[X]$ is monic and irreducible of degree n and suppose one root of f in $GF(p^n)$ is a generator of the multiplicative group of $GF(p^n)$. Show that every root of f in $GF(p^n)$ is a generator of the multiplicative group. Polynomials with this property are sometimes called primitive, although this conflicts with the meaning of "primitive" given in Section 4.11. Find a formula for the number of such primitive polynomials and compute the number for $p = 3$ and $n = 1, 2, 3, 4$.

6 Verify the correctness of the various calculations presented in the example involving computations in GF(8). Continue the example by constructing an array of structure constants for GF(8) as an algebra over $\mathbf{Z}_2$.

7 Using the description of GF(8) as $\mathbf{Z}_2[X]/\langle f \rangle$, where $f = 1 + X + X^3$, describe all the roots of f in GF(8) and also the roots of $1 + X^2 + X^3$.

8 Construct explicit descriptions of GF(9), GF(16), and GF(27) along the lines of the example in the text.

4. FACTORIZATION IN $\mathbf{Z}_p[X]$

If p is a prime, then $\mathbf{Z}_p[X]$ is a UFD. If f in $\mathbf{Z}_p[X]$ has positive degree n, then any irreducible factor of f must be one of the finitely many polynomials in $\mathbf{Z}_p[X]$ of degree at most n. Thus it is possible to factor f into a product of irreducible factors. However, this brute force algorithm is much too slow to be used when p^n is at all large. In 1970 E. R. Berlekamp showed in the paper listed in the bibliography that there are much faster algorithms for factoring in $\mathbf{Z}_p[X]$. This section is devoted to a discussion of variants of Berlekamp's algorithms.

THEOREM 1. Let A be a commutative algebra over $\mathbf{Z}_p$. The map $T{:}A \longrightarrow A$ defined by $xT = x^p$ is an algebra homomorphism. In particular, T is a linear transformation.

Proof. As in the proof of Theorem 3.3, the Binomial Theorem modulo p shows that $(x + y)^p = x^p + y^p$ for all x, y in A. Thus $(x + y)T = xT + yT$. If a is in $\mathbf{Z}_p$, then $a^p = a$, and so for x in A we have $(ax)^p = a^p x^p = ax^p$ and hence $(ax)T = a(xT)$. Therefore T is a linear transformation. Since $(xy)^p = x^p y^p$, it follows that T is an algebra homomorphism. □

The first factorization algorithm rests on the following curious result.

THEOREM 2. Let A be a finite dimensional commutative algebra over $\mathbf{Z}_p$ and let $T{:}A \longrightarrow A$ be the pth power map. Then A is a field if and only if both of the following conditions hold.

(a) T is a bijection.

(b) If x in A satisfies $xT = x$, then x has the form $a1$ for some a in $\mathbf{Z}_p$.

Proof. Recall that by Corollary 6.3.9, injectivity, surjectivity, and bijectivity are equivalent for linear transformations of A into itself.

Suppose first that A is a field. If x in A satisfies $xT = x^p = 0$, then $x = 0$, since fields are integral domains. Therefore T is injective and hence bijective. If $xT = x^p = x$, then x is a root of the polynomial $h = X^p - X$. Since A is a field, h has at most p roots in A. For each a in $\mathbf{Z}_p$ we have $(a1)^p = a^p 1 = a1$, and so the p elements $a1$ are the only elements x satisfying $xT = x$.

Now suppose conditions (a) and (b) hold. To show that A is a field, we must show that each nonzero x in A is invertible. Consider the ideal Ax of A. If $Ax = A$, then 1 is in Ax and so $1 = yx$ for some y in A and x is invertible. Thus we may assume $Ax \neq A$. The map $S{:}y \longmapsto yx$ of A onto

Ax is a linear transformation. Let W be the kernel of S. Since Ax is isomorphic as a vector space to A/W, we have

$$\dim_{\mathbf{Z}_p} W + \dim_{\mathbf{Z}_p} Ax = \dim_{\mathbf{Z}_p} A.$$

Suppose y is in $W \cap Ax$. Then $y = zx$ for some z in A and $y^2 = zxy = 0$. But then $y^p = 0$, and so $y = 0$ by condition (a). Therefore $W \cap Ax = \{0\}$. Since

$$\dim_{\mathbf{Z}_p}(W + Ax) + \dim_{\mathbf{Z}_p}(W \cap Ax) =$$
$$\dim_{\mathbf{Z}_p} W + \dim_{\mathbf{Z}_p} Ax = \dim_{\mathbf{Z}_p} A,$$

it follows that $\dim_{\mathbf{Z}_p}(W + Ax) = \dim_{\mathbf{Z}_p} A$, and so $W + Ax = A$. Therefore A is the internal direct sum $W \oplus Ax$. Thus there exist u in W and v in Ax such that $1 = u + v$. Then $1 = 1^p = u^p + v^p$. Now Ax is an ideal, so v^p is in Ax. It is also true that W is an ideal, for if $yx = 0$ and z is in A, then $(zy)x = z(yx) = 0$. Therefore u^p is in W. But 1 can be written as a sum of something in W and something in Ax in only one way. Thus $u^p = u$ and $v^p = v$. If $u = 0$, then $v = 1$ is in Ax, which we are assuming is not the case. If $v = 0$, then $1 = u$ is in W and so $x = ux = 0$, again contrary to our assumption. Thus u and v are both nonzero. Since u is in W and v is not in W, it follows that u and v are linearly independent. But, by condition (b), u and v must each be of the form $a1$ for some a in $\mathbf{Z}_p$. This is again a contradiction, so A is a field. □

Conditions (a) and (b) of Theorem 2 are not difficult to check provided we can find the matrix Q for T with respect to some basis $x_1, \ldots, x_n$ of A. Condition (a) is equivalent to $\det Q \neq 0$. If Y is the coordinate vector of an element y of A, then $yT = y$ if and only if $YQ = Y$ or $Y(Q - I) = 0$, where I is the n-by-n identity matrix. Thus $yT = y$ if and only if Y is a solution of the homogeneous linear system $(Q^t - I)Y = 0$. Since the dimension of the set of solutions of this system is $n - r$, where r is the rank of $Q^t - I$, condition (b) is equivalent to $r = n - 1$.

Let us use Theorem 2 to decide whether $f = 1 + 2X + X^2 + 2X^3 + 2X^4 + X^6$ and $g = 2 + 2X + X^2 + X^3 + 2X^4 + 2X^5 + X^6$ are irreducible in $\mathbf{Z}_3[X]$. By Corollary 5.4.4, f is irreducible if and only if $A = \mathbf{Z}_3[X]/\langle f \rangle$ is a field. We will work in A using the *ZNXF* procedures.

```
⎕IO←1
N←3
ZNXFINIT F←1 2 1 2 2 0 1
```

If

```
        ⎕←I←(⍳6)∘.=⍳6
 1 0 0 0 0 0
 0 1 0 0 0 0
 0 0 1 0 0 0
 0 0 0 1 0 0
 0 0 0 0 1 0
 0 0 0 0 0 1
```

then the rows of *I* correspond to the basis $1, X, X^2, X^3, X^4, X^5$ of A. If

```
        ⎕←Q←I ZNXFPOWER 3
 1 0 0 0 0
 0 0 0 1 0
 2 1 2 1 1
 2 0 0 2 0
 1 2 1 1 2
 2 2 1 0 2
```

then each row of *Q* is the cube of the corresponding row of *I*. Since the rows of *Q* are all elements of the subspace of A spanned by $1, X, X^2, X^3, X^4$, the cube map T does not map A onto itself, so A is not a field. Thus f is reducible. How can we find a factorization of f? Since T is not bijective, there is a nonzero element y of A such that $y^3 = 0$. To find such an element, we must solve the homogeneous linear system with matrix ⍉*Q*. The rows of the matrix W̲ obtained by

```
        (⍉Q) ZNLSYS 5⍴0
 0 0 0 0 0 0
        W̲
 1 1 0 1 0 0
 0 1 1 0 1 0
 2 2 1 0 0 1
```

are a basis for the set of solutions of this system. The first row of W̲ corresponds to the polynomial $h = 1 + X + X^3$. Since f divides h^3 but f does not divide h, the greatest common divisor of h and f must be a proper nontrivial factor of f. From

```
        ⎕←U←W̲[1;] ZNXGCD F
 1 1 0 1
        F ZNXQUOT U
 1 1 0 1
```

we find the factorization $f = (1 + X + X^3)^2$. Now let us turn to g. Proceeding as before,

```
        ZNXFINIT G←2  2  1  1  2  2  1
        ⎕←Q←I ZNXFPOWER 3
 1  0  0  0  0  0
 0  0  0  1  0  0
 1  1  2  2  1  1
 2  1  1  1  0  2
 2  1  2  0  2  0
 2  1  0  1  1  0
        ZNDET Q
1
```

we find this time that T is bijective. We must now consider condition (b) of Theorem 2. As noted, we want to know the solutions of the homogeneous linear system with matrix `⍉Q-I`. From the calculation

```
        (⍉Q-I) ZNLSYS 6⍴0
0  0  0  0  0  0
        W
 1  0  0  0  0  0
 0  1  2  1  0  1
```

We see that $h = X + 2X^2 + X^3 + X^5$ is an element of A that is not of the form $a1$ for some a in $\mathbf{Z}_3$ and that satisfies

$$0 = h^3 - h = h(h-1)(h-2).$$

Thus condition (b) of Theorem 2 fails and g is not irreducible. To find a factorization of g, we note that one of the elements h, $h - 1$, or $h - 2$ must not be invertible. Therefore the greatest common divisor of g and one of h, $h - 1$, or $h - 2$ must be a proper factor of g. Computing the greatest common divisors,

```
        G ZNXGCD H←0  1  2  1  0  1
1
        G ZNXGCD H ZNXDIFF 1
1  0  2  1
        G ZNXGCD H ZNXDIFF 2
2  2  0  1
        G ZNXQUOT 1  0  2  1
2  2  0  1
```

we find that $g = (2 + 2X + X^3)(1 + 2X^2 + X^3)$.

In the second example we see one drawback of this approach. If condition (a) holds but condition (b) fails, then to find a factorization we may need to compute p greatest common divisors. If p is large, this will be time-consuming. Berlekamp suggested the use of a random or probabilistic procedure in case p is large. A random procedure is one in which random choices are made. The probability that a particular choice will yield a solution is good. However, if one is extremely unlucky, a great many choices will have to be made before the solution is obtained. One such probabilistic algorithm for factoring in $\mathbf{Z}_p[X]$ is based on the following theorem.

THEOREM 3. Let f and g be relatively prime polynomials in $\mathbf{Z}_p[X]$, p a prime. Then

$$\mathbf{Z}_p[X]/<fg> \cong (\mathbf{Z}_p[X]/<f>) \oplus (\mathbf{Z}_p[X]/<g>),$$

where $\cong$ denotes isomorphism of $\mathbf{Z}_p$-algebras.

Proof. This theorem is essentially the same as Corollary 3.6.2. We can map $\mathbf{Z}_p[X]$ into $A = (\mathbf{Z}_p[X]/<f>) \oplus (\mathbf{Z}_p[X]/<g>)$ by mapping a polynomial h to $([h]_f, [h]_g)$, where $[h]_f$ denotes the coset $h + <f>$. If h is in the kernel of this map σ, then f and g each divide h. Since $\gcd(f,g) = 1$, this means fg divides h. Thus the kernel of σ is $<fg>$. By the First Isomorphism Theorem, $\mathbf{Z}_p[X]/<fg>$ is isomorphic to the image of σ. But $\mathbf{Z}_p[X]/<fg>$ has dimension $\deg(fg)$ and A has dimension $\deg(f) + \deg(g)$. Since $\deg(fg) = \deg(f) + \deg(g)$, it follows that $\mathbf{Z}_p[X]/<fg>$ and A have the same dimension. Therefore σ maps $\mathbf{Z}_p[X]$ onto A and $\mathbf{Z}_p[X]/<fg>$ is isomorphic to A. □

COROLLARY 4. Suppose f in $\mathbf{Z}_p[X]$ is equal to $f_1f_2 \ldots f_m$, where the f_i are distinct monic irreducible polynomials of the same degree n. Then $\mathbf{Z}_p[X]/<f>$ is isomorphic to the direct sum of m copies of $\mathrm{GF}(p^n)$.

Proof. By Theorem 3, $\mathbf{Z}_p[X]/<f>$ is isomorphic as a $\mathbf{Z}_p$-algebra to the direct sum of the algebras $\mathbf{Z}_p[X]/<f_i>$, which are all isomorphic to $\mathrm{GF}(p^n)$. □

To formulate our random factorization algorithm, we must distinguish the two cases $p \geq 3$ and $p = 2$. The case $p \geq 3$ is slightly simpler, and we consider it first.

THEOREM 5. Let $f = f_1 \ldots f_m$ in $\mathbf{Z}_p[X]$, where p is an odd prime, the f_i are distinct monic irreducible polynomials of the same degree n, and $m > 1$. Let $e = (p^n - 1)/2$. Suppose a polynomial g of degree less than mn is chosen randomly in $\mathbf{Z}_p[X]$. Then the probability that $\gcd(f, g^e - 1)$ is a proper factor of f is

$$q = 1 - \frac{1}{2^m}\left[\left(1 + \frac{1}{p^n}\right)^m + \left(1 - \frac{1}{p^n}\right)^m\right],$$

which is at least 4/9.

Proof. In choosing a random element g in $\mathbf{Z}_p[X]$ of degree less than

mn we are choosing a random element of $\mathbf{Z}_p[X]/<f>$, which by Corollary 4 is isomorphic to the direct sum A of m copies of GF(p^n). Every element of GF(p^n) is a root of

$$X^{p^n} - X = X(X^{p^n - 1} - 1) = X(X^e - 1)(X^e + 1).$$

Thus for e elements a of GF(p^n) we have $a^e = 1$, for e elements we have $a^e = -1$, and for one element, $a = 0$, we have $a^e = 0$. If $x = (a_1, \ldots, a_m)$ is an element of A, then $x^e = (a_1^e, \ldots, a_m^e)$ has only 0, 1, and -1 as components. The number of x such that x^e has no component equal to 1 is $(e + 1)^m$, and so the number of x such that x^e has at least one component equal to 1 is $p^{mn} - (e + 1)^m$. Of these, the number of x such that x^e has all components 1 is e^m. Thus the number of x in A such that x^e has at least one component equal to 1 and at least one component different from 1 is $p^{mn} - (e + 1)^m - e^m$. For any such x the element $x - 1$ is a nonzero noninvertible element of A. If g in $\mathbf{Z}_p[X]$ corresponds to one of these x's, then $g^e - 1$ is neither relatively prime to f nor divisible by f, so gcd(f, $g^e - 1$) is a proper factor of f. The probability of this occurring is

$$\begin{aligned} q = \frac{p^{mn} - (e + 1)^m - e^m}{p^{mn}} &= 1 - \left(\frac{e + 1}{p^n}\right)^m - \left(\frac{e}{p^n}\right)^m \\ &= 1 - \left(\frac{p^n + 1}{2p^n}\right)^m - \left(\frac{p^n - 1}{2p^n}\right)^m \\ &= 1 - \frac{1}{2^m}\left[\left(1 + \frac{1}{p^n}\right)^m + \left(1 - \frac{1}{p^n}\right)^m\right]. \end{aligned}$$

Now $1 - 1/p^n \leq 1$ and $1 + 1/p^n \leq 4/3$ as $p^n \geq 3$. Thus q is at least

$$1 - \frac{1}{2^m}\left[\left(\frac{4}{3}\right)^m + 1\right] = 1 - \left(\frac{2}{3}\right)^m - \frac{1}{2^m}.$$

This is an increasing function of m and, for $m = 3$, its value is

$$1 - \frac{8}{27} - \frac{1}{8} = \frac{125}{216} > \frac{4}{9}.$$

If $m = 2$, then

$$\begin{aligned} q = 1 - \frac{1}{4}\left[2 + \frac{2}{p^{2n}}\right] &\geq 1 - \frac{1}{4}\left(2 + \frac{2}{9}\right) \\ &= \frac{4}{9}. \quad \square \end{aligned}$$

If in the situation of Theorem 5 we were to choose 50 elements g at random, then the probability that no proper factor of f would be found would be at most $(5/9)^{50}$, or about 1.7×10^{-13}. To compute $\gcd(f, g^e - 1)$, we first raise g to the power e modulo f using the binary power algorithm, subtract 1, and compute the greatest common divisor with f.

Let us apply Theorem 5 to an example. Let $p = 2281$ and $f = 1815 + 1152X + 1804X^2 + X^3$ in $\mathbf{Z}_p[X]$. It is a fact that f is the product of three distinct monic polynomials of degree 1. To find them, we initialize to work in $\mathbf{Z}_p[X]/<f>$.

```
      ⎕IO←0
      N←2281
      ZNXFINIT F←1815 1152 1804 1
      E←(N−1)÷2
```

We choose a random polynomial g of degree at most 2

```
      ⎕←G←?3ρN
765 551 1313
```

and compute g^e modulo f.

```
      ⎕←H←G ZNXFPOWER E
1755 1344 251
```

Next, we compute $\gcd(f, g^e - 1)$.

```
      F ZNXGCD H ZNXDIFF 1
734 487 1
      F ZNXQUOT F1←734 487 1
1317 1
```

Here we have obtained the partial factorization $f = (X + 1317)f_1$, where $f_1 = 734 + 487X + X^2$. To complete the factorization, we repeat the process with f_1.

```
      ZNXFINIT F1
      G←?2ρN
      H←G ZNXFPOWER E
      F1 ZNXGCD H ZNXDIFF 1
1
```

Here we were unlucky and did not find a proper factor. Trying once more,

```
      G←?2ρN
      H←G ZNXFPOWER E
      F1 ZNXGCD H ZNXDIFF 1
847 1
      F1 ZNXQUOT 847 1
1921 1
```

we obtain the factorization $f = (847 + X)(1317 + X)(1921 + X)$.

In order to use Theorem 5 with a polynomial f about whose factorization we have no information, we must consider possible factors one degree at a time, starting with degree 1. Suppose we know that f has no nonconstant factors of degree less than m. Then any factor of degree m is irreducible and so is also a factor of $X^{p^m} - X$. Thus $g = \gcd(f, X^{p^m} - X)$ is the product of the distinct monic irreducible factors of f of degree m. We can factor g using Theorem 5 and remove all irreducible factors of degree m from f and then proceed to degree $m + 1$.

Theorem 5 is based on the factorization $X^{p^n} - X = X(X^e - 1)(X^e + 1)$, where $e = (p^n - 1)/2$. This factorization is valid only for odd primes p. If $p = 2$, then there is another factorization of $X^{p^n} - X$ that we can use to get a result similar to Theorem 5.

THEOREM 6. Let h be the polynomial

$$X + X^2 + X^4 + \ldots + X^{2^{n-1}}$$

in $\mathbf{Z}_2[X]$. Then $X^{2^n} - X = h(h + 1)$.

Proof. We have

$$h(h+1) = h + h^2 = X + X^2 + \ldots + X^{2^{n-1}} + X^2 + X^4 + \ldots + X^{2^n}$$
$$= X + X^{2^n} = X^{2^n} - X,$$

since squaring is a linear transformation and $1 = -1$ in $\mathbf{Z}_2$. □

Using Theorem 6, we can prove the following.

THEOREM 7. Let $f = f_1 \ldots f_m$ in $\mathbf{Z}_2[X]$, where the f_i are distinct irreducible polynomials of the same degree n and $m > 1$. Let $h = X + X^2 + \ldots + X^{2^{n-1}}$. Suppose a polynomial g of degree less than mn is chosen randomly in $\mathbf{Z}_2[X]$. Then the probability that $\gcd(f, h(g))$ is a proper factor of f is

$$q = 1 - \frac{1}{2^{m-1}}.$$

Proof. See Exercise 4. □

The procedure *ZNXFACTOR* factors a polynomial f by removing factors of degree m for $m = 1, 2, \ldots$. The product g of the distinct monic irreducible factors of f of degree m is computed. If $\deg(g) = rm$ with $r \geq 2$, then g is factored using Theorem 5 or Theorem 7 if p^m is large and by dividing g by all monic polynomials of degree m if p^m is small.

```
      N←2281
      F
1815 1152 1804 1
      ZNXFACTOR F
  847      1
 1317      1
 1921      1
```

Here we have recomputed the factorization of $1815 + 1152X + 1804X^2 + X^3$ in $\mathbf{Z}_p[X]$, $p = 2281$.

EXERCISES

1 For each of the following primes p and polynomials f in $\mathbf{Z}_p[X]$, factor f into irreducible factors using the algorithm based on Theorem 2.

(a) $p = 2, f = 1 + X^4 + X^5$.
(b) $p = 3, f = 1 + X + 2X^2 + 2X^5 + X^6$.
(c) $p = 2, f = 1 + X + X^2 + X^4 + X^8$.
(d) $p = 5, f = 1 - X - X^2 + 2X^4 - X^5 + X^6$.

2 The polynomial $f = 2 + 11X + 8X^2 + 4X^4 + 22X^5 + X^6$ in $\mathbf{Z}_{31}[X]$ is a product of two distinct monic irreducible polynomials of degree 3. Factor f using the method of Theorem 5.

3 The polynomial $g = 1 + X + X^5 + X^6 + X^7 + X^8 + X^9 + X^{10} + X^{12}$ in $\mathbf{Z}_2[X]$ is a product of two distinct irreducible polynomials of degree 6. Factor g using the method of Theorem 7.

4 Prove Theorem 7.

5 State and prove versions of Theorems 5 and 7 that describe how to find a nonzero noninvertible element of a commutative algebra over $\mathbf{Z}_p$ that is known to be isomorphic to a direct sum of m copies of $\mathrm{GF}(p^n)$, where $m \geq 2$.

8

LINEAR TRANSFORMATIONS

This chapter is devoted to the study of elements of the ring $\text{End}_F(V)$, where F is a field and V is a finite dimensional vector space over F. By Corollary 5.3.7, $\text{End}_F(V)$ is isomorphic to $M_n(F)$, the ring of n-by-n matrices over F, where $n = \dim_F(V)$. We know that the group $\text{GL}_n(F)$ of units of $M_n(F)$ is the set of n-by-n matrices with determinant different from 0. One of our goals will be to describe the equivalence classes of elements of $M_n(F)$ under an equivalence relation that extends the notion of conjugacy in $\text{GL}_n(F)$ to all of $M_n(F)$.

1. SIMILARITY

Throughout this section V will be a vector space of dimension n over a field F. Let A be the algebra $\text{End}_F(V)$. If v is in V and T is in A, then vT is, as usual, the image of v under T. This map from $V \times A$ to V makes V a right A-module. Fix an element T of A. The evaluation map taking f in $F[X]$ to $f(T)$ is an F-algebra homomorphism of $F[X]$ into A. This homomorphism allows us to consider V to be an $F[X]$-module. If v is in V and f is in $F[X]$, then vf will mean $vf(T)$. To emphasize the point that different choices of T lead to different $F[X]$-module structures on V, we will write V_T for V considered as an $F[X]$-module. We would like to be able to say when two linear transformations S and T in A yield isomorphic $F[X]$-modules V_S and V_T. We will use the complete description of finitely generated modules over the Euclidean domain $F[X]$ given in Sections 5 and 6 of Chapter to solve this problem.

In Section 3.10 we showed that conjugation by an element of a group G is an automorphism of G. The next theorem is a generalization of this result.

THEOREM 1. Let u be a unit of a ring A. The map $\sigma: A \longrightarrow A$ such that $x\sigma = u^{-1}xu$ is an automorphism of A. If A is an algebra over a commutative ring R, then σ is an R-algebra automorphism of A and, for all x in A and all f in $R[X]$, we have $f(x)\sigma = f(x\sigma)$.

Proof. For all x and y in A we have

$$(xy)\sigma = u^{-1}xyu = u^{-1}xuu^{-1}yu = (x\sigma)(y\sigma)$$

and

$$(x+y)\sigma = u^{-1}(x+y)u = u^{-1}xu + u^{-1}yu = (x\sigma) + (y\sigma).$$

Since $1\sigma = u^{-1}u = 1$, we see that σ is a ring homomorphism. If $x\sigma = 0$, then $0 = u^{-1}xu$, and so $x = u0u^{-1} = 0$. Therefore σ is injective. Also, since $(uxu^{-1})\sigma = u^{-1}uxu^{-1}u = x$, we see that σ is surjective and therefore σ is a ring automorphism. If A is an R-algebra for some commutative ring R, then for all a in R we have $(ax)\sigma = u^{-1}(ax)u = a(u^{-1}xu) = a(x\sigma)$ and hence σ is an automorphism of A as an R-algebra. Suppose $f = a_0 + a_1X + \ldots + a_nX^n$ is in $R[X]$. Then

$$\begin{aligned} f(x\sigma) &= a_0 + a_1(x\sigma) + \ldots + a_n(x\sigma)^n \\ &= (a_0 + a_1x + \ldots + a_nx^n)\sigma \\ &= f(x)\sigma. \quad \square \end{aligned}$$

Theorem 1 gives us our first answer to the question of when V_S and V_T are isomorphic $F[X]$-modules.

THEOREM 2. Let S and T be in $\text{End}_F(V)$. Then V_S and V_T are isomorphic as $F[X]$-modules if and only if there is a unit P of $\text{End}_F(V)$ such that $T = P^{-1}SP$.

Proof. Suppose first that V_S and V_T are isomorphic and let $P: V_S \longrightarrow V_T$ be an $F[X]$-isomorphism. Then P is certainly an F-isomorphism, that is, an isomorphism of vector spaces, and so P is a unit in $\text{End}_F(V)$. For all f in $F[X]$ and all v in V_S we have $vf(S)P = vPf(T)$. Take $f = X$. Then $vSP = vPT$ for all v in V. Therefore $SP = PT$ or $T = P^{-1}SP$.

Conversely, suppose $T = P^{-1}SP$ for some unit P of $\text{End}_F(V)$. Suppose f is in $F[X]$. It follows from Theorem 1 that $f(T) = f(P^{-1}SP) = P^{-1}f(S)P$ or $Pf(T) = f(S)P$. Then, for all v in V_S, we have $vf(S)P = vPf(T)$, so P is an $F[X]$-isomorphism of V_S onto V_T. $\square$

If S and T are in $\text{End}_F(V)$, we say S and T are *similar* if there is a unit P of $\text{End}_F(V)$ such that $T = P^{-1}SP$. Theorem 2 says that V_S and V_T are isomorphic if and only if S and T are similar. We define similarity of matrices in an analogous manner. Matrices A and B in $M_n(F)$ are similar if there is an invertible matrix Q such that $B = Q^{-1}AQ$. It is easily checked (see Exercise 1) that similarity is an equivalence relation for both linear transformations and matrices. The following theorems show the relation between these two notions of similarity.

THEOREM 3. Let T be in $\text{End}_F(V)$ and let $v_1, \ldots, v_n$ and $w_1, \ldots,$

w_n be bases of V. If A and B are the matrices for T with respect to $v_1, \ldots,$ v_n and $w_1, \ldots, w_n$, respectively, then A and B are similar.

Proof. Let Q be the matrix such that

$$w_i = \sum_j Q_{ij} v_j.$$

Then Q is in $\mathrm{GL}_n(F)$ and, by Corollary 5.3.10, $B = QAQ^{-1} = (Q^{-1})^{-1}AQ^{-1}$. Thus A and B are similar. □

THEOREM 4. Let S and T be elements of $\mathrm{End}_F(V)$. The following are equivalent.

(a) S and T are similar.

(b) For every basis $v_1, \ldots, v_n$ of V the matrix of S with respect to $v_1, \ldots, v_n$ is similar to the matrix for T with respect to the same basis.

(c) There exist bases $v_1, \ldots, v_n$ and $w_1, \ldots, w_n$ of V such that the matrix of S with respect to $v_1, \ldots, v_n$ is the same as the matrix of T with respect to $w_1, \ldots, w_n$.

Proof. (a) ⇒ (b). Suppose $T = P^{-1}SP$ for some unit P in $\mathrm{End}_F(V)$. Let A, B, and Q be the matrices of S, T, and P, respectively, with respect to the basis $v_1, \ldots, v_n$. Then $B = Q^{-1}AQ$, and A and B are similar.

(b) ⇒ (c). Let $w_1, \ldots, w_n$ be any basis of V and let A and B be the matrices of S and T with respect to $w_1, \ldots, w_n$. By (b), there is a matrix Q such that $B = QAQ^{-1}$. By Corollary 5.3.10, QAQ^{-1} is the matrix of S with respect to the basis $v_1, \ldots, v_n$, where

$$v_i = \sum Q_{ij} w_j.$$

(c) ⇒ (a). Suppose $v_1, \ldots, v_n$ and $w_1, \ldots, w_n$ are bases of V such that the matrix A of S with respect to $v_1, \ldots, v_n$ is the same as the matrix of T with respect to $w_1, \ldots, w_n$. Thus

$$v_i S = \sum_j A_{ij} v_j,$$

$$w_i T = \sum_j A_{ij} w_j.$$

There is a unique element P of $\mathrm{End}_F(V)$ such that $v_i P = w_i$. Since P is clearly surjective, P is a unit of $\mathrm{End}_F(V)$ and $w_i P^{-1} = v_i$. Thus

$$w_i P^{-1} SP = v_i SP = \left(\sum_j A_{ij} v_j\right) P$$
$$= \sum_j A_{ij} v_j P$$
$$= \sum_j A_{ij} w_j = w_i T$$

Thus $P^{-1}SP$ and T agree on $w_1, \ldots, w_n$, and so they agree on all of V. Therefore $T = P^{-1}SP$. □

Let us consider an example in which F is $\mathbf{Z}_7$ and V is the set of polynomials in $\mathbf{Z}_7[X]$ of degree at most 3. Let $D: V \longrightarrow V$ be the restriction of the derivative map to V. Thus

$$(a_0 + a_1 X + a_2 X^2 + a_3 X^3)D = a_1 + 2a_2 X + 3a_3 X^2.$$

Since

$$1D = 0,$$
$$XD = 1,$$
$$X^2 D = 2X,$$
$$X^3 D = 3X^2,$$

the matrix of D with respect to the basis $1, X, X^2, X^3$ of V is

```
      ⎕←A←4 4ρ0 0 0 0 1 0 0 0 0 2 0 0 0 0 3 0
0 0 0 0
1 0 0 0
0 2 0 0
0 0 3 0
```

Now define $T: V \longrightarrow V$ by

$$(a_0 + a_1 X + a_2 X^2 + a_3 X^3)T = a_3 X + 3a_0 X^2 + 2a_2 X^3.$$

Since

$$1T = 3X^2,$$
$$XT = 0,$$
$$X^2 T = 2X^3,$$
$$X^3 T = X,$$

the matrix for T with respect to $1, X, X^2, X^3$ is

```
      ⎕←B←4ρ0 0 3 0 0 0 0 0 0 0 0 2 0 1 0 0
0 0 3 0
0 0 0 0
0 0 0 2
0 1 0 0
```

If Q is the matrix given by

```
      ⎕IO←0
      ⎕←Q←(ι4)∘.=1 3 2 0
0 0 0 1
1 0 0 0
0 0 1 0
0 1 0 0
```

then Q is invertible with inverse QI obtained as follows.

```
      N←7
      ⎕←QI←ZNMATINV Q
0 1 0 0
0 0 0 1
0 0 1 0
1 0 0 0
```

The calculation

```
      Q ZNMATPROD A ZNMATPROD QI
0 0 3 0
0 0 0 0
0 0 0 2
0 1 0 0
```

shows modulo 7 that B is `Q+.×A+.×QI`, so A and B are similar matrices and hence D and T are similar linear transformations. The matrix Q is the transition matrix from the basis $1, X, X^2, X^3$ of V to the basis $X^3, 1, X^2, X$. Therefore B is the matrix for D with respect to this nonstandard basis.

If T is in $\text{End}_F(V)$, then the set J of polynomials g in $F[X]$ such that $g(T) = 0$ is an ideal of $F[X]$. Now $\text{End}_F(V)$ is isomorphic to $M_n(F)$, which has dimension n^2 over F. Thus $1, g, g^2, \ldots, g^{n^2}$ cannot be linearly independent over F, so there exist $a_0, \ldots, a_{n^2}$ in F, not all 0, such that $a_0 + a_1 g + \ldots + a_{n^2} g^{n^2} = 0$. Thus $J \neq \{0\}$, and so J consists of all multiples of a unique monic polynomial f. We call f the *minimal polynomial* of T. By the preceding argument, $\deg(f) \leq n^2$. In fact, as we will show in the next section, $\deg(f) \leq n$. We can also define the minimal polynomial of a matrix

A in $M_n(F)$ to be the monic polynomial f of smallest degree such that $f(A) = 0$.

In Section 7.1 we defined minimal polynomials of elements in extension fields. In that situation minimal polynomials are always irreducible. However, the minimal polynomials of linear transformations and matrices need not be irreducible. For example, the matrix

```
      ⎕←A←2 2 ρ1 0 0 2
1 0
0 2
```

in $M_2(\mathbf{Q})$ has minimal polynomial $f = X^2 - 3X + 2 = (X - 1)(X - 2)$. We can verify that $f(A) = 0$ as follows.

```
      ⎕←I←(ι2)∘.=ι2
1 0
0 1
      (A+.×A)+(¯3×A)+2×I
0 0
0 0
```

Therefore the minimal of A divides f. Since A is not a rational multiple of I, the minimal polynomial of A cannot have degree 1, and so the minimal polynomial is f, which is certainly not irreducible.

THEOREM 5. Let T be in $\mathrm{End}_F(V)$ and let A be the matrix of T with respect to a basis $v_1, \ldots, v_n$ of V. Then T and A have the same minimal polynomial.

Proof. The map that takes T in $\mathrm{End}_F(V)$ to its matrix A in $M_n(F)$ with respect to $v_1, \ldots, v_n$ is an F-algebra isomorphism. Thus if $a_0, \ldots, a_m$ are in F, then $a_0 + a_1T + \ldots + a_m T^m$ is 0 in $\mathrm{End}_F(V)$ if and only if $a_0 + a_1A + \ldots + a_m A^m$ is 0 in $M_n(F)$. Therefore T and A satisfy the same polynomials and so have the same minimal polynomial. □

We close this section with a simple but useful observation.

THEOREM 6. If A and B in $M_n(F)$ are similar, then $\det A = \det B$.

Proof. If A and B are similar, then $B = Q^{-1}AQ$ for some Q in $\mathrm{GL}_n(F)$. Thus

$$\det B = (\det Q^{-1})(\det A)(\det Q) = \det A,$$

since $\det Q^{-1} = (\det Q)^{-1}$. □

EXERCISES

1 Let R be a ring and let x and y be in R. Define x to be similar to

y if there is a unit u in R such that $y = u^{-1}xu$. Show that similarity is an equivalence relation on R.

2 Suppose S and T are similar linear transformations of the vector space V into itself. Show that the kernels of S and T have the same dimension and that the images of S and T have the same dimension.

3 Generalize Exercise 2 by showing that for all polynomials f in $F[X]$ the kernels of $f(S)$ and $f(T)$ have the same dimensions and the images of $f(S)$ and $f(T)$ have the same dimensions.

4 Let A, B, and Q be in $M_n(F)$ with A and B fixed and Q variable. Show that the matrix equation $QB = AQ$ is a system of linear equations on the entries of Q. Show that A and B are similar if and only if there is a Q such that $QB = AQ$ and $\det Q \neq 0$. (The condition $\det Q \neq 0$ is *not* a linear condition on the entires of Q.)

5 In $M_2(\mathbf{Z}_3)$, let

$$A = \begin{bmatrix} 0 & 1 \\ -1 & 0 \end{bmatrix} \quad \text{and} \quad B = \begin{bmatrix} 1 & 1 \\ 1 & -1 \end{bmatrix}.$$

Show that A and B are similar by solving the linear system $QB = AQ$ and exhibiting an invertible solution Q.

6 In $M_2(\mathbf{Q})$, let

$$A = \begin{bmatrix} 1 & 1 \\ 5 & 5 \end{bmatrix} \quad \text{and} \quad B = \begin{bmatrix} 2 & 4 \\ 2 & 4 \end{bmatrix},$$

let $V = \mathbf{Q} \oplus \mathbf{Q}$, and let S and T be the elements $\text{End}_{\mathbf{Q}}(V)$ whose matrices with respect to the standard basis of V are A and B, respectively. Show that the kernel and image of both S and T have dimension 1. Find vectors v_1 and w_1 spanning the kernels of S and T, respectively, and find vectors v_2 and w_2 spanning the images of S and T, respectively. Show that v_1, v_2 and w_1, w_2 are each bases of V. Compute the matrix of S with respect to v_1, v_2 and the matrix of T with respect to w_1, w_2. Are A and B similar?

7 Show that the matrices

$$A = \begin{bmatrix} 1 & 0 \\ 2 & 1 \end{bmatrix} \quad \text{and} \quad B = \begin{bmatrix} 1 & 0 \\ 4 & 1 \end{bmatrix}.$$

are similar in $M_2(\mathbf{Q})$ but not in $M_2(\mathbf{Z})$. That is, there does not exist a unit Q of $M_2(\mathbf{Z})$ such that $B = Q^{-1}AQ$, but there is a unit of $M_2(\mathbf{Q})$ with this property.

8 Prove that similar matrices have the same minimal polynomials.

9 In $M_2(\mathbf{Z}_5)$ let

$$A = \begin{bmatrix} 2 & 3 \\ 1 & 2 \end{bmatrix}.$$

Let c_0, c_1, c_2 be in $\mathbf{Z}_5$. Show that the matrix equation $c_0 + c_1A + c_2A^2 = 0$ corresponds to a system of homogeneous linear equations in the c_i. Describe all solutions of this system and compute the minimal polynomial of A.

2. RATIONAL CANONICAL FORM

In the previous section we defined similarity of linear transformations and of matrices and we showed how to reduce the problem of deciding when two linear transformations are similar to the problem of deciding when two matrices are similar. In the exercises several approaches to deciding similarity of matrices were introduced. These approaches are useful in special cases but are not adequate in general. In this section we present an algorithm for deciding when two n-by-n matrices A and B over a field F are similar and, when they are, for finding an element Q of $\mathrm{GL}_n(F)$ such that $B = Q^{-1}AQ$.

Let V be an n-dimensional vector space over F and let T be in $\mathrm{End}_F(V)$. The module V_T is a finitely generated $F[X]$-module; therefore, by the results of Sections 6.5 and 6.6, there are unique monic polynomials $f_1, \ldots, f_r$ of positive degree in $F[X]$ with f_i dividing f_{i+1}, $1 \leq i < r$, and a nonnegative integer m such that V_T is isomorphic to

$$(F[X]/\langle f_1 \rangle) \oplus \ldots \oplus (F[X]/\langle f_r \rangle) \oplus F[X]^m .$$

Since V is finite dimensional over F and $F[X]$ is not finite dimensional, m must be 0. Therefore V_T is isomorphic to

$$(F[X]/\langle f_1 \rangle) \oplus \ldots \oplus (F[X]/\langle f_r \rangle).$$

Since $f_1, \ldots, f_r$ determine V_T up to module isomorphism, by Theorem 1.2 the f_i determine T up to similarity. Thus the problem is how to determine the f_i.

The standard way of describing T is to choose a basis $v_1, \ldots, v_n$ of V and give the matrix A of T with respect to $v_1, \ldots, v_n$. Now $v_1, \ldots, v_n$ span V over F, and so $v_1, \ldots, v_n$ certainly generate V_T as an $F[X]$-module. Thus the map of $F[X]^n$ to V_T taking $(g_1, \ldots, g_n)$ to $v_1g_1 + \ldots + v_ng_n$ is a surjective $F[X]$-homomorphism. Let M be the kernel of this map. Then V_T is isomorphic to $F[X]^n/M$. Since M is a submodule of $F[X]^n$, we know

M can be generated by n elements. If we knew a matrix C over $F[X]$ such that the rows of C generated M, we could determine the f_i by reducing C over $F[X]$ using row and column operations as described in Section 6.5. The next theorem tells us one such matrix.

THEOREM 1. Let T be in $\text{End}_F(V)$, let A be the matrix of T with respect to the basis $v_1, \ldots, v_n$ of V, and let I be the n-by-n identity matrix. Then V_T is isomorphic to $F[X]^n/M$, where M is the submodule of $F[X]^n$ generated by the rows of $A - XI$.

Proof. As written, the matrix $A - XI$ is not in $M_n(F[X])$ but, indeed, in $M_n(F)[X]$. However, we remarked in Section 4.5 that these rings are isomorphic, and we agreed to identify them.

Let M be the kernel of the map τ of $F[X]^n$ onto V_T mapping $(g_1, \ldots, g_n)$ to $v_1 g_1 + \ldots + v_n g_n$. We will show that M is generated by the rows of $A - XI$.

The ith row of $A - XI$ is the vector

$$(A_{i1}, \ldots, A_{i,i-1}, A_{ii} - X, A_{i,i+1}, \ldots, A_{in}).$$

The image under τ of this vector is

$$v_1 A_{i1} + \ldots + v_{i-1} A_{i,i-1} + v_i(A_{ij} - T) + v_{i+1} A_{i,i+1} + \ldots + v_n A_{in},$$

which is

$$\sum_j A_{ij} v_j - v_i T.$$

However, by the definition of A,

$$v_i T = \sum_j A_{ij} v_j$$

so τ maps each row of $A - XI$ to 0. Therefore the rows of $A - XI$ are in M.

Let N be the submodule of $F[X]^n$ generated by the rows of $A - XI$. Then $N \subseteq M$. To show that $N = M$, it is sufficient to show that the dimension of $F[X]^n/N$ is the same as the dimension of $F[X]^n/M$, that is, n. (Why?) Let D be the matrix that is reduced over $F[X]$ and equivalent to $A - XI$. Then

$$D = \begin{bmatrix} g_1 & & 0 \\ & \ddots & \\ 0 & & g_n \end{bmatrix},$$

where the elementary divisors g_i are monic polynomials and g_i divides g_{i+1}, $1 \le i < n$. There exist units P and Q of $M_n(F[X])$ such that

$$D = P(A - XI)Q.$$

Therefore

$$\det D = (\det P)[\det(A - XI)](\det Q).$$

But $\det P$ and $\det Q$ are units in $F[X]$, that is, they are nonzero elements of F. Therefore $\det D$ and $\det(A - XI)$ have the same degree. Now $A - XI$ looks like

$$\begin{bmatrix} A_{11}-X & A_{12} & \cdots & & \\ A_{21} & A_{22}-X & A_{23} & & \cdots \\ & \cdot & & & \\ & \cdot & & & \\ & \cdot & & & \\ & \cdots & & & A_{nn}-X \end{bmatrix}.$$

Of the $n!$ terms in $\det(A - XI)$, every one has degree less than n except

$$(A_{11} - X)(A_{22} - X) \ldots (A_{nn} - X),$$

which has degree n and leading coefficient $(-1)^n$. Therefore $\det(A - XI)$ has degree n and hence so does $\det D$. By Corollary 6.5.3, the module $F[X]^n/N$ is isomorphic to $F[X]^n/L$, where L is the submodule of $F[X]^n$ generated by the rows of D, and $F[X]^n/L$ is isomorphic to

$$(F[X]/\langle g_i \rangle) \oplus \ldots \oplus (F[X]/\langle g_n \rangle).$$

The dimension of $F[X]/\langle g_i \rangle$ is $\deg(g_i)$, so $F[X]^n/N$ has dimension

$$\sum_i \deg(g_i) = \deg(g_1 \ldots g_n) = \deg(\det D) = n. \quad \square$$

COROLLARY 2. If A and B are in $M_n(F)$, then A and B are similar if and only if $A - XI$ and $B - XI$ are equivalent over $F[X]$.

Proof. Let $V = F^n$ and let S and T be the elements of $\mathrm{End}_F(V)$ defined by $vS = vA$ and $vT = vB$. Then A and B are the matrices of S and T, respectively, with respect to the standard basis of V. Now A and B are similar if and only if S and T are similar, which is true if and only if V_S and V_T are isomorphic. By Theorem 1, V_S is isomorphic to $F[X]^n/M$ and V_T is isomorphic to $F[X]^n/N$, where M and N are submodules of $F[X]^n$ generated by the rows of $A - XI$ and $B - XI$, respectively. By the results of Sections 6.5 and 6.6, $F[X]^n/M$ and $F[X]^n/N$ are isomorphic if and only if $A - XI$ and $B - XI$ are equivalent. $\square$

In the proof of Theorem 1 the determinant of $A - XI$ played an important role. This element of $F[X]$ is called the *characteristic polynomial* of A. We will encounter characteristic polynomials several times in this section and again in Section 3. The characteristic polynomial of

$$A = \begin{bmatrix} 1 & 3 & -1 \\ 2 & 0 & 1 \\ -3 & 1 & 2 \end{bmatrix}$$

is

$$\det \begin{bmatrix} 1-X & 3 & -1 \\ 2 & -X & 1 \\ -3 & 1 & 2-X \end{bmatrix} = -24 + 8X + 3X^2 - X^3.$$

In *CLASSLIB* there are procedures for reduction of matrices over $F[X]$, where F is $\mathbf{Z}_p$, p a prime. Let

```
      N←7
      ⎕←A←3 3ρ5 4 4 6 3 4 1 3 2
5 4 4
6 3 4
1 3 2
      ⎕←B←3 3ρ5 3 0 4 3 5 0 3 1
5 3 0
4 3 5
0 3 1
      ⎕←C←3 3ρ1 4 2 2 3 2 0 0 6
1 4 2
2 3 2
0 0 6
```

Let us use Corollary 2 and the procedure *ZNXREDUCE* to test the similarity of *A*, *B*, and *C* considered as elements of $GL_3(\mathbf{Z}_7)$. The matrix $A - XI$ is represented by *A*1, defined as follows.

```
      ⎕←I←(ι3)∘.=ι3                      DAZV A1←(3 3 1ρ,A),-I
1 0 0                              5 ¯1    4  0    4  0
0 1 0                              6  0    3 ¯1    4  0
0 0 1                              1  0    3  0    2 ¯1
```

[The array 3 3 1ρ ,*A* represents *A* considered as an element of

$M_3(\mathbf{Z}_3[X])$.]

Reducing *A*1,

```
      DAZV ZNXREDUCE A1
1 0 0    0 0 0    0 0 0
0 0 0    1 1 0    0 0 0
0 0 0    0 0 0    2 3 1
```

we see that $A - XI$ is equivalent to

$$D = \begin{bmatrix} 1 & 0 & 0 \\ 0 & 1+X & 0 \\ 0 & 0 & 2+3X+X^2 \end{bmatrix}.$$

Similarly,

```
      DAZV ZNXREDUCE (3 3 1⍴,B),-I
1 0 0 0    0 0 0 0    0 0 0 0
0 0 0 0    1 0 0 0    0 0 0 0
0 0 0 0    0 0 0 0    2 3 5 1
      DAZV ZNXREDUCE (3 3 1⍴,C),-I
1 0 0    0 0 0    0 0 0
0 0 0    1 1 0    0 0 0
0 0 0    0 0 0    2 3 1
```

we find that $B - XI$ is equivalent to

$$E = \begin{bmatrix} 1 & 0 & 0 \\ 0 & 1 & 0 \\ 0 & 0 & 2+3X+5X^2+X^3 \end{bmatrix},$$

while $C - XI$ is equivalent to D. Thus A and C are similar in $M_3(\mathbf{Z}_3)$ and B is not similar to either A or C.

It is convenient to be able to choose one representative from each similarity class of matrices in $M_n(F)$. Such sets of representatives are called *canonical forms.* We will now describe one canonical form.

If R is a commutative ring and J is any ideal of R, then $M = R/J$ is an R-module and multiplication of elements of M by a fixed element a of R is an R-endomorphism. Let us consider the case in which $R = F[X]$ and $J = \langle f \rangle$, where f is a monic polynomial of positive degree n. Multiplication by X is an $F[X]$-endomorphism T of $M = F[X]/J$ and so, in particular, T is in $\mathrm{End}_F(M)$. Let $v_i = X^i + J$, $0 \le i < n$. Then $v_0, \ldots, v_{n-1}$ is a basis of M. What is the matrix of T with respect to $v_0, \ldots, v_{n-1}$? If $0 \le i < n-1$, then $v_iT = (X^i + J)X = X^{i+1} + J = v_{i+1}$. Suppose $f = a_0 + a_1X + \ldots + a_{n-1}X^{n-1} + X^n$; then

$$v_{n-1}T = (X^{n-1} + J)X = X^n + J = -a_0 - a_1X - \ldots - a_{n-1}X^{n-1} + J$$
$$= -a_0v_0 - a_1v_1 - \ldots - a_{n-1}v_{n-1}.$$

Thus the matrix of T is

$$C = \begin{bmatrix} 0 & 1 & 0 & \cdot & \cdot & \cdot \\ & 0 & 1 & 0 & \cdot & \cdot & \cdot \\ & & & \cdot \\ & & & & \cdot \\ & & & & & \cdot \\ & & & & & 0 & 1 \\ -a_0 & -a_1 & & \cdot & \cdot & \cdot & -a_{n-1} \end{bmatrix}.$$

We call C the *companion matrix* of f. For example, if $f = 2 - 3X + X^2 - 4X^3 + X^4$ in $\mathbf{R}[X]$, then the companion matrix of f is

$$C = \begin{bmatrix} 0 & 1 & 0 & 0 \\ 0 & 0 & 1 & 0 \\ 0 & 0 & 0 & 1 \\ -2 & 3 & -1 & 4 \end{bmatrix}.$$

THEOREM 3. Let f be a monic polynomial of positive degree n in $F[X]$ and let C be the companion matrix of f. Then the minimal polynomial of C is f and the characteristic polynomial of C is $(-1)^n f$.

Proof. As noted, C is the matrix for the linear transformation T of $V = F[X]/\langle f \rangle$ defined by multiplication by X with respect to the basis $1 + J$, $X + J, \ldots, X^{n-1} + J$, where $J = \langle f \rangle$. By Theorem 8.1.5, C and T have the same minimal polynomial g. Now $g(T)$ is the linear transformation of V defined by multiplication by g. Since $g(T) = 0$, we must have

$$0 = (1 + J)g = g + J,$$

so g is in J. That is, f divides g. But $(X^i + J)f = X^i f + J = J$, so $f(T) = 0$. Therefore g divides f. Since f and g are monic, they must be equal.

Now let D be the reduced matrix over $F[X]$ equivalent to $C - XI$. Let M be the submodule of $F[X]^n$ generated by the rows of D. Then $F[X]^n/M$ is isomorphic to $V = F[X]/\langle f \rangle$, so

$$D = \begin{bmatrix} 1 & & & & 0 \\ & \cdot \\ & & \cdot \\ & 0 & & \cdot & 1 \\ & & & & & f \end{bmatrix}.$$

As noted in the proof of Theorem 1, the determinant of $C - XI$ is a multiple of $\det D$ by an element of F. Since $\det D = f$ is monic and $\det(C - XI)$ has leading coefficient $(-1)^n$, we have $\det(C - XI) = (-1)^n f$. □

The matrix

$$C = \begin{bmatrix} 0 & 1 & 0 & 0 \\ 0 & 0 & 1 & 0 \\ 0 & 0 & 0 & 1 \\ 6 & 3 & 1 & 2 \end{bmatrix}$$

in $M_4(\mathbf{Z}_{11})$ is the companion matrix of $f = 5 + 8X + 10X^2 + 9X^3 + X^4$ in $\mathbf{Z}_{11}[X]$. By Theorem 3, the minimal and characteristic polynomials of C are both f.

Let V and W be vector spaces over F and suppose S is in $\mathrm{End}_F(V)$ and T is in $\mathrm{End}_F(W)$. Let $S \oplus T$ denote the map of $V \oplus W$ into itself defined by $(v, w)(S \oplus T) = (vS, wT)$. Then for v, v' in V and w, w' in W, we have

$$\begin{aligned} [(v, w) + (v', w')](S \oplus T) &= (v + v', w + w')(S \oplus T) \\ &= ((v + v')S, (w + w')T) \\ &= (vS + v'S, wT + w'T) \\ &= (vS, wT) + (v'S, w'T) \\ &= (v, w)(S \oplus T) + (v', w')(S \oplus T). \end{aligned}$$

Similarly, $[a(v, w)](S \oplus T) = a[(v, w)(S \oplus T)]$ for all a in F. Thus $S \oplus T$ is in $\mathrm{End}_F(V \oplus W)$.

Let $v_1, \ldots, v_m$ be a basis of V and let $w_1, \ldots, w_n$ be a basis of W. Then it is easy to show (see Exercise 13) that $(v_1, 0), \ldots, (v_m, 0)$, $(0, w_1), \ldots, (0, w_n)$ is a basis of $V \oplus W$. Let A be the matrix of S with respect to $v_1, \ldots, v_m$ and let B be the matrix of T with respect to $w_1, \ldots, w_n$. Then

$$\begin{aligned} (v_i, 0)(S \oplus T) = (v_i S, 0) &= \left(\sum_j A_{ij} v_j, 0 \right) \\ &= \sum_j A_{ij}(v_j, 0). \end{aligned}$$

Similarly,

$$(0, w_i)(S \oplus T) = \sum_j B_{ij}(0, w_j),$$

so the matrix of $S \oplus T$ with respect to the basis $(v_1, 0), \ldots, (v_m, 0), (0, w_1), \ldots, (0, w_n)$ of $V \oplus W$ is the block matrix

$$C = \begin{bmatrix} A & 0 \\ 0 & B \end{bmatrix}.$$

We often write $C = A \oplus B$.

THEOREM 4. Suppose A is in $M_m(F)$ and B is in $M_n(F)$. Let $C = A \oplus B$. Then the minimal polynomial of C is the monic least common multiple of the minimal polynomials of A and B. The characteristic polynomial of C is the product of the characteristic polynomials of A and B.

Proof. Let f be in $F[X]$. Since

$$C = \begin{bmatrix} A & 0 \\ 0 & B \end{bmatrix},$$

it is easy to see that $f(C)$ is the block matrix

$$\begin{bmatrix} f(A) & 0 \\ 0 & f(B) \end{bmatrix}.$$

Thus $f(C) = 0$ if and only if both $f(A)$ and $f(B)$ are 0. This is equivalent to saying $f(C) = 0$ if and only if f is divisible by the minimal polynomial of A and by the minimal polynomial of B. Thus the minimal polynomial of C is the monic least common multiple of the minimal polynomials of A and B.

The characteristic polynomial of C is $\det(C - XI)$. Now

$$C - XI = \begin{bmatrix} A - XI_m & 0 \\ 0 & B - XI_n \end{bmatrix},$$

where I_m and I_n are the m-by-m and n-by-n identity matrices, respectively. By Exercise 4.7.18, $\det(C - XI) = [\det(A - XI_m)]\,[\det(B - XI_n)]$. □

Now we are in a position to be able to describe the canonical forms mentioned previously. Let A be in $M_n(F)$, let $V = F^n$, and let $T: V \longrightarrow V$ be defined by $vT = vA$. Then T is in $\text{End}_F(V)$ and A is the matrix of T with respect to the standard basis of V. Let $A - XI$ be equivalent over $F[X]$ to the reduced matrix

$$D = \begin{bmatrix} 1 & & & & & & 0 \\ & \ddots & & & & & \\ & & 1 & & & & \\ & & & f_1 & & & \\ & & & & \ddots & & \\ 0 & & & & & & f_r \end{bmatrix},$$

where $f_1, \ldots, f_r$ are monic polynomials of positive degree and f_i divides f_{i+1}, $1 \leq i < r$. Then the module V_T is isomorphic to

$$M = (F[X]/\langle f_1 \rangle) \oplus \ldots \oplus (F[X]/\langle f_r \rangle)$$

and the action of T on V_T corresponds to multiplication by X on M. Now we can choose a basis of $F[X]/\langle f_i \rangle$ such that the matrix for multiplication by X is the companion matrix C_i of f_i. Putting these bases together, we get a basis of M with respect to which the matrix for multiplication by X is

$$C = C_1 \oplus \ldots \oplus C_r = \begin{bmatrix} C_1 & & & 0 \\ & \cdot & & \\ & & \cdot & \\ 0 & & \cdot & C_r \end{bmatrix}.$$

Since V_T is isomorphic to M, we can choose a basis of V relative to which the matrix for T is C. Thus A is similar to C. We call C the *rational canonical form* for A. By Theorem 4, the minimal polynomial of A is f_r, since f_i divides f_r for $1 \leq i \leq r$, and the characteristic polynomial of A is

$$(-1)^n f_1 \ldots f_r .$$

As an obvious consequence of this result, we obtain the following theorem.

THEOREM 5 ***(Cayley-Hamilton Theorem).*** For any matrix A in $M_n(F)$, the minimal polynomial of A divides the characteristic polynomial of A.

Proof. In the preceding notation, this is simply the observation that f_r divides $(-1)^n f_1 \ldots f_r$. □

Theorem 5 is often stated in the following form. Every matrix in $M_n(F)$ satisfies its characteristic polynomial. Let us verify this in a special case with $n = 3$ and $F = \mathbf{Z}_5$. Let

```
      N←5
      ⎕←B←3 3⍴4 2 1 4 1 4 2 4 3
4 2 1
4 1 4
2 4 3
      ⎕←I←(⍳3)∘.=⍳3
1 0 0
0 1 0
0 0 1
```

The characteristic polynomial of B, considered as an element of $M_3(\mathbf{Z}_5)$, is

```
      ZNXDET (3 3 1⍴,B),-I
4 2 3 4
```

$4 + 2X + 3X^2 + 4X^3 = -1 + 2X - 2X^2 - X^3$. To show that $f(B)$ is the zero matrix, we compute the powers of B

```
      ⎕←B2←B ZNMATPROD B
 1 4 0
 3 0 0
 0 0 2
      ⎕←B3←B2 ZNMATPROD B
 0 1 2
 2 1 3
 4 3 1
```

and compute $f(B)$.

```
      N|(4×I)+(2×B)+(3×B2)+4×B3
 0 0 0
 0 0 0
 0 0 0
```

Now let us find the canonical form for the matrix A in $M_4(\mathbf{Z}_{17})$ represented by

```
      ⎕←A←4 4⍴8 6 1 11 15 7 11 12 6 5 3 12 2 0 6 2
  8  6  1 11
 15  7 11 12
  6  5  3 12
  2  0  6  2
```

Reducing $A - XI$ as before,

```
     N←17
     DAZV D←ZNXREDUCE A1←(4 4 1⍴,A),-(⍳4)∘.=⍳4
 1 0 0 0    0 0 0 0     0 0 0 0     0 0  0 0
 0 0 0 0    1 0 0 0     0 0 0 0     0 0  0 0
 0 0 0 0    0 0 0 0    15 1 0 0     0 0  0 0
 0 0 0 0    0 0 0 0     0 0 0 0    12 9 16 1
```

we see that $A - XI$ is equivalent to the reduced matrix

$$D = \begin{bmatrix} 1 & 0 & 0 & 0 \\ 0 & 1 & 0 & 0 \\ 0 & 0 & 15+X & 0 \\ 0 & 0 & 0 & 12+9X+16X^2+X^3 \end{bmatrix},$$

so the canonical form for A is the direct sum of the companion matrices of $15 + X$ and $12 + 9X + 16X^2 + X^3$; that is,

$$C = \begin{bmatrix} 2 & 0 & 0 & 0 \\ 0 & 0 & 1 & 0 \\ 0 & 0 & 0 & 1 \\ 0 & 5 & 8 & 1 \end{bmatrix}.$$

Now A is similar to C, so there is a matrix Q in $\mathrm{GL}_3(\mathbf{Z}_{17})$ such that $C = QAQ^{-1}$. We will use Theorem 6.6.10 to find one such Q.

The procedure *ZNXREDUCE* computed two arrays R and S.

```
          DAZV R
  3   0   0      0   0   0      0   0   0      0   0   0
  0   0   0      0   0   0      0   0   0      9   0   0
 14   0   0      0   0   0      7   0   0      8   7   0
 14  15   0      9   0   0      8  11   0     14  16  16
          DAZV S
  0   0   0      1   0   0     14   0   0      9   5   0
  1   0   0     10   3   0      1   8   0     16   5  15
  0   0   0      0   0   0      1   0   0      8   7   0
  0   0   0      0   0   0      0   0   0      1   0   0
```

These arrays represent the following matrices over $\mathbf{Z}_{17}[X]$.

$$R = \begin{bmatrix} 3 & 0 & 0 & 0 \\ 0 & 0 & 0 & 9 \\ 14 & 0 & 7 & 8+7X \\ 14+15X & 9 & 8+11X & 14+16X+16X^2 \end{bmatrix},$$

$$S = \begin{bmatrix} 0 & 1 & 14 & 9+5X \\ 1 & 10+3X & 1+8X & 16+5X+15X^2 \\ 0 & 1 & 1 & 8+7X \\ 0 & 0 & 0 & 1 \end{bmatrix}.$$

The product $R(A - XI)S$ is D. According to Theorem 6.6.10, we must compute $U = S^{-1}$. To find U, we use *ZNXMATINV*.

```
      DAZV U←ZNXMATINV S
 7 14     1  0     3  0    16  0
 1  0     0  0     3  0     1  8
 0  0     0  0     1  0     9 10
 0  0     0  0     0  0     1  0
```

and find that

$$U = \begin{bmatrix} 7+14X & 1 & 3 & 16 \\ 1 & 0 & 3 & 1+8X \\ 0 & 0 & 1 & 9+10X \\ 0 & 0 & 0 & 1 \end{bmatrix}$$

Let $V = (\mathbf{Z}_{17})^4$ and let T be the endomorphism of V given by multiplication on the right by A. The $\mathbf{Z}_{17}[X]$-module V_T is isomorphic to

$$(\mathbf{Z}_{17}[X]/\langle f\rangle) \oplus (\mathbf{Z}_{17}[X]/\langle g\rangle),$$

where $f = 15 + X$ and $g = 12 + 9X + 16X^2 + X^3$. Let v_1, v_2, v_3, v_4 be the standard basis of V and for $1 \leq i \leq 4$ define

$$w_i = \sum_{j=1}^{4} v_j U_{ij}$$

By Theorem 6.6.10, w_1 and w_2 must be 0 and V_T is the direct sum of the $\mathbf{Z}_{17}[X]$ submodules generated by w_3 and w_4, which are isomorphic to $\mathbf{Z}_{17}[X]/\langle f\rangle$ and $\mathbf{Z}_{17}[X]/\langle g\rangle$, respectively.

Now $v_j U_{ij}$ means the product of v_j and U_{ij} evaluated at A. Thus we must evaluate all the entries of U at A. Here is one way to do this. Since each U_{ij} has degree at most 1, we combine the identity matrix and A into a single rank 3 array.

```
      ⎕←H←2 4 4⍴(,(⍳4)∘.=⍳4),,A
 1  0  0  0
 0  1  0  0
 0  0  1  0
 0  0  0  1

 8  6  1 11
15  7 11 12
 6  5  3 12
 2  0  6  2
```

Next we form

```
      K←U ZNMATPROD H
```

The array K has rank 4 and `K[I;J;;]` is the value of the polynomial with coefficients `U[I;J;]` at the matrix A. For example,

```
      ⎕IO←1
      U[3;4;]
9 10
      17|(9×H[1;;])+10×H[2;;]
  4   9 10   8
 14  11  8   1
  9  16  5   1
  3   0  9  12
      K[3;4;;]
  4   9 10   8
 14  11  8   1
  9  16  5   1
  3   0  9  12
```

For any matrix B in $M_4(\mathbf{Z}_{17})$, the product v_jB is just the jth row of B. Thus, to compute the v_jU_{ij}, we want a rank 3 array L such that `L[I;J;]` is `K[I;J;J;]`. This array is

```
      L←1 2 2 3⍉K
```

Finally, we must sum on J to get the w_i, which are given by the rows of

```
      ⎕←W←17|+/[2]L
  0   0   0   0
  0   0   0   0
  3   0  10  12
  0   0   0   1
```

As promised, $w_1 = w_2 = 0$. Now the $\mathbf{Z}_{17}[X]$-submodule M of V_T generated by w_3 is one dimensional. The submodule N generated by w_4 has dimension 3 and a $\mathbf{Z}_{17}$-basis w_4, w_4A, and w_4A^2. Relative to this basis, the companion matrix for g is the matrix for T restricted to N. Thus w_3, w_4, w_4A, w_4A^2 is a basis of V and relative to this basis C is the matrix for T. The new basis is given by the rows of Q constructed as follows.

```
      Q←4 4⍴0
      Q[1 2;]←W[3 4;]
      Q[3;]←W[4;] ZNMATPROD A
      Q[4;]←Q[3;] ZNMATPROD A
      Q
  3   0  10  12
  0   0   0   1
  2   0   6   2
  5   8  15  13
```

This is the matrix we have been looking for!

```
        ⎕←QI←ZNMATINV Q
14   9   5   0
 0  16   5  15
 1   8   7   0
 0   1   0   0
        Q ZNMATPROD A ZNMATPROD QI
2 0 0 0
0 0 1 0
0 0 0 1
0 5 8 1
```

EXERCISES

1 Let C be in $M_n(F[X])$ and let N be the submodule of $F[X]^n$ generated by the rows of C. Suppose $f = \det C$ is not 0. Prove that $F[X]^n/N$ has dimension $\deg(f)$.

2 Let A be in $M_n(F)$ and let $f = c_n - c_{n-1}X + c_{n-2}X^2 - \ldots + (-1)^{n-1}c_1X^{n-1} + (-1)^n X^n$ be the characteristic polynomial of A. Show that c_i is the sum of determinants of the i-by-i diagonal submatrices of A, that is, the i-by-i submatrices obtained by selecting a set of i row indices and the same set of i column indices. In particular, $c_n = \det A$ and c_1 is the *trace* $A_{11} + \ldots + A_{nn}$ of A.

3 Show directly from the definitions that similar matrices have the same characteristic polynomials.

4 In $M_3(\mathbf{Z}_{13})$ let

$$A = \begin{bmatrix} 5 & 12 & 9 \\ 2 & 10 & 6 \\ 5 & 7 & 4 \end{bmatrix}, \quad B = \begin{bmatrix} 3 & 0 & 6 \\ 11 & 10 & 0 \\ 2 & 12 & 5 \end{bmatrix},$$

$$C = \begin{bmatrix} 5 & 0 & 9 \\ 5 & 9 & 8 \\ 6 & 2 & 4 \end{bmatrix}, \quad D = \begin{bmatrix} 8 & 2 & 6 \\ 7 & 0 & 7 \\ 9 & 3 & 11 \end{bmatrix}.$$

Decide whether A, B, C, D are similar to each other.

5 Write down the companion matrices of the polynomials $7 - 3X + 4X^2 + X^3$ and $1 - 5X + X^3 + X^5$ in $\mathbf{Q}[X]$.

6 Let A be a matrix in $M_6(\mathbf{R})$ such that $A - XI$ is equivalent to

$$D = \begin{bmatrix} 1 & & & & & 0 \\ & 1 & & & & \\ & & 1 & & & \\ & & & 1 & & \\ & 0 & & & f & \\ & & & & & g \end{bmatrix},$$

where $f = X^2 + 1$ and $g = (X-1)^2\,(X^2+1)$. What is the rational canonical form for A? What are the minimal and characteristic polynomials of A?

7 Determine the rational canonical form C for the matrix

$$A = \begin{bmatrix} 1 & -1 & -1 & -1 \\ 1 & -1 & 0 & -1 \\ -1 & 1 & -1 & 0 \\ 1 & 1 & -1 & -1 \end{bmatrix}$$

in $M_4(\mathbf{Z}_3)$.

8 In Exercise 7, find Q in $\mathrm{GL}_4(\mathbf{Z}_3)$ such that $C = QAQ^{-1}$.

9 A matrix A in $M_{10}(\mathbf{R})$ has minimal polynomial $X^3(X+1)^2$ and characteristic polynomial $X^6(X+1)^4$. What are the possible rational canonical forms for A?

10 Show that the matrices

$$A = \begin{bmatrix} 11 & 14 & 1 & 8 \\ 18 & 5 & 17 & 6 \\ 2 & 10 & 10 & 9 \\ 5 & 14 & 16 & 0 \end{bmatrix} \quad \text{and} \quad B = \begin{bmatrix} 16 & 6 & 3 & 3 \\ 7 & 15 & 10 & 3 \\ 10 & 13 & 11 & 11 \\ 6 & 18 & 5 & 3 \end{bmatrix}$$

in $M_4(\mathbf{Z}_{19})$ are similar and find Q in $\mathrm{GL}_4(\mathbf{Z}_{19})$ such that $B = Q^{-1}AQ$.

11 Let f and g be monic and relatively prime in $F[X]$. Prove that the companion matrix of fg is similar to the direct sum of the companion matrix of f and the companion matrix of g.

12 Let A be in $M_m(F)$ and B in $M_n(F)$. Show that

$$C = \begin{bmatrix} A & 0 \\ 0 & B \end{bmatrix} \quad \text{and} \quad D = \begin{bmatrix} B & 0 \\ 0 & A \end{bmatrix}$$

are similar and find Q in $\mathrm{GL}_{m+n}(F)$ such that $D = Q^{-1}AQ$.

13 Let V and W be vector spaces over the same field and let $v_1, \ldots,$

v_m and $w_1, \ldots, w_n$ be bases of V and W, respectively. Show that $(v_1, 0), \ldots, (v_m, 0), (0, w_1), \ldots, (0, w_n)$ is a basis of $V \oplus W$.

3. EIGENVALUES AND EIGENVECTORS

In Theorem 7.4.2 we encountered the problem of deciding, for a certain linear transformation T on a vector space V, which elements v of V satisfy $vT = v$. In this section we will discuss a slightly more general question.

Let V be a vector space over a field F and let T be in $\text{End}_F(V)$. An *eigenvector* for T is a nonzero element v of V such that $vT = \lambda v$ for some λ in F. We call λ the *eigenvalue* of T associated with v. (The adjective "eigen" in German means "own" or "characteristic of". The terms "characteristic value" and "proper value" are sometimes used instead of "eigenvalue".) Note that λ may be 0 but v must be nonzero and that v determines λ but λ does not usually determine v. If A is in $M_n(F)$, then an eigenvector of A is a nonzero element x of F^n such that $xA = \lambda x$ for some λ in F. As should be expected, the two types of eigenvectors are closely related.

THEOREM 1. Let V be a vector space over a field F, let T be in $\text{End}_F(V)$, and let A be the matrix of T with respect to a basis $v_1, \ldots, v_n$ of V. Then v in V is an eigenvector of T if and only if the coordinate vector x for v relative to $v_1, \ldots, v_n$ is an eigenvector of A.

Proof. The coordinate vector for vT is xA and v is nonzero if and only if x is nonzero. The coordinate vector for λv is λx. Thus $vT = \lambda v$ if and only if $xA = \lambda x$. □

If A is the matrix

```
      ⎕←A←3 3⍴6 ¯8 13 4 ¯9 19 0 ¯3 8
6  ¯8  13
4  ¯9  19
0  ¯3   8
```

in $M_3(\mathbf{Q})$, then `1 ¯1 1` is an eigenvector of A with eigenvalue 2.

```
      1 ¯1 1+.×A
2 ¯2 2
```

THEOREM 2. Let A be in $M_n(F)$ and let λ be in F. Then λ is an eigenvalue of A if and only if λ is a root of the characteristic polynomial of A.

Proof. Suppose x in F^n is an eigenvector of A with λ as its associated eigenvalue. Then $xA = \lambda x = x(\lambda I)$, where I is the n-by-n identity matrix. Thus $x(A - \lambda I) = 0$. Since $x \neq 0$, the matrix $A - \lambda I$ cannot be invertible, so

$\det(A - \lambda I) = 0$. Therefore λ is a root of the characteristic polynomial $f = \det(A - XI)$ of A. Conversely, if $f(\lambda) = 0$, then $\det(A - \lambda I) = 0$, and $A - \lambda I$ has rank less than n. Therefore the linear system $x(A - \lambda I) = 0$ has a nonzero solution. □

Note that the linear system $x(A - \lambda I) = 0$ in the proof of Theorem 2 would normally be written $(A^t - \lambda I)x = 0$. The set W of solutions of this system, including the zero vector, is called the λ-*eigenspace* of A. Let

```
      ⎕←A←4 4⍴3 1 4 0 2 2 2 1 2 4 1 4 3 0 3 1
 3 1 4 0
 2 2 2 1
 2 4 1 4
 3 0 3 1
      I←(⍳4)∘.=⍳4
```

in $M_4(\mathbf{Z}_5)$. From

```
      N←5
      ZNDET A1←A-2×I
0
```

we see that 2 is an eigenvalue of A. To find the eigenvectors, we solve the homogenous linear system with matrix ⍉A1.

```
      (⍉A1) ZNLSYS 4⍴0
0 0 0 0
      W
 1 1 1 0
 0 1 0 1
```

The eigenvectors of A that have 2 as their associated eigenvalue are the nonzero elements of the subspace of $(\mathbf{Z}_5)^4$ spanned by the rows of W.

It often happens that F is not a splitting field for the characteristic polynomial of an element A of $M_n(F)$. For example, the real matrix

$$A = \begin{bmatrix} 0 & 1 \\ -1 & 0 \end{bmatrix}$$

has characteristic polynomial

$$\det \begin{bmatrix} -X & 1 \\ -1 & -X \end{bmatrix} = 1 + X^2$$

and $1 + X^2$ has no roots in $\mathbf{R}$. However, if we consider A to be in $M_2(\mathbf{C})$, then $(i, 1)$ and $(-i, 1)$ are eigenvectors of A with associated eigenvalues i and $-i$, respectively.

If A and B are similar matrices, then A and B have the same characteristic polynomials and therefore the same eigenvalues. They do not necessarily have the same eigenvectors. However, for each eigenvalue λ, the λ-eigenspaces of A and B have the same dimension.

THEOREM 3. Let A be in $M_n(F)$. Then every eigenvalue of A is a root of the minimal polynomial of A.

Proof. Let x be an eigenvector with eigenvalue λ. Then $xA = \lambda x$, $xA^2 = (xA)A = (\lambda x)A = \lambda^2 x$ and, by induction, we have $xA^i = \lambda^i x$ for all nonnegative integers i. Thus if f is in $F[X]$, then $xf(A) = f(\lambda)x$. If f is the minimal polynomial of A, then $f(A) = 0$, so $f(\lambda)$ must be 0. □

We could also prove Theorem 3 using the description of the characteristic polynomial immediately preceding Theorem 2.5. Theorem 3 shows it is impossible for a matrix to have minimal polynomial $X(X - 1)$ and characteristic polynomial $X^2(X - 1)^3(X + 2)$.

A *diagonal matrix* is a square matrix whose entries off the main diagonal are 0. For example,

$$\begin{bmatrix} 1 & 0 & 0 & 0 \\ 0 & 1 & 0 & 0 \\ 0 & 0 & -2 & 0 \\ 0 & 0 & 0 & 3 \end{bmatrix}.$$

is a diagonal matrix. Linear transformations are considered particularly simple if some matrix for them is diagonal. This suggests the problem of deciding whether a given square matrix is *diagonalizable*, that is, similar to a diagonal matrix.

THEOREM 4. Let $\lambda_1, \ldots, \lambda_r$ be distinct elements of F and let D be the n-by-n diagonal matrix

$$\begin{bmatrix} \lambda_1 & & & & & & \\ & \ddots & & & & 0 & \\ & & \lambda_2 & & & & \\ & & & \ddots & & & \\ & 0 & & & \lambda_r & & \\ & & & & & \ddots & \end{bmatrix},$$

where λ_i occurs n_i times on the diagonal. Then the minimal polynomial of D is $(X - \lambda_1) \ldots (X - \lambda_r)$ and the characteristic polynomial of D is $(\lambda_1 - X)^{n_1} \ldots (\lambda_r - X)^{n_r}$.

Proof. The matrix $D - XI$ is a diagonal matrix with n_i diagonal entries $\lambda_i - X$, $1 \le i \le r$. Thus the characteristic polynomial of D, which is $\det(D - XI)$, is $(\lambda_1 - X)^{n_1} \ldots (\lambda_r - X)^{n_r}$. Each λ_i is an eigenvalue of D and so by Theorem 3 the minimal polynomial of D is divisible by $f = (X - \lambda_1) \ldots (X - \lambda_r)$. Now $f(D)$ is $(D - \lambda_1 I) \ldots (D - \lambda_r I)$. Since the first n_1 diagonal entries of $D - \lambda_1 I$ are 0, the next n_2 diagonal entries of $D - \lambda_2 I$ are 0, and so on, $f(D) = 0$. Therefore the minimal polynomial of D divides f. Since f is monic, f must be the minimal polynomial. □

We can now describe a necessary and sufficient condition for a square matrix to be diagonalizable.

THEOREM 5. Let A be in $M_n(F)$. Then A is diagonalizable in $M_n(F)$ if and only if the minimal polynomial of A has the form $(X - \lambda_1) \ldots (X - \lambda_r)$, where $\lambda_1, \ldots, \lambda_r$ are distinct elements of F.

Proof. If A is diagonalizable, then A is similar in $M_n(F)$ to a diagonal matrix D, which by Theorem 4 has minimal polynomial $(X - \lambda_1) \ldots (X - \lambda_r)$, where $\lambda_1, \ldots, \lambda_r$ are the distinct eigenvalues of D. Since A and D are similar, A and D have the same minimal polynomial.

Now suppose A has the minimal polynomial $f = (X - \lambda_1) \ldots (X - \lambda_r)$. Let $V = F^n$ and let T be the element of $\mathrm{End}_F(V)$ with $vT = vA$ for all v in V. The matrix $A - XI$ is equivalent over $F[X]$ to a matrix

$$\begin{bmatrix} f_1 & & & 0 \\ & \cdot & & \\ & & \cdot & \\ 0 & & & f_n \end{bmatrix},$$

where the f_i are monic polynomials and f_i divides f_{i+1}, $1 \le i < n$. Moreover, $f_n = f$. The $F[X]$-module V_T is isomorphic to the direct sum of the modules $F[X]/\langle f_i \rangle$, $1 \le i \le n$. If $1 \le i \le n$, then f_i divides f, and so $f_i = (X - \mu_1) \ldots (X - \mu_s)$, where the μ_j are distinct elements of F. By Theorem 7.4.3, $F[X]/\langle f_i \rangle$ is isomorphic to the direct sum of the modules $F[X]/\langle X - \mu_j \rangle$, $1 \le j \le s$. This means that V_T is a direct sum of one-dimensional submodules, $V = V_1 \oplus \ldots \oplus V_n$. If v_k is a nonzero element of V_k, then v_k is an eigenvector of T, and $v_1, \ldots, v_n$ form a basis of V. With respect to this basis, the matrix D of T is diagonal. Since A and D are similar, A is diagonalizable. □

The matrix

```
      ⎕←A←4 4⍴0 4 7 5 1 9 7 1 6 3 4 0 3 1 5 7
 0 4 7 5
 1 9 7 1
 6 3 4 0
 3 1 5 7
```

in $M_4(\mathbf{Z}_{11})$ has characteristic polynomial $3 + 10X + 2X^3 + X^4$,

```
      N←11
      ZNXDET (4 4 1⍴,A),-I←(⍳4)∘.=⍳4
3 10 0 2 1
```

which, in $\mathbf{Z}_{11}[X]$, is $(X - 2)^2\ (X - 8)^2$. If $f = (X - 2)\ (X - 8)$, then $f(A)$ is 0.

```
      (A-2×I) ZNMATPROD A-8×I
 0 0 0 0
 0 0 0 0
 0 0 0 0
 0 0 0 0
```

By Theorem 5, A is diagonalizable. As a check, we can compute the eigenspaces for $\lambda = 2$ and $\lambda = 8$.

```
      (⍉A-2×I) ZNLSYS 4⍴0
0 0 0 0
      ⎕←U←W
 4 2 1 0
 6 9 0 1
      (⍉A-8×1) ZNLSYS 4⍴0
0 0 0 0
      ⎕←V←W
 3 7 1 0
 2 2 0 1
```

The rows of U are a basis for the 2-eigenspace and the rows of V are a basis for the 8-eigenspace. If

```
      ⎕IO←1
      ⎕←P←U,[1]V
 4 2 1 0
 6 9 0 1
 3 7 1 0
 2 2 0 1
```

then P has nonzero determinant,

```
      ZNDET P
5
```

and so the rows of P are a basis for $(\mathbf{Z}_{11})^4$. A diagonal matrix similar to A is

```
      ⎕←D←P ZNMATPROD A ZNMATPROD ZNMATINV P
 2 0 0 0
 0 2 0 0
 0 0 8 0
 0 0 0 8
```

The diagonal entries of D are just the eigenvalues associated with the rows of P.

In Section 2 we described one canonical form for matrices under similarity, the rational canonical form. There is another widely used canonical form, the Jordan canonical form, which we can now describe. It is based on the following simple observation.

LEMMA 6. Let λ be an element of F and let I be the ideal of $F[X]$ generated by $(X-\lambda)^m$. Then the matrix for multiplication by X on $F[X]/I$ with respect to the basis $1+I, (X-\lambda)+I, \ldots, (X-\lambda)^{m-1}+I$ is

$$J = \begin{bmatrix} \lambda & 1 & & & & \\ & \lambda & 1 & & 0 & \\ & & & \cdot & & \\ & & & & \cdot & \\ & 0 & & & \cdot\ \lambda & 1 \\ & & & & & \lambda \end{bmatrix}.$$

Proof. The following identity is obvious.

$$(X-\lambda)^i X = \lambda(X-\lambda)^i + (X-\lambda)^{i+1}.$$

Thus for $0 \leq i < m-1$ we have

$$[(X-\lambda)^i + I]X = \lambda[(X-\lambda)^i + I] + [(X-\lambda)^{i+1} + I].$$

Since $(X-\lambda)^m$ is in I,

$$[(X-\lambda)^{m-1} + I]X = \lambda[(X-\lambda)^{m-1} + I]. \quad \square$$

The matrix J in Lemma 6 is called a *Jordan block.* [French mathematician Camille Jordan (1838-1922) made important contributions in algebra and in other areas of mathematics.] The minimal polynomial of J is $(X - \lambda)^m$. The only eigenvalue of J is λ, and the λ-eigenspace of J is one dimensional.

THEOREM 7. Let A be in $M_n(F)$ and assume all the roots of the characteristic polynomial of A lie in F. Then A is similar to a direct sum of Jordan blocks, which are unique up to order.

Proof. Let $A - XI$ be equivalent over $F[X]$ to the reduced matrix

$$\begin{bmatrix} f_1 & & & 0 \\ & \cdot & & \\ & & \cdot & \\ 0 & & \cdot & f_n \end{bmatrix}$$

and let $\lambda_1, \ldots, \lambda_r$ be the distinct roots of f_n in F. Then

$$f_i = (X - \lambda_1)^{m_{i1}} \ldots (X - \lambda_r)^{m_{ir}}$$

and $m_{ij} \le m_{i+1,j}$ for $1 \le i < n$. If $V = F^n$ and T in $\mathrm{End}_F(V)$ is multiplication by A, then V_T is isomorphic to the direct sum of the modules $F[X]/<f_i>$, which by Theorem 7.4.3 are isomorphic to the direct sums of the modules $F[X]/<(X - \lambda_j)^{m_{ij}}>$. Thus we can write V_T as a direct sum of modules V_{ij}, $1 \le i \le n$, $1 \le j \le r$, such that V_{ij} is isomorphic to $F[X]/<(X - \lambda_j)^{m_{ij}}>$. By Lemma 6, we can choose a basis for V_{ij} such that the matrix for $T|_{V_{ij}}$ is the m_{ij}-by-m_{ij} matrix

$$J_{ij} = \begin{bmatrix} \lambda_j & 1 & & & \\ & \cdot & & 0 & \\ & & \cdot & & \\ & & & \cdot & \\ 0 & & & \lambda_j & 1 \\ & & & & \lambda_j \end{bmatrix}.$$

Putting these bases together, we get a basis for V such that the matrix for T is the direct sum of the J_{ij}.

Now let us prove the uniqueness of the Jordan blocks. Suppose A is similar to a direct sum of m_{ij}-by-m_{ij} Jordan blocks J_{ij} having λ_j on the diagonal, $1 \le i \le n$, $1 \le j \le r$, where $\lambda_1, \ldots, \lambda_r$ are distinct elements of F and $m_{ij} \le m_{i+1,j}$, $1 \le i < n$. Then the module V_T is isomorphic to the direct sum of submodules V_{ij} isomorphic to $F[X]/<(X - \lambda_j)^{m_{ij}}>$. By Theorem 7.4.3,

$$V_{i1} \oplus V_{i2} \oplus \ldots \oplus V_{ir} \cong F[X]/<f_i>,$$

where

$$f_i = (X - \lambda_1)^{m_{i1}} \ldots (X - \lambda_r)^{m_{ir}}.$$

Thus V_T is isomorphic to $F[X]/S(D)$, where

$$D = \begin{bmatrix} f_1 & & & 0 \\ & \cdot & & \\ & & \cdot & \\ 0 & & \cdot & f_n \end{bmatrix}$$

and $A - XI$ is equivalent to D. Now each f_i is monic and f_i divides f_{i+1} for $1 \le i < n$. Therefore D is reduced over $F[X]$ and the f_i are the elementary divisors of $A - XI$. Changing either the eigenvalues λ_j or the sizes m_{ij} of the blocks yields a different sequence $f_1, \ldots, f_n$. Thus the Jordan blocks are unique up to order. □

A matrix that is the direct sum of Jordan blocks is said to be in *Jordan canonical form.*

Let us find the Jordan canonical form for the matrix

```
      ⎕←A←3 3ρ10 9 7 7 8 4 11 11 11
 10  9  7
  7  8  4
 11 11 11
```

in $M_3(\mathbf{Z}_{13})$. The characteristic polynomial of A is $1 - 3X + 3X^2 - X^3 = (1 - X)^3$.

```
      N←13
      ZNXDET (3 3 1ρ,A),-I←(ι3)∘.=ι3
1 10 3 12
```

Thus the only eigenvalue of A is 1. Since A is not equal to I, we know that A is not diagonalizable, so its Jordan canonical form has a block of size greater than 1. Since the square of $A-I$ is the zero matrix,

```
      (A-I) ZNMATPROD A-I
 0 0 0
 0 0 0
 0 0 0
```

the minimal polynomial of A is $(X - 1)^2$, and the Jordan canonical form for A has no blocks of size 3. Thus the Jordan canonical form must be

```
      ⎕←B←3 3⍴1 0 0 0 1 1 0 0 1
1 0 0
0 1 1
0 0 1
```

EXERCISES

1 Let T be an element of $\mathrm{End}_F(V)$, where V is finite dimensional over F. Let $\lambda_1, \ldots, \lambda_r$ be the distinct eigenvalues of T in F and let $V_i = \{v \in V \mid vT = \lambda_i v\}$. Show that $V_1 + V_2 + \ldots + V_r$ is $V_1 \oplus V_2 \oplus \ldots \oplus V_r$.

2 For each field F and matrix A over F find the eigenvalues of A and for each eigenvalue find a basis for the corresponding eigenspace.

(a) $F = \mathbf{Z}_5$,

$$A = \begin{bmatrix} 0 & 1 & 1 \\ 2 & 1 & 4 \\ 2 & 4 & 1 \end{bmatrix}.$$

(b) $F = \mathbf{Q}$,

$$A = \begin{bmatrix} -5 & -2 & 4 \\ 5 & 6 & -5 \\ -2 & 2 & 1 \end{bmatrix}.$$

(c) $F = \mathbf{Z}_{31}$,

$$A = \begin{bmatrix} 12 & 11 & 5 & 5 & 15 \\ 15 & -6 & 10 & 15 & 15 \\ 14 & 4 & 2 & 13 & 13 \\ -4 & 7 & 2 & 13 & 0 \\ -10 & 2 & 1 & 14 & -2 \end{bmatrix}.$$

3 Which of the matrices in Exercise 2 are diagonalizable?

4 Let J be a Jordan block in $M_m(F)$ with eigenvalue λ. For $i = 1, 2, \ldots,$ let V_i be the set of vectors x in F^m such that $x(J - \lambda I)^i = 0$. Show that V_i has dimension i for $1 \le i \le m$ and dimension m for $i > m$.

5 Let A be in $M_n(F)$ and assume all eigenvalues of A lie in F. For each eigenvalue λ and each positive integer i, let $V_i(\lambda)$ be the set of vectors x in F^n such that $x(A - \lambda I)^i = 0$. Show that the number of Jordan blocks with eigenvalue λ in the Jordan canonical form for A having size at least m in $\dim_F V_m(\lambda) - \dim_F V_{m-1}(\lambda)$. [Set $V_0(\lambda) = \{0\}$.]

6 For each field F and matrix A over F find the Jordan canonical form for A.

(a) $F = \mathbf{Z}_7$,

$$A = \begin{bmatrix} 6 & 6 & 3 \\ 5 & 6 & 1 \\ 4 & 1 & 4 \end{bmatrix}.$$

(b) $F = \mathbf{Z}_{11}$,

$$A = \begin{bmatrix} 3 & 7 & 1 & 8 & 8 \\ 5 & 7 & 2 & 1 & 10 \\ 9 & 10 & 1 & 7 & 5 \\ 7 & 8 & 2 & 4 & 4 \\ 9 & 8 & 1 & 10 & 3 \end{bmatrix}.$$

(c) $F = \mathbf{Q}$,

$$A = \begin{bmatrix} -2 & 0 & -1 \\ -8 & 2 & -2 \\ 1 & 0 & 0 \end{bmatrix}.$$

7 For each matrix A in Exercise 6 find an invertible matrix Q such that QAQ^{-1} is in Jordan canonical form.

Appendix 1
THE APL LANGUAGE

This appendix concentrates mainly on APL as a formal system of mathematical notation. The aspects of the language related to its implementation in terminal systems are discussed primarily in Appendix 2. We will, however, give a brief description of the operation of an APL terminal system so that readers can follow the sample dialogues presented in the text and reproduce them at a terminal. We will also show how to combine APL statements into procedures that can be executed on a terminal system.

The treatment of APL presented in this book is self-contained and goes deeply into the parts of the language that are likely to be of use to readers in solving the algebraic problems discussed. Certain aspects of APL are not covered. These include the system functions, some of the system variables, terminal input and output, the *CONTINUE* workspace, and accessing files. Individuals wishing to fill in these gaps should consult one or more of the books on APL listed in the bibliography.

It is now possible to work with APL on small, one-user computers. However, we will assume here that the APL system is running on a large computer that serves many individuals simultaneously. Before attempting to use a particular terminal system, one must obtain an account number for that system. Armed with an account number and having found a terminal appropriate to the system, one must connect the terminal to the computer and identify oneself by the account number. This is called *signing on.* The details of the sign-on process vary, depending on the version of the terminal system and the type of terminal being used. The computer center providing the APL service can supply information about signing on.

Once sign-on has been completed, the carriage or cursor of the terminal is positioned six spaces to the right of the left margin. At this point, one is free to type in an APL statement describing a computation to be performed. The statement is terminated by a carriage return. Now the computer begins to process the statement. In some cases results of the computation will be printed at the terminal. If the statement contains an error that makes complete processing impossible, an error message is printed. Any typing initiated by the computer begins at the left margin. When processing of the first statement has finished, the carriage or cursor is indented six spaces on a new line. A second APL statement may now be entered.

APL KEYBOARD

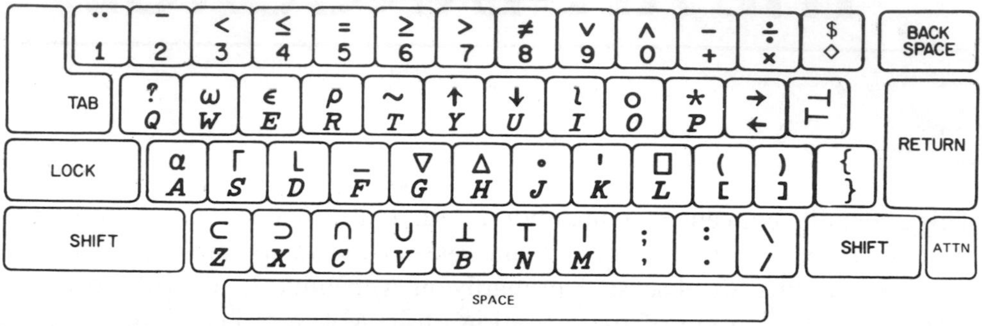

The layout of the keyboard varies from terminal to terminal. The layout of one terminal is given here. Readers should familiarize themselves with the keyboard of the terminal available to them.

In the description of the APL language that follows, no serious attempt has been made to anticipate every possible question. In almost all cases the best way to obtain the answer to "What would happen if . . .?" is simply to enter the particular construction at a terminal and see what happens. Beginning students of APL should let their terminal systems be the final arbiter concerning all questions about the APL language.

1. A SAMPLE TERMINAL SESSION

Let us imagine that we are seated in front of an APL terminal and that we have completed the sign-on process. We try a few simple arithmetic problems, entering the problems one at a time, signaling the end of the line with a carriage return and waiting for the computer's response. (To avoid wasting space on the page, the following listing has been broken in the middle and is presented in two columns. At the terminal, it would appear as a single long column.)

```
      4+3
7
      4-3
1
      4×3
12
      4÷3
1.333333333
      4*3
64
      3*4
81
```

It should be clear that the symbols +, -, ×, and ÷ represent the usual arithmetic operations and that ⋆ is used for the operation of exponentiation or raising to a power. Here are some more examples using two operations in the same statement.

```
      4+3-2
5
      4+3×2
10
```

```
      4×3+2
20
      4-3+2
¯1
```

The results of the last two computations may seem incorrect. This is due to the order in which the operations are performed. Consider the third statement, `4×3+2`. In normal mathematical notation, multiplication is said to take precedence over addition. Thus the product of 4 and 3 would be computed first and the result added to 2, giving 14. However, in APL there are so many basic or *primitive* operations that it was found to be impossible to construct useful precedence rules. Instead, the following simple rule was adopted.

The Right-to-Left Rule. Unless otherwise indicated by parentheses, the operations in an APL statement are performed from right to left, each operation taking as its right argument everything to the right of it.

In the statement `4×3+2` the rightmost operation `+` is performed first. Thus 3 and 2 are added and the result multiplied by 4, giving 20. The order of performing the operations can be changed using parentheses in the usual way.

```
      (4×3)+2
14
```

Now consider the other troublesome statement, `4-3+2`. Again, the rightmost operation is performed first. Thus 3 and 2 are added and the result subtracted from 4. The expected result is `-1` but we find `¯1` instead. The Right-to-Left Rule forces the symbol `¯` used when writing a negative constant to be different from the symbol `-` used to denote the operations of subtraction and negation. Note the difference in the following two statements.

```
      -4+3
¯7
```

```
      ¯4+3
¯1
```

In the first case we wish the negative of `4+3`, and in the second we are asking for the sum of `¯4` and `3`.

The definition of the APL division operation ÷ differs slightly from the usual definition of division of real numbers. Ordinarily, division by 0 is never permitted. However, in APL the expression 0÷0 is allowed and is given the value 1.

```
      0÷0
1
```

The many primitive APL operations can be classified into two types, according to the number of arguments they require. An operation with one argument is called *monadic* or *unary* and the symbol for that operation is always placed to the left of its argument. The only monadic operation we have encountered so far is the negation operation in the statement -4+3. An operation with two arguments is called *dyadic* or *binary* and the symbol for the operation is placed between the arguments. We have seen five dyadic operations: addition, subtraction, multiplication, division, and exponentiation. In order to keep the number of symbols required to write APL statements to a minimum, the same symbol is often used to represent a monadic operation and also a related dyadic operation. No ambiguity can arise, since a symbol being used monadically will have either nothing or another operation to its left. Thus in the statement

```
      -1--2
¯3
```

the first and third minus signs represent monadic negation while the second minus sign represents dyadic subtraction.

It is often convenient to save the result of a computation in the computer instead of having it printed at the terminal and forgotten. This is done by assigning the result to a variable using the assignment arrow ←.

```
      X←3+4
      X
7
      Y←X*0.5
      Y
2.645751311
```

```
      Q←X×X+Y
      Q
67.52025918
      X←X+1
      X
8
```

Typing the name of a variable and a carriage return causes the current value of that variable to be typed at the terminal. The Right-to-Left Rule applies to ← so that in the preceding example X←X+1, 1 is added to the current value of *X* and then this new value is assigned to *X*.

Often we will want to assign a value to a variable and, at the same time, display that value at the terminal. This can all be done in one statement.

```
      ⎕←Z←2+11×7-4
35
```

Assigning a value to the "quad" symbol ⎕ indicates that the value is to be printed at the terminal.

Any relatively short sequence of letters, underlined letters, and digits is a valid variable name provided that the sequence starts with a letter or an underlined letter. No blank spaces are permitted within a variable name. Thus

```
      ABC                     S̲U̲M̲
      X1                      C12B3
      SUM                     GREATESTCOMMONDIVISOR
```

are all valid names. A letter and the same letter underlined are considered to be different symbols, and so *SUM* and *S̲U̲M̲* are different names.

Several assignments may appear in one APL statement. The Right-to-Left Rule governs the order in which these assignments are made.

```
      I←J←K←0                       M←(A←4)×B←5
      I                             B
0                             5
      J                             A
0                             4
      K                             M
0                             20
```

Numbers with very large or very small absolute values may be entered using scientific notation.

```
      AVOGADRO←6.022E23             PLANCK←6.626E¯34
```

Here we have entered Avogadro's number 6.022×10^{23} and Planck's constant 6.626×10^{-34}. Scientific notation may be used by the computer for output.

```
      7*31
1.57775382E26
```

EXERCISES

1 Carry out at a terminal all of the dialogues presented in this section.

2 Evaluate each of the following APL expressions on paper and then check your answers by entering them at a terminal.

- **(a)** `7+12`
- **(b)** `8-3`
- **(c)** `2-9`
- **(d)** `-5`
- **(e)** `0÷0`
- **(f)** `2*4`
- **(g)** `4*2`
- **(h)** `0*0`
- **(i)** `2*0.5`
- **(j)** `¯3*¯3`
- **(k)** `6+4-2`
- **(l)** `6-4+2`
- **(m)** `6+¯4+2`
- **(n)** `(6-4)+2`
- **(o)** `6-4-2`
- **(p)** `8×4÷2`
- **(q)** `8÷4×2`
- **(r)** `8÷4÷2`
- **(s)** `0*0*0`
- **(t)** `¯5-¯2`
- **(u)** `5-4-3-2`

3 What would be the final values of *X*, *Y*, and *Z* after the statement

```
Z←(Y←2+Z)×X←3+Z←4×X←1
```

is entered at a terminal?

2. ARRAYS

An APL statement may involve three kinds of objects: arrays, primitive operations, and defined procedures. This section is devoted to arrays. Defined procedures are discussed in Section 4; the remaining sections of this appendix deal with the various types of primitive operations.

The three most common types of arrays in APL are *scalars*, *vectors*, and *matrices*. All of the constants and variables that we used in the previous section were scalars, that is, single real numbers. A variable may also be a vector.

```
      ⎕←V←1 2 3 5 8 13
1 2 3 5 8 13
```

The components of a vector of constants are separated by spaces. The *I*th component of *V* is denoted `V[I]`, and *I* is referred to as an *index*.

```
      V[2]
2
      V[5]
8
```

```
      V[1 3 5]
1 3 8
      V[2 4 6]
2 5 13
```

In the last two examples a vector was used as an index to obtain a vector of components of *V*.

The dyadic primitive operation ρ lets us reshape a vector into a matrix.

```
      ⎕←A←2 3ρV
1  2  3
5  8 13
```

```
      ⎕←B←3 2ρV
1  2
3  5
8 13
```

In each of the preceding examples the first argument of ρ is a vector giving the number of rows and columns the matrix is to have. The matrix is built up from the second vector argument a row at a time. This reshape operation is discussed more fully in Section 5.

Two indices are required to specify a particular entry in a matrix. (One usually refers to the components of a vector. The term "entry" is used with matrices and more complicated arrays and may also be used with vectors.)

```
      A[1;2]
2
```

```
      B[3;1]
8
```

The two indices of a matrix are separated by a semicolon. An entire row or column of a matrix may be obtained by leaving the appropriate index unspecified.

```
      A[2;]
5 8 13
```

```
      B[;1]
1 3 8
```

If two vectors are used as indices of a matrix, then the result is a matrix.

```
      A[1 2;2 3]
2  3
8 13
```

Entries in an array can be changed using the assignment arrow.

```
      B
1  2
3  5
8 13
      B[1;2]←¯5
```

```
      B[2 3;1]←1 0
      B
1  ¯5
1   5
0  13
```

The *rank* of an array is the number of indices required to specify a particular entry in the array. Thus scalars, vectors, and matrices have rank

0, 1, and 2, respectively. APL permits the definition of arrays of any finite rank. For example:

```
      ⎕←C←2 2 2⍴1 2 3 4 5 6 7 8
 1 2
 3 4

 5 6
 7 8
      C[1;2;1]
3
      C[;1;]
 1 2
 5 6
```

Here C is an array of rank 3. Note that C is displayed at the terminal as the two matrices `C[1;;]` and `C[2;;]`. If

```
      ⎕←I←1
1
      ⎕←J←1⍴1
1
```

```
      ⎕←K←1 1⍴1
1
      ⎕←L←1 1 1⍴1
1
```

then each of the arrays I, J, K, and L has one entry, but I is scalar, J is a vector of length 1, K is a 1-by-1 matrix, and L has rank 3.

The *shape* of an array is a vector that gives the number of different values each index of the array may have. The monadic operation ⍴ gives the shape of its argument.

```
      ⍴V
6
      ⍴A
2 3
      ⍴B
3 2
```

```
      ⍴C
2 2 2
      ⍴A[;1]
2
      ⍴V[1 3 5]
3
```

The components of the shape of X are sometimes called the *dimensions* of X. The length of ⍴X is the rank of X, which is therefore ⍴⍴X.

```
      ⍴⍴V
1
      ⍴⍴A
2
      ⍴⍴B
2
```

```
      ⍴⍴C
3
      ⍴⍴1
0
```

If X is scalar, then X has rank 0 and ρX is a vector of length 0, that is, a vector with no components.

```
      ρ1
                    ←———— (blank line "printed" at the terminal)
```

The vector V has six components and we have been indexing V by elements of the set {1, 2, 3, 4, 5, 6}. There are times when it is more convenient to index V by elements of the set {0, 1, 2, 3, 4, 5}. The smallest value an index may have is called the *index origin* and is denoted ⎕*IO*. The index origin is one of a class of variables called *system variables*, which are described in Section A2.2. The origin may be 1 or 0 and can be changed by a normal APL assignment. During the sign-on process of an APL terminal system, the origin is automatically set to 1.

```
      V
1 2 3 5 8 13
      V[2]
2
      V[1 3 5]
1 3 8
```

```
      ⎕IO←0
      V[2]
3
      V[1 3 5]
2 5 13
```

The origin affects the indexing of matrices and other arrays as well.

```
      A
 1  2  3
 5  8 13
      ⎕IO←1
```

```
      A[1;2]
2
      ⎕IO←0
      A[1;2]
13
```

The monadic operation ι is called the *index generator*. If N is a nonnegative integer, then ιN is the vector of possible indices for a vector of length N. This vector depends on the index origin.

```
      ⎕IO←1
      ι3
1 2 3
      ι5
1 2 3 4 5
```

```
      ⎕IO←0
      ι3
0 1 2
      ι5
0 1 2 3 4
```

The vector ι0 is the empty vector.

Most of the arrays we will be using will have entries that are real numbers. However, APL does allow the formation of *character arrays*, that is, arrays whose entries may be any symbol on the APL keyboard. Individual characters and vectors of characters are enclosed in single quotation marks when entered at a terminal. For example:

```
      ⎕←E←'='
=
      ρρE
0
      ⎕←NAME←'NAME'
NAME
```

```
      ρNAME
4
      ρρNAME
1
      NAME[1 2 3 0]
AMEN
```

Here *E* is a scalar and *NAME* is a vector whose components are the letters *N*, *A*, *M*, and *E*.

An array of any rank may be used as an index provided that all the entries in the array are valid indices.

```
      ρC
2 2 2
      P←1 2 1
      ⎕←Q←2 2ρ1 2 2 1
 1 2
 2 1
      R←2 1 2ρ1 1 1 2
      ρR
2 1 2
```

```
      ⎕IO←1
      D←C[P;Q;R]
      ρD
3 2 2 2 1 2
      D[3;2;1;1;1;2]
3
      C[P[3];Q[2;1];R[1;1;2]]
3
```

EXERCISES

1 Carry out at a terminal all of the dialogues presented in this section.

2 Let `A←2 3 5 7 11` and `B←3 1 4`. Evaluate each of the following expressions on paper, first assuming the index origin is 1 and then assuming the origin is 0. Check your answers at a terminal.

(a) `A[4]`	**(d)** `ι7`	**(g)** `ριl0`
(b) `B[2 1]`	**(e)** `ρA`	**(h)** `ρρA[2]`
(c) `A[B]`	**(f)** `ρρB`	**(i)** `A[1+⎕IO]`

3 Suppose the statements

```
      E←ι5
      E[5 3 1 3 5]←ι5
```

are entered at a terminal with `⎕IO` set to 1. What would you expect the resulting vector *E* to be? Check your answer at a terminal. (Different APL systems may give different results.)

4 Enter the following matrices at a terminal.

```
      A                          X
2  7  ¯1                     1  1  0
3 ¯2   5                     0  1  1
      B                      0  0  0
6  2   0
4  1   3
```

5 Let *A*, *B*, and *X* be as in Exercise 4. Evaluate each of the following expressions on paper, first assuming an origin of 1 and then an origin of 0. Check your answers at a terminal.

(a) `A[1;1]`	**(d)** `X[;2]`	**(g)** `ρA[1;]`
(b) `X[2;1]`	**(e)** `ρB`	**(h)** `ρρB[;1]`
(c) `B[1;]`	**(f)** `ρρX`	**(i)** `ρ3 4ρι12`

6 Let `⎕IO←1` and `C←'ABEFGILNRSU. '`. Evaluate the following.

(a) `ρC`
(b) `ρρC`
(c) `C[1 7 5 3 2 9 1 13 6 10 13 4 11 8 12]`

7 Suppose that the shapes of the arrays *A*, *B*, *C*, *D*, are `3 2 5`, `2 3`, `4 7 3`, and `2 4 6 8`, respectively. For each of the following arrays, give the rank and the shape, assuming each is defined.

(a) `A[B;C;D]`	**(d)** `C[B;1;A]`	**(g)** `B[B;]`
(b) `A[B;D;]`	**(e)** `C[B;C;A]`	**(h)** `B[;B]`
(c) `B[A;D]`	**(f)** `D[;;;]`	**(i)** `A[;C[D;B;A];A]`

3. PRIMITIVE SCALAR OPERATIONS

There are many primitive operations in APL. However, we will require only a small fraction of these operations in our study of algebra. In addition, the APL operations fall naturally into classes consisting of operations with similar properties, making the job of learning about the operations relatively easy. In this section we will describe some *scalar operations*, operations that take one or two scalar arguments and produce a scalar result.

First, we will discuss a class of operations that includes the familiar operations of arithmetic we used in Section 1. Table 1 lists all of the monadic arithmetic operations.

Symbol	Name
-	Negation
|	Absolute value
⌊	Floor
⌈	Ceiling
×	Signum
÷	Reciprocal
!	Factorial
*	Exponential
⍟	Natural logarithm
○	π times

Table 1. The Monadic Arithmetic Operations

The first two operations, negation and absolute value, should be familiar.

```
      -1
¯1
      -¯2
2
```

```
      |¯3
3
      |3
3
```

The *floor* ⌊X of a real number X is the largest integer less than or equal to X.

```
      ⌊2.1
2
      ⌊¯2.1
¯3
```

```
      ⌊¯3
¯3
      ⌊0
0
```

The *ceiling* ⌈X of a real number X is the smallest integer greater than or equal to X.

```
      ⌈2.1
3
      ⌈¯2.1
¯2
```

```
      ⌈¯3
¯3
      ⌈0
0
```

The *signum* ×X of a real number X is 1, 0, or ¯1, according to whether X is positive, zero, or negative.

```
      ×2
1
      ×0
0
```

```
      ×¯2
¯1
      ×3.5
1
```

The *reciprocal* ÷X of X is 1÷X.

```
      ÷4                                ÷¯3
0.25                                ¯0.3333333333
```

If N is a positive integer, then the *factorial* !N of N is the product of the integers 1, 2, . . . N. In addition, !0 is defined to be 1.

```
      !3                                !6
6                                   720
      !4                                !0
24                                  1
```

The symbol ! is made by overstriking the single quote ' with a period.

The last three operations in Table 1 are not required in this book but are included for completeness. The value of ⋆X is e raised to the power X. The value of ⍟X is the natural logarithm of X. The symbol ⍟ is made by overstriking ○ and ⋆. The value of ○X is π times X.

```
      ⋆1                                ⍟⋆3
2.718281828                         3
      ⋆2                                ○1
7.389056099                         3.141592654
      ⍟10                               ○2
2.302585093                         6.283185307
```

The dyadic arithmetic operations are listed in Table 2.

Symbol	Name
+	Addition
−	Subtraction
×	Multiplication
÷	Division
⋆	Exponentiation
⌊	Minimum
⌈	Maximum
|	Remainder
!	Binomial coefficient
⍟	Logarithm
○	Trigonometric, etc.

Table 2. The Dyadic Arithmetic Operations

We have already encountered the first five dyadic arithmetic operations in Section 1. If X and Y are real numbers, then X⌊Y is the smaller of the two and X⌈Y is the larger.

```
      3⌊4
3
      ¯3⌊¯4
¯4
      1.2⌊0.7
0.7
```

```
      3⌈4
4
      ¯3⌈¯4
¯3
      1.2⌈0.7
1.2
```

The definition of the *remainder* operation is a little more complicated. Let X and Y be real numbers. The X remainder of Y, denoted by X|Y, is a number of the form Y+N×X, where N is an integer. If X is 0, then Y+N×X is Y for any N and so 0|Y is Y. If X is not 0, we choose the correct number of the form Y+N×X as follows. If X is positive, then X|Y is the smallest number of the form Y+N×X that is greater than or equal to 0. If X is negative, then X|Y is the largest number of the form Y+N×X that is less than or equal to 0.

```
      3|4
1
      4|3
3
      3|11
2
      ¯3|11
¯1
```

```
      0|3
3
      0|¯3
¯3
      1.5|5
0.5
      ¯1.5|5
¯1
```

Let M and N be nonnegative integers with M less than or equal to N. The *binomial coefficient* M!N is (!N)÷(!N-M)×!M and is the coefficient of $X^{N-M}Y^{M}$ in the polynomial $(X+Y)^{N}$. It is also the number of M-element subsets in a set with N elements. The traditional notation for M!N is

$$\binom{N}{M}.$$

```
      2!6
15
      3!8
56
```

```
      0!8
1
      8!8
1
```

The value of X⍟Y is the logarithm of Y to the base X.

```
      10⍟100                    4⍟32
2                         2.5
```

The dyadic operation ○ takes as a left argument an integer I, which must have absolute value at most 7. The value of I○X is given in the following table.

I	I○X	(-I)○X
0	(1-X*2)*0.5	(1-X*2)*0.5
1	Sin X	Arcsin X
2	Cos X	Arccos X
3	Tan X	Arctan X
4	(1+X*2)*0.5	(¯1+X*2)*0.5
5	Sinh X	Arcsinh X
6	Cosh X	Arccosh X
7	Tanh X	Arctanh X

We will now consider two other classes of scalar APL operations. In carrying out an algorithm it is often necessary to do different things, depending on whether a particular condition is true or false. In order to be able to handle such decisions within APL, the numbers 1 and 0 are used to represent "true" and "false", respectively. Several primitive operations have been included in the language to allow the formulation of conditions whose truth or falsity can be tested. The use of numbers to represent logical values permits arithmetic operations to be performed with them. The usefulness of this will be seen later.

The six relational operations <, ≤, =, ≥, >, and ≠ each take two real arguments and have the value 1 or 0, according to whether or not the indicated relation holds. Thus X≥Y has the value 1 if X is greater than or equal to Y and 0 otherwise. Here are some examples.

```
      4<5                       7≥6
1                         1
      4≤3                       10≠10
0                         0
      10=11                     5<6<7
0                         0
```

The last example may be puzzling. The Right-to-Left Rule applies to all APL statements and all operations. Thus the expression 6< 7 is evaluated first. Its value is 1. Then 5< 1 is evaluated, giving 0.

There are several logical operations that take only 0 and 1 as arguments.

The most important are ~, ∧, ∨, the *not*, *and*, and *or* operations, respectively. The first is monadic while the other two are dyadic.

```
      ~0                          1∧1
1                           1
      ~1                          1∨0
0                           1
      ~5=3                        0∨0
1                           0
      1∧0                         (5<6)∧6<7
0                           1
```

Sometimes ~*X* is referred to as the *logical negation* of *X*.

There are two other dyadic logical operations. They are ⍲ and ⍱, the *nand* and *nor* operations, respectively. The definition of *X*⍲*Y* is ~*X*∧*Y*, while *X*⍱*Y* means ~*X*∨*Y*. The symbols for these operations are made by overstriking ∧ and ∨, respectively, with ~.

There is one more monadic scalar operation that may be of use. It is denoted by ? and is called *roll*. If *N* is a positive integer, then ?*N* is a randomly chosen integer from ⍳*N*.

```
      ?4                          ?10
1                           5
      ?10                         ?4
8                           3
```

Note that repeated evaluations of ?*N* with the same value of *N* produce different results.

All of the primitive scalar operations in APL are extended to arrays of arbitrary rank in a very natural way. In calculus we learn to add vectors by adding corresponding components. This is the way vectors are added in APL.

```
      X←2 3 1 7 5 10                   X+Y
      Y←1 3 4 5 8 10             3 6 5 12 13 20
```

In fact, every scalar operation is defined for vectors in this componentwise manner.

```
      X-Y
1 0 ¯3 2 ¯3 0
      X×Y
2 9 4 35 40 100
      -X
¯2 ¯3 ¯1 ¯7 ¯5 ¯10
      X⌊Y
1 3 1 5 5 10
      X|Y
1 0 0 5 3 0
      !X
2 6 1 5040 120 3628800
```

```
      ?X
1 1 1 5 5 4
      ⎕←C←X=Y
0 1 0 0 0 1
      ⎕←D←Y<X×X
1 1 0 1 1 1
      ~C
1 0 1 1 1 0
      C∨D
1 1 0 1 1 1
      C∧D
0 1 0 0 0 1
```

If f is a monadic scalar operation and g is a dyadic scalar operation, then for any arrays A and B the expressions fA and AgB are evaluated entry by entry. Here are some examples with matrices.

```
      ⎕←A←2 2⍴¯1 3 2 ¯4
¯1  3
 2 ¯4
      ⎕←B←2 2⍴⍳4
1 2
3 4
      |A
1 3
2 4
      A+B
0 5
5 0
      A×B
 ¯1   6
  6 ¯16
      A<B
1 0
1 1
      A*B
 ¯1              9
  8            256
```

Note that *A*×*B* is not the matrix that is usually called the matrix product of *A* and *B*. The way to describe the matrix product in APL will be discussed in Section 7.

In general, *AgB* is defined only when *A* and *B* have the same rank and the same shape. There is one exception, however. If one of the arrays has just a single entry, then that array is automatically expanded to an array of rank and shape agreeing with the other argument.

```
      X
2 3 1 7 5 10
      X+1
3 4 2 8 6 11
      X*2
4 9 1 49 25 100
      X≥4
0 0 0 1 1 1
```

```
      X-ι1
1 2 0 6 4 9
      2×A
 ¯2  6
  4 ¯8
```

The roll operation is also extended to arrays in an entry-by-entry manner.

```
      ?2 4 6 8
1 3 2 4
```

The name of the roll operation comes from the fact that it is possible to simulate the rolling of dice by executing ?6 6.

```
      ?6 6
1 5
```

```
      ?6 6
3 4
```

One can create random matrices of zeros and ones as follows.

```
      ⎕IO←0
      ⎕←D←2 3ρ2
 2 2 2
 2 2 2
```

```
      ?D
 0 0 1
 1 1 0
```

EXERCISES

1 Carry out at a terminal all of the dialogues presented in this section.

2 Evaluate each of the following APL expressions on paper and then check your answers by entering the expressions at a terminal.

(a) `×¯3.7`
(b) `÷0.05`
(c) `|¯3.7`
(d) `⌊¯3.7`
(e) `⌈¯3.7`
(f) `!4`
(g) `2.9⌊3.1`
(h) `¯2.9⌈¯3.1`
(i) `4|7`
(j) `4|¯7`
(k) `¯4|7`
(l) `2.5|7.1`
(m) `2!4`
(n) `3⌊4⌈2`
(o) `3⌈4⌊2`
(p) `÷÷2`
(q) `8*÷3`
(r) `3+×¯4`
(s) `2|4|6`
(t) `5||¯7`
(u) `2!!3`

3 Show that for any real number *X* the value of `|X` is the same as `X××X`.

4 Explain why forming `⌊X+0.5` amounts to rounding *X* to the nearest integer.

5 Evaluate the following expressions as in Exercise 2.

(a) `3<4`
(b) `1≤¯1`
(c) `7=8`
(d) `2≥2`
(e) `5>6`
(f) `3≠¯2`
(g) `1∨0`
(h) `0∧1`
(i) `∼1∧1`
(j) `∼3=¯2`
(k) `(1>0)∨6<7`
(l) `(3≠3)∧1>0`
(m) `0=4|12`
(n) `(5*2)=5×5`
(o) `120=!5`
(p) `1=2×÷2`
(q) `1>0∧6<7`
(r) `2*1+0∨¯2+⌊3.5`

6 Let `A←2 3 5 7 11`, `B←3 1 4`, and `⎕IO←1`. Evaluate each of the following expressions on paper and check your answers at a terminal.

(a) `A+2`
(b) `A×⍳5`
(c) `B<2`
(d) `B⌈2 4 6`
(e) `1+⍳1`
(f) `⍴1+⍳1`
(g) `1+⍳0`
(h) `⌈(⍳8)÷2`
(i) `0=(⍳12)|12`

7 Construct a vector *T* whose components are the first 100 odd numbers in order.

4. DEFINED PROCEDURES

Not all computations can be performed easily by entering APL statements one at a time at the terminal. Often we will want to do the same basic calculation many times with different values of one or more of the variables. In this case it is more convenient to give the system a complete list of all the statements to be executed and refer to the entire collection by a single short name. This is known as defining a procedure. (The word normally used in books on APL is "function". However, some APL procedures are not functions as we define the term in Chapter 1.)

We begin with a very elementary example. Suppose we have the following right triangle.

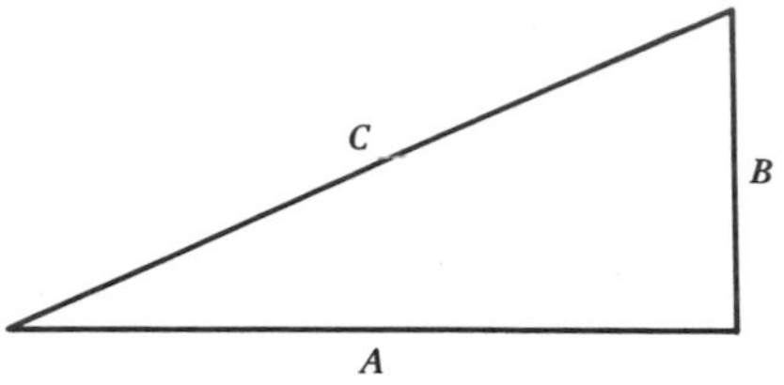

If *A* and *B* are known, then by the Pythagorean Theorem the value of *C* may be computed by

```
C←((A*2)+B*2)*0.5
```

Assuming that we have to calculate the lengths of the hypotenuses of many right triangles, it is convenient to define a procedure called *HYPOT* as follows.

```
      ∇C←A HYPOT B
[1]   C←((A*2)+B*2)*0.5
[2]   ∇
      3 HYPOT 4
5
```

```
      X←12 HYPOT 5
      X
13
      12 HYPOT 3 HYPOT 4
13
```

We have been in what is called the *execution mode* in which statements are executed as soon as they are entered. The symbol ∇ (read "del") indicates that we wish to enter the *definition mode* in which statements are saved for execution at some later time. The remainder of the line following the ∇ is called the *header* of the procedure. It gives the following information.

1. The name of the procedure. (The rules for naming procedures are the same as those for naming variables.)
2. The names of the variables used within the procedure to represent the arguments, if any.
3. The name of the variable used within the procedure to represent the value, if any, returned by the procedure.

In our example the name of the procedure is *HYPOT* and it requires two arguments, *A* and *B*. Following the convention described earlier for primitive APL operations, the name of the procedure is placed between its two arguments. The procedure is to return a value that is referred to within the procedure by the name *C*.

After the header has been entered, the system prompts by typing [1], indicating that the first line of the procedure is to be entered. After that line has been typed, the system prompts with [2] for the second line.

In our case there is no second line, so we type another ∇ to tell the system that the definition is complete and we wish to return to the execution mode. The procedure is now available for use. Note that the Right-to-Left Rule applies to defined procedures as well as to primitive operations.

A defined procedure may have zero, one, or two arguments and may or may not return an explicit value. Thus there are six basic forms for the header of a procedure.

```
      ∇PROC                        ∇Z←PROC
      ∇PROC Y                      ∇Z←PROC Y
      ∇X PROC Y                    ∇Z←X PROC Y
```

Here the name of the procedure is *PROC*, the variables *X* and *Y* represent arguments, and the variable *Z* represents a value returned by the procedure.

The following dialogue illustrates how the variables used in the header have meaning only within the procedure.

```
      A←10+C←¯7
      B
VALUE ERROR
      B
      ∧
      D←8 HYPOT 15
      D
17
```

```
      A
3
      C
¯7
      B
VALUE ERROR
      B
      ∧
```

Here we define two variables *A* and *C* and check that *B* has no value assigned to it. (When we enter the name *B*, the computer responds with the message *VALUE ERROR*, retypes the line, and types a caret below *B*.) Next we use *HYPOT* to compute *D*. Within *HYPOT* the values of the arguments 8 and 15 are assigned to variables *A* and *B* and the value 17 is assigned to *C*. However, after the calculation of *D*, we find that the values of *A* and *C* are unchanged from their values prior to the use of *HYPOT* and that no value is assigned to *B*. The variables *A*, *B*, and *C* in the header of *HYPOT* are treated as separate variables, unrelated to other variables with the same names that may be defined.

Variables that have meaning only within a procedure are called *local variables.* They may have the same names as variables already existing outside the procedure. The arguments of a procedure and the value returned by the procedure are local variables. Additional local variables may be used within a procedure. They are listed in the header preceded by a semicolon and separated by semicolons. Procedures may also use *global variables*, variables that may have been assigned values before execution of the procedure begins and that continue to exist after completion of the procedure. A procedure may even create new global variables.

The following examples illustrate the use of local and global variables. Suppose we need to evaluate the expression

$$(x^3 + a^2)^3 + (x^3 + a^2)^2 + 3$$

for various values of a and x. We start by defining a procedure F to do the evaluation. The expression $x^3 + a^2$ occurs twice so it is reasonable to compute it first.

```
      ∇Y←A F X;B
[1]   B←(X*3)+A*2
[2]   Y←3+(B+1)×B*2
[3]   ∇
      1 F 2
813
      1 F 3
22739
      10 F 4
4437843
      10 F 5
11441253
```

The variable B used in F is a local variable.

Suppose, however, that we knew the value of a will remain the same for a large number of values of x. We might then proceed differently, defining the procedure G as follows.

```
      ∇Y←G X;B
[1]   B←(X*3)+A*2
[2]   Y←3+(B+1)×B*2
[3]   ∇
      A←1
      G 2
813
      G 3
22739
      A←10
      G 4
4437843
      G5
11441253
      G 100
1.000301030E18
```

The variable A in G is now a global variable whose value must be assigned before G is used.

Once a procedure has been defined on a terminal system, the procedure may be listed or modified. Information concerning the editing of defined procedures may be found in Section A2.1 and in APL reference manuals.

A defined procedure may use or *invoke* other procedures, as illustrated in the following somewhat artificial example.

```
      ∇Z←X SUM Y
[1]   Z←X+Y
[2]   ∇
      ∇Z←X PROD Y
[1]   Z←X×Y
[2]   ∇
```

```
      ∇C←A MAX B
[1]   C←(A SUM B)⌈A PROD B
[2]   ∇
      1 MAX 2
3
      3 MAX 4
12
```

In fact, as we will soon see, a procedure may even invoke itself!

The lines of a procedure are normally executed in the order of their line numbers. Sometimes it is necessary to change this order by *branching.* A branch is indicated by a right-pointing arrow: →. The two basic kinds of branches are illustrated in the following procedure *SUMOFSQUARES*, which computes the sum of the squares of the first *N* positive integers.

```
      ∇S←SUMOFSQUARES N;I
[1]   S←0
[2]   I←1
[3]   →(N<I)/0
[4]   S←S+I*2
[5]   I←I+1
[6]   →3
[7]   ∇
```

```
      SUMOFSQUARES 3
14
      SUMOFSQUARES 10
385
      SUMOFSQUARES 0
0
```

This procedure is intended only as an example of the use of branching and not as a model for constructing other APL procedures. In spirit, it is really a FORTRAN program. As we will learn in Section 6, the whole procedure *SUMOFSQUARES* can be replaced by a single short APL statement. Also, it is well known that the expression `(N×(N+1)×1+2×N)÷6` gives the value of the sum of the first *N* squares.

The procedure *SUMOFSQUARES* contains two branches. Line 6 indicates an *unconditional branch* to line 3, while line 3 contains a *conditional branch.* Although it is possible to indicate conditional branches in many ways in APL, perhaps the most common is the following.

→(condition)/number

Here "condition" represents an expression with a logical value, 1 for true or 0 for false, and "number" is a line number. The statement specifies that a branch is to be made only if the condition is true. Thus line 3 of

SUMOFSQUARES

may be read "if *N* is less than *I*, then branch to line 0". Of course, there is no line 0. A branch to a nonexistant line terminates execution of the procedure.

The variable I runs through thc integers 1 , 2 , . . . , N. In line 4 we add I*2 to S. Since S is initially given the value 0, the value of S after a given execution of line 4 is the sum of the first I squares. Execution continues until, at line 5, the value of I becomes N+1. After the branch back to line 3, execution is stopped by the conditional branch.

The procedure SUMOFSQUARES illustrates the three aspects of a *loop*. The loop variable I is initialized in line 2, incremented in line 5, and tested in line 3. It is common practice in APL programming to combine the incrementing and the testing steps.

```
      ∇S←SUMSQ N;I                       SUMSQ 3
[1]   S←I←0                        14
[2]   →(N<I←I+1)/0                       SUMSQ 10
[3]   S←S+I*2                      385
[4]   →2                                 SUMSQ 0
[5]   ∇                            0
```

Because I is incremented before the test is made even on the first pass through the loop, it is necessary to initialize I with one less than the first value actually desired, that is, with 0.

There is another approach possible to the problem of computing the sum of the first N squares.

```
      ∇S←RSUMSQ N                        RSUMSQ 3
[1]   →(N>0)/3                     14
[2]   →S←0                               RSUMSQ 10
[3]   S←(RSUMSQ N-1)+N*2           385
[4]   ∇                                  RSUMSQ 0
                                   0
```

The idea behind RSUMSQ is simple. In order to compute the sum of the first N squares, it is enough to add N*2 to the sum of the first N-1 squares. The sum of the first N-1 squares is computed by adding (N-1)*2 to the sum of the first N-2 squares, and so on. Obviously, this method is not going to work unless there is some value of N for which we already know the sum of the first N squares. In RSUMSQ the sum of the first 0 squares is taken to be 0. (This is consistent with the standard mathematical convention that defines a sum with no summands to be 0.) For any positive integer N the value of RSUMSQ N is the sum of N*2 and RSUMSQ N-1.

The use by a procedure of itself is called *recursion*. There is a close connection between recursive procedures and proofs by induction. An induction proof that some mathematical object exists can often be converted into a recursive procedure for constructing the object.

There are some practical difficulties in using recursive APL procedures. Each time one procedure uses another procedure, additional memory space in the computer is required. Since memory is limited, this means that procedures cannot continue invoking other procedures indefinitely.

It is possible to write APL procedures that will never terminate by themselves. Perhaps the simplest is the following.

```
      ∇INFINITELOOP
[1]   →1
[2]   ∇
```

Here the single line of the procedure consists of a branch to itself that continues indefinitely until there is some outside intervention. Procedures of this type are said to contain an *infinite loop*. Infinite loops often occur unintentionally. If a procedure seems to be taking longer than expected to finish, one should signal attention by hitting the attention or break key on the terminal. This causes the procedure to be suspended between two lines. One may now look at the values of the variables in the procedure to see if everything is going as intended. Computation may be resumed if desired by typing → followed by the number of the next line to be executed.

Additional information concerning suspended procedures can be found in Sections A2.3 and A2.5 and in APL reference manuals. A word of caution is in order, however. Beginning students of APL often forget to terminate suspended procedures when they are no longer needed. This can cause problems, especially when local variables having the same names as global variables are involved. It is a good rule to terminate suspended procedures as soon as possible by typing → followed by a carriage return. Within a procedure, a right arrow by itself aborts the entire calculation currently under way.

There is one more point concerning branching that needs to be mentioned. The editing of a procedure may cause line numbers to change. If the line number for a particular statement changes, then any branches to that statement will also have to be changed. Since procedures often go through many versions before they reach their final form, a great deal of work can be saved by the use of statement labels. The following version of *SUMSQ* uses a label *LOOP* on line 2.

```
      ∇S←SUMSQX N;I
[1]   S←I←0
[2]   LOOP:→(N<I←I+1)/0
[3]   S←S+I*2
[4]   →LOOP
[5]   ∇
```

Statement labels are local variables that are automatically given a value equal to the number of the line on which they occur. A label is separated from the remainder of the statement by a colon. It is good programming practice to use statement labels whenever possible.

Loops need not involve loop variables that count passes through the loop.

```
      ∇Y←SQRT X
[1]   →(0>X)/0
[2]   →(0=Y←X)/0
[3]   BACK:→(X≠Y×Y←(Y+X÷Y)÷2)/BACK
[4]   ∇
      SQRT 43.56
6.6
      SQRT 0
0
      I←SQRT ¯1
VALUE ERROR
      I←SQRT ¯1
        ∧
```

The procedure *SQRT* computes the square root *Y* of its argument *X*. For nonnegative values of *X* we take *X* to be a first approximation to *Y*. If *X* is 0, we stop immediately. Otherwise, given any value for *Y*, we replace *Y* by the average of *Y* and *X*÷*Y* and continue until *X* and *Y*×*Y* agree to the accuracy specified by the comparison tolerance. (See Section A2.2.) If *X* is negative, then the procedure is terminated in line 1 before any value is assigned to *Y*. This accounts for the error message printed in the previous example.

EXERCISES

1 Enter the definitions of each of the procedures presented in this section and try them out.

2 Define and use a procedure called *AVERAGE* that returns the average of its two arguments.

3 Define and use a procedure *POWERS* with one argument *X* and no explicit result that creates three global variables *X*2, *X*3, and *X*4, which are the square, cube, and fourth power of *X*, respectively.

4 Define and use a procedure *PARITY* with a single argument *N*, assumed to be an integer, that returns the value 1 if *N* is odd and the value 0 if *N* is even.

5 Write a procedure *POWERSUM* modeled after *SUMSQ* that has two arguments *M* and *N* and that returns the sum of the *M*th powers of 1, 2, . . . , *N*.

6 Write a recursive version *RPOWERSUM* of *POWERSUM* based on *RSUMSQ*.

7 Write a recursive procedure *FACT* for computing factorials that does not involve the use of the primitive operation !. Use the identity (!*N*) = *N*×!*N*−1 and the starting value of 1 for !0.

8 The *Fibonacci sequence* is the sequence 1, 2, 3, 5, 8, 13, . . . , in which each term after the second is the sum of the two terms preceding it. Construct a recursive procedure *FIB* such that *FIB I* is the value of the *I*th term in the Fibonacci sequence.

9 Write a procedure *FIBSUM* such that *FIBSUM N* in the sum of the first *N* terms of the Fibonacci sequence. Do not use recursion or the procedure *FIB* of Exercise 8.

10 Any positive integer n can be written uniquely in the form $2^r m$, where r and m are integers and m is odd. Call m the *odd part* of n. Write a procedure *ODDPART* such that *M*←*ODDPART N* makes *M* the odd part of *N*. Use a loop similar to the one in *SQRT*. This can be done by taking *M* to be *N* initially and then replacing *M* by *M*÷2 as long as *M* is even.

5. PRIMITIVE MIXED OPERATIONS

In this section we consider a class of primitive APL operations that are not applied entry by entry to their arguments. The operations in this class are usually called *mixed*.

Table 1 lists the monadic mixed operations. Note that there are three different ways of denoting the operation reverse.

Symbol	Name
⍴	Shape
⍳	Index generator
⌽ ⊖ ⌽[*I*]	Reverse
⍋	Grade up
⍒	Grade down
⍉	Transpose
,	Ravel
⌹	Matrix inversion
⍎	Execute
⍕	Standard format

Table 1. The Monadic Mixed Operations

We have already seen the shape and index generator operations in Section 2. The operations ⌹, ⍎, and ⍕ are described in Section 8. We will discuss the remaining five operations in Table 1 one at a time. The symbols ⌽, ⊖, ⍉, ⍋, ⍒, ⍎, ⍕, and ⌹ are entered at a terminal by overstriking ○, ○, ○, Δ, ∇, ⊥, ⊤, and ⎕ by |, -, \, |, |, ∘, ∘, and ÷, respectively.

If *V* is a vector, then ⌽*V* is the vector obtained from *V* by *reversing* the order of the components.

```
      ⌽V←5 3 2 1
1 2 3 5
```

```
      ⌽⍳4
4 3 2 1
```

With matrices, the symbol ⌽ indicates the reversal of the rows, while ⊖ indicates reversal of the columns.

```
      ⎕←A←2 3⍴⍳6
 1 2 3
 4 5 6
      ⌽A
 3 2 1
 6 5 4
```

```
      ⊖A
4 5 6
1 2 3
```

If *C* is an array of rank greater than 2, we can specify reversal along any of the "axes" of *C* by using the indexed form ⌽[*I*].

```
      ⎕←C←2 2 2⍴⍳8
1 2
3 4

5 6
7 8
      ⌽[1]C
5 6
7 8

1 2
3 4
```

```
      ⌽[2]C
3 4
1 2

7 8
5 6
      ⌽[3]C
2 1
4 3

6 5
8 7
```

The symbol ⌽ always denotes reversal along the last axis, while ⊖ always denoted reversal along the first axis. The indexing in ⌽[I] is subject to the index origin, just as in the case of array indexing.

```
      ⌽[1]A
4 5 6
1 2 3
      ⎕IO←0
```

```
      ⌽[1]A
3 2 1
6 5 4
      ⎕IO←1
```

The *grade up* operation ⍋ takes a vector V as an argument and produces a vector P that is a permutation of ⍳⍴V such that the components of V[P] are in increasing order.

```
      V
5 3 2 1
      ⍋V
4 3 2 1
      V[⍋V]
1 2 3 5
```

```
      ⍋W←2 1 7 4
2 1 4 3
      W[⍋W]
1 2 4 7
```

If the components of V are not distinct, then P←⍋V is chosen so that whenever I < J and V[P[I]] = V[P[J]], then P[I] < P[J].

```
      ⍋1 2 3 1 2 3
1 4 2 5 3 6
```

This determines *P* uniquely. The result of ⍋ is origin dependent.

```
      ⍋V
4 3 2 1
      ⎕IO←0
```

```
      ⍋V
3 2 1 0
      ⎕IO←1
```

The *grade down* operation ⍒ is defined very much like ⍋. The components of V[⍒V] are in decreasing order, however.

```
      ⍒W
3 4 1 2
```

```
      W[⍒W]
7 4 2 1
```

The *transpose* operation ⍉, when applied to a matrix, gives the usual matrix transpose.

```
      A
 1 2 3
 4 5 6
```

```
      ⍉A
 1 4
 2 5
 3 6
```

If *X* is any array, then ⍉X is the array obtained by reversing the order of the indices. If *X* has rank 4 and Y←⍉X, then Y[I;J;K;L] is X[L;K;J;I]. If *X* is a scalar or a vector, then ⍉X is the same as *X*.

```
      ⍴D←2 3 4⍴⍳24
2 3 4
```

```
      ⍴⍉D
4 3 2
```

The *ravel* ,X of an array *X* is the vector that lists the entries of *X* in the order they would be printed at a terminal.

```
      A
 1 2 3
 4 5 6
      ,A
1 2 3 4 5 6
      V
5 3 2 1
      ,V
5 3 2 1
```

```
      ,1
1
      ⍴,1
1
      ,2 2 2⍴⍳8
1 2 3 4 5 6 7 8
```

The dyadic mixed operations are given in Table 2.

Symbol	Name
⍴	Reshape
⍳	Index of
∊	Membership
↓	Drop
↑	Take
/ ⌿ /[*I*]	Compress
\ ⍀ \[*I*]	Expand
⌽ ⊖ ⌽[*I*]	Rotate
, ,[*I*]	Catenate, laminate
⍉	Generalized transpose
?	Deal
⊥	Decode
⊤	Encode
⌹	Matrix division
⍕	Format

Table 2. The Dyadic Mixed Operations

The last four operations–decode, encode, matrix division, and format–will be described in Section 8. The other operations are discussed in the remainder of this section.

The reshape operation was introduced in Section 2. In general, *V*⍴*W* is defined whenever *V* and *W* are vectors or scalars and the entries of *V* are nonnegative integers. If none of the components of *V* is 0, then *W* must be nonempty. The result is an array whose shape is ,*V* and whose entries are taken from *W*.

```
      ⎕←B←3 4⍴⍳20
 1  2  3  4
 5  6  7  8
 9 10 11 12
      2 3⍴⍳4
1 2 3
4 1 2
      5⍴1
1 1 1 1 1
      ⍴2 2 2 2⍴0
2 2 2 2
      ⍴3 0 2 0⍴1
3 0 2 0
```

If *W* has too many entries, then only enough entries are taken from the beginning of *W* to form the desired array. If *W* has too few entries, then the entries of *W* are repeated.

In the expression *V*⍳*X* the array *V* must be a vector, but *X* may have any rank. The result is an array *Y* with the same shape as *X*. Each entry in *Y* is the index in *V* of the first occurrence of the corresponding entry in *X*. If some entry of *X* does not occur as a component of *V* at all, then one more than the largest valid index for *V* is used for the entry in *Y*. The result of *V*ι*X* is origin dependent.

```
      V
5 3 2 1
      Vι3
2
      W
2 1 7 4
      VιW
3 4 5 5
```

```
      WιV
5 5 1 2
      A
 1 2 3
 4 5 6
      VιA
 4 3 2
 5 1 5
```

The APL membership operation ∊ is closely related to the usual notion of set membership. If *A* and *B* are arrays, then *C*←*A*∊*B* defines *C* to be the logical array with the same shape as *A* such that an entry of *C* is 1 if and only if the corresponding entry of *A* occurs among the entries of *B*.

```
      A
 1 2 3
 4 5 6
      W
2 1 7 4
```

```
      A∊W
 1 1 0
 1 0 0
      W∊A
1 1 0 1
```

If *V* is a vector and *N* is a scalar or a vector of length 1, then *N*↓*V* is the vector obtained by *dropping* or omitting the first *N* components of *V* if *N*>0 or the last |*N* components of *V* if *N*< 0.

```
      V
5 3 2 1
      2↓V
2 1
```

```
      ¯2↓V
5 3
      ρ5↓V
0
```

Rows and columns may be dropped from a matrix.

```
      B
 1  2  3  4
 5  6  7  8
 9 10 11 12
      1 1↓B
 6  7  8
10 11 12
```

```
      ¯1 ¯1↓B
 1 2 3
 5 6 7
      ¯1 2↓B
 3 4
 7 8
```

The *take* operation ↑ is defined in much the same way as ↓, but now the first argument indicates the part of the second argument to be kept or taken.

```
      V
5 3 2 1
      3↑V
5 3 2
      ¯3↑V
3 2 1
```

```
      B
1  2  3  4
5  6  7  8
9 10 11 12
      2 2↑B
1 2
5 6
```

If we try to take more than is present, then zeros are added at the end or the beginning, depending on the sign of the first argument.

```
      V
5 3 2 1
      7↑V
5 3 2 1 0 0 0
```

```
      ¯6↑V
0 0 5 3 2 1
```

In the case of character arrays, spaces are used instead of zeros.

Let V be a vector and let X be a logical vector of length ρV. The *compression* X/V of V by X is the vector consisting of the components of V corresponding to the 1's in X.

```
      V
5 3 2 1
      X←1 1 0 1
      X/V
5 3 1
```

```
      (~X)/V
2
      (V>2)/V
5 3
```

An array may be compressed along any axis. The symbols /, ⌿, and /[I] are used to denote compression along the last, first, and Ith axes, respectively.

```
      B
 1  2  3  4
 5  6  7  8
 9 10 11 12
      X
1 1 0 1
      X/B
 1  2  4
 5  6  8
 9 10 12
      1 0 1⌿B
 1  2  3  4
 9 10 11 12
```

```
      1 0 1/[1]B
1  2  3  4
9 10 11 12
      ⎕IO←0
      X/[1]B
1  2  4
5  6  8
9 10 12
      ⎕IO←1
```

As we saw with the reverse operation, the indexing in /[*I*] is the subject to the index origin.

The operation of *expansion* is almost inverse to that of compression. Let *V* be a vector and let *X* be a logical vector with exactly ρ*V* entries equal to 1. The expansion *X**V* of *V* by *X* is the vector of length ρ*X* with the components of *V* in the positions corresponding to ones in *X* and 0's elsewhere. (If *V* is a character vector, then the 0's in *X* correspond to spaces in *X**V*.)

```
      V                              X\V
5 3 2 1                        5 0 3 2 0 1
      X←1 0 1 1 0 1
```

An array may be expanded along any one of its axes, with \\, ⍀, and \\[*I*] being used completely analogously to /, ⌿, and /[*I*].

```
      B                            1 0 1 1⍀B
 1  2  3  4                    1  2  3  4
 5  6  7  8                    0  0  0  0
 9 10 11 12                    5  6  7  8
      X\B                      9 10 11 12
 1  0  2  3  0  4                  X\[2]B
 5  0  6  7  0  8              1  0  2  3  0  4
 9  0 10 11  0 12              5  0  6  7  0  8
                               9  0 10 11  0 12
```

If *V* is a vector and *N* is a positive integer, then *N*⌽*V* is the vector obtained by *rotating* *V* to the left *N* places. Any components shifted off the left end reappear at the right. If *N*<0, then *N*⌽*V* is the result of rotating *V* to the right |*N* places.

```
      V                             ¯1⌽V
5 3 2 1                        1 5 3 2
      2⌽V                             7⌽V
2 1 5 3                        1 5 3 2
```

Arrays may be rotated along any axis, with ⌽, ⊖, and ⌽[*I*] being used to indicate which axis.

```
      B                             ¯1⊖B
 1  2  3  4                    9 10 11 12
 5  6  7  8                    1  2  3  4
 9 10 11 12                    5  6  7  8
      1⌽B                            2⌽[2]B
 2  3  4  1                    3  4  1  2
 6  7  8  5                    7  8  5  6
10 11 12  9                   11 12  9 10
```

It is possible to rotate the rows or columns of a matrix by different amounts.

```
      B                         0 2 ¯1 2⊖B
 1  2  3  4                 1 10 11 12
 5  6  7  8                 5  2  3  4
 9 10 11 12                 9  6  7  8
      0 1 2⌽B
 1  2  3  4
 6  7  8  5
11 12  9 10
```

The components of the first argument give the amount the rows or columns are to be rotated.

If *V* and *W* are vectors, then *V*,*W* is the vector obtained by following the components of *V* with the components of *W* and is called the *catenation* of *V* and *W*.

```
      V                      V,W
5 3 2 1                 5 3 2 1 2 1 7 4
      W                      W,V
2 1 7 4                 2 1 7 4 5 3 2 1
```

Two arrays *A* and *B* of the same positive rank may be catenated along any axis, provided their dimensions along the other axes agree. The expression *A*,*B* denotes catenation along the last axis. If some other axis is desired, then the form *A*,[*I*]*B* must be used.

```
      A                      A,[1]A
1 2 3                   1 2 3
4 5 6                   4 5 6
      A,A               1 2 3
1 2 3 1 2 3             4 5 6
4 5 6 4 5 6
```

The catenation *A*,*B* is also defined if the ranks of *A* and *B* differ by 1 or if one array has positive rank and the other is a scalar. Here are a few examples.

```
      A                      A,[1]⍳3
1 2 3                   1 2 3
4 5 6                   4 5 6
      A,⍳2              1 2 3
1 2 3 1                      7,A
4 5 6 2                 7 1 2 3
      (⍳2),A            7 4 5 6
1 1 2 3
2 4 5 6
```

Two arrays A and B of the same shape may be *laminated* to form an array of rank 1+ρρA in which the dimension along the new axis is 2. If A and B are scalars, the result is written A,B and is a vector of length 2. If A and B have positive rank, then one must write A,[I]B, where I is not an integer. This indicates that the new axis is to be created after the (⌊I)th axis and before the (⌈I]th axis.

```
      1,2
1 2
      V
5 3 2 1
      W
2 1 7 4
      V,[0.5]W
 5 3 2 1
 2 1 7 4
```

```
      V,[1.5]W
5 2
3 1
2 7
1 4
      ⎕IO←0
      V,[¯0.5]W
5 3 2 1
2 1 7 4
      ⎕IO←1
```

The most common use of the *generalized transpose* operation is in obtaining the main diagonal of a matrix. If B is a matrix, the *main diagonal* of B is the vector whose Ith component is B[I;I].

```
      B
 1  2  3  4
 5  6  7  8
 9 10 11 12
      1 1⍉B
1 6 11
```

```
      ⎕IO←0
      0 0⍉B
1 6 11
      ⎕IO←1
```

In origin 1 the main diagonal of B is 1 1⍉B and in origin 0 it is 0 0⍉B. If X is any array and V is a suitable vector of length ρρX, then Y←V⍉X is the array in which the Ith index of X is taken to be the V[I]th index of Y. The required condition on V is that the set of components of V be precisely the same as the set of components of ιM for some integer M.

```
      D
 1  2  3  4
 5  6  7  8
 9 10 11 12

13 14 15 16
17 18 19 20
21 22 23 24
```

```
      ⎕←F←1 2 1⍉D
 1  5  9
14 18 22
      F[2;3]
22
      D[2;3;2]
22
```

The expression 1 2 1⍉D denotes the array obtained from D by setting the first and third indices equal, that is, `F[I;J]` is `D[I;J;I]`. In origin 0 we would have to write `0 1 0⍉D`. Here are some additional examples.

```
      1 1 1⍉D
1 18
      1 1 2⍉D
  1  2  3  4
 17 18 19 20
```

```
      2 1 2⍉D
1 14
5 18
9 22
```

If `M` and `N` are nonnegative integers with `M≤N`, then `M?N` is a vector of length `M` whose components are chosen randomly from `⍳N` without repetitions. In particular, `N?N` is a random permutation of `⍳N`.

```
      3?10
9 3 5
      3?10
5 8 9
```

```
      10?10
4 3 1 5 9 8 10 6 7 2
```

This dydaic operation `?` is called *deal.*

EXERCISES

In the following exercises assume that the index origin is 1. Exercises 1 and 2 refer to the following arrays.

```
      X
1 0 1 1 0 1
      Y
1 1 0 1 0 0 1 1 1
      Z
4 2 1 2 3 4
```

```
      R
1 0 0 0
      E
 1 3 5
 2 4 6
```

1 Evaluate each of the following APL expressions on paper and check your answers by entering the expressions at a terminal.

- **(a)** `⌽Z`
- **(b)** `⊖E`
- **(c)** `⌽[2]E`
- **(d)** `⍋Z`
- **(e)** `⍒Z`
- **(f)** `Z[⍋Z]`
- **(g)** `⍉E`
- **(h)** `⍉Z`
- **(i)** `,E`

2 Evaluate the following expressions as in Exercise 1.

- **(a)** `2 4⍴Z`
- **(b)** `22⍴1`
- **(c)** `Z⍳E`
- **(d)** `E∊Z`
- **(e)** `X/Z`
- **(f)** `¯3⌽Z`
- **(g)** `3 3⍴R`
- **(h)** `¯3↓Z`
- **(i)** `7↑Z`
- **(j)** `Y\Z`
- **(k)** `1⌽E`
- **(l)** `X,Y`
- **(m)** `1 2⌽E`
- **(n)** `E,0`
- **(o)** `E,[1]E`
- **(p)** `Z,[0.5]X`
- **(q)** `Z,[1.5]X`
- **(r)** `1 1⍉E`

3 Suppose G is an array of rank 4. Write an expression defining H to be an array of rank 3 such that `H[I;J;K]` is `G[I;K;J;I]`.

4 For what vectors V are `⍒V` and `⌽⍋V` the same?

5 Let V be a vector and let X and Y be logical vectors such that `X\V` and `Y/V` are defined. Describe the vectors `X/X\V` and `Y\Y/V`.

6. REDUCTION AND SCAN

In traditional mathematical notation the sum of the components of a vector $\nu = (\nu_1, \ldots, \nu_n)$ is written

$$\sum_{i=1}^{n} \nu_i$$

and the product of the components is written

$$\prod_{i=1}^{n} \nu_i .$$

In APL, the sum and product of the components of the vector V are denoted `+/V` and `×/V`, respectively.

```
      V←2 3 5 7
      +/V
17
```

```
      ×/V
210
```

More generally, if f is any dyadic scalar primitive operation, then f/`V` denotes the value of the APL expression

$$V[1]fV[2]f\ldots fV[N],$$

where N is the length of V and, for purposes of illustration, we are assuming the index origin is 1. We call f/`V` the *f-reduction* of V. Here are some examples.

```
      2⌈3⌈5⌈7
7
      ⌈/V
7
```

```
      2|3|5|7
0
      |/V
0
```

The ⌈-reduction `⌈/V` gives the maximum component of V while `⌊/V` is the minimum component.

If X is a logical vector, then `∨/X` is 1 provided at least one component of X is 1 and `∧/X` is 1 only when all the components of X are 1.

```
      X←1 1 0 1 0
      ∨/X
1
```

```
      ∧/X
0
```

Thus, to test equality of vectors V and W of the same length, we form

```
∧/V=W.
```

```
      V
2 3 5 7
      W←2 3 6 7
```

```
      V=W
1 1 0 1
      ∧/V=W
0
```

If A and B are arrays of the same shape, we check equality of A and B with the statement `∧/,A=B`.

```
      ⎕←A←B←2 3⍴⍳6
 1 2 3
 4 5 6
```

```
      ∧/,A=B
1
```

The operation of f-reduction satisfies the condition that f/V,W is the same as f/V,f/W for any nonempty vectors V and W.

```
      V
2 3 5 7
      W
2 3 6 7
```

```
      +/V,W
35
      +/V,+/W
35
```

In order for this condition to hold when W is ⍳0, it is necessary to define f/⍳0 to be a right identity element for f, provided such an identity element exists.

```
      +/⍳0
0
      ×/⍳0
1
      -/⍳0
0
```

```
      ⌈/⍳0
¯7.237005577E75
      ⌊/⍳0
7.237005577E75
```

The identity element for ⌈ should probably be denoted $-\infty$, but ⌊/⍳0 is defined to be the smallest number representable on the particular terminal system being used. Similarly, ⌊/⍳0 is the largest number on the system. The f-reduction of a vector of length 1 is the single component of that vector.

```
      +/,6                          ×/,6
6                             6
```

Reduction can be applied to arrays of any rank. The *f*-reduction of a scalar *S* is just *S*. If *A* is a matrix, then *f*/*A* is the vector whose *I*th entry is *f*/*A*[*I*;], the *f*-reduction of the *I*th row of *A*. Reduction along the columns is denoted *f*⌿*A*.

```
      A                             +⌿A
 1 2 3                        5 7 9
 4 5 6
      +/A
6 15
```

The indexed form *f*/[*I*] may also be used.

```
      +/[2]A                        +/[1]A
6 15                          5 7 9
```

Reductions of arrays of rank 3 and larger are similarly defined.

The *f-scan* operation, which is denoted *f*\, is closely related to *f*-reduction. Although the result of the *f*-scan is independent of the index origin, it is easier to describe with origin 1 indexing. If *V* is a vector, then *W*←*f**V* is the vector with the same length as *V* such that *W*[*I*] is *f*/*I*↑*V*.

```
      V                             ×\V
2 3 5 7                       2 6 30 210
      +\V                           ∧\1 1 1 0 1 0
2 5 10 17                     1 1 1 0 0 0
```

The components of +*V* are often called the partial sums of *V* and the components of ×*V* are called the partial products.

For any array *A* the *f*-scan *f**A* is an array with the same shape as *A*. The axis along which the scan is performed is indicated in the usual way. A scan along the last axis is *f*\, along the first axis is *f*⍀. and along the *I*th axis is *f*\[*I*].

```
      A                             +⍀A
 1 2 3                         1 2 3
 4 5 6                         5 7 9
      +\A                           +\[2]A
  1  3  6                       1  3  6
  4  9 15                       4  9 15
```

EXERCISES

1 Let *V* be a vector of nonzero numbers. Describe `-/V` and `÷/V` in traditional notation.

2 For any matrix *A* the numbers `+/+/A` and `+/,A` are equal. For what other scalar dyadic operations *f* are `f/f/A` and `f/,A` the same?

3 How is it possible to test two arrays *A* and *B* for equality without running the risk of producing an error message when *A* and *B* may have different ranks or different shapes?

4 Let *A* be a matrix. Is it possible to say which is larger, `⌈/⌊/A` or `⌊/⌈/A`?

5 Let *A* be a matrix. Show that `+\+⍀A` and `+⍀+\A` are always the same and describe this matrix.

6 Write an APL expression for the sum of the squares of the integers from 1 to *N*. Compare this expression to the procedure *SUMSQX* in Section 4.

7. INNER AND OUTER PRODUCTS

Suppose $x = (x_1, x_2, x_3)$ and $y = (y_1, y_2, y_3)$ are three-dimensional real vectors. Then the inner product of x and y is defined to be $x_1y_1 + x_2y_2 + x_3y_3$. Using APL notation, we can write the inner product of *X* and *Y* as `+/X×Y`. We can also write it as `X+.×Y`.

```
      X←2 ¯1 3
      Y←1 4 ¯2
      +/X×Y
¯8
```

```
      X+.×Y
¯8
```

The operation `+.×` is one example of an APL inner product. The inner products form the largest single class of APL primitive operations. Suppose *f* and *g* are two dyadic scalar primitive operations and *X* and *Y* are vectors of the same length. The inner product `Xf.gY` is defined to be `f/XgY`.

```
      X×.+Y
9
      ×/X+Y
9
```

```
      X+.⌊Y
¯2
      +/X⌊Y
¯2
```

If one of the operands *X* and *Y* has length 1, it is expanded to match the length of the other operand.

```
      Z←,2                              X+.×2 2 2
      X+.×Z                     8
8
```

There are hundreds of possible inner products, and a surprisingly large number of them have useful applications. The reader should experiment with a variety of inner products to get a feel for their utility.

It is possible to form the inner product of arrays of any rank. If one of the operands is a scalar, then it is converted to a vector of length 1 before proceeding. If `A` and `B` are arrays of positive rank, then `Af.gB` is defined only if `A` and `B` are *conformable* in the sense that the last dimension `¯1↑⍴A` of `A` is equal to the first dimension `1↑⍴B` of `B` or one of these dimensions is equal to 1. If `A` and `B` are conformable, then `Af.gB` has rank two less than the sum of the ranks of `A` and `B` and has shape `(¯1↓⍴A),1↓⍴B`. The entries of `Af.gB` are calculated by forming all possible inner products of vectors along the last axis of `A` with vectors along the first axis of `B`. For example, if `A` and `B` are matrices, then `C←Af.gB` defines `C` to be the matrix such that `C[I;J]` is `A[I;]f.gB[;J]`, the inner product of the `I`th row of `A` and the `J`th column of `B`.

```
      ⎕←A←2 2⍴1 ¯2 3 0
  1 ¯2
  3  0
      ⎕←B←2 2⍴2 1 ¯1 3
  2  1
 ¯1  3
      ⎕←C←A+.×B
  4 ¯5
  6  3
      A[1;]+.×B[;1]
4
      C[1;1]
4
      A[2;]+.×B[;1]
6
      C[2;1]
6
```

Here we see that the `+.×` inner product for matrices is just the usual matrix product.

Another important inner product is `∨.∧`, which is defined only for logical arrays.

```
      ⎕←R←3 2⍴1 0 0 1 1 1
 1 0
 0 1
 1 1
      ⎕←S←2 2⍴1 1 0 1
 1 1
 0 1
      ⎕←T←R∨.∧S
 1 1
 0 1
 1 1
      R[2;]∨.∧S[;1]
0
      T[2;1]
0
```

If *A* is a matrix and *V* is a vector, then *Af.gV* is the vector whose *I*th component is *A*[*I*;]*f.gV*. Similarly, *Vf.gA* is the vector whose *I*th component is *Vf.gA*[;*I*].

```
      A
 1 ¯2
 3  0
      V←2 ¯1
```

```
      A+.×V
4 6
      V+.×A
¯1 ¯4
```

As a final example, let us consider the case in which *A* is a matrix and *B* has rank 3. If *C* is *Af.gB*, then *C* has rank 3 and *C*[*I*;*J*;*K*] is *A*[*I*;]*f.gB*[;*J*;*K*].

Besides the inner products, APL also contains operations called outer products, one for each dyadic scalar primitive operation *f*. If *V* and *W* are vectors, then the outer product *V*∘.*fW* is the matrix *A* such that *A*[*I*;*J*] is *V*[*I*]*fW*[*J*].

```
      V←2 7
      W←¯3 4
      V∘.+W
 ¯1  6
  4 11
      V∘.×W
 ¯6  8
¯21 28
```

```
      V∘.|W
1 0
4 4
      (⍳3)∘.=⍳3
1 0 0
0 1 0
0 0 1
```

In general, the outer product *A*∘.*fB* is formed by applying *f* to all possible pairs of an entry in A and an entry in B. The rank of A∘.fB is the sum of the ranks of A and B and the shape of A∘.fB is (⍴A),⍴B.

EXERCISES

1 Assume V←4 ¯3 1 and W←2 1 ¯1. Evaluate the inner product of V and W for each of the following inner product operations by hand. Check yourself by entering the appropriate expression at a terminal.

(a) +.×	**(d)** ∧.=	**(g)** *.∣
(b) -.+	**(e)** ∨.<	**(h)** +.⌊
(c) ⌈.÷	**(f)** +.≠	**(i)** ⌊.⌈

2 Let V and W be as in Exercise 1. Evaluate the outer product of V and W for each of the following outer product operations by hand. Check your answers at a terminal.

(a) ∘.+	**(c)** ∘.=	**(e)** ∘.⌈
(b) ∘.×	**(d)** ∘.∣	**(f)** ∘.÷

3 Show that whenever A, B, and C are conformable arrays with 2≤⍴B, then A+.×B+.×C is the same as (A+.×B)+.×C. Give an example showing that this is not always true when B is a vector.

*4 Determine which inner products satisfy the associativity property described in Exercise 3.

8. SOME ADDITIONAL OPERATIONS

In this section we will investigate six more APL operations—the decode, encode, matrix inverse, matrix division, execute, and format operations—whose definitions are somewhat more involved than the definitions of the operations we have discussed so far.

The *decode* operation ⊥ and the *encode* operation ⊤ are useful for working with radix notation using bases other than 10, for "packing" many small integers into a single large integer to save space, and for certain computations with polynomials. If B and U are vectors of the same length, then B⊥U is defined to be W+.×U, where W is the *weighting vector* of length ⍴B defined as follows: W[⍴W] is 1 and W[I-1] is B[I]×W[I]. For example, if B is 10 5 3 2, then W is 30 6 2 1.

```
      B←10 5 3 2
      W←30 6 2 1
      U←4 3 7 1
      V←2 ¯1 3 1.5
      B⊥U
153
      W+.×U
153
      B⊥V
61.5
      W+.×V
61.5
```

If B is a scalar, then B is extended to form a vector of length ρU.

```
      10⊥U
4371
      11⊥U
5765
```

Here we see that B⊥U gives the decimal representation for the integer whose digits in radix notation using the base B are given in the vector U.

The decode operation can be used to evaluate polynomials. For example, to evaluate $3X^3 - X^2 + 2X + 5$ at $X = 7$, we form

```
      7⊥3 ¯1 2 5
999
```

Note that the coefficients must be listed in the order of decreasing powers of X.

Decode is extended to arrays of any rank in a manner similar to inner products. If either B or U is a scalar, then it is converted into a vector of length 1 before beginning to evaluate B⊥U. If B and U are conformable for inner products, then B⊥U is obtained by forming all possible numbers C⊥V, where C is a vector along the last axis of B and V is a vector along the first axis of U.

```
      ⎕←R←2 3⍴4 5 3 2 6 ¯1
 4 5  3
 2 6 ¯1
      ⎕←S←3 2⍴6 ¯1 3 4 ¯2 2
 6 ¯1
 3  4
¯2  2
      ⎕←T←R⊥S
  97   ¯1
 ¯41    4
      R[1;]⊥S[;2]
¯1
      T[1;2]
¯1
```

The encode operation can be used to compute the vector of digits in the representation of an integer in any specified base.

```
      10 10 10 10⊤2573
2 5 7 3
      (5⍴8)⊤2573
0 5 0 1 5
```

The entries in the arguments of ⊤ do not have to be positive integers.

```
      2.5 1.7 ¯4.8 3.1 ⊤ 7.3
1.5 0.7 ¯2.8 1.1
```

If *B* is a vector and *X* is a scalar, then *U*←*B*⊤*X* is defined as follows.

1. If *B* is empty, then *U*←⍳0.
2. If 0 = ¯1↑*B*, then *U*←(-⍴*B*)↑*X*.
3. If 0 ≠ *D*←¯1↑*B*, then *U* is defined recursively to be

   ((¯1↓*B*)⊤*Q*),*Y*,

   where *Y*←*D*|*X* and *Q*←(*X*-*Y*)÷*D*.

Here are some additional examples.

```
      ⍴(⍳0)⊤7
0
      5 6 7 0⊤13
0 0 0 13
      B←2.5 1.7 ¯4.8 3.1
      X←7.3
      ⎕←Y←3.1|X
1.1
      ⎕←Q←(X-Y)÷3.1
2
      ((¯1↓B)⊤Q),Y
1.5 0.7 ¯2.8 1.1
      B⊤X
1.5 0.7 ¯2.8 1.1
```

Encode is extended to arrays of any rank in a manner similar to outer products. If *B* and *X* are any arrays, then *B*⊤*X* is obtained by computing all possible vectors *C*⊤*Y*, where *C* is a vector along the first axis of *B* and *Y* is an entry in *X*.

```
      10 10 10⊤932 17 485
9 0 4
3 1 8
2 7 5
      ⎕←B←3 2ρ10 7
10  7
10  7
10  7
      B⊤325
3 6
2 4
5 3
      X←126 37 98
      ρU←B⊤X
3 2 3
      U[;2;3]
2 0 0
      B[;2]⊤X[3]
2 0 0
```

The encode operation can be used to construct matrices of characteristic vectors.

```
      ⎕IO←0                          ⎕IO←1
      ⎕←V←2 2 2⊤ι8
 0 0 0 0 1 1 1 1
 0 0 1 1 0 0 1 1
 0 1 0 1 0 1 0 1
```

The columns of V are the characteristic vectors for the eight subsets of ι3.

Decode and encode can also be used to pack a vector of small integers into a single large integer in order to save space and speed up certain types of comparisons. If

```
      ⎕←A←3 4ρ6 1 4 2 3 5 1 0 3 2 6 4
 6 1 4 2
 3 5 1 0
 3 2 6 4
```

then all of the entries in A are nonnegative integers less than 7. We can pack A into the vector

```
      ⎕←W←7⊥A
318 86 209 102
```

Each component of *W* describes a column of *A*. We can recover *A* from *W*, if needed, as follows.

```
      7 7 7⊤W
 6 1 4 2
 3 5 1 0
 3 2 6 4
```

The next two operations are represented by the symbol ⌹, which is called domino and is formed by overstriking ⎕ with ÷. The monadic operation is called *matrix inverse*, and the dyadic operation is called *matrix division.*

If *A* is a nonsingular square matrix, then ⌹*A* is the inverse of *A*.

```
      ⎕←A←2 2 ρ3 5 2 3
 3 5
 2 3
      ⎕←C←⌹A
  ¯3   5
   2  ¯3
      A+.×C
 1.000000000E0        7.993605777E¯15
¯4.440892099E¯16     1.000000000E0
      C+.×A
 1.000000000E0        6.883382753E¯15
 0.000000000E0        1.000000000E0
```

The result of ⌹*A* on an APL system is actually only an approximation to the inverse of *A*. In the preceding example, the matrix printed for *C* is the exact inverse of *A*, but the matrix *A*+.×*C* is not quite the identity matrix. This happens because some rounding of the entries in *C* takes place when *C* is displayed. The value of *C* in the system is not exactly the inverse of *A*.

If *B* is a matrix or a vector, then *X*←*B*⌹*A* is (⌹*A*)+.×*B*, which is the unique solution of the linear equation ∧/,*B*=*A*+.×*X*. This equation would be written $AX = B$ in conventional notation.

The expressions ⌹*A* and *B*⌹*A* are defined more generally when *A* is a matrix whose columns are linearly independent, which implies that *A* has at least as many rows as columns. Suppose the columns of *A* are linearly independent and *B* is a vector of length 1↑ρ*A*. There need not be any vector *X* such that A+.×X is equal to B, but there is a unique X that minimizes the expression +/(B-A+.×X)*2, which is the sum of the squares of the components of the difference B-A+.×X. This vector X is the result of B⌹A and is called the least-squares solution of the equation ∧/B=A+.×X. The vector A+.×X is in the vector space *W* spanned by the columns of A and is, in fact, the vector in *W* closest to B. We call A+.×X the *orthogonal projection* of B onto *W*. The vector B-A+.×X is orthogonal to *W*. If D is a matrix, then D⌹A is the matrix whose Ith column is D[;I]⌹A. The value of ⌹A is M⌹A, where M is the identity matrix of size 1↑⍴A.

```
      ⎕←A←3 2⍴1 1 2 ¯1 1 2
1  1
2 ¯1
1  2
      ⎕←D←3 2⍴8 5 2 2 1 ¯4
8  5
2  2
1 ¯4
      D⌹A
2  1
1 ¯1
      D[;2]⌹A
1 ¯1
```

The symbol ⍎, made by overstriking ⊥ with ∘, represents the *execute operation*, which allows an APL procedure to construct an APL statement as a vector of characters and then execute that statement. If S is a vector of characters that is a well-formed APL statement, then the expression ⍎S causes S to be executed.

```
      ⍎'3+4'
7
      ⎕←I←⍎'2×5'
10
      ⍎'J←I+3'
      J
13
```

The last operation in this section is represented by the symbol ⍕, which is made by overstriking ⊤ with ∘. The operation ⍕ is called *format*, and it gives the user more control over the way output is printed at the terminal than is possible using the system variable ⎕PP, which is discussed in Section A2.2. If A is an array, then ⍕A is a character array corresponding to the way A would be displayed at the terminal.

```
      ⎕←A←2 2ρ17 ¯3 4 19
 17 ¯3
  4 19
      R←⍕A
      ρR
2 8
      R
 17 ¯3
  4 19
      R[1;7]
¯
```

Format may be used to control the number of decimal places printed for entries in an array. For example, let

```
      ⎕←T←(2 3ρι6)*0.5
 1           1.414213562 1.732050808
 2           2.236067977 2.449489743
```

To print out T with its entries rounded to two decimal places, we enter 2⍕T.

```
      2⍕T
 1.00 1.41 1.73
 2.00 2.24 2.45
```

We can spread out the entries by asking that six spaces be used for each entry.

```
      6 2⍕T
  1.00  1.41  1.73
  2.00  2.24  2.45
```

Different spacing and different precision can be used in each column.

```
      4 2 6 3 4 1⍕T
1.00 1.414 1.7
2.00 2.236 2.4
```

Here four spaces and two places after the decimal point are specified for the first column, six spaces and three decimal places for the second column, and four spaces and one decimal place for the third column. Additional information about ⍕ can be found in the APL reference books.

EXERCISES

1 Use ⊥ to determine the decimal representations for the integers whose base 8 representations are 123, 4113, and 327614.

2 Use ⊤ to determine the base 7 representations of the integers 53_{10}, 972_{10}, and 27631_{10}.

3 Evaluate $3X^3 - 2X^2 + 4X + 1$ for $X = 3, -2$, and 1.5 using ⊥.

4 Construct a matrix whose rows are the characteristic vectors for the 16 subsets of ⍳4.

5 Compute the inverse of the matrix

$$\begin{bmatrix} 2 & -1 & 4 \\ 0 & 6 & -2 \\ 5 & 3 & 1 \end{bmatrix}$$

using ⌹.

6 What would be the result of executing the statement

```
⍎'3×',5⍴'2+'
```

on an APL system? Check your answer at a terminal.

7 Let *B* be a vector, let *X* be a scalar, and let *U*←*B*⊤*X*. Show that *B*|*U* is *U*. Suppose *I* is an integer from 1 to ⍴*B*, and *C*←*I*↓*B* and *V*←*I*↓*U*. Show that *C*⊥*V* is (×/*C*)|*X*.

8 Let *A* be a matrix whose columns are linearly independent and let *B* and *C* be vectors of length 1↑⍴*A*. Show that (*B*+*C*)⌹*A* is equal to (*B*⌹*A*)+*C*⌹*A*.

Appendix 2
APL SYSTEMS

This appendix provides information about the way the APL language described in Appendix 1 has been implemented in computer systems. The main topics covered are editing of APL statements, modification of procedures, system variables, system commands, workspaces, debugging techniques, and programming efficiency. APL systems differ in many ways from one another. Your system may not behave exactly as described here. Contact your local installation for help. Additional material can be found in APL reference manuals.

1. EDITING

One might get the impression from reading the APL dialogues in this book that no one ever makes a mistake typing an APL statement and that procedures always work correctly when they are entered and therefore never need to be changed. This is not the case. The purpose of this section is to show how to correct mistakes on the line currently being typed and how to modify procedures when they do not work as intended. First, however, it is necessary to discuss the types of terminals commonly used with APL systems.

There are two basic standards governing the way terminals communicate with computers. These standards are referred to by their acronyms: EBCDIC (Extended Binary Coded Decimal Interchange Code) and ASCII (American Standard Code for Information Interchange). Terminals manufactured by IBM are generally EBCDIC terminals, while most other terminals use the ASCII standard. It is not hard to distinguish between EBCDIC and ASCII terminals. On an ASCII terminal keyboard there is a key marked "CONTROL" (often abbreviated "CNTRL"). On an EBCDIC terminal there is no such key. On an ASCII terminal the key that interrupts a computation in progress is marked "BREAK" or "BRK", while on an EBCDIC terminal this key is marked "ATTN".

If an error is discovered on a line before the RETURN key is hit, it is possible to backspace to the error and then retype the line correctly from that point. On an EBCDIC terminal one backspaces to the error and hits the ATTN key. The system causes the terminal to space downward one line and print the symbol ∨. The terminal spaces down another line, and the corrected version of the line may be entered. The following example illustrates the correction of an error while typing the statement 2×3×5×7 on an EBCDIC terminal.

```
      2×3+5+
          ∨
          ×5×7
210
```

On an ASCII terminal one backspaces to the error and hits the LINEFEED or LF key. The terminal spaces down one line, and the correction is entered. The previous example would look like this on an ASCII terminal.

```
      2×3+5+
          ×5×7
210
```

The characters of an APL statement may be typed in any order. For example, it is possible to enter the statement 2×3×5×7 by typing 2 3 5 7 and backspacing to fill in the missing times signs.

The remainder of this section deals with methods of displaying and changing procedure definitions. Let us assume that we have entered the definition of a procedure *PROC*. To display this definition, we type ∇ to enter the definition mode, the name of the procedure, the symbols [⎕], which indicate that the entire procedure is to be displayed, and another ∇ to leave the definition mode.

```
      ∇PROC [⎕]∇
    ∇ X PROC Y
[1]   SIM←X+Y
[2]   DIFF←X-Y
[3]   QUOT←X÷Y
[4]   MOD←X|Y
    ∇
```

Now suppose that we decide to change *PROC* by entering a new line at the end. We enter ∇*PROC* and the system prompts for line [5], which we enter.

```
      ∇PROC
[5]   EXP←X*Y
[6]
```

The system prompts for the next line but, at this point, we realize that we forgot a line between lines [2] and [3]. We ask to be prompted for line [2.5] and, when we are, we type the new line.

```
        [2.5]
[2.5]   PROD←X×Y
[2.6]
```

The system prompts for line [2.6], but we do not wish to insert any more lines. Instead, we decide to delete line [4], which is not really needed. We enter [∆4].

```
        [∆4]
[4]
```

The system prompts for the line we deleted. (Some systems prompt for the next line.) Now we see that the variable *SIM* in line [1] should really be *SUM*. We decide the quickest way to make the change is to retype the line. We ask to be prompted for line [1] and reenter the line.

```
        [1]
[1]     SUM←X+Y
[1.1]
```

At this point we would like to look at the current definition of *PROC*.

```
        [⎕]
      ∇ X PROC Y
[1]     SUM←X+Y
[2]     DIFF←X-Y
[2.5]   PROD←X×Y
[3]     QUOT←X÷Y
[5]     EXP←X*Y
      ∇
[6]     ∇
```

We are satisfied with *PROC* and leave the definition mode. The system now renumbers the lines of the procedure with consecutive positive integers. This can be seen by displaying the procedure again.

```
      ∇PROC [⎕]∇
    ∇ X PROC Y
[1]   SUM←X+Y
[2]   DIFF←X-Y
[3]   PROD←X×Y
[4]   QUOT←X÷Y
[5]   EXP←X*Y
    ∇
```

It is possible to request the definition of a procedure be displayed from line [*N*] on. This is done by entering [⎕*N*]. To have just line [*N*] printed, enter [*N*⎕]. The system will type the line and then prompt for that same line. For the purposes of editing only, the header of the procedure is considered to be line [0].

It would be inefficient to modify a long line in a procedure by retyping it entirely. There is another method which can be used instead. The following example uses this method to change the header of our procedure *PROC* to read *SUM*←*PROC X*.

```
      ∇PROC [0⎕8]
[0]   X PROC Y
      //4    /
[0]   SUM←PROC X∇
```

The expression [0⎕8] indicates that we want to begin modifying line [0] at approximately position 8, counted from the left margin. Line [0] is printed, the terminal spaces down one line and the carriage or cursor stops under position 8. A / typed under a character indicates that character is to be deleted. A number from 1 to 9 typed under a character indicates that the specified number of blanks is to be inserted before the character. (More than 9 blanks can be requested by typing a letter. The *I*th letter of the alphabet requests 5×*I* blanks.) After a carriage return, the line is retyped with the deletions and inserted blanks. The cursor is positioned at the first inserted blank. The line may now be edited in the usual manner. In particular, the blanks may be filled in.

Additional characters may be added at the end of line [*N*] by entering [*N*⎕0]. Line [*N*] is typed, and the cursor is positioned at the end of the line.

2. SYSTEM VARIABLES

In an APL system there is a class of special variables, called *system variables*, that describe the environment in which the APL programs run. For us the most important system variable is the index origin ⎕*IO*. However, there are several other system variables that may occasionally be of use. These are `⎕CT`, `⎕PP`, `⎕PW`, `⎕AI`, and `⎕WA`, the comparison tolerance, the printing precision, the printing width, the accounting information, and the work area available, respectively.

When computations are performed in an APL system, the results are often only approximations to the "true" values, that is, the values obtained using exact calculations in the field of real numbers. For example, if we let `X←2*0.5`, we should not be surprised if `X*2` is not exactly 2. For this reason, when values are compared in an APL system, they are not checked for exact equality but, instead, for equality within the tolerance specified by the system variable `⎕CT`. That is, `A` and `B` are considered equal if the absolute value of their difference does not exceed `⎕CT` multiplied by the larger of the absolute values of `A` and `B`. Thus `A=B` is equivalent to

```
(|A-B)≤⎕CT×(|A)⌈|B
```

The value of `⎕CT` may be any number from 0 to 1. The normal value is `1E¯13`, that is, 10^{-13}. Setting `⎕CT` to 0 produces exact comparisons.

```
      ⎕←X←2÷3
0.6666666667
      Y←3×X
      Y-2
¯2.220446049E¯16
      ⎕CT←0
```

```
      2=Y
0
      ⎕CT←1E¯13
      2=Y
1
```

Besides the relational operations `<`, `≤`, `=`, `≥`, `>`, and `≠`, the operations `⌈`, `⌊`, `∊`, and `⍳` are affected by the value of `⎕CT`.

```
      ⎕CT←0
      ⌊Y
1
      1 2⍳Y
3
```

```
      ⎕CT←1E¯13
      ⌊Y
2
      1 2⍳Y
2
```

The system variable `⎕PP` indicates the number of significant figures that are to be printed when numeric values are displayed at the terminal. Typically, `⎕PP` can vary from 1 to 16, with 10 as the normal value.

```
      ⎕PP←16
      ⎕←X←2*0.5
1.414213562373095
      2*35
34359738368
```

```
      ⎕PP←10
      X
1.414213562
      2*35
3.435973837E10
```

Terminals differ in the number of characters that can be printed on a line. The system variable ⎕*PW* can be assigned a convenient value less than the width of the terminal line to assist the system in formatting output for the terminal.

```
      ⎕PW←40
      ⍳20
1 2 3 4 5 6 7 8 9 10 11 12 13
      14 15 16 17 18 19 20
      ⎕PW←80
      ⍳20
1 2 3 4 5 6 7 8 9 10 11 12 13 14 15 16 17 18 19 20
```

In order to determine the amount of CPU time used in a particular calculation, we can use the system variable ⎕*AI*, which is a vector of length 4. The only component of ⎕*AI* that we need to be concerned with is the second, which tells the amount of CPU time used since sign-on. The units used depend on the particular APL system. Older systems may use 1/60-ths of a second, while newer systems usually use milliseconds. To time how long it takes to square a 10-by-10 integer matrix, we may enter the following.

```
      A←?10 10⍴100                      B←A+.×A
      ⎕AI                               ⎕AI
1001 74 223187 209210            1001 96 246006 230023
```

On this system CPU time is measured in milliseconds, and computing *B* took 96 – 74 = 22 milliseconds.

The active workspace, the portion of the computer's memory containing the variables and procedures currently in use, has only a limited amount of space. The amount of unused space is given by the system variable ⎕*WA*. The units as well as the maximum value of ⎕*WA* depend on the particular system you are using.

```
      ⎕WA                               ⎕WA
124280                           120264
      Z←⍳1000
```

Here 4016 units (in this case, bytes) of memory are required to store the array *Z*. Four bytes are needed to store each entry of *Z*. The remaining bytes are used to store other information, such as the name and the shape of *Z*. (On some APL systems the vectors ⍳*N* are stored using an amount of space that does not depend on *N*.) It is not possible to assign a value to ⎕*WA*.

In addition to system variables there are also special functions called *system functions.* These functions, which can be used within defined procedures, provide alternatives to some of the system commands described in Section 3. Two system functions permit a procedure to modify another procedure and to define new procedures. Thus APL programs can write and run other APL programs. Information about system functions can be found in APL reference books.

3. WORKSPACES AND SYSTEM COMMANDS

The variables and procedures with which we are currently working are stored in a portion of the computer's memory called the *active workspace.* These variables and procedures are lost if we sign off without first saving the active workspace into our library. Each user has a library in which inactive workspaces may be saved for future use. The number of workspaces that can be stored in the library is set by the system manager.

The APL system contains certain system commands that facilitate workspace management. Some of these commands allow us to save the active workspace into the library or bring a workspace, or part of a workspace, from the library into the active workspace. Other commands permit us to determine the status of our active workspace and modify that status if desired. All system commands start with a right parenthesis.

To find out what workspaces are in our library, we issue the `)LIB` command.

```
      )LIB
CLASSLIB
EXAMPLES
DEMO
```

Here we see that in addition to the two workspaces `CLASSLIB` and `EXAMPLES`, which are used throughout this book, there is a workspace `DEMO`. To see what is in `DEMO`, we must first put a copy of it into the active workspace with the `)LOAD` command.

```
      )LOAD DEMO
SAVED  10.39.26 08/09/79
```

The `)LOAD` command does not affect the contents of the library. It merely makes a copy of the workspace and places the copy in the active workspace, destroying anything that may have been in the active workspace.

We can find out the names of the variables in `DEMO` with the `)VARS` command.

```
      )VARS                          ρρM
A        M       X              0
      ρA                             ρX
25 40                           37
```

There are three variables in *DEMO*, a matrix *A*, a scalar *M*, and a vector *X*.

To see what procedures are in *DEMO*, we use the)*FNS* command. (Remember that APL procedures are usually called functions.)

```
      )FNS
PROC1    PROC2
      ∇PROC1 [⎕]∇
      ∇Z←X PROC1 Y;U;V
[1]   U←3 PROC2 X+Y
[2]   V←0.5 PROC2 X-Y
[3]   Z←U PROC2 V
    ∇
```

In *DEMO* there are two procedures *PROC*1 and *PROC*2, the first of which we have listed. If we try to execute *PROC*1, we run into trouble.

```
      4 PROC1  7
DOMAIN ERROR
PROC2[1] Z←Y*X
           ∧
```

At this point, computation is halted. We can get a better idea of where we are with the)*SI* command.

```
      )SI
PROC2[1] *
PROC1[2]
```

The)*SI* command lists the *state indicator*, which contains a list of the procedures currently halted along with the corresponding line numbers, most recently halted first. In our example we are stopped at line [1] of *PROC*2, which was called from line [2] of *PROC*1. It is now possible to examine the variables in *PROC*2 to see what is wrong. This will be discussed in Section 5. To clear the state indicator we enter →.

```
      →
      )SI
```

It is possible to bring part or all of a workspace from the library into the active workspace without destroying the contents of the active workspace. This is done with the *)COPY* command.

```
      )COPY CLASSLIB ZGCD ZLCM EXPAND
SAVED  16.59.26 06/19/79
      )FNS
EXPAND   PROC1    PROC2    ZGCD    ZLCM
      35 ZGCD 56
7
      35 ZLCM 56
280
```

Here we have copied the procedures *ZGCD*, *ZLCM*, and *EXPAND* from the workspace *CLASSLIB* into the active workspace. Omitting the list of objects causes the entire workspace to be copied.

We can delete variables and procedures from the active workspace with the *)ERASE* command.

```
      )ERASE A ZLCM
      )VARS
M       X          R̲        S̲
      )FNS
EXPAND   PROC1   PROC2   ZGCD
```

The variables R̲ and S̲ were created by *ZGCD*.

If we happen to forget the name of the active workspace, we can find out what it is with the *)WSID* command.

```
      )WSID
DEMO
```

Suppose now we wish to save the active workspace in the library without affecting the workspace *DEMO* still in the library. To do this, we must save the workspace with a new name.

```
       )SAVE DEMO1
   10.55.07 08/09/79
```

The *)SAVE* command places a copy of the active workspace into the library but does not change the active workspace. To wipe the active workspace clean, we issue the *)CLEAR* command.

```
      )CLEAR
CLEAR WS
```

Clear workspaces do not have a name associated with them. If at some later time we decide we no longer need the workspace *DEMO* in the library, we can drop it.

```
      )DROP DEMO
  10.55.35 08/09/79
      )LIB
CLASSLIB
EXAMPLES
DEMO1
```

In order to facilitate the copying of several related objects from a workspace in the library to the active workspace, it is possible to collect together a number of arrays and procedures and give a name to the collection. The term normally used for such a collection is "group", but these groups have nothing to do with the groups of Chapter 3. To illustrate the formation and modification of groups, let us reload the workspace *DEMO*1.

```
      )LOAD DEMO1
SAVED  10.55.07 08/09/79
      )VARS
M       X         R̲         S̲
      )FNS
EXPAND  PROC1     PROC2     ZGCD
```

To form a group named *INTEGER* consisting of the variables R̲ and S̲ and the procedures *ZGCD* and *EXPAND*, we enter

```
        )GROUP INTEGER R̲ S̲ ZGCD EXPAND
```

To get a list of the groups currently defined, we use the)*GRPS* system command.

```
      )GRPS
INTEGER
```

To recall what is in a group, we use the)*GRP* command.

```
      )GRP INTEGER
EXPAND  ZGCD      R̲      S̲
```

We may add new members to the group as follows.

```
      )GROUP INTEGER INTEGER M
      )GRP INTEGER
EXPAND  M         ZGCD      R̲         S̲
```

If we erase a group, we destroy the definition of the group and also erase the objects in the group. To disperse a group, that is, remove its definition without affecting its members, use the *)GROUP* command with an empty list of objects.

```
)GROUP INTEGER
)GRPS
```

4. ERROR MESSAGES

From time to time the system may be unable to complete a computation or carry out a system command. When this happens, an error message or trouble report is printed at the terminal. The following is a brief description of the more common error messages.

DEFN ERROR	Misuse of ∇ or ⎕.
DOMAIN ERROR	Argument not in the domain of the primitive operation.
ENTRY ERROR	An invalid character has been transmitted, perhaps an improper overstrike.
INDEX ERROR	Index value out of range.
LENGTH ERROR	Shapes of the arguments not conformable.
RANK ERROR	Rank of an argument of a primitive operation is wrong.
RESEND	Either an error in transmission or too many characters on a line.
SI DAMAGE	One of the procedures in the state indicator has been adversely affected by editing or by a *)COPY* or *)ERASE*.
SYNTAX ERROR	The statement is not formed correctly; for example, a procedure is used without the proper arguments, unmatched parenthesis.
SYMBOL TABLE FULL	Too many names in use.
SYSTEM ERROR	This is serious. Send as complete a description as possible of what happened to the system manager.
VALUE ERROR	Name does not have a value, procedure did not return a value, or constant has a value too large or too small for internal representation.
WS FULL	The space available in the active workspace is

too small to complete the task in progress. Check)*SI* and clear halted procedures if possible.

5. DEBUGGING

Frequently, a newly defined procedure does not work as intended. This section describes some simple techniques for finding out what is wrong. As a start, let us go back to the problem we had in Section 3 with the procedures *PROC*1 and *PROC*2.

```
      ∇PROC1 [⎕]∇
    ∇ Z←X PROC1 Y;U;V
[1]   U←3 PROC2 X+Y
[2]   V←0.5 PROC2 X-Y
[3]   Z←U PROC2 V
    ∇
      4 PROC1 7
DOMAIN ERROR
PROC2[1] Z←Y*X
             ∧
```

The error message tells us that one of the variables *X* or *Y* must be outside the domain of the exponential operation. We can examine the values of these variables.

```
      X                          Y
0.5                       ¯3
```

Now we see what is wrong. The procedure *PROC*2 is trying to take the square root of a negative number. Since we have no information about what *PROC*1 and *PROC*2 are supposed to do, we cannot tell if the problem is in the definition of *PROC*1, in the definition of *PROC*2, or in our choice of the arguments 4 and 7. Before leaving this example, however, it is useful to remark that we can list and edit *PROC*2 at this point.

```
      ∇PROC2 [⎕]∇
    ∇ Z←X PROC2 Y
[1]   Z←Y*X
    ∇
```

However, without first clearing the state indicator, we cannot list or edit *PROC* 1.

```
      ∇PROC1 [⎕]∇
DEFN ERROR
      ∇PROC1
            ∧
      →
```

```
          ∇PROC1 [⎕]∇
        ∇ Z←X PROC1 Y;U;V
[1]       U←3 PROC2 X+Y
[2]       V←0.5 PROC2 X-Y
[3]       Z←U PROC2 V
        ∇
```

Only the most recently halted procedure can be edited.

Sometimes a procedure does not produce any error messages but seems to be running a long time. We can temporarily stop execution and check on how things are going. In Section A1.4 we defined a procedure *SUMSQX* for computing sums of squares.

```
      ∇SUMSQX [⎕]∇
    ∇ S←SUMSQX N;1
[1]   S←I←0
[2]   LOOP:→(N<I←I+1)/0
[3]   S←S+I*2
[4]   →LOOP
    ∇
```

Suppose we decide to compute the sum of the squares of the first 10^6 positive integers using *SUMSQX*.

```
      SUMSQX 1E6
```

We wait a few seconds, but there is no response from the system. Curious, we hit the ATTN or BRK key to stop execution and look at the current value of *I* and *S*.

```
      SUMSQX[3]
      I
1110
```

```
      S
455261135
```

Everything seems to be all right, so we resume execution at the line where we stopped.

```
      →3
```

We wait a while longer and decide to check our progress again.

```
      SUMSQX[2]
      I
2163
```

```
      S
3375587894
```

By now we realize that it will take too long to complete the calculation and we abandon the effort by clearing the state indicator.

```
      →
```

It is possible to keep track of the execution of a procedure without continually stopping it and restarting it. This is done with the trace facility. To trace the execution of *SUMSQX*, we create a special variable *T∆SUMSQX* that lists the line numbers at which we want tracing information.

```
      T∆SUMSQX←1 2 3 4
```

Here we are requesting tracing information at every line. We now begin execution of the procedure.

```
      SUMSQX 3
SUMSQX[1] 0
SUMSQX[2] →3
SUMSQX[3] 1
SUMSQX[4] →2
SUMSQX[2] →3
SUMSQX[3] 5
SUMSQX[4] →2
SUMSQX[2] →3
SUMSQX[3] 14
SUMSQX[4] →2
SUMSQX[2] →0
14
```

For each line number in *T∆SUMSQX* the system prints the procedure name, the line number, and the result of that line. Branches are indicated by a branch arrow followed by the number of the next line to be executed. Tracing is terminated by assigning ⍳0 to *T∆SUMSQX*.

```
      T∆SUMSQX←⍳0
      SUMSQX 3
14
```

In addition to the trace facility there is also a stop facility. If we want *SUMSQX* to stop every time it is about to execute line 4, we assign 4 to the special variable *S∆SUMSQX*.

```
      SΔSUMSQX←4                          →4
      SUMSQX 3
                                   SUMSQX[4]
SUMSQX[4]                                S
      S                            14
1                                         →4
      →4                           14
                                          SΔSUMSQX←⍳0
SUMSQX[4]
      S
5
```

In this example we have examined the value of *S* each time the procedure stopped and then restarted it again.

There is yet another way to observe the execution of a procedure, but this requires editing the procedure to temporarily insert the characters ⎕← to indicate that results are to be printed at the terminal.

```
      ∇SUMSQX [3⎕7]                     SUMSQX 4
[3]   S←S+I*2                      1
      2                            5
[3]   ⎕←S←S+I*2                    14
[4]   ∇                            30
                                   30
```

The modified *SUMSQX* prints out each value of *S* as it is computed in line [3]. The final 30 in the display is the result that would be printed without any tracing.

6. PROGRAMMING EFFICIENCY

A given computation can usually be formulated in many ways in APL. Some formulations may execute faster than others, and some may require less space than others. Usually the initial concern is to find something that works. However, if the computation is to be repeated often, it becomes important to devise a method of performing the computation that is efficient in its use of both time and space. In this section we present some techniques for efficient APL programming that have been used in the procedures in *CLASSLIB*.

We start with a brief discussion of the way APL stores entries in arrays. Let us define five vectors of length 1000 and use the system variable ⎕*WA* to find out how much space was used to store each vector.

```
      ⎕WA
91032
      A←1000⍴1
      ⎕WA
90888
      B←1000⍴2
      ⎕WA
86872
      C←1000⍴2.5
```

```
      ⎕WA
78856
      D←1000⍴1E10
      ⎕WA
70840
      E←1000⍴'X'
      ⎕WA
69824
```

The number of bytes required to store *A*, *B*, *C*, *D*, and *E* is 144, 4016, 8016, 8016, and 1016, respectively. What caused the difference? The answer is that APL systems have at least four different methods of storing arrays. Different methods are used for logical arrays, for arrays of small integers, for arrays of fractions and large integers, and for arrays of characters. Let us consider another example.

```
      ⎕WA
69824
      F←1000⍴4÷2
```

```
      ⎕WA
61808
```

This example seems to contradict what was just said. The entries in *F* seem to be the same as the entries in *B*, but roughly twice as much space was used for *F* as was used for *B*. The problem is that APL systems usually consider the result of any division to be a fraction, even if it happens to be an integer. We can force the system to recognize `4÷2` as an integer by the use of the floor operation.

```
      ⎕WA
61808
      G←1000⍴⌊4÷2
```

```
      ⎕WA
57792
```

The foregoing discussion can be summarized as follows. Be aware of the different methods that APL systems use to store arrays and of techniques for helping the system find the best way for your specific needs.

As an illustration of how a different formulation of a computation can reduce the time required to perform it, let us look at the problem of constructing the inverse *Q* of the permutation *P*. Probably the most natural way to obtain *Q* is the following.

```
      Q←⍋P
```

However, if *P* is very long, then there is another method that requires less computer time on most systems.

```
      Q←(⍴P)⍴2
      Q[P]←⍳⍴P
```

The use of 2 in the first line, instead of 0 or 1, is to get the system to allocate the right amount of space for Q the first time.

Some of the APL operations are somewhat slow when used with large arrays. These include ⍋, ⍒, ∊, and dyadic ⍳. These operations should be used only when there is no faster alternative available.

Here is another example of how a little ingenuity can help a great deal. Suppose A is a 4-by-4 matrix and X and Y are vectors of length 1000 with entries in ⍳4. We wish to compute the vector Z such that Z[I] is A[X[I];Y[I]]. Let us assume ⎕IO is 0. Then the "obvious" way to get Z is

```
Z←0 0⍉A[X;Y]
```

However, A[X;Y] is a matrix with 1 million entries, and trying to compute it will clearly cause a WS FULL condition on all but the largest APL systems. We could write a procedure containing a loop that computes the entries in Z one at a time. This would be efficient as far as space is concerned, but it would be quite slow.

There is another approach that uses the ravel ,A. To illustrate this method, let us consider a specific example.

```
      ⎕←A←4 4⍴2 ¯1 3 0 1 4 ¯2 ¯3 0 ¯4 1 2 3 2 4 1
2 ¯1  3  0
1  4 ¯2 ¯3
0 ¯4  1  2
3  2  4  1
      ⎕IO←0
      X←?1000⍴4
      Y←?1000⍴4
```

We begin by forming the ravel of A.

```
      B←,A
```

Our next task is to find a quick method of locating an entry in B given the row and column indices of the corresponding entry in A. The index in B of A[I;J] is J+4×I.

```
      A[2;3]
2
      B[3+4×2]
2
```

```
      A[1;2]
¯2
      B[2+4×1]
¯2
```

From this we see that the vector Z can be constructed very easily.

```
      Z←B[Y+4×X]                       A[X[237];Y[237]]
      Z[237]                    ¯3
¯3
```

An equivalent construction, which might be more convenient with arrays of higher rank, is the following.

```
      ρW←X,[¯0.5]Y                     Z[237]
2 1000                          ¯3
      Z←B[4⊥W]
```

In order to determine which of several alternative approaches to a given computation is the fastest, each alternative may be timed using the system variable ⎕AI as described in Section 2.

Appendix 3
THE SUPPLEMENTAL WORKSPACES

Two APL workspaces have been created to supplement this book. The workspace *CLASSLIB* contains procedures for carrying out various algebraic computations. All of the procedures referred to in the text are in *CLASSLIB*. The workspace *EXAMPLES* contains arrays that describe various algebraic objects. These arrays appear in the sample dialogues and in some of the exercises. All of the dialogues assume that the contents of both *CLASSLIB* and *EXAMPLES* are present in the active workspace. One way to arrange this is to execute the following system commands.

```
)LOAD CLASSLIB
)COPY EXAMPLES
```

However, if space is limited, it may be necessary to copy only the procedures and arrays that are needed at the moment.

What follows is a brief introduction to the organization of *CLASSLIB* and *EXAMPLES*. More complete information, including procedure listings, can be found in the *CLASSLIB* user's manual, which is published separately.

1. *CLASSLIB*

The workspace *CLASSLIB* contains over 200 procedures related to algebraic computation. These procedures are named using a special naming convention that will be described shortly. Illustrations in the text of the use of individual procedures may be located using the index. There are also four global variables, two of general importance and two used by only one procedure.

The global variables are:

EPSILON	A scalar, normally set to 10^{-13}, which controls the extent to which small entries in an array of real or complex numbers are set to 0.
NOTEST	A logical scalar, normally set to 0. When *NOTEST* is set to 1, certain validity checking of arguments is skipped, thereby speeding up some computations.
BIGPRIMES	A vector listing the large primes used by *MPZDET*.
BIGINV	A vector of positive integers used by *MPZDET*. For *I* > 1 the value of *BIGINV*[*I*] is the inverse modulo *BIGPRIMES*[*I*] of ×/(*I*-1)↑*BIGPRIMES*.

With seven exceptions, the names of the procedures in *CLASSLIB* consist of two parts, a prefix indicating the algebraic system in which the computation is to be performed and a suffix describing the nature of the computation. For example, the prefix *RX* refers to **R**[*X*] and the suffix *SUM* means addition in a ring. Thus *RXSUM* is used to add arrays of real polynomials.

The seven exceptions to the naming convention are *DAQ*, *DARV*, *DAZV*, *DERR*, *EXPAND*, *EXPANDV*, and *TRAV*. The first three take a single argument and construct a character matrix that, when displayed at the terminal, makes clear the algebraic object represented by the argument. The procedure *TRAV* performs the analogue of the monadic transpose ⍉ on arrays that are considered to arrays of vectors, such as arrays of complex numbers or arrays of polynomials. The remaining procedures, *DERR*, *EXPAND*, and *EXPANDV*, carry out specialized tasks for other procedures and are not meant to be called directly.

The following is a list of the prefixes and the algebraic systems to which they refer. Some of the systems are defined by global variables that must be created before the procedures with those prefixes can be used. References are given to sections in the text where details may be found concerning the way individual objects in the systems are represented.

C	The field of complex numbers. (4.3)
FR	The finite ring described by the variables *FRPLUS*, *FRTIMES*, *FRNEG*, and *FRINV*. (4.3)
FRX	The ring of polynomials over the ring for *FR*. (4.4)
GAUSS	The ring of Gaussian integers. (4.2, 4.9)
GP	The symmetric group on ⍳*N*, where *N* is the last dimension of the argument(s). (3.7, 3.8)

GT	The finite group described by the variables *GTABLE*, *GTINV*, and *GTIO*. (3.2)
MPZ	The ring of integers with multiple precision. (2.5)
Q	The field of rational numbers with exact computation. (4.3)
R	The field of real numbers. (4.3)
RA	The real algebra described by the structure constants *RASC*. (5.4)
RX	The ring of real polynomials. (4.4)
S	Sets, mostly sets of nonnegative integers. (Chapter 1)
Z	The ring of integers. (Chapter 2, 4.3)
ZA	The integer algebra described by the structure constants *ZASC*, (5.4)
ZN	The ring of integers module $\underline{N}$. Some procedures require $\underline{N}$ to small and/or prime. (4.3)
ZNA	The algebra over the ring for *ZN* described by the structure constants *ZNASC*. (5.4)
ZNX	The ring of polynomials over the ring for *ZN*. (4.4)
ZX	The ring of integer polynomials. (4.4)

The following is a list of the suffixes, the operations to which they refer, and references to sections containing examples.

ALLORB	Compute all orbits of a permutation group. (3.8)
CHREM	Solve simultaneous congruences, as in the Chinese Remainder Theorem. (2.4)
CHV	Compute characteristic vectors of sets. (1.3)
CONJ	Compute complex conjugate. (4.3)
CYCIN	Convert a permutation from cycle to vector notation. (3.7)
CYCOUT	Convert a permutation from vector to cycle notation. (3.7)
DEGREE	Compute the degree of a polynomial. (4.4)
DET	Compute the determinant of a square matrix. (4.6)
DIFF	Compute the difference in a ring. (2.5, 4.3, 4.4)
EQREL	Test whether a square logical matrix defines an equivalence relation. (1.2)
EVAL	Evaluate a polynomial at a point in the coefficient ring. (4.4)

FACTOR	Factor an element of a UFD. (2.4, 4.10, 7.4)
FEL	Select the first element in a set of nonnegative integers. (1.4)
GCD	Compute a greatest common divisor. (2.2, 4.10)
INTERP	Interpolate a polynomial. (4.12)
INV	Compute an inverse in the group of units of a ring. (4.3)
IRRED	Find "small" irreducible elements in certain Euclidean domains. See also *PRIMES*. (4.10)
LCM	Compute a least common multiple. (2.2)
LCON	Compute the characteristic matrix for left congruence modulo a subgroup. (3.3)
LSYS	Solve a linear system of equations. (6.4, 6.7)
MAG	Compute the magnitude of a real or complex number. (2.5, 4.3)
MATINV	Compute the inverse of a matrix. (4.7, 5.3, 6.2)
MATPROD	Compute the product of two matrices. (4.5)
NEG	Compute the negative in a ring. (4.3)
NORM	Compute the norm of a complex number. (4.3)
ORBIT	Compute a single orbit of a permutation group. (3.8)
POWER	Compute a nonnegative power in a ring. (2.5, 4.3)
PRIMES	Find "small" prime elements in certain Euclidean domains. See also *IRRED*. (2.4)
PROD	Compute a product in a group or a ring. (2.5, 3.2, 4.3, 4.4)
QUOT	Compute a quotient in a Euclidean domain or other ring. (2.1, 4.3, 4.9)
RCON	Compute the characteristic matrix for right congruence modulo a subgroup. (3.3)
REDUCE	Reduce a matrix over a Euclidean domain using row and column operations. (6.5)
REM	Compute a remainder in a Euclidean domain or in certain other rings. (2.1, 2.5, 4.9)
ROWREDUCE	Row reduce a matrix over a Euclidean domain. (6.2)
SGN	Compute the signum of a real number or the sign of a permutation. (2.5, 3.8)
SGP	Determine all elements in a subgroup from a set of generators. (3.3, 3.8)
SORT	Sort the entries of a vector into increasing order and

	remove duplicates. (1.1)
SUB	Construct all subsets of a given size in ι*N*. (1.1)
SUM	Compute a sum in a ring. (2.5, 4.3, 4.4)
SYMG	Construct all elements of a symmetric group. (3.7)

In addition to the suffixes just listed, there are four suffixes that describe operations of a somewhat different type. These are the suffixes *TEST*, *NRMLZ*, *INIT*, and *CHECK*. Procedures ending in *TEST* are used to decide whether a given APL array is a valid representation for an array with entries in a particular algebraic system. A value of 0 is returned if the APL array does represent an array over the system, and a value of 1 is returned when the APL array is not a valid representation. For some algebraic systems, such as the integers modulo *N̲*, a given array may have many representations by APL arrays. Procedures ending in *NRMLZ* test for validity and, when the argument is valid, return the standard representation. For example, *QNRMLZ* reduces fractions to lowest terms.

Several of the prefixes refer to algebraic systems defined by global variables. Procedures with the suffix *INIT* may be used to initialize these global variables. Procedures with the suffix *CHECK* can be used to decide if an array or pair of arrays really defines a system of the appropriate type. For example, *GTCHECK* checks whether an array is a group table.

The procedures in *CLASSLIB* print error messages if they detect errors. In addition, if the system decides that something is wrong, the system will print an error message. One frequent cause of difficulty is the absence from the active workspace of a necessary procedure because it was erased or not copied in order to save space. It is a good practice to have all procedures with a given prefix in the active workspace if you expect to use any one of them. The procedures *DERR*, *EXPAND*, and *EXPANDV* should also be present. If procedures with the prefix *ZN*, *ZNX*, or *Q* are being used, then the procedures with the prefix *Z* should be in the active workspace too.

2. *EXAMPLES*

The workspace *EXAMPLES* contains copies of the arrays *BIGPRIMES* and *BIGINV* that are normally in *CLASSLIB*. In addition, there are arrays representing various algebraic objects. The first part of the name indicates the type of object according to the following scheme.

B	Binary operation.
DESIGN	Block design.
GP	Set of permutations.
E	Equivalence relation.

G	Group table.
H	Homomorphism.
INV	List of group inverses.
PLUS	Addition table of a finite ring.
TIMES	Multiplication table of a finite ring.
MPLUS	Addition table for a module.
MTIMES	Module action for a finite ring.

BIBLIOGRAPHY

Other introductory algebra texts

Garrett Birkhoff and Saunders MacLane, *A Survey of Modern Algebra*, 4th ed. New York: Macmillan, 1977.

I. N. Herstein, *Topics in Algebra*, 2nd ed. New York: Wiley, 1975.

More advanced algebra texts

Nathan Jacobson, *Basic Algebra One and Basic Algebra Two*. San Francisco: W. H. Freeman, 1974, 1980.

Serge Lang, *Algebra*. Reading, Mass.: Addison-Wesley, 1965.

B. L. Van der Waerden, *Algebra*, 2 vols, 7th ed. New York: Frederick Ungar, 1970.

APL reference

Leonard Gilman and Allen G. Rose, *APL: An Interactive Approach*, 2nd rev. ed. New York: Wiley, 1976.

In a class by itself

Donald E. Knuth, *The Art of Computer Programming*, Vols. 1 and 2, 2nd ed. Reading, Mass.: Addison-Wesley, 1974 and 1981.

Articles referred to

Elwyn R. Berlekamp, "Factoring Polynomials over Large Finite Fields." *Math. of Comp.*, *24* (1970), 713-735.

John Brillhart, "Note on Representing a Prime as a Sum of Two Squares." *Math. of Comp.*, *26* (1972), 1011-1013.

John D. Dixon, "The Probability of Generating the Symmetric Group." *Math. Zeit.*, *110* (1969), 199-205.

Daniel Gorenstein, "The Classification of Finite Simple Groups." *Bull. Amer. Math. Soc.* (New Series), *1* (1969), 43-199.

Theodore S. Motzkin, "The Euclidean Algorithm." *Bull. Amer. Math. Soc.*, *55* (1949), 1142-1146.

Charles C. Sims, "Group-Theoretic Algorithms, a Survey." *Proc. International Congress of Mathematicians*, Helsinki (1978), 979-985.

INDEX